Methods in Molecular Biology

Series Editor
John M. Walker
School of Life and Medical Sciences
University of Hertfordshire
Hatfield, Hertfordshire, AL10 9AB, UK

For further volumes:
http://www.springer.com/series/7651

Computational Stem Cell Biology

Methods and Protocols

Edited by

Patrick Cahan

Department of Biomedical Engineering, Johns Hopkins University, Baltimore, MD, USA

Humana Press

Editor
Patrick Cahan
Department of Biomedical Engineering
Johns Hopkins University
Baltimore, MD, USA

ISSN 1064-3745 ISSN 1940-6029 (electronic)
Methods in Molecular Biology
ISBN 978-1-4939-9223-2 ISBN 978-1-4939-9224-9 (eBook)
https://doi.org/10.1007/978-1-4939-9224-9

This Humana Press imprint is published by the registered company Springer Science+Business Media, LLC, part of Springer Nature.
The registered company address is: 233 Spring Street, New York, NY 10013, U.S.A.

Preface

The title of this book, *Computational Stem Cell Biology*, bears some explanation. First, let me propose a definition. Computational stem cell biology (CSCB) is the field concerned with the generation and application of computational approaches to understand the unique properties of stem cells. One of the unique features of stem cells is, of course, their fate potency, a term that refers to the ability of a cell to select from and to become one of several distinct cell types. The other unique property of a stem cell is its ability to maintain its fate potency over time, often through many rounds of cell division—a process called self-renewal. Now, we have a slightly more expanded definition of CSCB: the development and application of computation to understand the basis of fate potency, cell fate decision-making, and self-renewal.

Why do we need computational tools to understand stem cells? There are at least two distinct reasons. The first is that in the absence of sufficient empirical data, it is difficult to pose, reason about, and evaluate competing hypotheses. Computer models have proven to be an invaluable aid in these kinds of situations, where rigorous methods are needed to generate and explore nontrivial hypotheses. In fact, in what might be considered one of the earliest CSCB studies, in 1952, Alan Turing used computational modeling to describe the theory of reaction diffusion as a mechanism to explain embryonic patterning, the process by which a collection of indistinguishable progenitor cells becomes organized into spatially and functionally distinct groups. We are no longer in a data-poor era, yet computer modeling is even more valuable now than in Alan Turing's day. This is due, in part, to the fact that the models needed to explain stem cell behavior are unquestionably complex, and the availability of more data is only increasing this complexity. Therefore, computer models can help us to explore and understand unanticipated aspects of our models. In this book, we have eight chapters that describe theory-grounded and practical modeling of stem cell behavior, ranging from stem-to-niche interactions to clonal dynamics to gene regulatory networks that contribute to multi-stability. Taken together, this section will provide a student or practitioner with the tools needed to jump in and use a range of modern computational modeling approaches customized for the analysis of stem cell behavior.

About 10 years ago, technologies to perform genome-scale measurements of single cells started to emerge, leading to the current single-cell OMICS revolution. Having a single-cell perspective of stem cells is incredibly valuable because of their very defining property: the ability to become something else. Rather than looking at a population of cells in which heterogeneity in state and fate tendency is obscured, single-cell approaches allow us to quantify this heterogeneity, infer dynamics, and identify regulators of fate decisions. In the second section of this book, we have five chapters that describe distinct methods to analyze single-cell genome-scale measurements in the context of stem cells: one chapter is an example of the rapidly proliferating "trajectory inference" methodologies which organize snapshots of single-cell profiles into dynamic biological trajectories, two chapters take different approaches to reconstructing gene regulatory networks (one in pluripotent stem cells, the second in hematopoietic stem cells), one chapter details a sophisticated tool for exploring cell fate decisions from scRNA-seq data, and one chapter describes how to identify stem cells from a population and to infer their lineage potential. Undoubtedly, as single-cell OMICS expand rapidly in the near future, we can expect more innovation in the development of computational methods, including in the areas of data integration, imputation, and cell-to-cell interactions.

The metabolic state of the cell is indelibly linked to stem cell behavior. Relatively speaking, the study of the metabolome, or the complete metabolite profile of a cell, in stem cells has been given relatively less attention as compared to transcriptomics and proteomics. This is now changing, and our third section includes two chapters dealing with metabolomics in stem cells. The first chapter describes how to model the dynamic metabolic networks of stem cells, whereas the second chapter gives a broader overview of metabolomics methods and applications to stem cells. Understanding stem cell metabolism is essential because, among other reasons, it allows us to design cell culture media that may be tailored to undifferentiated and stable growth, or to differentiation to selected lineages, or to maturation of the final stage cell population. As our need to culture stem cells in precisely defined media increases, we anticipate the development of new computational methods that distill insight from metabolomics data and that integrate it with other genome-scale measurements.

Since 2006, when Shinya Yamanaka demonstrated that pluripotent stem cells could be derived from somatic sources through the process of cellular reprogramming, the stem cell field has undergone a radical transformation and expansion. Much of this change has been driven by the promise of using pluripotent stem cells to model diseases in a dish, to discover and determine the safety of drugs, and ultimately to replace diseased or otherwise damaged tissues. However, a prerequisite for any of these applications is that pluripotent stem cell-derived cells are faithful reflections of the cell type that they are meant to represent or replace. The final section of this book deals with computational methods to assess the quality and fidelity of the so-called engineered cells and to devise improved cell fate engineering protocols. The first chapter describes a platform to select factors for cell fate reprogramming. The second describes an engineering approach to program cell fate with synthetic circuits. The third chapter describes a clever approach to infer DNA copy number variation in stem cells, and the fourth describes our approach to assess the fidelity of engineered populations for use in disease modeling studies.

In the near future, we expect that we will witness more attempts to bridge across the themes defined by the four sections of this book. Already, transcriptional and metabolic networks are being integrated to uncover the interplay between these two molecular readouts of cell state. It will be very exciting to see how computational modeling, such as the agent-based approach used in Chapter 1, might be adapted to capitalize on single-cell RNA-seq data, especially in applications to assess and improve cell fate. We hope that the methods provided here will be useful as you explore stem cells, and we hope that they will inspire you to make your own contributions to this nascent field.

Baltimore, MD, USA ***Patrick Cahan***

Contents

Contributors

HUSSEIN M. ABDALLAH • *Department of Electrical Engineering and Computer Science, Massachusetts Institute of Technology, Cambridge, MA, USA*

ANN C. BABTIE • *Department of Life Sciences, Centre for Integrative Systems Biology and Bioinformatics, Imperial College London, London, UK*

IPSITA BANERJEE • *Department of Chemical Engineering, University of Pittsburgh, Pittsburgh, PA, USA*

PATRICK CAHAN • *Institute for Cell Engineering, Johns Hopkins University, Baltimore, MD, USA; Department of Biomedical Engineering, Johns Hopkins University, Baltimore, MD, USA*

THALIA E. CHAN • *Department of Life Sciences, Centre for Integrative Systems Biology and Bioinformatics, Imperial College London, London, UK*

SRIRAM CHANDRASEKARAN • *Department of Biomedical Engineering, University of Michigan, Ann Arbor, MI, USA*

CAMDEN CHEEK • *Department of Biomedical Engineering, University of Michigan, Ann Arbor, MI, USA*

KAI-YUAN CHEN • *Department of Biomedical Engineering, Duke University, Durham, NC, USA; Center for Genomic and Computational Biology, Duke University, Durham, NC, USA*

J. A. CORNWELL • *Department of Life Sciences, Faculty of Dentistry, University of Sydney, Westmead Centre for Oral Health, Westmead Hospital, Westmead, NSW, Australia*

ANTONIO DEL SOL • *Computational Biology Group, Luxembourg Centre for Systems Biomedicine (LCSB), University of Luxembourg, Esch-Belval, Luxembourg*

DOMITILLA DEL VECCHIO • *Department of Mechanical Engineering, Massachusetts Institute of Technology, Cambridge, MA, USA*

SARA-JANE DUNN • *Microsoft Research, Cambridge, UK*

BERTHOLD GÖTTGENS • *Department of Haematology, Wellcome-MRC Cambridge Stem Cell Institute, University of Cambridge, Cambridge, UK*

JULIAN GOUGH • *Medical Research Council (MRC) Laboratory of Molecular Biology, Cambridge Biomedical Campus, Cambridge, UK*

PHILIP GREULICH • *Mathematical Sciences, University of Southampton, Southampton, UK; Institute for Life Sciences, University of Southampton, Southampton, UK*

DOMINIC GRÜN • *Max-Planck Institute of Immunobiology and Epigenetics, Freiburg, Germany*

FIONA K. HAMEY • *Department of Haematology, Wellcome-MRC Cambridge Stem Cell Institute, University of Cambridge, Cambridge, UK*

ANDRÁS HARTMANN • *Computational Biology Group, Luxembourg Centre for Systems Biomedicine (LCSB), University of Luxembourg, Esch-Belval, Luxembourg*

UMA S. KAMARAJ • *Program in Cardiovascular and Metabolic Disorders, Duke-NUS Medical School, Singapore, Singapore*

HU LI • *Department of Molecular Pharmacology and Experimental Therapeutics, Center for Individualized Medicine, Mayo Clinic, Rochester, MN, USA*

EMILY K. W. LO • *Institute for Cell Engineering, Johns Hopkins University, Baltimore, MD, USA; Department of Biomedical Engineering, Johns Hopkins University, Baltimore, MD, USA*

EDROALDO LUMMERTZ DA ROCHA • *Stem Cell Transplantation Program, Division of Pediatric Hematology and Oncology, Boston Children's Hospital and Dana-Farber Cancer Institute, Boston, MA, USA; Department of Biological Chemistry and Molecular Pharmacology, Harvard Medical School, Boston, MA, USA; Harvard Stem Cell Institute, Cambridge, MA, USA; Manton Center for Orphan Disease Research, Boston, MA, USA*

MOHAN MALLESHAIAH • *Division of Systems Biology, Montreal Clinical Research Institute, Montreal, QC, Canada*

DAVID MENN • *School of Biological and Health Systems Engineering, Arizona State University, Tempe, AZ, USA*

ROBERT MINES • *Department of Biomedical Engineering, Duke University, Durham, NC, USA*

R. E. NORDON • *Graduate School of Biomedical Engineering, University of New South Wales, Sydney, NSW, Australia*

JOHN F. OUYANG • *Program in Cardiovascular and Metabolic Disorders, Duke-NUS Medical School, Singapore, Singapore*

JOSE M. POLO • *Department of Anatomy and Developmental Biology, Monash University, Clayton, VIC, Australia; Development and Stem Cells Program, Monash Biomedicine Discovery Institute, Clayton, VIC, Australia; Australian Regenerative Medicine Institute, Monash University, Clayton, VIC, Australia*

OWEN J. L. RACKHAM • *Program in Cardiovascular and Metabolic Disorders, Duke-NUS Medical School, Singapore, Singapore*

MARK A. RAGAN • *Institute for Molecular Bioscience, The University of Queensland, Brisbane, QLD, Australia*

SRIKANTH RAVICHANDRAN • *Computational Biology Group, Luxembourg Centre for Systems Biomedicine (LCSB), University of Luxembourg, Esch-Belval, Luxembourg*

SAGAR • *Max-Planck Institute of Immunobiology and Epigenetics, Freiburg, Germany*

NATHAN SALOMONIS • *Division of Biomedical Informatics, Cincinnati Children's Hospital Medical Center, Cincinnati, OH, USA; Department of Pediatrics, University of Cincinnati School of Medicine, Cincinnati, OH, USA*

XILING SHEN • *Department of Biomedical Engineering, Duke University, Durham, NC, USA; Center for Genomic and Computational Biology, Duke University, Durham, NC, USA*

FANGZHOU SHEN • *Department of Biomedical Engineering, University of Michigan, Ann Arbor, MI, USA*

MICHAEL P. H. STUMPF • *Department of Life Sciences, Centre for Integrative Systems Biology and Bioinformatics, Imperial College London, London, UK*

ZHEN SUN • *Department of Basic Medical Sciences, Center for Stem Cell and Regenerative Medicine, The First Affiliated Hospital, Institute of Hematology, School of Medicine, Zhejiang University, Zhejiang, Hangzhou, China*

ATEFEH TAHERIAN FARD • *Australian Institute for Bioengineering and Nanotechnology, The University of Queensland, Brisbane, QLD, Australia*

KEITH TASK • *Department of Chemical Engineering, University of Pittsburgh, Pittsburgh, PA, USA*

XIAO WANG • *School of Biological and Health Systems Engineering, Arizona State University, Tempe, AZ, USA*

URI WEISSBEIN • *Department of Genetics, The Azrieli Center for Stem Cells and Genetic Research, Silberman Institute of Life Sciences, The Hebrew University, Jerusalem, Israel*
HUA YU • *Department of Basic Medical Sciences, Center for Stem Cell and Regenerative Medicine, The First Affiliated Hospital, Institute of Hematology, School of Medicine, Zhejiang University, Zhejiang, Hangzhou, China*
ZHONGDA ZENG • *Dalian ChemDataSolution Information Technology Co. Ltd., Dalian, China*
CHENYANG ZHANG • *Department of Basic Medical Sciences, Center for Stem Cell and Regenerative Medicine, The First Affiliated Hospital, Institute of Hematology, School of Medicine, Zhejiang University, Zhejiang, Hangzhou, China*
JIN ZHANG • *Department of Basic Medical Sciences, Center for Stem Cell and Regenerative Medicine, The First Affiliated Hospital, Institute of Hematology, School of Medicine, Zhejiang University, Zhejiang, Hangzhou, China*
JING ZHAO • *Department of Basic Medical Sciences, Center for Stem Cell and Regenerative Medicine, The First Affiliated Hospital, Institute of Hematology, School of Medicine, Zhejiang University, Zhejiang, Hangzhou, China*

Part I

Modeling

Chapter 1

Agent-Based Modelling to Delineate Spatiotemporal Control Mechanisms of the Stem Cell Niche

Robert Mines, Kai-Yuan Chen, and Xiling Shen

Abstract

Agent-based modelling (ABM) offers a framework to realistically couple subcellular signaling pathways to cellular behavior and macroscopic tissue organization. However, these models have been previously inaccessible to many systems biologists due to the difficulties with formulating and simulating multi-scale behavior. In this chapter, a review of the Compucell3D framework is presented along with a general workflow for transitioning from a well-mixed ODE model to an ABM. These techniques are demonstrated through a case study on the simulation of a Notch-Delta Positive Feedback, Lateral Inhibition (PFLI) gene circuit in the intestinal crypts. Specifically, techniques for gene circuit-driven hypothesis formation, geometry construction, selection of simulation parameters, and simulation quantification are presented.

Key words Stem cell niche, Cell fate differentiation, Biological control mechanisms, Agent-based modelling, Multi-scale modelling

1 Introduction

The human body is thought to be composed of roughly 37.2 trillion cells [1] that can be categorized into 200–2260 cell types [2] based on the differential expression of approximately 20,000 protein-coding genes [3, 4]. These changes in gene expression are driven by differential chromatin accessibility [5, 6], DNA methylation [6, 7], and transcription factor [8], which form intra- and intercellular feedback loops that control stem cell differentiation and cell fate decisions [9–11]. Therefore, tissue regeneration and homeostasis are complex processes [12–17].

The adult stem cell niches of various biological tissues could be considered a prime example of dynamical complexity. To replace dying cells and to recover from injury, most adult tissues in invertebrate and vertebrate animals contain specialized regions or compartments known as stem niches [18–21]. Peter Schofield was the first to develop the concept of the stem niche when he noticed that there was a relationship between the relative spatial positions of

Patrick Cahan (ed.), *Computational Stem Cell Biology: Methods and Protocols*, Methods in Molecular Biology, vol. 1975, https://doi.org/10.1007/978-1-4939-9224-9_1,

hematopoietic stem cells and colony-forming immune cells in the spleen of *Drosophila* that correlated with their differentiation state and the age structure of the population [22]. He hypothesized that when stem cells left local microenvironments known as the "stem niche," they would lose the signaling cues necessary to maintain stemness and undergo differentiation [22]. Further research has subsequently identified stem cell niches for hematopoietic stem cells in the bone marrow, bulge cells in the hair follicle, LGR5+ stem cells (ICSCs) in the crypts of the intestinal epithelium, muscle satellite cells in the basal lamina, and neural stem cells in the subventricular zone (SVZ) and for other cell types in diverse tissues [20, 21]. Currently, a stem niche is loosely defined as a tissue microenvironment that promotes the growth, altered gene expression and metabolism, and symmetric division of stem cells by providing spatially patterned gradients of transcription factors and other signaling molecules, an underlying layer of mesenchymal support cells and extracellular matrix that provides signaling cues and macroscopic structure, significant vascularization, and an interface for neural input [18–21, 23]. As stem cells leave their niche, they differentiate into one of the many cell types that surround their niche, providing signaling cues back to the niche to form feedback loops without central control and typically self-organizing into well-defined patterns.

To illustrate these points, this chapter will focus on the stem niches at the bottoms of the intestinal crypts of Lieberkühn [24]. The crypts (depicted in Fig. 1) are located at the bases of the intestinal villi and are composed of roughly 2000 cells in three distinct regions: the stem niche where roughly 15 LGR5+ stem cells and 15 Paneth cells reside in mosaic/checkerboard patterns, the transit amplifying (TA) region with 6–8 generations of rapidly dividing progenitor cells, and then the terminally differentiated region which contains absorptive enterocytes, mucous and digestive enzyme-secreting goblet cells, and other rare cell types like enteroendocrine and tuft cells [24–27]. Excluding the Paneth cells, the intestinal epithelium is largely replaced every 3–5 days as the mitotic pressure generated by cell division lower in the crypt drives the upper cells to undergo anoikis as they are pushed from the tip of the villus [26]. The differentiation events, total crypt cell count, length of the crypt, and patterning of the stem niche are maintained via an interconnected set of signaling pathways including Notch, Wnt, Ephrin B, BMP, and Hedgehog [24, 26, 28]. Most notably, Wnt proteins are secreted locally within a one-cell radius by Paneth cells, while the underlying mesenchyme also provides a gradient that is classically depicted as linearly decreasing from the crypt base to the villus [29]. Wnt upregulation of ß-catenin subsequently drives expression of Hes1 and homogeneously promotes stemness [11]. In contrast, Notch signaling upregulates stemness, but it is more commonly associated with

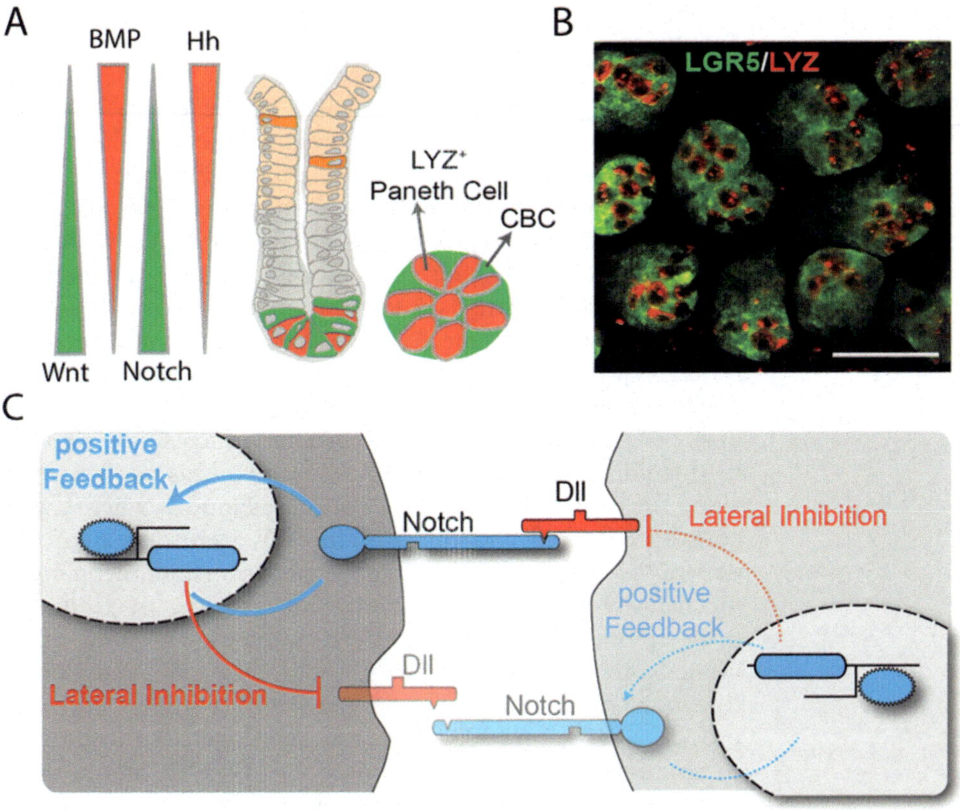

Fig. 1 (**a**) Schematic of a longitudinal cross section of the intestinal crypt. Paneth cells are represented in red, LGR5+ stem cells (CBC) in green, transit amplifying (TA) cells in gray, absorptive enterocytes in tan, and goblet cells in orange. The underlying gradients of Notch, Wnt, BMP, and Hedgehog (Hh) signaling are shown to the left. A schematic of the transverse cross section depicting the characteristic mosaic/checkerboard pattern of the crypt base is also shown. (**b**) Fluorescent micrograph of crypt bases of murine intestinal crypts with Paneth cells stained with an anti-Lysozyme antibody (red) and stem cells labeled with an LGR5-EGFP fusion protein (green). (**c**) Schematic illustration of a lateral inhibition, positive feedback mechanism for Notch-Delta signaling where opacity represents the strength of expression/activity. (Figures reprinted with permission from Chen et al., 2017, [9])

differentiation due to its juxtacrine lateral inhibition mechanism [9, 30]. Stem cells have Notch Receptors 1 and 2 (Notch1/Notch2) which can be activated by Paneth cell's Delta-like ligands 1 and 4 (DLL1/4), Jagged 1 (Jag1), or Atoh1 triggering upregulation of Hes1 in the stem cell and inhibition of its expression of Delta-like ligands preventing activation of Notch signaling in the Paneth cells [9]. In addition, all other secretory cells including goblet and enteroendocrine cells are also in a low Notch, high Delta state [31]. BMP forms a gradient that runs in the opposite direction from villus to crypt and is associated with cell cycle arrest and terminal differentiation of TA cells into absorptive enterocytes and goblet cells [32, 33]. Ephrin B is used to attract cells from the

+4 stem cell position to the crypt base after damage occurs to the Paneth cells, but features of its pathway and that of Hedgehog signaling are less well characterized [26].

Given the remarkable regularity of the crypt structure and given that a cell's position in a crypt is related to its age due to the mitosis-driven convection effect, significant amounts of experimental data exist on the intestinal crypt, and numerous modelling attempts have been made to describe them. Arguably, the earliest attempts to model tissue morphogenesis and its subsequent dysregulation in cancer date back to the pioneering work of Alan Turing on reaction-diffusion mechanisms as the source of biological patterns [34]. Early models of the intestinal crypts produced in the late 1980s and early 1990s by Loeffler and Potten [35, 36], Meinzer et al. [37], and Finney et al. [38] attempted to explain the variation in thymidine labeling of mitotically active cells and growth rates along the crypt villus axis using grid-based models where each point constituted a cell and the growth rates were dependent on the generation of the cell, without considering an underlying molecular mechanism driving differentiation or growth. In 1996, Collier et al.'s pioneering work on pattern formation via lateral inhibition in the Notch/Delta circuit, while topological in formulation, began to connect cell differentiation behavior to underlying signaling processes [30]. In 2003, Lee et al. developed the first rigorously parameterized, computational model of Wnt/ß-catenin signaling, accounting for all mechanistic steps, and utilized it to explain the significance of the core destruction complex proteins via parameter sensitivity analyses [39]. Lee's model has been the basis of almost all subsequent models of Wnt signaling, whether ODE or agent based (ABM), and has been subsequently modified to study the effects of cancer-associated APC mutations on ß-catenin levels [40], shuttling of an APC-ß-catenin complex to the nucleus [41], and the interaction between membrane-bound and cytosolic ß-catenin [42]. While bifurcation, robustness, and sensitivity analyses could be used to develop hypotheses about why certain signaling motifs were utilized, the predicted molecular phenotypes still could not be directly related to the macroscopic phenotypes observed in mammalian tissues.

To finally bridge the gap between the modelling approaches, the agent-based modelling paradigm was adapted from economics, epidemiology, and the social sciences in order to study the relationship between subcellular signaling and macroscopic tissue organization. This was first achieved for the intestinal stem cell niche in 2009 by Van Leeuwen et al. who developed an agent-based model in the *Cancer, Heart, and Soft Tissue Environment* (CHASTE) [43]. Using a hexagonal Voronoi tessellation where cells were connected by springs of varying stiffness on a square domain with periodic boundary conditions, they coupled the cellular growth rate, rate of mitotic events, and strength of cellular adhesion [44] to a simplified model of the Wnt dynamics proposed by Mirams et al. [45] and were able to accurately reproduce data from the previously described labeling experiments and was later used to study tumor initiation after mutations to the Wnt pathway [46]. Subsequently, many alternative multi-scale and agent-based frameworks became available linking the subcellular signaling to macroscopic behavior. Murray et al. created a continuum PDE model that accurately reproduced the results of Van Leeuwen [47] and used it to study invasion of other crypts by mutated cells [48]. Zhang et al. created another PDE model and utilized it to describe the crypt invagination process during development [49]. A number of other lattice-free methods were developed in three-dimensional geometries. Buske et al. created an elastic sphere model of intestinal crypts where cell fate decisions were determined using an active rule system (instead of a rigorous molecular modelling approach) that incorporated Wnt, Notch, and Ephrin B signaling to study cell fate decisions [50]. Du et al. used a subcellular element (modified FEM) approach to study the effects of Wnt diffusivity and degradation along with BMP negative regulation on the size of the stem niche [31]. Pin et al. used a granular TETRIS model to study regeneration of the crypt from a single cell using a rule-based approach to model differentiation [51, 52]. A summary of these models is included in Table 1.

In particular, Compucell3D offers a fundamentally distinct approach to modelling tissue morphogenesis and spatial organization [52]. This software package (available at http://www.compucell3d.org/) was developed by James Glazier's group to implement the Glazier-Graner-Hogeweg (GGH model, also known as a Cellular Potts Model) in a highly fast and efficient manner [53–56]. Unlike the deterministic, lattice-free methods, the GGH model places cells on a lattice of pixels and calculates an energy Hamiltonian based on their interactions at their borders and other user-defined energy corrections related to cell behavior [52, 53]. The energy landscape is then stochastically explored using a kinetic Monte Carlo algorithm to advance the model in time [53]. Additionally, this software package has many other convenient features including cross-platform availability, a graphical user interface to monitor simulations in real time, a wizard to automatically construct the energy correction terms in Python and add them into the base model, and the capacity to simultaneously solve SBML models for all cells in the simulation and use the resulting intracellular information to modify the behaviors of the cells as if they had differentiated into new cell types with unique macroscopic properties [52]. Given the adaptability and ease of use of this software, it has been employed in simulating the intestinal stem niche [9], tumor cell-type stratification [57], epithelial to mesenchymal transition [58], angiogenesis [59], xenobiotic

Table 1
Summary of the relative benefits and disadvantages of various multi-scale modelling frameworks along with references to papers using these approaches

Model Type	Benefits	Disadvantages	Refs.
Partial Differential Equation (PDE)	• Classical stability analysis and bifurcation methods can be directly employed. • Faster simulation times allow higher resolution parameter sweeps.	• Typically formulated in terms of cell densities instead of particulate cells. • Difficult to consider the effects of cell growth and division. • Simulating the motion of non-diffusing species such as cell bound receptors is problematic.	[47–49]
Rule-Based Off Lattice ABM	• Only requires a Boolean or topological understanding of the signaling pathway which is more appropriate for genome level models. • Easy to incorporate numerous cell types. • Less computationally intensive than coupling deterministic kinetics to cell behaviors. • Not affected by lattice anisotropy effects.	• Requires significant assumptions about the shapes of chemical gradients which are generally assumed to be static. • Cannot link the signaling pathway kinetics to the macroscopic behavior because they are not modeled. • No standard frameworks exist, and subcellular element methods to control cell shape and motion are both mathematically and computationally complex.	[31, 50–52]
Kinetic Off Lattice ABM (CHASTE)	• Most popular framework for modelling the intestinal crypt and cardiac tissues. • Provides deterministic simulation of tissues without a lattice via Voronoi tessellation or overlapping spheres. • Has support for PDE solution on irregular and evolving geometries. • Can also perform cellular automaton or Cellular Potts simulations, but these features are less frequently used.	• Primarily developed for Linux. MacOS version is also available, but only partial support for Windows exists. • SBML integration is still under development. • Voronoi tessellation methods are not as accurate as subcellular elements. • Dynamical models need to be developed relating the macroscopic behavior of the simulated cells to their intracellular and extracellular chemical environments.	[43, 46]
Glazier-Granner-Hogeweg ABM (Compucell3D)	• Most widely used framework for studying tissue morphogenesis. • Simulations are inherently stochastic allowing them to capture these effects. • Cell behavior is described via energy/Hamiltonian methods which are easier to modify than dynamical equations.	• The GGH model can exhibit some non-biological behaviors depending on the choice of simulation parameters and the formulation of the Hamiltonian. • Stochastic simulations are significantly more computationally intensive than deterministic or rule-based models.	[9, 57–61]

(continued)

Table 1 (continued)

Model Type	Benefits	Disadvantages	Refs.
	• Currently supports SBML integration, has a built in PDE solver, and cross-platform availability. • Built in wizards allow model development without requiring the user to write any code in some cases.	• Direct comparison of the model parameters and experimental data is non-trivial. • Model can experience lattice anisotropy.	

clearance in the liver [60], and fusion of the secondary palate bones [61].

Since there are many diverse approaches to agent-based modelling even for the stem niche, it is impossible to write a general protocol to agent-based modelling of biological self-organization and patterning problems since different problems require different levels of mechanistic detail and the capacity to model different kinds of interactions. Because Compucell3D is a relatively easy framework for beginners to convert a subcellular signaling model into an agent-based simulation, the subsequent protocols will focus on Compucell3D and how it can be used to study stem cell dynamics in the intestinal niche. Specifically, this chapter will present the basics of formulating a gene circuit and converting it to a model in SBML, setting macroscopic cell behavior parameters in Compucell3D, initializing the stem niche geometry, and quantifying the results of an agent-based simulation.

2 Glazier-Graner-Hogeweg Model and Its Implementation in Compucell3D

2.1 Development of the Cellular Potts and GGH Hamiltonians

As previously stated, the GGH model is fundamentally distinct from the deterministic, lattice-free models since it is based on the Ising model from statistical mechanics rather than the traditional biomechanical models used to describe tissues [53–55, 62]. The original Ising model was used to describe continuous, second-order phase transitions that lead to the development of permanent magnetic fields in ferromagnetic materials (like iron) as they are cooled below their Curie temperature (T_C) leading to spontaneous alignment of the quantum mechanical spins (σ) of their constituent atoms [53, 63]. In the Ising model, the atoms are placed at regularly spaced lattice points in one-, two-, or three-dimensional Euclidean space, spins can only be oriented up or down ($\sigma = 1, -1$), atoms only experience nearest neighbor effects, and classical

Boltzmann statistics can be used to derive the probability of a configuration (Eq. 1) [53, 63].

$$P(\{\sigma_i\}) = \frac{1}{Z}e^{-\frac{H(\{\sigma_i\})}{kT}}$$
$$Z = \sum_{\sigma_i} e^{-\frac{H(\{\sigma_i\})}{kT}} \quad (1)$$

In the Hamiltonian for the Ising model, all spins have the same magnitude J, but the sign of the energy contribution is determined by the nature of the interactions. Favorable homotypic interactions with aligned spins contribute $-J$, while heterotypic interactions contribute $+J$. This can be expressed concisely as shown in Eq. 2 [53, 63].

$$H_{\text{Ising}} = -\frac{J}{2} \times \sum\nolimits_{(i,j)\in \text{Adj}} \sigma(i)\sigma(j) \quad (2)$$

Renfrey Potts later generalized the Ising model to accept multiple degenerate spins and reformulated it to only consider energy contributions from heterotypic interactions that occur on the borders of regions with the following Hamiltonian in Eq. 3 [53, 64].

$$H_{\text{Potts}} = J \times \sum\nolimits_{(i,j)\in \text{Adj}} (1 - \delta(\sigma(i)\sigma(j))) \quad (3)$$

While this Hamiltonian allowed adhesion between boundaries, it still lacked the generality to describe numerous distinct cell types and was restricted to a single homogeneous contact energy. Additionally, there was no way to address size and shape constraints on the regions [53]. (At this time, an algorithm to generate stochastic trajectories, known as the Metropolis Algorithm, had already been developed. However, it will be introduced after discussion of the GGH Hamiltonian).

Weaire and Komode came up with an extension of the Potts Hamiltonian to study growth of bubbles in soap froths [65, 66]. The Potts Hamiltonian inherently leads to domain coarsening and the formation of large regions to minimize the surface area to volume ratio, thus minimizing the boundary energy to volume ratio [53]. However, to make the process occur more smoothly, they imposed an additional elastic constraint to drive the regions' volumes ($v(\sigma)$) to a given target volume ($V_t(\sigma)$) and scaled this new energetic term with the inverse compressibility of the gas (λ). The updated Hamiltonian is presented in Eq. 4 [53, 65, 66].

$$H_{\mathrm{WH}} = J \times \sum_{(i,j)\in \mathrm{Adj}} (1 - \delta(\sigma(i)\sigma(j))) + \lambda \sum\nolimits_{\sigma} (v(\sigma) - V_t(\sigma))^2 \quad (4)$$

These volume constraints are incredibly useful in cellular agent-based modelling since it prevents uncontrolled growth but does not require exact modelling of the intrinsic growth pathways.

In 1991, Glazier and Graner proposed that the Weaire-Komode formulation of the Potts Hamiltonian could potentially be used to test the differential adhesion hypothesis (DAH) of embryogenesis [53]. This hypothesis claims that underlying patterns of gene expression result in changes in the strength of cell-cell adhesion and that the cells in a tissue will reorganize like a highly viscous fluid leading to the emergence of structures that minimize the tissue's total free energy [67, 68]. Specifically, Glazier and Graner demonstrated that cells could be discretized onto a lattice of pixels and that the Weaire-Komode Hamiltonian could be used to describe them if they assigned each pixel of a cell an index $\sigma_i \in \{1, \ldots, N\}$ corresponding to the cell it was a part of and introduced the concept of cell types $\tau(\sigma_i)$ [69, 70]. Last, they generalized the contact energy J to be a symmetric function of the adjacent cell types $J_{ij} = J(\tau(\sigma_i), \tau(\sigma_j))$. The resulting cellular Potts model (CPM) Hamiltonian is shown in Eq. 5 [69, 70].

$$H_{\mathrm{CPM}} = \sum_{(i,j)\in \mathrm{Adj}} J_{ij}(1 - \delta(\sigma(i)\sigma(j))) + \lambda \sum\nolimits_{\sigma} (v(\sigma) - V_t(\sigma))^2 \quad (5)$$

While incredibly flexible, this formulation of cell behavior has a few limitations. First, while bacteria can experience thermal membrane motion, mammalian cells tend to experience long-term, motor-driven changes in their cytoskeletons that are not describable by equilibrium Boltzmann statistics [53, 54]. Second, at sufficiently high temperatures, membrane regions can disconnect from the cells and form "blebs" which rapidly dissipate but can confound calculations of cell morphology and volume [53, 54]. Third, researchers tend to use positive values for the contact energies J_{ij} since this inherently minimizes boundary areas. However, cells tend to be stabilized by their contacts with other cells relative to the surrounding media. In the positive contact energy formulation, smaller cells will be able to undergo higher penalty boundary fluctuations than larger cells and diffuse faster, which is erroneous since larger cells move faster biologically [53, 54]. While the results are still relatively accurate and recapitulate numerous emergent features of biological systems, two corrections are possible: First, one can use negative contact energies and introduce a surface area

(3D) or circumference (2D) constraint on the cells. Second, one can add an additional dissipative term to the Hamiltonian to associate costs with all cell movements during the time step (which will be described in greater detail in the next section) [53, 54].

Finally, Paulien Hogeweg proposed a series of extensions to the Hamiltonian and other non-Hamiltonian extensions to the CPM model that later became known as the GGH model [71–73]. For instance, an elastic constraint for surface area ($s(\sigma)$) was proposed (Eq. 5) [53].

$$H_{\mathrm{Surf}} = \sum\nolimits_{\sigma} \lambda_s(\tau)(s(\sigma) - S_t(\tau(\sigma)))^2 \tag{6}$$

Other modifications include introducing equations that adjust the target volumes in response to concentrations of growth factors [71–73], introduction of fields of diffusing chemicals [74], and mitotic events in cells [75]. Given the incredible flexibility of the Hamiltonian formulation of these models, it would be impossible to cover all possible extensions of the model, and they will be addressed as necessary in the model formulation step of the protocol. For a complete list of model extensions, we direct the reader to the Compucell3D documentation for further information including representative code examples [76]. Additionally, many of these features can be directly incorporated into the model using Compucell3D's built-in wizard shown in Fig. 2 [52, 77].

2.2 Stochastic Simulation via the Metropolis Algorithm

As alluded to previously, instead of simply using the Hamiltonian to calculate the equilibrium statistics of the system, the GGH model utilizes the Metropolis Algorithm to simulate the system's approach to equilibrium or its evolution in general (if nonrandom behaviors like growth and mitosis are included) [52, 53, 77]. In statistical mechanics, the bulk macroscopic properties of a material are inferred from the molecular behavior of its constituent parts which evolve according to a simple Hamiltonian. Specifically, these bulk properties are calculated via ensemble averaging over the Boltzmann distribution of the configuration space (Eq. 1) as shown in Eq. 7 [78].

$$\begin{aligned} A &= \sum\nolimits_{\sigma_i} A(\sigma_i) P(\{\sigma_i\}) \\ &= \frac{1}{Z} \sum\nolimits_{\sigma_i} A(\sigma_i) \mathrm{e}^{-\frac{H(\{\sigma_i\})}{kT}} = \frac{\sum_{\sigma_i} A(\sigma_i) \mathrm{e}^{-\frac{H(\{\sigma_i\})}{kT}}}{\sum\limits_{\sigma_i} \mathrm{e}^{-\frac{H(\{\sigma_i\})}{kT}}} \end{aligned} \tag{7}$$

The Metropolis Algorithm was originally developed to rapidly and efficiently perform the summation or integration over the configuration spaces of many particle, molecular systems to relate the microscopic Hamiltonians to bulk chemical properties via the

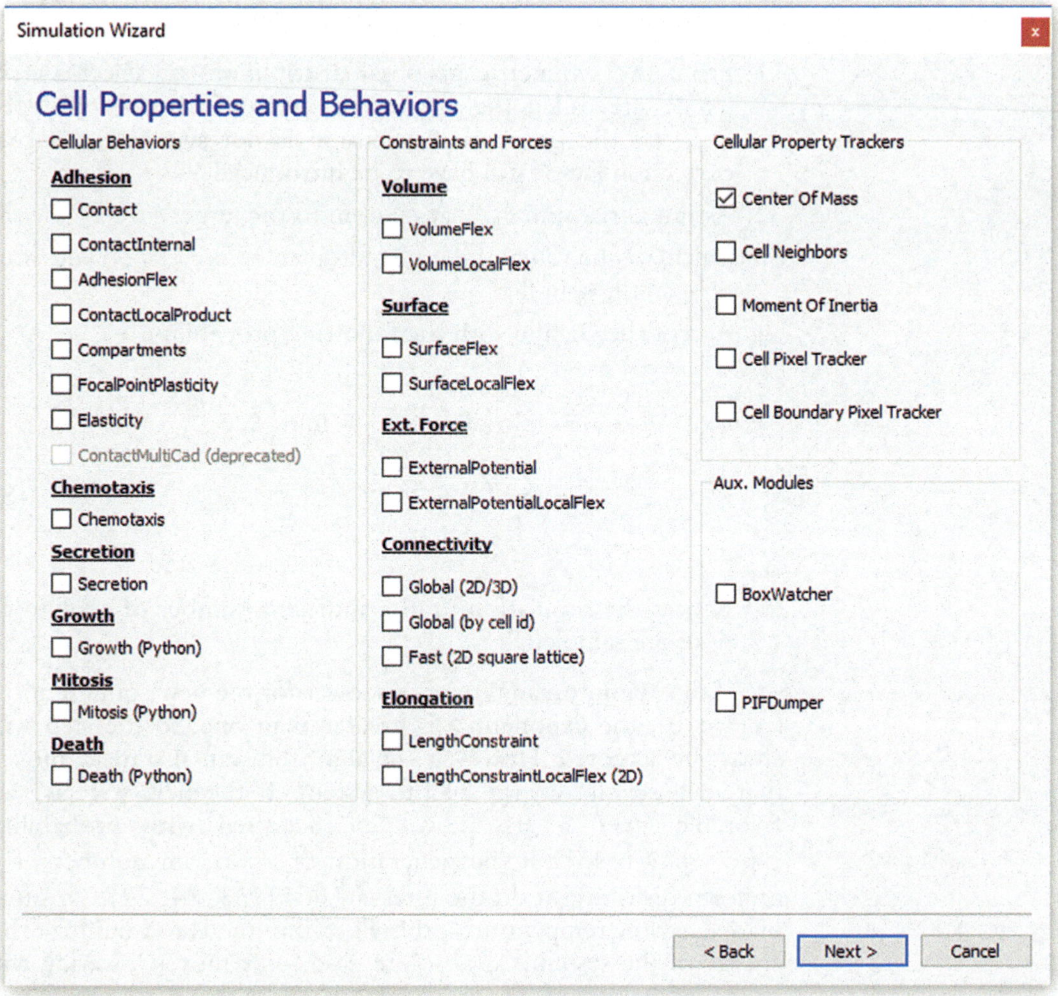

Fig. 2 Compucell3D (CC3D) includes a program called Twedit++ which functions as an editor for CC3D XML (.cc3d) files or Python scripts for the underlying simulation. One of the most useful features of Twedit++ is the simulation wizard which provides a graphical interface for users to select features that they want to add to the model Hamiltonian and specify representative parameter values. This window shows pre-defined Hamiltonian extensions that the user can select from with no programming experience required [12]

ensemble averaging [79, 80]. In many particle systems, the summation or integrable in the normalization constant Z becomes computationally intractable since the Hamiltonians represent the summation of the molecular potentials of hundreds of particles, and the integral becomes at least 3 N dimensional (where N is the number of particles in a three-dimensional box). However, the Metropolis Algorithm avoids this dimensionality problem by considering the relative probability of moving to a new state versus remaining in the current state by taking a ratio that eliminates the normalization constant [79–81].

In the context of the Ising or Potts models, the Metropolis Algorithm works as described below [53, 79–81]:

1. From the symmetric proposal distribution τ_{ij}, select a new lattice site called the target site, and record its target spin σ_{Old}. (If the proposal distribution is not symmetric, a bias correction factor will have to be introduced.)
2. Assign a trial spin σ_{New} at random to the target site.
3. Calculate the value of the Hamiltonian before (H_{Old}) and after (H_{New}) the spin flip.
4. Accept the change with the following probability:

$$p(\sigma_{\text{Old}} \to \sigma_{\text{New}}) = \min\left(1, \frac{P_{\text{New}}}{P_{\text{Old}}}\right) = \min\left(1, e^{-\frac{H(\sigma_{\text{New}})-H(\sigma_{\text{Old}})}{kT}}\right)$$
$$= \min\left(1, e^{-\frac{\Delta H}{kT}}\right) \tag{8}$$

5. Repeat the simulation until a sufficient number of configurations are sampled.

If the Hamiltonian (energy) is lower for the new configuration ($\Delta H < 0$), the exponential is greater than one, so the step will always be accepted. However, the algorithm can also make moves that increase the energy of the system. If the new state is less favorable ($\Delta H > 0$), it will be accepted with probability $\exp\left(-\frac{\Delta H}{kT}\right)$ based on the generation of a random number of a uniform distribution on the interval [0, 1] [53, 54, 79]. To summarize, at low temperatures, this algorithm tends to minimize the energy of the system. At absolute zero, only moves lowering the energy of the system are permitted until the system gets trapped in a local or global minimum. At high temperatures or in the limit of infinite temperature, all moves become equally likely.

2.3 Implementation of the Metropolis Algorithm in Compucell3D

The basic Metropolis Algorithm is employed with little modification to simulating pixel copy attempts in GGH model simulations. The application of the algorithm to a single pixel copy is presented in Fig. 3 [52, 76, 77]. During a single Monte Carlo Step (MCS), CC3D attempts a pixel copy at every lattice point. If a pixel of a cell is contacting a neighboring cell, the neighboring cell can attempt to copy the pixel. The basic CPM Hamiltonian and all Hamiltonian extension terms are calculated. If the pixel flip lowers the energy of the system, it is immediately accepted. If not, the ratio of the acceptance probability is calculated and compared to a random number to determine if the transition to a higher energy state is accepted [52, 76, 77]. After iterating over all lattice sites, the changes are implemented, and the results for the MCS are displayed

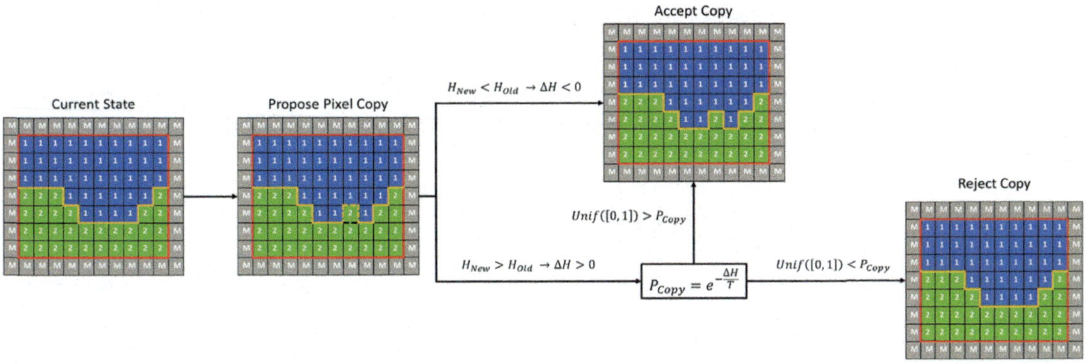

Fig. 3 The procedure for a pixel copy attempt in the GGH model implemented in Compucell3D. Blue squares represent cell type 1, green squares represent cell type 2, and gray squares represent the tissue culture media. Two interaction energies are present indicated by the yellow and red borders. The interaction energy between cell types 1 and 2 is less than their interaction energy with the surrounding media. For each pixel flip attempt, a pixel is chosen at random, and the Hamiltonian is recalculated. If the new Hamiltonian value is lower, the pixel copy is automatically accepted. If it is higher, a random number is generated from a uniform distribution on the interval [0, 1] and compared to the flip probability P_{Flip} to determine if the flip occurs

in the user interface. As stated previously, Hogeweg suggested a modification to the acceptance probability that adds in dissipative effects (Eq. 9) [53, 54, 62].

$$P_{\text{Accept}} = \mathrm{e}^{-\frac{\Delta H - \Delta E_0}{kT}} \tag{9}$$

However, this term is not always included and may not be necessary if all cells have roughly the same volume.

3 Methods

3.1 Formulating a Mathematical Model in SBML

One feature of the Compucell3D platform is the capacity to directly integrate mathematical models of subcellular signaling and metabolic networks into the agent-based simulation by reading it in through a Systems Biology Markup Language (.sbml) file [82]. The original goal of SBML was to provide a format for biological models to increase reproducibility, readability, and reusability of biological models by providing an easily editable format to develop models in [82]. Today, hundreds of software packages are available to automatically simulate SBML files as ODEs, PDEs, or stochastically and to conduct other high-level analyses on the models (*see* **Note 1**). Additionally, thousands of models are stored in an SBML showcase that is publicly available. Additional information on the available software and models can be found at the official SBML site (http://sbml.org/) [83].

While numerous software packages exist to formulate SBML models, tellurium is recommended for use with Compucell3D [76, 84]. Users can simply specify the reactions they desire to simulate in a simple notation along with the parameter values and initial conditions, and tellurium will automatically simulate it deterministically or stochastically and convert the model to the significantly longer SBML format [84]. As an example, consider the Positive Feedback, Lateral Inhibition (PFLI) model of Notch-Delta signaling that our lab developed for use with Compucell3D [9].

$$
\begin{aligned}
\frac{\mathrm{d}[N_i]}{\mathrm{d}t} &= \beta_{N0} + \frac{\beta_N (D_j[N_i])^p}{K_N^p + (D_j[N_i])^p} - \alpha_N[N_i] - k_c[N_i][D_i] - k_t[N_i]D_j \\
\frac{\mathrm{d}[D_i]}{\mathrm{d}t} &= \beta_{D0} + \frac{\beta_D K_D^h}{K_D^h + (D_j[N_i])^h} - \alpha_D[D_i] - k_c[N_i][D_i] - k_t[N_i]D_j \\
\frac{\mathrm{d}[R_i]}{\mathrm{d}t} &= k_t[N_i]D_j - \alpha_R[R_i]
\end{aligned}
\tag{10}
$$

In the equations above, N, D, and R are the concentrations of Notch Receptor, Delta-like ligands, and Notch Intracellular Domain in cell i and neighboring cells j. The rate of receptor ligand interaction that produces R is proportional to $[N_i][D_i]$ for cis interactions and proportional to $[N_i]D_j$ for trans interactions, where D_j indicates the average concentration of Delta in the neighboring cells. Expression of Notch is autocatalyzed by NICD, so it is modelled as an activating Hill function of $[N_i]D_j$. In contrast, Delta is suppressed by downstream targets of NICD, so it is modelled as a repressing Hill function of $[N_i]D_j$ [9]. The first reaction can be modelled in tellurium as follows [84]:

$$
\begin{aligned}
\$X \Rightarrow N;\ &BN0 + BN^{*}(\mathrm{Davg}^{*}N)\hat{}p/(KN\hat{}p + (\mathrm{Davg}^{*}N)\hat{}p) \\
&- AN^{*}N - Kc^{*}N^{*}D - Kt^{*}N^{*}\mathrm{Davg};
\end{aligned}
$$

In this notation, "\$" denotes that the line of code is a reaction. "$X \Rightarrow N$;" indicates that Notch is increased by this reaction, but we are not directly coupling its stoichiometry to any of the other species. The final part of the line is just the rate law for Notch from Eq. 10. After specifying parameter values and initial conditions further down in the code, the equations can be simulated and plotted with two lines of code or written to a *.sbml* file. The complete tellurium *.py* file is presented in Fig. 4 along with representative simulation results.

```
import tellurium as te
import roadrunner

r = te.loada('''
// Created by libAntimony v2.9.4
model *cell()

  // Compartments and Species:
  compartment compartment_;
  species $Davg in compartment_, $X in compartment_, $Navg in compartment_;
  species D in compartment_, N in compartment_, B in compartment_, R in compartment_;

  // Reactions:
   $X => N; BN0 + BN*(Davg*N)^p/(KN^p + (Davg*N)^p) - AN*N - Kc*N*D - Kt*N*Davg;
   $X => D; BD0 + BD*KD^h/(KD^h + (Davg*N)^h) - AD*D - Kc*N*D - Kt*D*Navg;
   $X => B; 0;
   $X => R; Kt*Davg*N - AR*R;

  // Species initializations:
  Davg = 0;
  X = 0;
  Navg = 0;
  D = 0;
  N = 0;
  B = 0;
  R = 0;

  // Compartment initializations:
  compartment_ = 1;

  // Variable initializations:
  BN0 = 10;
  BN = 10;
  p = 3;
  KN = 0.0042;
  AN = 1;
  Kc = 0;
  Kt = 1;
  BD0 = 0;
  BD = 100;
  KD = 0.015;
  h = 3;
  AD = 1;
  AR = 1;

  // Other declarations:
  const compartment_, BN0, BN, p, KN, AN, Kc, Kt, BD0, BD, KD, h, AD, AR;
end
''')

result = r.simulate(0, 40, 500)
r.plot(result, xtitle = 'Time', ytitle = 'Concentrations (nM)')

sbml_file = open('C:\CompuCell3D-64bit\Projects\CC3D_Crypt\SBML_Models\Tellurium_Test.sbml' , 'w')
print >> sbml_file,r.getSBML()
sbml_file.close()
```

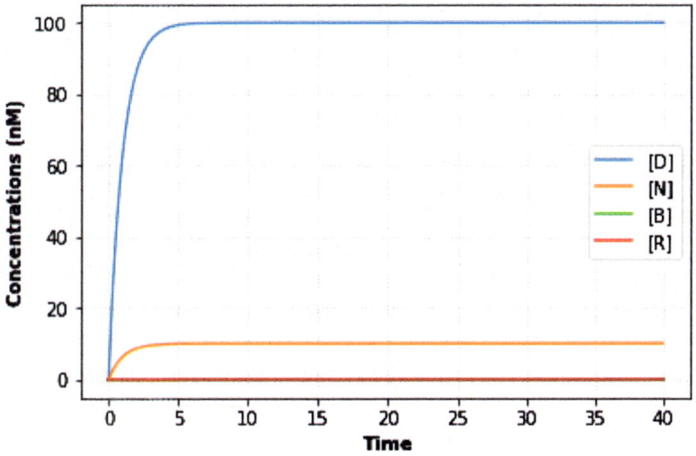

Fig. 4 Representative tellurium file written in Python. Reactions are represented by an $\Rightarrow$ symbol while parameters are specified with $=$. The model is simulated in the first block of text after the commented region and written into an SBML file in the last block. The inset figure represents the results of the simulation

3.2 Developing Hypotheses for Simulation via Stability and Bifurcation Analysis

Fundamentally, agent-based modelling should be treated as a computational experiment instead of a mathematical model. In GGH models, the cells are assigned random shapes and random initial conditions and allowed to evolve stochastically. Therefore, replicate simulations and error or uncertainty analysis must be performed. Additionally, depending on the number of cells and complexity of the SBML model, a single simulation run could take on the order of minutes to hours. Last, agent-based simulations can introduce artifacts derived from non-biological behavior or due to changes in physical dimensionality (e.g., diffusion constants are altered in different dimensions) [53, 54, 62]. Therefore, thorough investigation of the underlying dynamical properties of the gene circuit, in a single-cell or two-cell model, should be performed prior to its integration into an agent-based model.

Ideally, any analysis of a dynamical system should begin with calculation of steady states. In this terminology, steady states refer to points where all derivatives in the system of ODEs go to zero, but being at steady state does not mean that the system is stable with respect to perturbations from the steady state [85, 86]. For instance, perturbations from steady state could cause the system to transition to a stable steady state, to diverge to infinity, or to permanently circle the steady state in a limit cycle. Steady states should be found analytically whenever possible, but in many cases, the models are too complex for simple algebraic methods. Under these circumstances, the problem can be formulated as a least squares minimization problem and solved using built-in tools in MATLAB, Python, or Mathematica, or numeric integration can be performed to determine the long-term behavior of the system starting from various initial conditions [9]. This approach was utilized to determine the number of steady states and their values for the intracellular concentration of Delta D_i as a function of $\langle D_j \rangle$ in a one-cell model while considering the effects of changing the Hill Function coefficients p and h revealing bistable behavior as the functions became increasingly nonlinear [56].

After identification of the steady states, the stability of these points should be determined. For one-dimensional, first-order autonomous systems, this analysis proceeds as follows. Consider the following general ODE in Eq. 11 [85, 86].

$$\frac{dx}{dt} = f(x) \tag{11}$$

Defining the steady-state concentration as $\bar{x}$ such that $f(\bar{x}) = 0$, let us now transform the system to a new variable $y = x - \bar{x}$, and Taylor expand the function $f(x)$ around the steady state (Eq. 12) [85, 86].

$$
\begin{gathered}
\frac{\mathrm{d}y}{\mathrm{d}t} = \frac{\mathrm{d}y}{\mathrm{d}x}\frac{\mathrm{d}x}{\mathrm{d}t} = f(x) \\
\frac{\mathrm{d}y}{\mathrm{d}t} = f(\bar{x}) + f'(\bar{x})(\mathrm{x} - \bar{\mathrm{x}}) + \frac{1}{2}f''(\bar{x})\Delta(x - \bar{x})^2 + \dots \\
\frac{\mathrm{d}y}{\mathrm{d}t} = f(\bar{x}) + f'(\bar{x})\mathrm{y} + \frac{1}{2}f''(\bar{x})\mathrm{y}^2 + \dots \\
\frac{\mathrm{d}y}{\mathrm{d}t} \approx f'(\bar{x})y = ay \\
y = y_0\exp(at) \\
x = \bar{x} + (x_0 - \bar{x})\exp(at)
\end{gathered}
\tag{12}
$$

If $a < 0$, perturbations from steady state will decay until the system returns to the steady state. In contrast, if $a > 0$, perturbations from steady state can grow without bound. For an N-dimensional system, the vector equivalent of a is the Jacobian Matrix J, and the system can be linearized in terms of the Jacobian (Eq. 13) [85, 86].

$$
\begin{gathered}
J = \begin{bmatrix} \frac{\mathrm{d}f_1}{\mathrm{d}x_1} & \dots & \frac{\mathrm{d}f_n}{\mathrm{d}x_1} \\ \vdots & \ddots & \vdots \\ \frac{\mathrm{d}f_1}{\mathrm{d}x_n} & \dots & \frac{\mathrm{d}f_n}{\mathrm{d}x_n} \end{bmatrix} \\
\frac{\mathrm{d}\vec{x}}{\mathrm{d}t} \approx J(\vec{x} - \bar{x})
\end{gathered}
\tag{13}
$$

While one cannot directly exponentiate the matrix J, the matrix can be diagonalized in terms of its eigenvalues $\{\lambda_i\}$. This effectively decouples all the equations and allows us to understand the asymptotic behavior in terms of $\exp(\lambda_i t)$ [85, 86]. If all $\{\lambda_i\}$ are real, the linearized solution will diverge smoothly from steady state if any $\lambda_i > 0$, or it will converge smoothly toward it if all $\lambda_i < 0$. If any of the eigenvalues is complex, the solution will exhibit oscillations, but it will still converge or diverge to the steady state based on the real part of the eigenvalues. If the real parts of the eigenvalues are zero, the solution will be purely oscillatory and will converge permanently to a limit cycle around the steady state [85, 86].

At times, linearizing the system in terms of the Jacobian can be cumbersome. An alternative approach is to simply plot the null-clines (curves or surfaces where each ODE in the system equals zero) and overlay the vector field in a two- or three-dimensional phase plane to determine the stability of the steady states graphically [85, 86]. This approach is demonstrated in Fig. 5b revealing that the concentration of the Notch receptor in a two cell model functions as a bistable switch (three steady states, extreme points stable, central point unstable) when there is sufficient nonlinearity in the Hill coefficients [9].

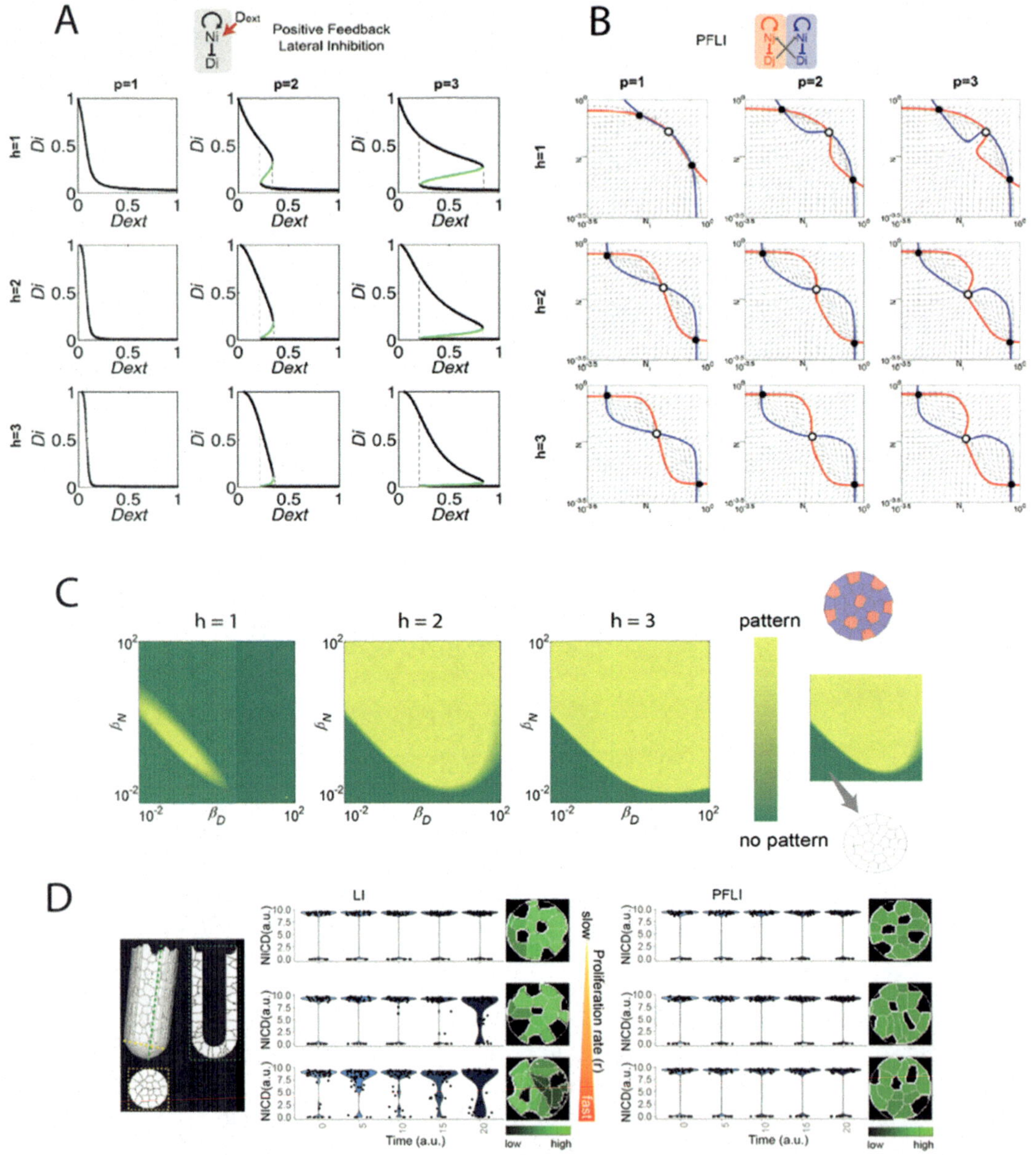

Fig. 5 (**a**) Steady states of intracellular Delta levels D_i as a function of external/average Delta concentration D_{ext} over various values of the Hill coefficients n and p. The black lines represent stable steady states, while the green represents an unstable steady state. The system shows bifurcation behavior as it transitions from monostability to bistability back to monostability. (**b**) Phase plane analysis of the Notch receptor steady state in a two-cell model while varying Hill coefficients h and p. (**c**) Bifurcation analysis of a multicell model of the gene circuit while varying the max promoter strengths β_N and β_D. The regions were determined by the maximum Lyapunov exponent method, and yellow regions correspond to the parameter sets that generated a pattern. The green regions did not produce a pattern. (**d**) Capacity of the agent-based model to produce patterning at the crypt bottom as a function of cell division rate and positive feedback strength. The left portion lacks feedback, while the right portion has the Notch autocatalysis mechanism indicating that positive feedback improves robustness. (Figure reprinted with permission from Chen et al., *Mol Sys Bio*, [9])

Finally, bifurcation analysis should be performed. Mathematically, bifurcation refers to a topological transition in the dynamical behavior of a system of differential equations in response to a smooth change made to a parameter value [85, 86]. For instance, a monostable solution could become bistable and function as a switch or begin to oscillate as the strength of a feedback loop was increased. Bifurcations are observed in Fig. 5a, b in response to changing the activation and repression Hill coefficients. Figure 5c shows an increase in the size of the region of bistability (yellow) of the PFLI in response to changing the promoter strengths for Notch and Delta (β_N, β_D) for a multicell simulation of the circuit [9]. To conduct a bifurcation analysis like these, one simply has to conduct a parameter sweep over a grid of two to three parameter values and report a variable of interest such as the number of steady states or the maximum Lyapunov coefficient to assess bistability or multistability. Alternatively, one could investigate the number of autocorrelation peaks or Fourier transform peaks to determine if the behavior was oscillatory. For high-dimensional parameter spaces ($N \geq 4$), systematic sweeps over parameter spaces become computational intensive and hard to visualize, so alternative Monte Carlo methods should be used for sampling (*see* **Note 2**).

After thoroughly investigating the circuit, hypotheses for simulation can finally be formed. For instance, the motivation behind the PFLI gene circuit was to demonstrate that the addition of a positive feedback to the existing lateral inhibition (LI) mechanism would increase the robustness of the gene circuit to perturbation [9]. Armed with the information of where bistability should occur within the parameter space, this circuit could be tested with and without the positive feedback term in a model of the crypt geometry while varying the proliferation rate. At higher proliferation rates, the PFLI circuit maintained a better differentiation pattern at the crypt base relative to the LI circuit supporting the prior bifurcation analysis as shown in Fig. 5d [9].

3.3 Setting Simulation Parameters and Constructing a Geometry in Compucell3D

Compucell3D projects at the bare minimum are composed of a main python script and a collection of Python steppables that update the Hamiltonian during the Monte Carlo step. A CC3D . xml file can also be included to rapidly access parameter values but is not necessary and can be constraining for advanced programming. The first parameters that need to be set in Compucell3D are the lattice type, lattice dimensions, temperature, neighbor order, cell types, and contact energies [52, 53, 76].

To begin, the lattice type must be set to rectangular or hexagonal. Hexagonal lattices benefit from reduced lattice anisotropy (*see* **Note 3**), but they are harder to specify custom geometries in [53, 76]. Accordingly, it is simpler to utilize rectangular lattices for complex geometries, but this should be compensated for by increasing the neighbor order. Neighbor order refers to the group of pixels from which average properties are calculated during a pixel copy attempt. First order refers to the pixels immediately to the left and right and top and bottom of the current pixel. Second order includes adjacent pixels that are diagonal to target pixel. Third order includes pixels in the second shell around the target pixel, and higher-order effects can also be considered [76]. While smoother boundaries and simulation results will be obtained with higher neighbor orders, it significantly increases the computational complexity of the simulation, so accuracy must be balanced with speed.

Temperature in Compucell3D does not refer to temperature in the classic thermodynamic sense, but it actually refers to the average fluctuation amplitude of the cell membrane in the simulations. Referring back to Eq. 12, the probability of accepting an unfavorable increase in the energy of the Hamiltonian is related to the temperature of the system [53, 54]. However, the GGH model can exhibit artificial phase transitions due to poor choices of temperature. At extremely low temperatures, the ratio $\frac{\Delta H}{kT}$ becomes much larger than 1, so the probability of accepting a move that increases energy approaches 0. In these cases, the system will "anneal" to a local minimum in the energy where the pixels align along the lattice in the most favorable manner freezing the system in an anisotropic state [53, 76]. In contrast, in the limit of infinite temperatures, the ratio approaches 0 driving the acceptance probability of all moves to 1 regardless of the energy penalty. This creates a scenario where cells (especially in 3D environments) spin off small packets of pixels known as "blebs" as if the cells were an evaporating liquid. The approximate range of temperatures contact energy J and n lattice site neighbors is given by Eq. 14 [53, 76].

$$0.2 \leq \frac{\Delta H}{T} \approx \frac{Jn}{2T} \leq 2 \tag{14}$$

Next, the cell types, contact energies, and initial configuration need to be specified. While Compucell3D has a built-in blob initializer for randomized and stratified tissue sheet starting conditions on hexagonal and rectangular lattices, the hallmark of the stem niche (including the hair follicle and intestinal crypts) is the unique test tube-like geometry that it exhibits (Fig. 6) [24, 52, 56]. Additionally, a mechanism for anoikis-mediated cell death as the mitotic pressure drives cells of the villus tip needs to be considered. In this section, a method to generate this geometry and constraint on cell death will be presented.

One limitation of the Compucell3D software relative to the more biomechanical ABM frameworks is that rigid structures cannot be directly generated. The creation of a rigid structure will be accomplished via definition of additional cell types. By default, Compucell3D starts every simulation with a "medium" cell type

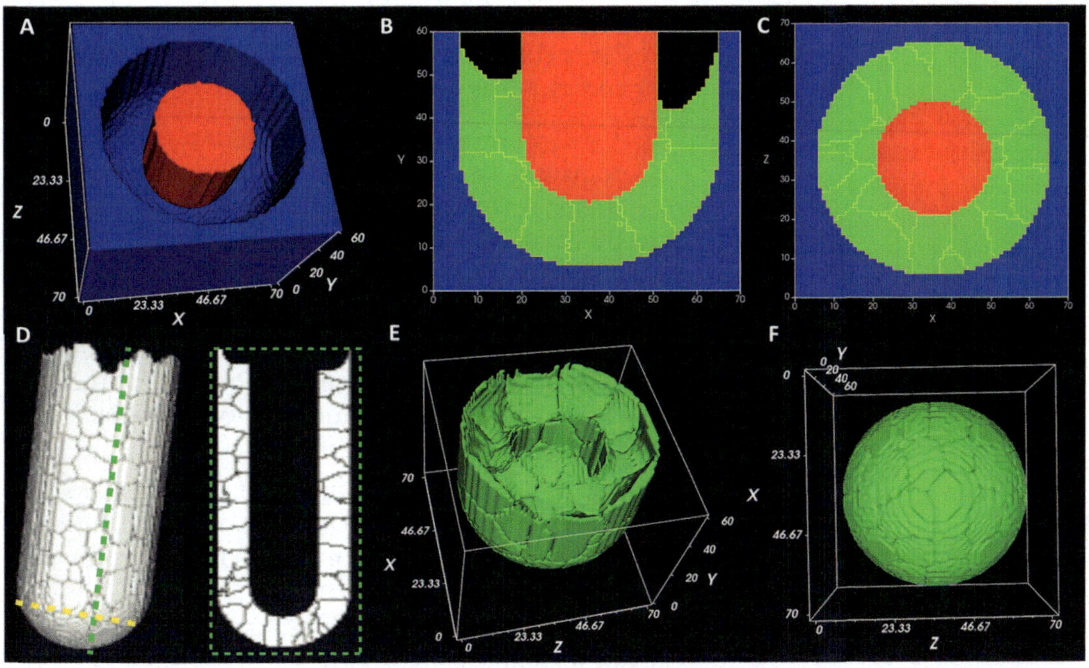

Fig. 6 (**a**) Central tube is indicated in red, and basal membrane is indicated in blue. (**b**) Longitudinal cross section with same color notation. Green region represents epithelial cells. (**c**) Transverse cross section of the crypt base. (**d**) 3D representation of the total crypt structure. (**e**) 3D representation of the stem niche. (**f**) 3D representation of the base of the stem niche [9]

(Type M/0) corresponding to the growth media the cells are residing in. At the same time, three new cell types will be defined "epithelial" (E/1), "central tube" (C/2), and "basal embrane" (B/3). The basal membrane will serve as the anchor for the epithelial cells to attach to and move along while the central tube cells will prevent the cells from detaching into the interstitial medium. The basic algorithm for establishing the geometry can be established using a modified mitosis steppable and goes as follows (*see* **Note 4**) [9].

1. Initialize a cubic region with the Compucell3D Uniform Initializer.
2. Define the radius of curvature r as half the width of the cubic region.
3. Generate central tube
   (a) If coordinate $y \geq r$ and if $(x - r)^2 + (z - r)^2 \leq \left(\frac{r-5}{2}\right)^2$, overwrite cell at this coordinate to be a central tube cell, and clear cell fragments to generate cylindrical region.
   (b) Else if $y < r$ and $(x - r)^2 + (y - r)^2 + (z - r)^2 \leq \left(\frac{r-5}{2}\right)^2$, overwrite cell at this coordinate to be a central tube cell, and clear cell fragments to generate hemispherical portion

of central tube. Assign cell volume to target volume, and set volume compressibility λ_V to 15,000 to create a sufficient energy penalty to freeze cells.

4. Generate basal membrane
   (a) If coordinate $y \geq r$ and if , overwrite cell at this coordinate to be a basal membrane cell, and clear cell fragments to generate complement of cylindrical region.
   (b) Define the number of segments that the tube is composed of. Calculate the angular position of the cells, and determine the segment number by dividing by the number of degrees in an arc. The segment number can be used for identification or as an alternative form of positional information.
   (c) If coordinate $y < r$ and if $(x-r)^2+(y-r)^2+(z-r)^2 \geq (r-5)^2$, overwrite cell at this coordinate to be a basal membrane cell, and clear fragments to generate complement of hemispherical region.
   (d) Set $(x, y, z) = (0, 0, 0)$ to satisfy criteria that the system contains medium cell type.$(x - r)^2 + (z - r)^2 \geq (r - 5)^2$
   (e) Freeze basal membrane cells analogously to central tube cells.
5. Kill cells crossing the top of the crypt and unattached to the basal membrane
   (a) Check vertical position relative to y dimension of basal membrane.
   (b) Identify cells not attached to basal membrane by checking neighbor list.
   (c) Append cell indices to list of cells to kill.
   (d) Set target volume of cells in kill list to 0 and clear indices.
6. Set temperature to 0, and let the system "anneal" by allowing remaining cells to reach target volume, and divide to fill the remaining space before starting simulation.

The cell killing program for Anoikis proceeds similarly to **step 5** but runs every Monte Carlo step and does not include annealing.

Last, appropriate selection of a hierarchy of contact energies is critical for efficient cell motion along the length of the crypt. Let the contact energies be symbolized as J_{ij} where the indices indicate the two cell types, and be aware that the contact energies are symmetric. The ideal ordering of contact energies is given in Eq. 15 [9].

$$0 = J_{\mathrm{MM}} = J_{\mathrm{CC}} = J_{\mathrm{BB}} < J_{\mathrm{EB}} = J_{\mathrm{EC}} < J_{\mathrm{EE}} < J_{\mathrm{EM}} < J_{\mathrm{BC}} = J_{\mathrm{BM}} = J_{\mathrm{CM}} = 10 \tag{15}$$

In this hierarchy, there is no energetic penalty for homotypic interactions between the medium (J_{MM}), the central tube (J_{CC}), and the basal membrane (J_{BB}) and themselves. There is a slight penalty for the epithelial cell contacting the basal membrane (J_{EB}) and central tube (J_{EC}) to ensure that they are not anchored in place. However, this penalty is much smaller than the penalty for epithelial cells touching themselves which drives them to spread out (J_{EE}). There is a high penalty for epithelial cells touching the medium (J_{EM}), but it is lower than the highest penalty for the interactions between the basal membrane (J_{BM}) or connecting tube (J_{CM}) and the medium [9] (*see* **Note 5** for a simpler example on contact energy hierarchies in two dimensions.)

3.4 Integrating SBML and Initializing the Simulation

The final part of the model is the SBML integration into the simulation. At this stage, cell behavior properties can be connected directly to the underlying reaction kinetics. Utilizing Compucell3D's built-in SBML features, a class can be defined containing a set of functions necessary to integrate the SBML model for each cell, to calculate the mean field properties of the neighboring cells for use in the integration, and to trigger mitosis when cells cross both size and intracellular chemical concentration thresholds [52, 82]. The possibilities for cell behaviors as a function of the underlying chemical networks are limitless, so the reader is directed to the official Compucell3D documentation (www.compucell3d.org/Manuals) for more information on scripting novel Python modules for use in custom simulations [76].

To give an illustrative example, the CC3D crypt model for the PFLI gene circuit works as follows [9]. Two classes are defined. The first performs numerical integration for each cell. The second updates the mitotic behaviors of the cells based on the associated chemical kinetics. Within the first class, the path to model file is presented, and the SBML model is associated to the epithelial cell type [9]. Subsequently, the initial conditions are set randomly selecting concentrations for each protein based on uniform initial distributions centered around their original values in the ODE model. These values are stored in a dictionary format associated with each cell. The next step is detection of nearest neighbors and calculation of average Delta and Notch concentrations on these neighbors [9]. The second class for mitotic events contains two functions. The first function of the mitosis class identifies cells that meet the growth threshold and Notch threshold to undergo mitosis since these correspond to stem cells (Notch High/Delta Low). The second function in this class divides the mitotic cells along their shortest axis and copies the concentration information to the daughter cells [9]. In this way, the complex spatial information needed to calculate the average external Delta and Notch concentrations is provided to the gene circuits, and the cell behaviors of the stem and Paneth cells are tied to gene circuit without making arbitrary assumptions about the growth and division rates.

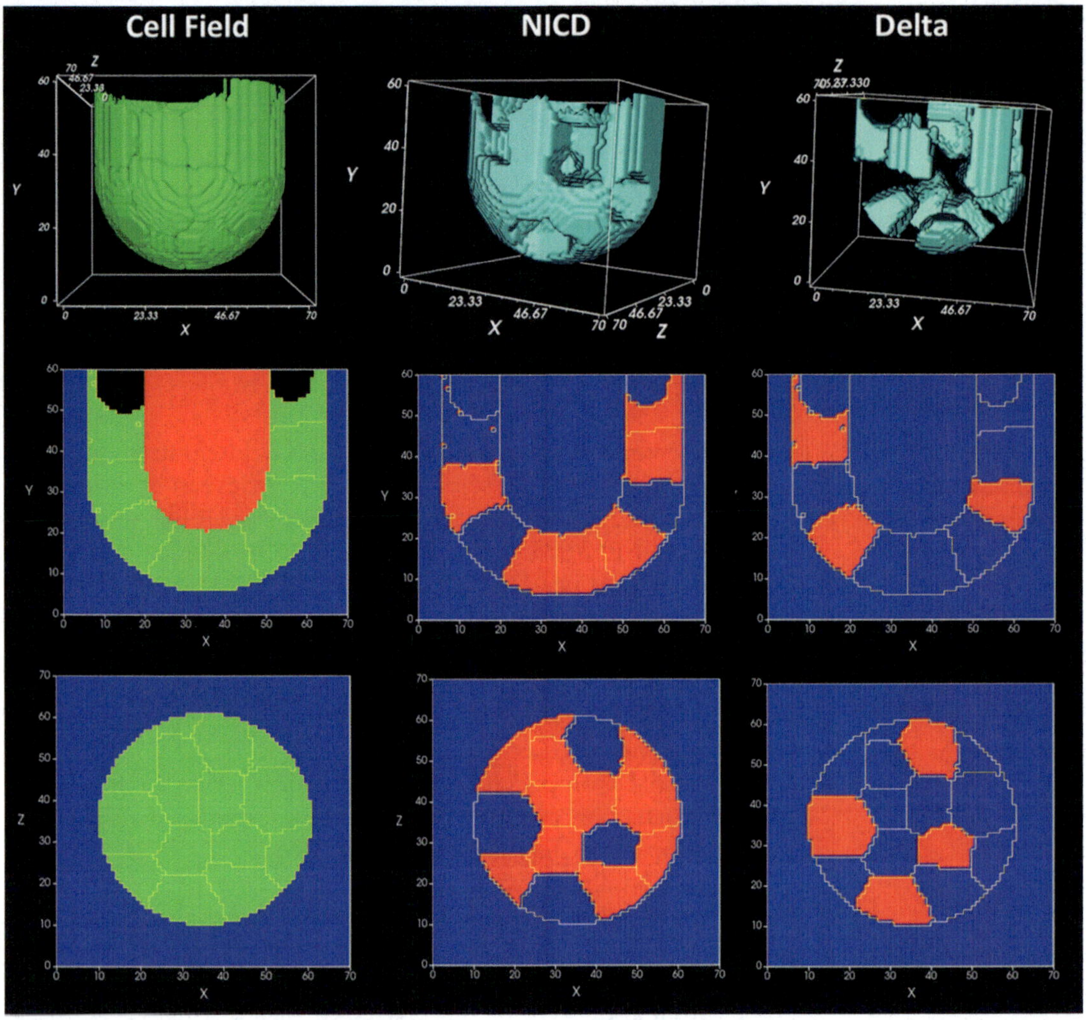

Fig. 7 (**Left**) Cell field of the crypt. (**Middle**) NICD concentrations predicted by PFLI model. (**Right**) Delta concentrations predicted by PFLI model. (**Top**) 3D representation of the cell or chemical field. (**Middle**) Longitudinal cross section of cell or chemical field. (**Bottom**) Transverse cross section of the crypt base. In the cell field column, green corresponds to epithelial cells, red corresponds to the central tube, and blue corresponds to the basal membrane. In the chemical field columns, red corresponds to high levels of the chemical, while blue indicates its absence. The lateral inhibition mechanism appears to be in effect since the NICD and Delta results are exact opposites

Results for a representative simulation run are presented in Fig. 7. The SBML model was integrated into the model appropriately, and the set of parameters that was previously utilized was able to generate a pattern as expected. In two-dimensional cross sections, a color map is available to visualize the concentrations showing that this circuit functions as a sharp switch with high quality as indicated by the complete absence or total saturation of cells with either NICD or Delta.

3.5 Analyzing the Simulation Results

Numerous methods are available to analyze the simulations. Compucell3D tracks the concentration values, locations of all cells, and the total energy of the system at every time point. Based on this information, numerous schemes can be developed to quantify the simulation results. As shown in Fig. 5d, violin plots can be constructed to study distributions of intracellular species as a function of time, promoter strength, position, or other parameters [9]. Each point in the violin plots represents results obtained from a single cell. Others have used agent-based models like these to replicate results from mitotic indexing experiments by keeping track of the number of cell divisions cells go through at specific points along the crypt or have estimated the velocity profile or cell flux as a function of position along the crypt axis [43]. Snippert et al. demonstrated that crypts are monoclonal by showing that stochastic variation in stem cell growth rates leads to the expulsion of all other stem cells from the crypt base except those in the fastest growing lineage [23].

Fundamentally, there is no limitation to the kind of experiments that can be simulated computationally in an agent-based modelling framework. However, the in silico experiments should be replicated, and the data should be handled to account for stochasticity. Specifically, efforts should be made to quantify the uncertainty in the simulations using standard statistical methods, and if possible, hypothesis testing should be performed to determine if varying model parameters or varying entire models result in a statistically significant change in some quantitative metric. Even if the underlying data are not Gaussian, tests such as the Kolmogorov-Smirnov test can be utilized to test for differences between data sets non-parametrically [87] (i.e., when the underlying distribution is not known), or the Chi-squared test can be used to assess differences in frequencies of categorical data such as variances in the relative numbers of different cell types [88].

4 Conclusion

In this chapter, we briefly reviewed the current state of agent-based modelling with an emphasis on understanding the spatiotemporal mechanisms underlying biological signaling pathways in the intestinal stem cell niche. The underlying theory of the Glazier-Graner-Hogeweg agent-based modelling framework was presented, and the algorithm behind a Monte Carlo step in the Compucell3D implementation of the GGH model was described. Then, methods for formulating a gene circuit in SBML, developing testable hypotheses using bifurcation and stability analyses, setting up a new Compucell3D simulation and constructing an initial geometry, integrating the SBML model into the cellular behavior, and analyzing the results were provided.

Fundamentally, agent-based modelling is poised to bridge the gaps in understanding between the dynamical properties of cellular gene circuits and cellular behaviors, especially in the context of tissue regeneration, homeostasis, and patterning. The development of software tools will make agent-based modelling more accessible to the general modelling community. For example, Compucell3D implementation of the GGH model by extending the Hamiltonian term to model new behaviors is significantly simpler than formulating new dynamical equations to describe the deterministic motion of cells as in other frameworks. In the end, agent-based modelling will never be a "one-size-fits-all" approach to integrating macroscopic and subcellular physiological models, but we hope that the techniques and insights outlined in this chapter may provide modellers with a hands-on tutorial to try this framework.

5 Notes/Troubleshooting

1. As efforts continue to make scientific work more reproducible, SBML will become more and more of a necessity since it standardizes mathematical models for reuse and simplifies error checking [82]. However, beyond this, SBML simplifies many dynamic analyses since it is compatible with a large number of software packages. Notably, it can be integrated into BRENDA and COBRA for metabolic pathway analyses; Gillespie2 for stochastic simulation; libSBML, SBML Toolbox, and SimBiology Toolbox for Matlab; MesoRD for stochastic reaction diffusion problems; Wolfram System Modeler for Mathematica integration; and XPPAUT for high-level bifurcation analyses. A full list of software is available at http://sbml.org/SBML_Software_Guide/SBML_Software_Matrix [83].
2. For more than three-dimensional parameter sweeps, systematic sampling of a parameter space becomes computational intensive and time-consuming, and simultaneous visualization of all data points becomes impossible. For higher-dimensional data sets, a better approach to sampling the parameter space is via Monte Carlo sampling. For an N-dimensional data set, the reader can generate a set of random numbers from N probability distributions corresponding to the acceptable or expected parameter ranges for the parameters of interest. The metric being used for the bifurcation analysis is calculated for each vector of N random parameters. Then, an $N \times N$ matrix of scatter plots is constructed where the on-diagonal plots are marginal histograms of the parameter values grouped by the bifurcation metric, and the off-diagonal scatter plots are related to the joint marginal densities of each combination of the two variables (i.e., the density of scatter plot points is proportional

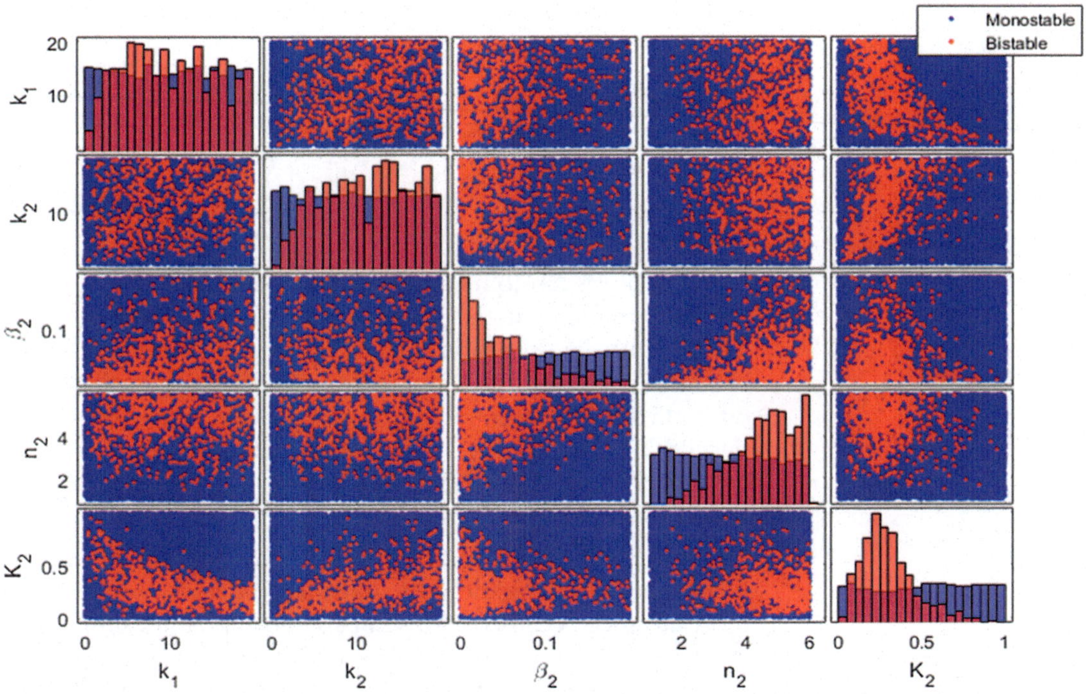

Fig. 8 Monte Carlo parameter sweep for a bistable switch with a single feedback loop described by five parameters. Red dots indicate bistable parameter sets, while blue dots indicate monostable results. The on-diagonal histograms are marginal distributions with respect to each variable, while the off diagonals represent joint probability distributions. Unlike traditional bifurcation analysis, this method introduces a certain amount of uncertainty into the results since it can only project the shape of the "edges" of the higher-dimensional bistable region onto the plane

to the joint, marginal probability density function) with each data point colored by its respective group (e.g., monostable or bistable). This technique is illustrated in Fig. 8 for a bistable switch composed of a single feedback loop and described by five parameters. Instead of getting sharply defined regions, one finds some overlap in the data. This occurs since the true bistable region is a high-dimensional space, and when it is projected on to the plane, only the "boundaries" or "edges" of this region are visible.

3. Anisotropy refers to the phenomena by which pattern forming or structured materials align along specific axes of some lattice to minimize their energy. These effects are significantly stronger in rectangular and cubic lattices than in hexagonal lattices. While counterintuitive, a rectangular coordinate system can be isomorphically mapped onto a hexagonal lattice, and a cubic coordinate system can be mapped onto a rhombic dodecahedron lattice. For the rectangular to hexagonal case, the mapping is given by the following equations where L is the

length between hexagonal pixels, and even and odd refer to the row number (Eq. 16) [76].

$$
\begin{aligned}
(x_{\text{hex}}, y_{\text{hex}}) &= \left(x_{\text{cart}} L, \frac{\sqrt{3}}{2} L y_{\text{cart}} \right) \text{ if } y \in \text{even} \\
(x_{\text{hex}}, y_{\text{hex}}) &= \left(\left(x_{\text{cart}} + \frac{1}{2} \right) L, \frac{\sqrt{3}}{2} L y_{\text{cart}} \right) \text{ if } y \in \text{odd}
\end{aligned}
\tag{16}
$$

The blob, random initializer in Compucell3D can automatically implement the hexagonal or rhombic dodecahedral lattices, but the user should be aware that the neighbor order should be decreased to 1 and that the diffusion operator works differently. For additional information, please review the Compucell3D Hexagonal Lattices Manual (http://www.compucell3d.org/BinDoc/cc3d_binaries/Manuals/Hexagonal Lattice.pdf) [76].

4. For full instructions on integrating an SBML model into a Compucell3D simulation or guidance on how to construct more advanced cell behavior steppables by updating simulation parameters during the simulation, looping over cell types, applying macroscopic forces, or varying the mitotic properties of cells, the reader should consult the Python Scripting Manual for Compucell3D Version 3.7.6 (https://media.readthedocs.org/pdf/pythonscriptingmanual/latest/pythonscriptingmanual.pdf) [76].
5. Before attempting to set up a contact energy hierarchy for a complex system, the best method to familiarize yourself with the contact energies is to develop a 2D cell sorting simulation using the Compucell3D simulation wizard. Specify two cell types so that you have one population of cells that will condense and one that will remain dispersed. Next try varying the six contact energies in the range from 0 to 10 to observe the variety of patterns that can be formed. Vary the temperature to observe freezing and evaporation. Try adjusting the relative ratio of cell types and using the engulfment starting condition. Last, try changing the contact energies to negative values. For a full list of exercises and explanations of the underlying phenomena, review the Compucell3D introductory manual (http://www.compucell3d.org/BinDoc/cc3d_binaries/Manuals/Introduction_To_CompuCell3D_v.3.7.4.pdf) [76].

References

1. Bianconi E, Piovesan A, Facchin F, Beraudi A, Casadei R, Frabetti F, Vitale L, Pelleri MC, Tassani S, Piva F, Perez-Amodio S, Strippoli P, Canaider S (2013) An estimation of the number of cells in the human body. Ann Hum Biol 40:463–471. https://doi.org/10.3109/03014460.2013.807878

2. Hatano A, Chiba H, Moesa HA, Taniguchi T, Nagaie S, Yamanegi K, Takai-Igarashi T, Tanaka H, Fujibuchi W (2011) CELLPEDIA: a repository for human cell information for cell studies and differentiation analyses. Database 2011:bar046. https://doi.org/10.1093/database/bar046
3. Clamp M, Fry B, Kamal M, Xie X, Cuff J, Lin MF, Kellis M, Lindblad-Toh K, Lander ES (2007) Distinguishing protein-coding and noncoding genes in the human genome. Proc Natl Acad Sci 104:19428–19433. https://doi.org/10.1073/pnas.0709013104
4. Church DM, Goodstadt L, Hillier LW, Zody MC, Goldstein S, She X, Bult CJ, Agarwala R, Cherry JL, DiCuccio M, Hlavina W, Kapustin Y, Meric P, Maglott D, Birtle Z, Marques AC, Graves T, Zhou S, Teague B, Potamousis K, Churas C, Place M, Herschleb J, Runnheim R, Forrest D, Amos-Landgraf J, Schwartz DC, Cheng Z, Lindblad-Toh K, Eichler EE, Ponting CP, Consortium TMGS (2009) Lineage-specific biology revealed by a finished genome assembly of the mouse. PLoS Biol 7:e1000112. https://doi.org/10.1371/journal.pbio.1000112
5. Tsompana M, Buck MJ (2014) Chromatin accessibility: a window into the genome. Epigenetics Chromatin 7:33. https://doi.org/10.1186/1756-8935-7-33
6. Kumar A, Garg S, Garg N (2014) Regulation of gene expression. In: Reviews in cell biology and molecular medicine. American Cancer Society, Atlanta, pp 1–59
7. Hernando-Herraez I, Garcia-Perez R, Sharp AJ, Marques-Bonet T (2015) DNA methylation: insights into human evolution. PLoS Genet 11:e1005661. https://doi.org/10.1371/journal.pgen.1005661
8. Lambert SA, Jolma A, Campitelli LF, Das PK, Yin Y, Albu M, Chen X, Taipale J, Hughes TR, Weirauch MT (2018) The human transcription factors. Cell 172:650–665. https://doi.org/10.1016/j.cell.2018.01.029
9. Chen K-Y, Srinivasan T, Tung K-L, Belmonte JM, Wang L, Murthy PKL, Choi J, Rakhilin N, King S, Varanko AK, Witherspoon M, Nishimura N, Glazier JA, Lipkin SM, Bu P, Shen X (2017) A Notch positive feedback in the intestinal stem cell niche is essential for stem cell self-renewal. Mol Syst Biol 13:927. https://doi.org/10.15252/msb.20167324
10. Borggrefe T, Lauth M, Zwijsen A, Huylebroeck D, Oswald F, Giaimo BD (2016) The Notch intracellular domain integrates signals from Wnt, Hedgehog, TGFβ/BMP and hypoxia pathways. Biochim Biophys Acta 1863:303–313. https://doi.org/10.1016/j.bbamcr.2015.11.020
11. Kay SK, Harrington HA, Shepherd S, Brennan K, Dale T, Osborne JM, Gavaghan DJ, Byrne HM (2017) The role of the Hes1 crosstalk hub in Notch-Wnt interactions of the intestinal crypt. PLoS Comput Biol 13:e1005400. https://doi.org/10.1371/journal.pcbi.1005400
12. Ladyman J, Lambert J, Wiesner K (2013) What is a complex system? Eur J Philos Sci 3:33–67. https://doi.org/10.1007/s13194-012-0056-8
13. Ross J, Arkin AP (2009) Complex systems: from chemistry to systems biology. Proc Natl Acad Sci 106:6433–6434. https://doi.org/10.1073/pnas.0903406106
14. Whitesides GM, Ismagilov RF (1999) Complexity in chemistry. Science 284:89–92. https://doi.org/10.1126/science.284.5411.89F
15. Weng G, Bhalla US, Iyengar R (1999) Complexity in biological signaling systems. Science 284:92–96. https://doi.org/10.1126/science.284.5411.92
16. Koch C, Laurent G (1999) Complexity and the nervous system. Science 284:96–98. https://doi.org/10.1126/science.284.5411.96
17. Cooke J (1973) Properties of the primary organization field in the embryo of Xenopus laevis: IV. Pattern formation and regulation following early inhibition of mitosis. Development 30:49–62
18. Lander AD, Kimble J, Clevers H, Fuchs E, Montarras D, Buckingham M, Calof AL, Trumpp A, Oskarsson T (2012) What does the concept of the stem cell niche really mean today? BMC Biol 10:19. https://doi.org/10.1186/1741-7007-10-19
19. Rojas-Ríos P, González-Reyes A (2014) Concise review: the plasticity of stem cell niches: a general property behind tissue homeostasis and repair. Stem Cells 32:852–859. https://doi.org/10.1002/stem.1621
20. Ferraro F, Celso CL, Scadden D (2010) Adult stem cells and their niches. Adv Exp Med Biol 695:155–168. https://doi.org/10.1007/978-1-4419-7037-4_11
21. Morrison SJ, Spradling AC (2008) Stem cells and niches: mechanisms that promote stem cell maintenance throughout life. Cell 132:598–611. https://doi.org/10.1016/j.cell.2008.01.038
22. Schofield R (1978) The relationship between the spleen colony-forming cell and the haematopoietic stem cell. Blood Cells 4:7–25

23. Snippert HJ, van der Flier LG, Sato T, van Es JH, van den Born M, Kroon-Veenboer C, Barker N, Klein AM, van Rheenen J, Simons BD, Clevers H (2010) Intestinal crypt homeostasis results from neutral competition between symmetrically dividing Lgr5 stem cells. Cell 143:134–144. https://doi.org/10.1016/j.cell.2010.09.016
24. Clevers H (2013) The intestinal crypt, a prototype stem cell compartment. Cell 154:274–284. https://doi.org/10.1016/j.cell.2013.07.004
25. Barker N (2014) Adult intestinal stem cells: critical drivers of epithelial homeostasis and regeneration. Nat Rev Mol Cell Biol 15:19–33. https://doi.org/10.1038/nrm3721
26. Meran L, Baulies A, Li VSW (2017) Intestinal stem cell niche: the extracellular matrix and cellular components. Stem Cells Int 2017:7970385. https://www.hindawi.com/journals/sci/2017/7970385/. Accessed 8 Apr 2018
27. Gjorevski N, Ordóñez-Morán P (2017) Intestinal stem cell niche insights gathered from both in vivo and novel in vitro models. Stem Cells Int 2017:8387297. https://www.hindawi.com/journals/sci/2017/8387297/. Accessed 8 Apr 2018
28. Durand A, Donahue B, Peignon G, Letourneur F, Cagnard N, Slomianny C, Perret C, Shroyer NF, Romagnolo B (2012) Functional intestinal stem cells after Paneth cell ablation induced by the loss of transcription factor Math1 (Atoh1). Proc Natl Acad Sci 109:8965–8970. https://doi.org/10.1073/pnas.1201652109
29. Farin HF, Jordens I, Mosa MH, Basak O, Korving J, Tauriello DVF, de Punder K, Angers S, Peters PJ, Maurice MM, Clevers H (2016) Visualization of a short-range Wnt gradient in the intestinal stem-cell niche. Nature 530:340–343. https://doi.org/10.1038/nature16937
30. Collier JR, Monk NAM, Maini PK, Lewis JH (1996) Pattern formation by lateral inhibition with feedback: a mathematical model of delta-notch intercellular signalling. J Theor Biol 183:429–446. https://doi.org/10.1006/jtbi.1996.0233
31. Du H, Nie Q, Holmes WR (2015) The Interplay between Wnt mediated expansion and negative regulation of growth promotes robust intestinal crypt structure and homeostasis. PLoS Comput Biol 11:e1004285. https://doi.org/10.1371/journal.pcbi.1004285
32. Qi Z, Li Y, Zhao B, Xu C, Liu Y, Li H, Zhang B, Wang X, Yang X, Xie W, Li B, Han J-DJ, Chen Y-G (2017) BMP restricts stemness of intestinal Lgr5+ stem cells by directly suppressing their signature genes. Nat Commun 8:13824. https://doi.org/10.1038/ncomms13824
33. Walton KD, Whidden M, Kolterud Å, Shoffner SK, Czerwinski MJ, Kushwaha J, Parmar N, Chandhrasekhar D, Freddo AM, Schnell S, Gumucio DL (2016) Vilification in the mouse: Bmp signals control intestinal villus patterning. Development 143:427–436. https://doi.org/10.1242/dev.130112
34. Turing AM (1952) The chemical basis of morphogenesis. Phil Trans R Soc Lond B 237:37–72. https://doi.org/10.1098/rstb.1952.0012
35. Loeffler M, Stein R, Wichmann H-E, Potten CS, Kaur P, Chwalinski S (1986) Intestinal cell proliferation. I. A comprehensive model of steady-state proliferation in the crypt. Cell Prolif 19:627–645. https://doi.org/10.1111/j.1365-2184.1986.tb00763.x
36. Potten CS, Loeffler M (1987) A comprehensive model of the crypts of the small intestine of the mouse provides insight into the mechanisms of cell migration and the proliferation hierarchy. J Theor Biol 127:381–391. https://doi.org/10.1016/S0022-5193(87)80136-4
37. Meinzer HP, Sandblad B, Baur HJ (2008) Generation-dependent control mechanisms in cell proliferation and differentiation—the power of two. Cell Prolif 25:125–140. https://doi.org/10.1111/j.1365-2184.1992.tb01486.x
38. Finney KJ, Appleton DR, Ince P, Sunter JP, Watson AJ (1988) Proliferative status of colonic mucosa in organ culture: 3H-thymidine-labelling studies and computer modelling. Virchows Arch B 56:397–405. https://doi.org/10.1007/BF02890043
39. Lee E, Salic A, Krüger R, Heinrich R, Kirschner MW (2003) The roles of APC and axin derived from experimental and theoretical analysis of the Wnt pathway. PLoS Biol 1:e10. https://doi.org/10.1371/journal.pbio.0000010
40. Cho K-H, Baek S, Sung M-H (2006) Wnt pathway mutations selected by optimal β-catenin signaling for tumorigenesis. FEBS Lett 580:3665–3670. https://doi.org/10.1016/j.febslet.2006.05.053
41. Schmitz Y, Wolkenhauer O, Rateitschak K (2011) Nucleo-cytoplasmic shuttling of APC can maximize β-catenin/TCF concentration. J Theor Biol 279:132–142. https://doi.org/10.1016/j.jtbi.2011.03.018
42. van Leeuwen IMM, Byrne HM, Jensen OE, King JR (2007) Elucidating the interactions between the adhesive and transcriptional functions of β-catenin in normal and cancerous

cells. J Theor Biol 247:77–102. https://doi.org/10.1016/j.jtbi.2007.01.019

43. Van Leeuwen IMM, Mirams GR, Walter A, Fletcher A, Murray P, Osborne J, Varma S, Young SJ, Cooper J, Doyle B, Pitt-Francis J, Momtahan L, Pathmanathan P, Whiteley JP, Chapman SJ, Gavaghan DJ, Jensen OE, King JR, Maini PK, Waters SL, Byrne HM (2009) An integrative computational model for intestinal tissue renewal. Cell Prolif 42:617–636. https://doi.org/10.1111/j.1365-2184.2009.00627.x
44. Meineke FA, Potten CS, Loeffler M (2001) Cell migration and organization in the intestinal crypt using a lattice-free model. Cell Prolif 34:253–266. https://doi.org/10.1046/j.0960-7722.2001.00216.x
45. Mirams GR, Byrne HM, King JR (2010) A multiple timescale analysis of a mathematical model of the Wnt/β-catenin signalling pathway. J Math Biol 60:131–160. https://doi.org/10.1007/s00285-009-0262-y
46. Osborne JM, Walter A, Kershaw SK, Mirams GR, Fletcher AG, Pathmanathan P, Gavaghan D, Jensen OE, Maini PK, Byrne HM (2010) A hybrid approach to multi-scale modelling of cancer. Philos Trans R Soc Lond Math Phys Eng Sci 368:5013–5028. https://doi.org/10.1098/rsta.2010.0173
47. Murray PJ, Walter A, Fletcher AG, Edwards CM, Tindall MJ, Maini PK (2011) Comparing a discrete and continuum model of the intestinal crypt. Phys Biol 8:026011. https://doi.org/10.1088/1478-3975/8/2/026011
48. Murray PJ, Kang J-W, Mirams GR, Shin S-Y, Byrne HM, Maini PK, Cho K-H (2010) Modelling spatially regulated beta-catenin dynamics and invasion in intestinal crypts. Biophys J 99:716–725. https://doi.org/10.1016/j.bpj.2010.05.016
49. Zhang L, Lander AD, Nie Q (2012) A reaction–diffusion mechanism influences cell lineage progression as a basis for formation, regeneration, and stability of intestinal crypts. BMC Syst Biol 6:93. https://doi.org/10.1186/1752-0509-6-93
50. Buske P, Galle J, Barker N, Aust G, Clevers H, Loeffler M (2011) A comprehensive model of the spatio-temporal stem cell and tissue organisation in the intestinal crypt. PLoS Comput Biol 7:e1001045. https://doi.org/10.1371/journal.pcbi.1001045
51. Pin C, Watson AJM, Carding SR (2012) Modelling the spatio-temporal cell dynamics reveals novel insights on cell differentiation and proliferation in the small intestinal crypt. PLoS One 7:e37115. https://doi.org/10.1371/journal.pone.0037115
52. Swat MH, Thomas GL, Belmonte JM, Shirinifard A, Hmeljak D, Glazier JA (2012) Multi-scale modeling of tissues using CompuCell3D. Methods Cell Biol 110:325–366. https://doi.org/10.1016/B978-0-12-388403-9.00013-8
53. Glazier JA, Balter A, Popławski NJ (2007) Magnetization to morphogenesis: a brief history of the glazier-Graner-Hogeweg model. In: Single-cell-based models in biology and medicine. Birkhäuser Basel, pp 79–9
54. Marée AFM, Grieneisen VA, Hogeweg P (2007) The cellular potts model and biophysical properties of cells, tissues and morphogenesis. In: Single-cell-based models in biology and medicine. Birkhäuser, Basel, pp 107–136
55. Savill NJ, Merks RMH (2007) The cellular potts model in biomedicine. In: Single-cell-based models in biology and medicine. Birkhäuser, Basel, pp 137–150
56. Chen N, Glazier JA, Izaguirre JA, Alber MS (2007) A parallel implementation of the Cellular Potts Model for simulation of cell-based morphogenesis. Comput Phys Commun 176:670–681. https://doi.org/10.1016/j.cpc.2007.03.007
57. Swat MH, Thomas GL, Shirinifard A, Clendenon SG, Glazier JA (2015) Emergent stratification in solid tumors selects for reduced cohesion of tumor cells: a multi-cell, virtual-tissue model of tumor evolution using CompuCell3D. PLoS One 10:e0127972. https://doi.org/10.1371/journal.pone.0127972
58. Abdulla T, Luna-Zurita L, de la Pompa JL, Schleich J-M, Summers R (2013) Epithelial to mesenchymal transition—The roles of cell morphology, labile adhesion and junctional coupling. Comput Methods Prog Biomed 111:435–446. https://doi.org/10.1016/j.cmpb.2013.05.018
59. Boas SEM, Merks RMH (2015) Tip cell overtaking occurs as a side effect of sprouting in computational models of angiogenesis. BMC Syst Biol 9:86. https://doi.org/10.1186/s12918-015-0230-7
60. Sluka JP, Fu X, Swat M, Belmonte JM, Cosmanescu A, Clendenon SG, Wambaugh JF, Glazier JA (2016) A liver-centric multiscale modeling framework for xenobiotics. PLoS One 11:e0162428. https://doi.org/10.1371/journal.pone.0162428
61. Hutson MS, Leung MCK, Baker NC, Spencer RM, Knudsen TB (2017) Computational model of secondary palate fusion and disruption. Chem Res Toxicol 30:965–979. https://doi.org/10.1021/acs.chemrestox.6b00350
62. Balter A, Merks RMH, Popławski NJ, Swat M, Glazier JA (2007) The glazier-Graner-

Hogeweg model: extensions, future directions, and opportunities for further study. In: Single-cell-based models in biology and medicine. Birkhäuser, Basel, pp 151–167

63. Ising E (1925) Beitrag zur Theorie des Ferromagnetismus. Z Für Phys 31:253–258. https://doi.org/10.1007/BF02980577
64. Potts RB (1951) The mathematical investigation of some cooperative phenomena. Thesis
65. Weaire D, Kermode JP (1983) Computer simulation of a two-dimensional soap froth. Philos Mag B 48:245–259. https://doi.org/10.1080/13642818308228287
66. Wealire D, Kermode JP (1984) Computer simulation of a two-dimensional soap froth II. Analysis of results. Philos Mag B 50:379–395. https://doi.org/10.1080/13642818408238863
67. Steinberg MS (1963) Reconstruction of tissues by dissociated CELLS. Science 141:401–408. https://doi.org/10.1126/science.141.3579.401
68. Steinberg MS (1970) Does differential adhesion govern self-assembly processes in histogenesis? Equilibrium configurations and the emergence of a hierarchy among populations of embryonic cells. J Exp Zool 173:395–433. https://doi.org/10.1002/jez.1401730406
69. Glazier JA, Graner F (1993) Simulation of the differential adhesion driven rearrangement of biological cells. Phys Rev E 47:2128–2154. https://doi.org/10.1103/PhysRevE.47.2128
70. Graner F, Glazier JA (1992) Simulation of biological cell sorting using a two-dimensional extended Potts model. Phys Rev Lett 69:2013–2016. https://doi.org/10.1103/PhysRevLett.69.2013
71. Marée AFM, Hogeweg P (2002) Modelling Dictyostelium discoideum morphogenesis: the culmination. Bull Math Biol 64:327–353. https://doi.org/10.1006/bulm.2001.0277
72. Marée AFM, Hogeweg P (2001) How amoeboids self-organize into a fruiting body: Multicellular coordination in Dictyostelium discoideum. Proc Natl Acad Sci 98:3879–3883. https://doi.org/10.1073/pnas.061535198
73. Savill NJ, Hogeweg P (1997) Modelling morphogenesis: from single cells to crawling slugs. J Theor Biol 184:229–235. https://doi.org/10.1006/jtbi.1996.0237
74. Merks RMH, Glazier JA (2006) Dynamic mechanisms of blood vessel growth. Nonlinearity 19:C1–C10. https://doi.org/10.1088/0951-7715/19/1/000
75. Mombach JCM, de Almeida RMC, Iglesias JR (1993) Mitosis and growth in biological tissues. Phys Rev E 48:598–602. https://doi.org/10.1103/PhysRevE.48.598
76. Manuals - CompuCell3D. http://www.compucell3d.org/Manuals. Accessed 10 Apr 2018
77. Izaguirre JA, Chaturvedi R, Huang C, Cickovski T, Coffland J, Thomas G, Forgacs G, Alber M, Hentschel G, Newman SA, Glazier JA (2004) CompuCell, a multimodel framework for simulation of morphogenesis. Bioinformatics 20:1129–1137. https://doi.org/10.1093/bioinformatics/bth050
78. Chandler D (1987) Introduction to modern statistical mechanics. Oxford University Press, Oxford
79. Metropolis N, Rosenbluth AW, Rosenbluth MN, Teller AH, Teller E (1953) Equation of state calculations by fast computing machines. J Chem Phys 21:1087–1092. https://doi.org/10.1063/1.1699114
80. Hastings WK (1970) Monte Carlo sampling methods using Markov chains and their applications. Biometrika 57:97–109. https://doi.org/10.1093/biomet/57.1.97
81. Hoff PD (2009) A first course in Bayesian statistical methods [electronic resource]. Springer, London, New York
82. Hucka M, Finney A, Sauro HM, Bolouri H, Doyle JC, Kitano H, Arkin AP, Bornstein BJ, Bray D, Cornish-Bowden A, Cuellar AA, Dronov S, Gilles ED, Ginkel M, Gor V, Goryanin II, Hedley WJ, Hodgman TC, Hofmeyr J-H, Hunter PJ, Juty NS, Kasberger JL, Kremling A, Kummer U, Le Novère N, Loew LM, Lucio D, Mendes P, Minch E, Mjolsness ED, Nakayama Y, Nelson MR, Nielsen PF, Sakurada T, Schaff JC, Shapiro BE, Shimizu TS, Spence HD, Stelling J, Takahashi K, Tomita M, Wagner J, Wang J, Forum SBML (2003) The systems biology markup language (SBML): a medium for representation and exchange of biochemical network models. Bioinforma Oxf Engl 19:524–531
83. Main Page - SBML.caltech.edu. http://sbml.org/Main_Page. Accessed 10 Apr 2018
84. Choi K, Medley JK, Cannistra C, Konig M, Smith L, Stocking K, Sauro HM (2016) Tellurium: a python based modeling and reproducibility platform for systems biology. bioRxiv:054601. https://doi.org/10.1101/054601

85. Ingalls BP (2013) Mathematical modeling in systems biology: an introduction. MIT Press, Cambridge
86. Logan JD (2013) Applied mathematics. John Wiley & Sons, Hoboken
87. Bianchi C, Cirillo P, Gallegati M, Vagliasindi PA (2007) Validating and calibrating agent-based models: a case study. Comput Econ 30:245–264. https://doi.org/10.1007/s10614-007-9097-z
88. Pope AJ, Gimblett R (2015) Linking Bayesian and agent-based models to simulate complex social-ecological systems in semi-arid regions. Front Environ Sci 3. https://doi.org/10.3389/fenvs.2015.00055

Chapter 2

Modeling Cellular Differentiation and Reprogramming with Gene Regulatory Networks

András Hartmann, Srikanth Ravichandran, and Antonio del Sol

Abstract

Gene expression regulation is a fundamental cellular process that enables robust functioning of cells. How different genes interact among themselves to coordinate and maintain the overall gene expression profile observed in a cell is a key question in cellular biology. However, the immense complexity arising due to the scale and the nature of gene-gene interactions often hinders obtaining a global understanding of gene regulation. In this regard, network models of gene regulation based on gene-gene interactions, commonly referred to as gene regulatory networks (GRNs), serve important purpose of describing the overall interactions within a cell and provide a systematic approach to study their global behavior. In particular, in the context of cellular differentiation and reprogramming, where regulated changes in gene expression play a crucial role, precise knowledge of a cell type-specific GRN can enable control of the eventual cell fates with potential clinical applications. In this chapter, we describe our computational methodologies that we have tailor-made with purpose of applications to cell fate control. Briefly, we introduce the process of cellular differentiation and reprogramming, describe GRNs and common strategies to model them, and, finally, introduce the concept of determinants of cellular reprogramming and differentiation. In the Methods section, we elaborate on the different steps involved in the computational pipeline, including initial gene expression data processing, characterization of prior knowledge network, algorithm to remove non-cell type-specific edges, topological characterization of the inferred network, and Boolean network simulations to mimic cellular transitions. Finally, we provide a strategy to identify determinants of cellular reprogramming and differentiation based on the proposed computational methods.

Key words Cellular differentiation, Reprogramming, Modeling, Gene regulatory networks, Boolean networks, Attractors

1 Introduction

Cellular differentiation is a key biological process through which a less specialized and immature cell evolves in to a specialized cell, reaching its fully mature form and function [1, 2]. Stem cells are crucial for developmental processes as they are pluripotent and have the ability to differentiate and transform into any cell type of the

András Hartmann and Srikanth Ravichandran contributed equally to this work.

Patrick Cahan (ed.), *Computational Stem Cell Biology: Methods and Protocols*, Methods in Molecular Biology, vol. 1975, https://doi.org/10.1007/978-1-4939-9224-9_2,

body. Importantly, this ability of stem cells elicit tremendous interest in cellular reprogramming, which is a process where a more differentiated cell type is transformed into a progenitor or pluripotent stem cell by targeted manipulation of specific transcription factors (TFs) [1].

Due to the clinical potential of reprogrammed stem cells, it is important to understand how stem cell fate decisions are determined. The potential cell fates of stem cells include self-renewal, differentiation and lineage specification, programmed cell death, and quiescence [2]. Further, these choices of cell fate identities also apply to the immediate progeny of stem cells, known as progenitor or transit-amplifying cells. A major challenge in the field of stem cell biology is to identify the key molecular factors that determine the different cell fate options confronting stem and progenitor cells. Understanding the mechanisms involved in controlling the stem cell fate has important applications for biomedical research in general, ranging from elucidating the causes of cancer to the use of stem cells in regenerative medicine.

1.1 Gene Regulatory Networks

Over the years, several approaches have been developed that enable the experimental or computational identification of interactions between genes and TFs within a given cellular context [3–6]. However, it is not straightforward to interpret and analyze these datasets directly due to the sheer volume of the data and its complexity. In this regard, computational network modeling is a powerful approach to overcome the inherent complexity of biological processes that arise due to several interacting genes and the nonlinear nature of their interactions [7, 8]. For instance, mammalian genome consists of approximately 30,000 genes, which will result in staggering 2^{30000} potential combinations of gene expression values when we consider only "ON" and "OFF" states for each gene [2]. However, in reality, gene expression is continuous; this makes the potential gene expression space even larger. It is difficult to comprehend how such a large repertoire of potential gene expression states can be constrained so that only several hundreds of different cell types, which together comprise an organism, are stably maintained. To achieve this, gene expression and gene-gene interactions must exhibit robustness with respect to varying non-physiological intrinsic and extrinsic signals: that is, the cell states must not be very sensitive to small fluctuations in these signals [2]. In other words, the systems must exhibit stable behavior under intrinsic and extrinsic noise. Robustness of biological systems acts as a strong constraint and greatly limits the number of solutions or "states" for the system (i.e., a cell), and such stable cellular phenotypes are included in the large space of all possible cellular states [9]. Therefore, despite a very large number of potential cellular states, only a small fraction of states robustly exists as stable cell types [2]. Uncovering the dynamic and intertwined nature of

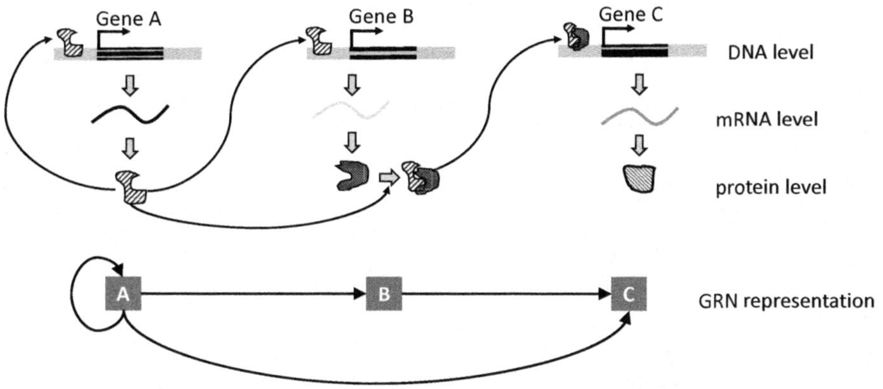

Fig. 1 Schematic of a GRN. The figure shows a toy representation of a gene regulatory network among three different genes. It can be seen that different levels of interactions, including protein DNA binding and protein-protein interactions, are modeled in the GRN

gene regulation that results in stable gene expression states corresponding to cell types requires integrative computational network approaches.

Gene expression patterns used to characterize distinct cellular states are tightly regulated by transcriptional and epigenetic mechanisms [10]. In the Waddington epigenetic landscape representation, cell types correspond to stable steady states known as "attractors," separated by a series of ridges depicting barriers between them [11]. These stable states represent stable gene expression patterns orchestrated by specific transcriptional programs. The process of cellular differentiation and reprogramming are precisely regulated by activation and inhibition of specific sets of TFs. The activation and inhibition interactions can be mapped as network of regulations among the interacting genes (Fig. 1). Such gene regulatory networks (GRNs) serve as abstract cellular models and enable us to study how a set of genes interact among each other to collectively determine the cell behavior. Processes, such as cellular differentiation or lineage commitment, can be understood as the transition from one attractor to another in this landscape, controlled by the underlying GRNs.

1.2 Boolean Modeling

A GRN topology specifies genes and proteins (TFs) that can interact with each other. Such a static representation of regulatory interactions in the GRNs allows for the network characterization based on topological features, such as centrality measures, network modularity, etc. Further, it also specifies the underlying regulatory logic (Fig. 2) that governs the interactions connecting individual genes. A given network topology, including the interaction rules, can be mathematically modeled, with solutions indicating whether a particular gene is active, or not, under a particular conditions. These models describe the time evolution of variables that are subject to fixed rules.

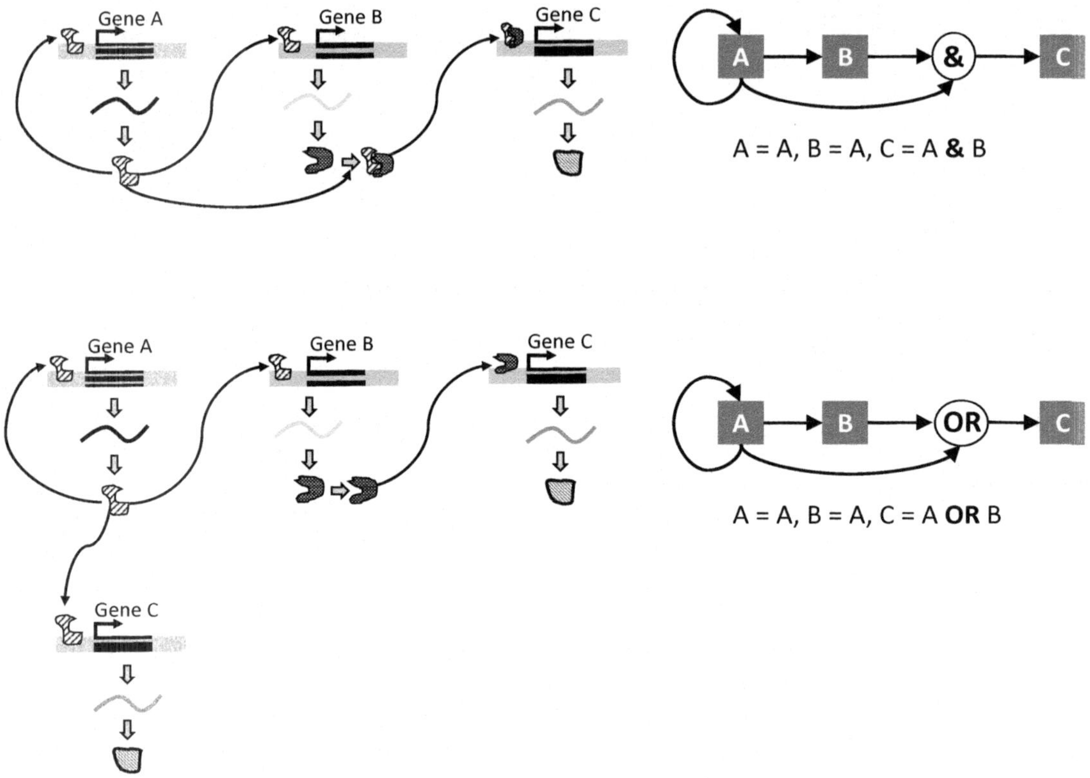

Fig. 2 Modeling of GRN. The schematic is same as Fig. 1 except that the mode of regulation of the target gene can be modeled differently depending upon biological knowledge. In case both A *and* B are known to form a protein complex and bind to the promoter of gene C, the GRN can be modeled with an AND gate. In case A *or* B both can independently activate the target, then GRN can be modeled with an OR gate

Various different mathematical formalisms exist to model GRN dynamics [8]. Two major formalisms are discrete Boolean modeling and continuous modeling based on differential equation. Differential equation-based dynamical models are one of the commonly employed formalisms to model GRNs [12]. In this formalism, a set of coupled ordinary differential equations (ODEs) define the dynamics of the GRN, where the activity of each gene is described by a differential equation modeling its activation and inhibition reactions. Although this formalism is powerful to study the system dynamics, a major bottleneck in building continuous differential equation-based models is the lack of availability rate parameters governing each of the reactions. In this context, Boolean modeling does not require information of reaction rate parameters and offers a simple yet powerful alternative to ODEs for modeling GRNs.

Kauffman initially proposed Boolean modeling framework for studying genetic regulatory networks [13]. In this framework, a gene can exist in only two states: active "ON" and inactive "OFF" (represented by Boolean 1 and 0, respectively) and the interaction between genes and proteins which is represented using Boolean logic functions, "AND," "OR," and "NOT." The simplification of Boolean models comes at the cost of discretization of the gene expression (or gene activity) to only two expression levels, namely, 1 or 0, which can denote either "ON" or "OFF" or active or inactive, respectively. Nevertheless, Boolean modeling can capture the discrete (pseudo-time) dynamics of the GRN, relying only on the topology and the regulatory logic, and has been successfully applied in the past to model various biological phenomena [7, 8].

1.3 Inference of GRNs

In order to accurately describe the gene regulation dynamics and identify genes that confer or perturb the GRN stability, accurate GRN inference is inevitable. An important challenge for regulatory network reconstruction is to build network models that can capture the changes in gene expression when cells are undergoing transitions from stable states. In general, GRNs can be reconstructed based on prior knowledge resources such as literature or specific interaction databases, the so-called prior knowledge networks (PKNs), and/or purely from specific experimental datasets by inferring interactions between statistically correlated levels of biological entities such as genes or proteins. The PKNs can be noisy and include information from different sources and combine interactions previously determined in different biological contexts such as different cell types, tissues, or experimental conditions in the attempt to collect all relevant biological events. Therefore, the PKN does not contain the information of context specificity, i.e., not specific for a cell type or specific condition. On the other hand, there are a number of methods to infer GRNs directly from data through reverse engineering based on statistical correlation or mutual information, such as the structural equation model (SEM) [14], the Gaussian graphical model (GGM) [15], the algorithm for the reconstruction of accurate cellular networks (ARACNE) [16], and context likelihood of relatedness (CLR) [17]. These methods in general produce GRNs highly contextualized for the cell types or conditions depending on the input experimental data. However, these networks generally lack directionality and usually yield a number of "false" interactions because of the statistical co-occurrence of events that are not due to cause-effect relationships and indirect correlations. Therefore, both PKN-based methods and network inference from experimental data do not result in networks that can be used directly to construct reliable dynamical models to describe or simulate biological processes [18], especially, when the purpose of the GRN is to model gene expression state as a stable phenotype and to predict input-output relationship such as responses to single or multiple perturbations.

In order to overcome some of abovementioned limitations and to account for building GRN models that can capture the distinct gene expression states, we had developed an algorithm that enables contextualization of a PKN to specific expression states or cellular phenotypes [18, 19]. Given a PKN and a training set of experimental data, the method attempts to infer subnetwork/s of the PKN that best describe the experimental data for an adopted Boolean dynamical model [19]. The resulting contextualized network(s) constitutes an optimized GRN models able to describe known phenotypes based on the gene expression states and can be used to simulate transitions between them in response to perturbations.

1.4 Determinants of Cellular Reprogramming and Differentiation

Despite the facts that several mechanistic details are not yet understood about the pathways involved in the regulation of stem and progenitor cell fates, it is now generally accepted that a few core TFs are crucial players in determining cell fate decisions [20–23]. Identification of these key genes has wide range of applications for cell fate engineering and regenerative medicine [24]. There is thus an extensive research effort devoted to design systematic data-driven approaches to predict drivers of cellular conversions.

We describe in detail the methods involved in our model-based approach for building context-specific GRNs and identification of reprogramming determinants.

2 Methods

The structure of this section is the following: After introducing the general pipeline for the prediction, we describe its building blocks in detail. Finally, we describe a specific example using the building blocks: a pipeline to predict cell fate determinants.

The methodology described in this section for the identification of reprogramming determinants is a model-based approach, where the Boolean network serves as modeling framework. A general pipeline is shown in (Fig. 3). The predictions are based on metrics derived either from the gene expression or from the network structure (topological features). Hence, it is merely important to acquire networks properly representing the modeled context. Our methodology involves a process to identify context-specific networks (*contextualization*). The input of the identification is a not context-specific prior knowledge network (PKN) and the context-specific gene expression information. Prior to the contextualization, the (accordingly pre-processed, e.g., normalized) continuous-domain gene expression has to be discretized into Boolean "ON"/ "OFF" values. This process is referred to as *booleanization*.

2.1 Discrete Model of Gene Regulatory Network

Discrete (Boolean) models provide an abstraction of gene regulatory circuits. A Boolean network is described with the triplet $B = (V, E, F)$, where $V = \{v_1, \ldots, v_n\}$ is the set of nodes and $E \subseteq V \times V \times \{+1, -1\}$ represents signed directed edges spanned between the nodes. The set $F = \{f_1, \ldots, f_n\}, f_i : \{0, 1\}^n \rightarrow \{0, 1\}$

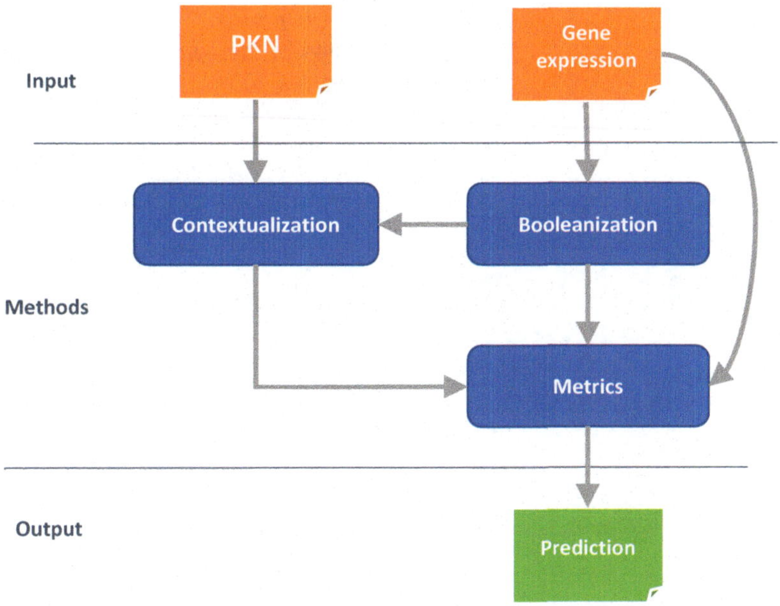

Fig. 3 General pipeline for the identification of reprogramming determinants. The pipeline consists from the following three steps: (1) Booleanization derives discrete "ON"/"OFF" values from the continuous-domain gene expression, (2) network identification (contextualization) creates context-specific network, (3) ranking of the putative drivers based on different metrics predicts the most relevant drivers of the cellular transition

represents the Boolean functions controlling the dynamics of the network. The gene expression is represented by the state vector $x = [x_1 x_2 \ \ldots \ x_n]^T$, and the time evolution of the states is $x_i(k+1) = f_i(x_i(k))$, where $x_i(k)$ denotes for the state of gene i at the discrete time-instance k.

Due to the exponential explosion of the potential Boolean functions, in most practically relevant cases, the data is not sufficient to infer unique Boolean functions for all the genes. In order to overcome this limitation, there exist mainly two simplified custom logic schemes that are then applied for each gene: majority rule (MR) and inhibition-dominant (ID) rule (*see* Table 1). The schemes are based on the number of active incoming activations and inhibitory edges:

$$\mathrm{Inh}_i(k) = \sum_{j=1}^{n} x_j(k), \text{ if } (v_j, v_i, -1) \in V, \quad (1)$$

$$\mathrm{Act}_i(k) = \sum_{j=1}^{n} x_j(k), \text{ if } (v_j, v_i, 1) \in V \quad (2)$$

Table 1
Formal description of the applied schemes for the time evolution of the applied schemes

Name	Rule
Majority rule (MR)	$\mathrm{Inh}_i(k) > \mathrm{Act}_i(k) \Rightarrow x_i(k+1) = 0$ $\mathrm{Inh}_i(k) < \mathrm{Act}_i(k) \Rightarrow x_i(k+1) = 1$ Otherwise $\Rightarrow x_i(k+1) = x_i(k)$
Inhibition dominant (ID)	$\mathrm{Inh}_i(k) > 0 \Rightarrow x_i(k+1) = 0$ $\mathrm{Inh}_i(k) = 0\ \&\ \mathrm{Act}_i(k) > 0 \Rightarrow x_i(k+1) = 1$ Otherwise $\Rightarrow x_i(k+1) = x_i(k)$

MR implements majority voting between activator and inhibitor edges. In the ID rules, the presence of an inhibition dominates over any number of activations

2.2 Booleanization

2.2.1 Relative Booleanization

Depending on the application, the networks may be specific for one context or to a cellular transition with starting and end phenotype. In the latter case, it is convenient to take the relative gene expression of the two phenotypes into account, e.g., genes are differently expressed between two conditions if their difference on logarithmic scale is larger than a threshold, (ϵ):

$$\left|\log_2\left(C^i\right) - \log_2\left(C^j\right)\right| > \epsilon \tag{3}$$

where C^i and C^j denote for the gene expression of a gene in context i and j, respectively. More sophisticated statistical methods for determining differentially gene expression, in case replicated measurements are available, are described in [25] and references therein.

2.2.2 Absolute Booleanization

If, however, there are only one or more than two phenotypes for which Boolean values should be acquired, relative gene expression is not suitable, and methods for absolute booleanization have to be applied. These methods usually anticipate a representative background distribution and discretize the gene expression based on that. Specifically, the barcode method of [26, 27] estimates a Gaussian mixture model for the not expressed and a uniform distribution for the expressed genes. A z-score is estimated for the sample to test based on the Gaussian model. The null hypothesis is that the gene is not expressed, and the gene is only considered to be expressed if the null hypothesis can be rejected. Hence the algorithm can only minimize the impact of false positives.

The recently developed RefBool algorithm [28] overcomes this limitation. For each gene, it aims to estimate the most probable expression level where the gene shifts from being not expressed to being expressed. The best separation between "OFF" and "ON" is associated with the steepest derivate of the cumulative distribution function (CDF) of the background gene expression. The CDF is approximated with a step function from both sides using bootstrapping method. This results in a threshold distribution, which is typically bimodal. Finally, z-scores are derived from the threshold distribution for both genes being expressed and being not expressed. In addition, z-scores can be derived for belonging to an intermediate state.

2.3 Network Reconstruction (Contextualization)

Our approach to network reconstruction is based on the non-context-specific prior knowledge network (PKN), which is then pruned according to the context-specific transcriptomic data. No new edges are introduced to the PKN during contextualization, but the edges that are in conflict with the context-specific knowledge and the applied ruleset are removed (pruned) from the prior knowledge.

2.3.1 Literature Knowledge

Prior knowledge network can be inferred from general information such as PPI (protein-protein interactions) or the correlation structure among the genes. In our applications, we consider PKNs derived from literature databases. The current knowledge about experimentally validated transcriptional interactions that are reported in the scientific literature is stored in databases like MetaCore or Pathway Studio [29]. Although these datasets are large-scale, and grow dinamically thanks to regular updates based on literature mining and/or hand-curation, the contained context information to the regulatory interactions is scarce or not accessible. There are efforts to create hand-curated context-specific networks [30], but currently they are mostly specific for tissue types rather than for cell types and cell lines. The advantage of algorithmically identified context-specific networks is that they only require computational time; in trade-off they might not be as reliable as manually curated networks. Certain edges in a PKN may not have sign information, meaning that they are compatible with both signs $E_{\mathrm{PKN}} \subseteq V \times V \times \{+1, -1, 0\}$.

2.3.2 Problem Formulation

Given a PKN and a training set of experimental data, find a subnetwork of the PKN, and infer the missing sign information within such that the identified network is consistent with the experimental data and the applied Scheme (MR or ID). The training set consists of at least one set of booleanized gene expression values.

2.3.3 Identification Using Genetic Algorithm (GA)

Genetic algorithm (GA) is a well-established, heuristic method based on the principles of evolution and natural selection [31] that is widely used to solve search and optimization problems. In our application, the method relies on the Boolean modeling formalism with synchronous updating scheme. The goal of the algorithm is to make the prior knowledge compatible with the context-specific transcriptional data, e.g., to identify GRNs that capture the steady-state behavior of the phenotypes in the training dataset.

The genetic algorithm is used for the identification applying the steps below in an iterative manner.

1. **Initialization:** An initial population of networks is generated from the PKN by randomly sampling edges to keep (e.g., with probability $P = 0.5$).
2. **Selection:** Survival of the fittest selection is applied based on the fitness function as defined in Eq. (4), representing the hamming distance between the simulated and the actual attractor states.
3. **Termination:** At this point, the algorithm verifies the stopping criteria: if the maximum number of iterations is reached or all the scores in the population of pruned networks are higher than a defined value (e.g. 90%). If this condition holds, the algorithm terminates and returns the last evaluated population. Otherwise the next population of pruned networks is generated.
4. **Genetic Operations:**
   (a) **Mutation:** This step alters the selected networks probabilistically, by removing or introducing interactions. Also, inferred signs are subject to probabilistic mutations, flipping signs from activation to inhibition, or vice versa.
   (b) **Crossover:** In order to produce new solutions, a pair of "parent" solutions is selected for recombination from the pool selected previously. One plausible crossover operation is to apply single-point crossover both to the interactions and the inferred signs. After recombination, two new individuals are obtained, of which the first one contains the first interaction subarray of the first individual and the second subarray of the second individual, whereas the other one contains the first subarray of the second and the second subarray of the first individual. We choose this crossover operation to ensure that each phenotype-specific network is treated independently during the pruning and sign estimation process.

The fitness function of a network m to a training set of M booleanized phenotypes $\{\sigma_1, \ldots, \sigma_M\}$ is

$$S_m = 1 - \frac{\sum_{i=1}^{M} \left|\sigma_i - \hat{\sigma}_{i,m}\right|}{M} \quad (4)$$

where $|\cdot|$ represents the normalized hamming distance, σ_i is the booleanized expression vector of phenotype i, and $\hat{\sigma}_{i,m}$ is the point attractor reached on network m when simulating starting from σ_i

according to the chosen ruleset. If no point attractor is reached, then the distance is considered to be maximal (*see* **Note 1**).

It has been observed that the GA algorithm converges reasonably fast to a solution; the method involves several user-defined parameters (*see* a complete list and recommended values for our application in **Note 2**). Some possible extensions of the method are described in **Notes 3** and **4**. A variant of the method that is more adequate to infer a set of networks is introduced in **Note 5**.

2.4 Metrics for Drivers of Cellular Conversions

To systematically predict driver genes of cellular cell fate determinants in all possible contexts is still an open problem. Reprogramming determinants (RDs) are usually identified based on the level of gene expression or on topological features of the GRN. The most comprehensive methodologies integrate more than one criteria to the final prediction (*see* a list of possible predictors below).

2.4.1 Cellular Decision Based on Differences in Expression

One view of cell identity is that it can be induced through the activity of just a small number of core TFs. The unique gene expression signature across various human cell types has been identified in order to build an atlas of candidate core TFs [21]. These candidate core TFs can serve as putative targets for cellular conversion.

2.4.2 Normalized Ratio Difference (NRD)

This measure has recently been introduced in [20] to predict TF pairs whose expression ratios show a significant change in daughter cells in comparison to the stem/progenitor cells. For a pair of gene expression C_i and C_j, the measure calculates the ratio

$$\mathrm{NRD}_{i,j} = \frac{\left(\frac{C_i^{\mathrm{Prog}}}{C_j^{\mathrm{Prog}}} - \frac{C_i^{\mathrm{Daughter}}}{C_j^{\mathrm{Daughter}}}\right)}{\left(\frac{C_i^{\mathrm{Prog}}}{C_j^{\mathrm{Prog}}}\right)} \tag{5}$$

2.4.3 Structural Patterns that Stabilize Phenotypes

Along the interpretation of the Waddington landscape, cellular phenotypes are represented by the stable steady states (i.e., attractors) of GRNs, separated by epigenetic barriers. In the Boolean modeling framework, the cellular transition can be modeled by changing the phenotype from one stable attractor state of the GRN to another. In this context, network motifs, such as positive circuits and strongly connected components (SCCs), play an important role. Positive circuits are motifs that stabilize point attractors and can be identified, e.g., using Jensen's algorithm, while SCCs are subnetworks where each node is accessible from any other node. When selecting candidate TFs, it is rational to prioritize the nodes in the network that are part of these motifs.

2.4.4 Ranking Based on Simulation

It is possible to directly simulate the changes of gene perturbations on the identified GRNs in order to discover the best RD candidates for cellular transition. This involves simulating and ranking all possible combinations of TF perturbations, e.g., using the fitness function as in Eq. (4). In case of n genes, finding the best RD set with cardinality l implies that the number of necessary simulations is $N_{\text{sim}} = \binom{n}{l} = \frac{n!}{l! * (n-l)!}$. Moreover, if $n \gg l$—which is typically the case—then the number of simulations necessary for the process is approximately exponential in the number of genes $N_{\text{sim}} \sim n^l$. Therefore, this kind of simulations is only numerically tractable on small-scale networks or on an appropriately reduced set of candidate genes.

2.5 Example

Below we detail the steps to reconstruct the computational pipeline described in [20]. The approach relies on the assumption that stem/progenitor cells correspond to stable gene expression states maintained by the balanced expression of cell fate determinants. The cell fate determinants are located in interconnected feedback loops (strongly connected components). In binary cell fate decisions, these strongly connected components consist of differentially expressed TFs between two daughter cell types from the stem/progenitor cells and stabilize the two stable gene expression states corresponding to these two daughter cell types. Upregulated TFs that cooperate within the same daughter cell type compete with those upregulated in the other daughter cell type.

1. Identify differentially expressed genes between the two daughter cell types (2.2.1).
2. Calculate the NRD ratios for both daughter cell types, and test for significance (2.4.2).
3. Identify a set of GRNs (2.3.3).
   (a) The training set consists of the differential expression of the genes that are also part of significant NRD pairs.
4. Identify the largest strongly connected component (SCC) in the set of GRNs (2.4.3).
5. Rank the differentially expressed significant NRD pairs with the following criteria:
   (a) Ratio of SCCs they are present in.
   (b) Ratio of SCCs they are directly connected.
   (c) Minimum out degree.

3 Notes

1. It would be more precise to take the distance to the closest attractor of the network instead of the simulation results. However, in practice, the enumeration of all attractor states of a network would be too time-consuming, and the distance to attractor reached by simulation serves as a reasonable approximation.
2. GA involves several user-defined parameters (*see* a complete list and recommended values in Table 2). As a rule of thumb, larger **population size** increases the likelihood of finding a globally optimal solution but requires more **iterations to converge**, thus increasing the computational load. The choices of the **mutation probability** and **recombination probability** are crucial for the performance of the GA. On one hand, high probabilities of genetic operations lead to deeper exploration of the search space. On the other hand, a low mutation probability allows the GA to converge faster toward an attractor state, even though it may be only a local optimum.
3. In the returned population of the GA, the best scoring solution might not be unique. In order to obtain the most likely GRN representation, it is beneficial to derive a consensus solution, e.g., taking the union of all the edges in the top-ranking solutions. In this case one has to verify that the elements of the training dataset still correspond to point attractors of the consensus solution.
4. The principle of elitism is often applied in GA. This means to propagate the "elite" best-fitting networks directly to the next generation.
5. It has been shown that GA can be applied in order to identify a contextualized solution. However, because of insufficient amount of training data, the problem might be underdetermined, the optimal solution is not unique, and it is impossible to decide which network instance to take based on the fitness

Table 2
List of GA parameters and recommendations for parameter values

Parameter	Recommended value
Population size	>5 × number of interactions
Number of maximal iterations	Between 100 and 1000
Mutation probability	0.01 (1% of bits are expected to mutate)
Probability of recombination	0.9 (90% of the population produce children)

function. If this is the case, it is better to consider a population of networks for the prediction.

A slight variant of GA, estimation of distribution algorithm (EDA) [32], is more adequate to retain a population of networks, because it better preserves the heterogeneity in the population. The main difference between EDA and GA is that while GA uses genetic operators on the network instances, EDA directly uses the distribution of edges.

Practically, this means that instead of the genetic operators in **step 4**, the next generation of the network is assembled as follows: For each interaction, an empirical probability is estimated based on the frequency of the interaction being present in the best-fitting (selected) networks. Then the networks in the next are sampled independently from these probabilities. In other words, the frequency of each interaction is taken as a measure of probability, providing a background for the next population random generation. Moreover, some noise can be introduced to the sampling process by sampling from truncated probability distributions, e.g., probability distribution values larger/smaller than an upper/lower threshold are truncated to the threshold value, respectively. Recommendation for upper/lower thresholds is 0.8/0.2.

References

1. Jaenisch R, Young R (2008) Stem cells, the molecular circuitry of pluripotency and nuclear reprogramming. Cell 132:567–582
2. Enver T, Pera M, Peterson C, Andrews PW (2009) Stem cell states, fates, and the rules of attraction. Cell Stem Cell 4:387–397
3. Stormo GD, Zhao Y (2010) Determining the specificity of protein-DNA interactions. Nat Rev Genet 11:751–760
4. Maerkl SJ, Quake SR (2007) A systems approach to measuring the binding energy landscapes of transcription factors. Science 315:233–237
5. Hallikas O, Palin K, Sinjushina N, Rautiainen R, Partanen J, Ukkonen E, Taipale J (2006) Genome-wide prediction of mammalian enhancers based on analysis of transcription-factor binding affinity. Cell 124:47–59
6. Inukai S, Kock KH, Bulyk ML (2017) Transcription factor-DNA binding: beyond binding site motifs. Curr Opin Genet Dev 43:110–119
7. Le Novere N (2015) Quantitative and logic modelling of molecular and gene networks. Nat Rev Genet 16:146–158
8. Karlebach G, Shamir R (2008) Modelling and analysis of gene regulatory networks. Nat Rev Mol Cell Biol 9:770–780
9. Lim WA, Lee CM, Tang C (2013) Design principles of regulatory networks: searching for the molecular algorithms of the cell. Mol Cell 49:202–212
10. Paulsson J (2004) Summing up the noise in gene networks. Nature 427:415–418
11. Waddington CH (1959) Canalization of development and genetic assimilation of acquired characters. Nature 183:1654–1655
12. Chen T, He HL, Church GM (1999) Modeling gene expression with differential equations. Pac Symp Biocomput:29–40
13. Kauffman S (1969) Homeostasis and differentiation in random genetic control networks. Nature 224:177–178
14. Liu B, de la Fuente A, Hoeschele I (2008) Gene network inference via structural equation modeling in genetical genomics experiments. Genetics 178:1763–1776
15. Ma S, Gong Q, Bohnert HJ (2007) An Arabidopsis gene network based on the graphical Gaussian model. Genome Res 17:1614–1625

16. Margolin AA, Nemenman I, Basso K, Wiggins C, Stolovitzky G, Dalla Favera R, Califano A (2006) ARACNE: an algorithm for the reconstruction of gene regulatory networks in a mammalian cellular context. BMC Bioinformatics 7(Suppl 1):S7
17. Faith JJ, Hayete B, Thaden JT, Mogno I, Wierzbowski J, Cottarel G, Kasif S, Collins JJ, Gardner TS (2007) Large-scale mapping and validation of Escherichia coli transcriptional regulation from a compendium of expression profiles. PLoS Biol 5:e8
18. Rodriguez A, Crespo I, Androsova G, del Sol A (2015) Discrete logic modelling optimization to contextualize prior knowledge networks using PRUNET. PLoS One 10:e0127216
19. Crespo I, Perumal TM, Jurkowski W, del Sol A (2013) Detecting cellular reprogramming determinants by differential stability analysis of gene regulatory networks. BMC Syst Biol 7:140
20. Okawa S, Nicklas S, Zickenrott S, Schwamborn JC, Del Sol A (2016) A generalized gene-regulatory network model of stem cell differentiation for predicting lineage specifiers. Stem Cell Reports 7:307–315
21. D'Alessio AC, Fan ZP, Wert KJ, Baranov P, Cohen MA, Saini JS, Cohick E, Charniga C, Dadon D, Hannett NM, Young MJ, Temple S, Jaenisch R, Lee TI, Young RA (2015) A systematic approach to identify candidate transcription factors that control cell identity. Stem Cell Reports 5:763–775
22. Rackham OJ, Firas J, Fang H, Oates ME, Holmes ML, Knaupp AS, Suzuki H, Nefzger CM, Daub CO, Shin JW, Petretto E, Forrest AR, Hayashizaki Y, Polo JM, Gough J (2016) A predictive computational framework for direct reprogramming between human cell types. Nat Genet 48:331–335
23. Cahan P, Li H, Morris SA, Lummertz da Rocha E, Daley GQ, Collins JJ (2014) CellNet: network biology applied to stem cell engineering. Cell 158:903–915
24. Cohen DE, Melton D (2011) Turning straw into gold: directing cell fate for regenerative medicine. Nat Rev Genet 12:243–252
25. Soneson C, Delorenzi M (2013) A comparison of methods for differential expression analysis of RNA-seq data. BMC Bioinformatics 14:91
26. McCall MN, Uppal K, Jaffee HA, Zilliox MJ, Irizarry RA (2011) The Gene Expression Barcode: leveraging public data repositories to begin cataloging the human and murine transcriptomes. Nucleic Acids Res 39: D1011–D1015
27. McCall MN, Jaffee HA, Zelisko SJ, Sinha N, Hooiveld G, Irizarry RA, Zilliox MJ (2014) The gene expression barcode 3.0: improved data processing and mining tools. Nucleic Acids Res 42:D938–D943
28. Jung S, Hartmann A, Del Sol A (2017) RefBool: a reference-based algorithm for discretizing gene expression data. Bioinformatics 33:1953–1962
29. Shmelkov E, Tang Z, Aifantis I, Statnikov A (2011) Assessing quality and completeness of human transcriptional regulatory pathways on a genome-wide scale. Biol Direct 6:15
30. Gross AM, Ideker T (2015) Molecular networks in context. Nat Biotechnol 33:720–721
31. Goldberg DE (1989) Genetic algorithms in search, optimization and machine learning. Addison-Wesley Longman Publishing Co., Inc., Boston
32. Hauschild M, Pelikan M (2011) An introduction and survey of estimation of distribution algorithms. Swarm and Evolutionary Computation 1:111–128

Chapter 3

Cell Population Model to Track Stochastic Cellular Decision-Making During Differentiation

Keith Task and Ipsita Banerjee

Abstract

Human pluripotent stem cells are defined by their potential to give rise to all of the lineages of an embryo proper. Guiding the differentiation of embryonic stem cells or induced pluripotent stem cells can be achieved by exposing them to a succession of signaling conditions meant to mimic developmental milieus. However, achieving a quantitative understanding of the relationship between proliferation, cell death, and commitment has been difficult due to the inherent heterogeneity of pluripotent stem cells and their differentiation. Here, we describe a computational modeling approach to track the dynamics of germ layer commitment of human embryonic stem cells. We demonstrate that simulations using this model yield specific hypotheses regarding proliferation, cell death, and commitment and that these predictions are consistent with experimental measurements.

Key words Human stem cell differentiation, Endoderm, Embryonic stem cells, Activin, Population modeling

1 Introduction

1.1 Background

Embryonic stem cells (ESCs) have gained much attention in recent years for their ability to differentiate into cells of any of the three primary germ layers (endoderm, mesoderm, and ectoderm) as well as to remain in a pluripotent state under appropriate conditions [1, 2]. Numerous types of endoderm-like cells emerging during gastrulation have been described, including primitive, visceral, parietal, and definitive endoderm [3]. The definitive endoderm, henceforth referred to as endoderm, gives rise to such tissues as liver and pancreas [4, 5]. A focus of stem cell study in recent years has been the directed differentiation of ESC into endoderm cells for subsequent differentiation into hepatic or pancreatic lineage. Extensive research has established the possibility of deriving endoderm following alternate routes. What is however lacking is a thorough mechanistic investigation of the dynamics of differentiation. For example, studies have shown Nodal, a key component in

Patrick Cahan (ed.), *Computational Stem Cell Biology: Methods and Protocols*, Methods in Molecular Biology, vol. 1975, https://doi.org/10.1007/978-1-4939-9224-9_3,

the Nodal signaling pathway which induces endoderm, can be mimicked in vitro by human Activin A [6, 7]. Endoderm formation with the addition of Activin A has been experimentally verified in numerous studies [8–10]. However, the routes by which pluripotent cells differentiate into the endoderm germ layer during Activin A exposure have been less studied. Furthermore, while there has been some interest in modeling the gene regulatory network of differentiating stem cells using population-averaged information [11, 12], the heterogeneity and stochasticity of the process of differentiation demands more careful analysis. Mechanistic studies will be beneficial in efficient derivation of mature, functional cell types, and mathematical models which incorporate differences at the cellular level can be used to elucidate these mechanisms.

1.2 Review of Mathematical Models of Stem Cells

Initial efforts toward a quantitative understanding of stem cells focused on gene regulatory networks. Several key genes have been identified which govern pluripotency and include Oct4, Nanog, and Sox2 [13–16]. Understanding how these genes interact can give insight into self-renewal behavior, and mathematical models have been utilized to gain a more quantitative understanding of the system. Notable studies of this network include the study by Chickarmane et al. [17] which reports identification of a bistable switch in the Oct4-Sox2-Nanog network leading to a binary decision of the cells to self-renew or differentiate. In a follow-up work [18], the authors further extend the model to incorporate lineage-specific differentiation, namely, to endoderm and trophectoderm. MacArthur et al. [19] also analyzed the Oct4-Sox2-Nanog network coupled with a lineage specification network to investigate the induction of pluripotent cells from somatic cells.

The mathematical analysis of stem cell gene regulatory networks has also been used to systematically assess the effectiveness of conversion between different cell types. CellNet [20] is an open-source platform which reconstructs gene regulatory networks from experimental gene expression profiles and compares these networks to those of target cells. In stem cell engineering, conversion between cell types has been accomplished via numerous routes, including reprogramming somatic cells to pluripotency, directed differentiation of ESC to somatic cells, and direct conversion between somatic cells. In this transformation, a key question is how similar the cells generated in vitro are to the target cell type. CellNet mathematically compares the reconstructed networks to those of target cells to answer this question. This tool has been extensively used for a wide range of cell types, including ESC, iPSC (induced pluripotent stem cells), and somatic cells (fibroblast, hepatocyte, etc.). Studies using CellNet have found that agreement between the networks of converted cells and their targeted cell types is stronger when directed differentiation of ESC is utilized as compared to direct conversion of somatic cells. However, in both cases, networks of the parent cell persist after differentiation/conversion. These findings and the CellNet platform can be used to improve the routes by which cells are transformed toward other cell types.

Mathematical approaches have also been utilized to understand signaling pathways involved in pluripotency and differentiation. Prudhomme et al. [21] performed a thorough systematic analysis of how the intracellular signaling relates to different extracellular cues during differentiation of mouse ESC. A partial least square multivariate model was built to show the role of signaling proteins in self-renewal, differentiation, and proliferation of stem cells. In a follow-up work, Woolf et al. [22] investigated the signaling network to determine the "cue-signal-response" interactions through a Bayesian network algorithm. The nodes of the network are assigned to be an extracellular stimulus, a signaling protein, or a cell response, following which the model identified interconnections between nodes. Mahdavi et al. [23] employed sensitivity analysis in the Stat3 pathway to predict how self-renewal in mouse ESC is controlled.

Beyond intracellular processes, there have been modeling efforts to describe population behavior during self-renewal and pluripotency and how this behavior is manifested from single-cell characteristics. Viswanathan et al. [24] proposed a single-cell model of the ESC system which would account for the heterogeneity in the cell population. They based their model on the number of ligands/receptors per cell, and it predicted the behavior of ESC self-renewal and differentiation and the system's response to different exogenous stimuli. Such analysis has potential use in selecting specific tuning parameters while guiding ESC toward a specific fate [24, 25]. Prudhomme et al. [26] developed an ordinary differential equation-based kinetic model to quantify the differentiation dynamics in response to combinations of different extracellular stimuli. Based on the experimental data of ESC response to different combinations of extracellular matrix and cytokines, the authors estimated kinetic rate constants for each culture condition.

ESCs are heterogeneous in nature, with high intercellular variability of mRNA and protein expression within a population, both during self-renewal and differentiation. Therefore, specific techniques are needed to model population heterogeneity. A common modeling technique to describe population dynamics in a heterogeneous system is the population balance equation (PBE), which has been used to model various systems, including adult stem cell behavior [27]. An alternate approach to capture the dynamics of a heterogeneous cellular population is the cellular ensemble model. Here individual cells are tracked with time, with the behavior of the individual cells dictated by rules or equations which are solved for each cell in the population [28]. Distributions and variability associated with the parameters of these rules and equations can capture the heterogeneity in the population. Glauche et al. have utilized this approach to describe lineage specification of hematopoietic stem cells, with cellular choices governed by a competition of different lineage propensities [29]. Through their model, various cellular populations are tracked with time, and insight into the differentiation process is obtained.

Following a similar conceptual approach, we have developed a stochastic model to track the early differentiation and germ layer commitment of human embryonic stem cells. In our model, individual cells in the population are stochastically evolved in time, following specific user-defined rules, through which the system dynamics is extracted. The model predicts three aspects of endoderm formation: total cell proliferation, cell death, and lineage commitment. In order to understand the mechanism favoring the process of stem cell differentiation, we simulate several alternate mechanisms and compare the simulated dynamics with our experimental data. Endoderm is experimentally induced in hESC through alternate pathways: addition of Activin A and Activin A supplemented with the growth factors basic fibroblast growth factor (FGF2) and bone morphogenetic protein 4 (BMP4) [9, 30, 31]. Differentiation dynamics of the cell population is experimentally tracked by analyzing percentage of cell population expressing endoderm-specific proteins: Sox17 and CXCR4 [32, 33]. Through agreement between the experimental data and simulated dynamics, we elucidate the mechanisms behind initial lineage commitment during definitive endoderm differentiation induction.

1.3 Directed Differentiation of Human Embryonic Stem Cells

Here, we provide some background on the experimental conditions used to generate the data in Fig. 1. Human embryonic stem cells (H1 cell line) were cultured under feeder-free conditions. 6-well culture dishes were incubated with Matrigel™ coating (hESC-qualified matrix) for 30 min. hESC colonies were plated onto the Matrigel layer with 1 mL mTeSR®1 hESC media and supplement. The cells were incubated at 37 °C in 5% CO_2, and the mTeSR®1 media was replaced daily. For the differentiation study, two alternate conditions were compared for endoderm induction: human Activin A (henceforth referred to as "Condition A") and human Activin A supplemented with the growth factors FGF2 and BMP4 ("Condition B"). To commence endoderm induction, DMEM/F12 (Invitrogen, Carlsbad, CA, USA), 1xB27® supplement (Invitrogen), and 0.2% bovine serum albumin (BSA) supplemented with 100 ng/mL human Activin A (Condition A) or 100 ng/mL human Activin A, 100 ng/mL FGF2, and 100 ng/mL BMP4 (Condition B) were used as differentiation media, which were replaced daily for a total of 5 days. Upon induction of differentiation, cells were harvested daily for subsequent analysis.

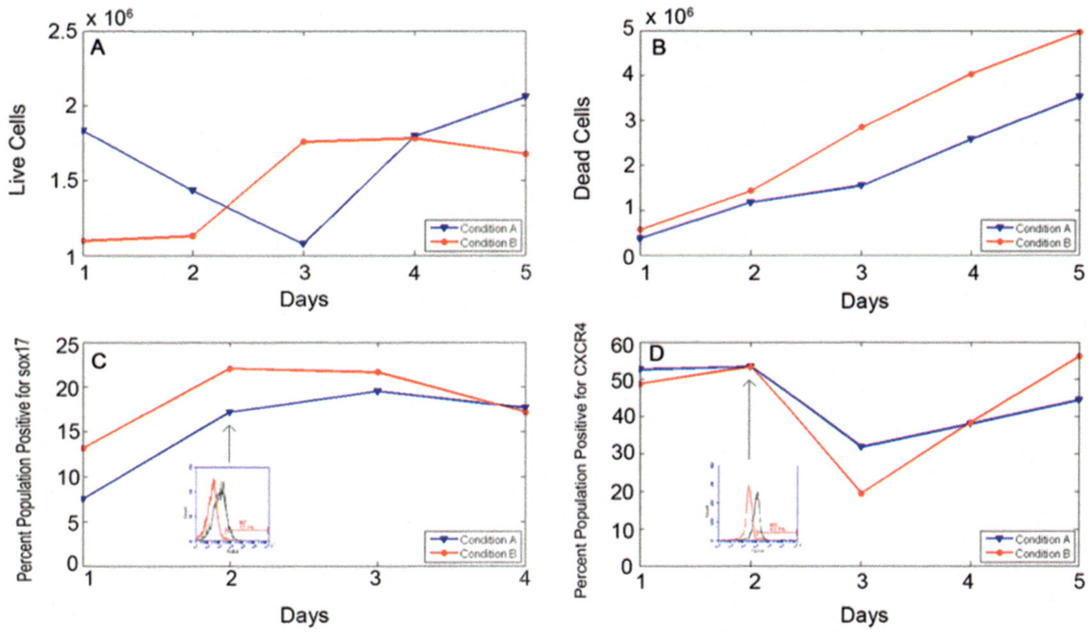

Fig. 1 Experimental results of cell behavior during endoderm induction. Cellular growth (**a**) and death (**b**) dynamics for conditions A and B. Temporal behavior of cellular population positive for Sox17 (**c**) and CXCR4 (**d**). Inset: representative output of flow cytometry data. Red histogram: secondary antibody only control. Black histogram: sample. Red gate denotes sample taken to be positive

For live/dead assay, the plated cells were dissociated with Trypsin+EDTA, trypan blue was added to distinguish live from dead, and the cells were counted using a standard hemacytometer.

For flow cytometry, harvested cells were first fixed for 15 min in 4% methanol-free formaldehyde in phosphate-buffered saline (PBS). Cells were washed twice and permeabilized in 0.1% Saponin +0.5% BSA in PBS for 30 min. To block non-specific binding, the cells were incubated in 3% BSA + 0.25% dimethyl sulfoxide (DMSO) + 0.1% Saponin in PBS for 30 min. A portion of cells were then set aside as the negative control (secondary antibody only without primary). The cells to be used as the positive samples were then incubated in blocking buffer with goat antihuman Sox17 and rabbit antihuman CXCR4 primary antibodies, 1:200 dilution, for 30 min. The cells were washed twice with blocking buffer, resuspended in the buffer, and incubated with donkey anti-goat APC (1:350 dilution) and donkey anti-rabbit FITC (1:200 dilution) for 30 min (both the samples and negative control). Two washings were followed by 10 min of 0.2% Tween 20 to further eliminate non-specific staining. Cells were washed and transferred to flow cytometry tubes. Accuri C6 © flow cytometer was used to quantify Sox17-APC and CXCR4-FITC expression. Cells stained with the secondary antibody only (without primary antibody) were first analyzed; this population was taken as the negative, and the gate was set beyond these cells to eliminate false positives due to autofluorescence and non-specific secondary antibody binding. The completely stained samples (primary and secondary antibody stained) were then analyzed, and the percentage of the population falling within the set gate was recorded as the positive sample for the respective antibody.

The hESC culture was analyzed for cellular growth and death dynamics during endoderm induction by both Activin A (Condition A) and Activin A/FGF2/BMP4 (Condition B) conditions as illustrated in Fig. 1. Cellular growth kinetics were found to exhibit nonlinear dynamics, while cell death remained predominantly linear over time. A proliferation lag time is exhibited in Condition A up until Day 3, during which time the number of live cells decreases because of cell death. Beyond this time, cells begin to proliferate in a roughly linear fashion. Interestingly, the majority of cell growth in Condition B occurs before Day 3.

Figure 1 also illustrates the dynamics of Sox17 (Fig. 1c) and CXCR4 (Fig. 1d) expression for both the experimental conditions, with the insets illustrating representative flow cytometry results. Overall the dynamics of differentiation, as judged by the fraction of cells positive for Sox17 and CXCR4, was similar for both conditions A and B, with some differences in the magnitude. The population of cells positive for Sox17 exhibits an initial increase followed by a saturation behavior. The fraction of cells positive for CXCR4 is relatively constant until the second day, after which there is a significant drop. Subsequently, there is an approximately linear increase in the CXCR4 population which is more prominent in Condition B. Understanding the cellular behavior in terms of these four output dynamics (cell growth, death, population positive for Sox17 and positive for CXCR4) was the focus of the study.

2 Methods

2.1 Development of a Mathematical Model

The system of stem cell differentiation to endoderm is modeled using a stochastic population-based model (*see* **Note 1**). The basic formulation of the model is based on earlier reports for hematopoietic stem cells [29, 34]. In order to adapt the model to pluripotent stem cell system, we incorporated specific modifications, followed by a stringent model analysis using parameter sensitivity and feasibility studies. Figure 2 illustrates the pseudo-code describing the implementation of the main simulation, parameter ensemble, and sensitivity analysis.

2.1.1 Signaling Regimes

The model is initiated by a population of cells, the properties of which are evolved by specific preassigned rules. The cells are primarily categorized into two signaling regimes: Ω and A; Ω can be considered as an active regime supporting cellular proliferation and

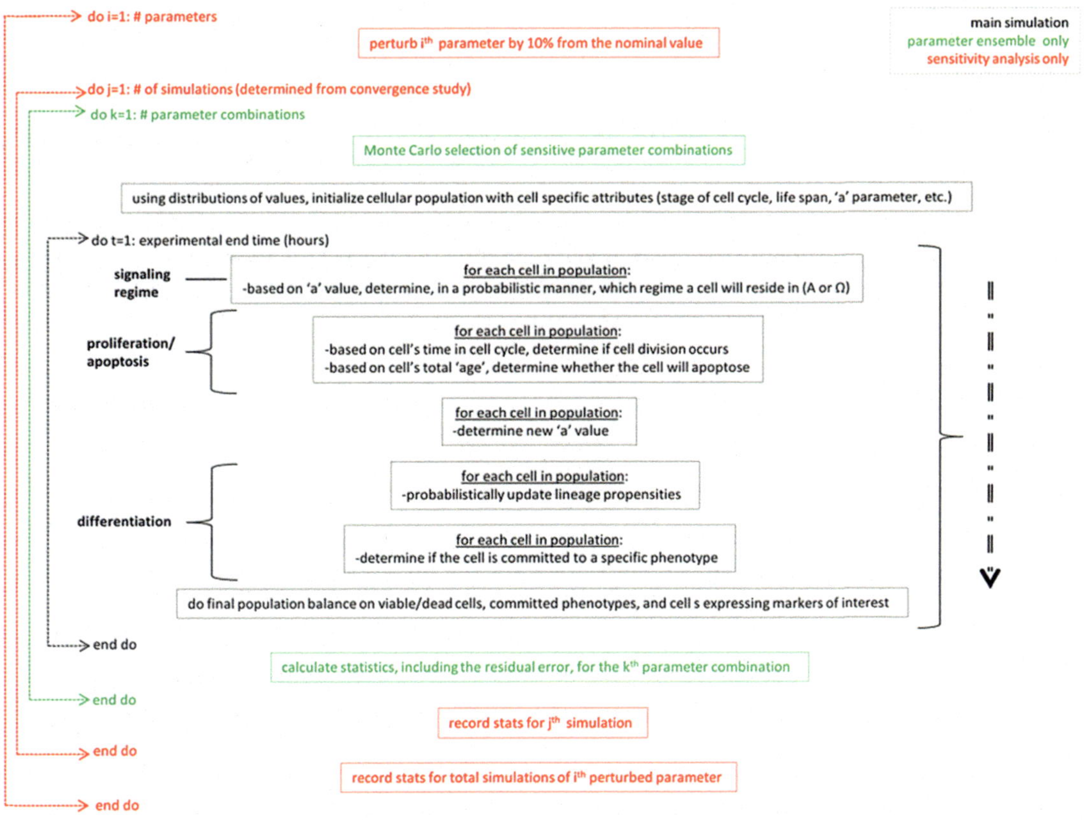

Fig. 2 Implementation of mathematical model. Pseudo-code describing flow of events in the population-based model. Black: events to simulate temporal behavior of cellular population (main routine). Green: model inclusions for parameter ensemble, which runs main routine using different parameter value combinations. Red: model inclusions for sensitivity analysis, which runs main routine 4000 times (replications determined by convergence study) for each perturbed parameter value, the output being the parameter sensitivities

differentiation, while A is a more dormant regime where cells are quiescent and prone to dedifferentiation. The cells can transfer in between these two regimes, an event decided upon primarily by a cell-specific parameter, termed as "a" value. This parameter "a" is randomly assigned to each cell at the beginning of simulation and is updated at each time step. This "a" value is unaltered in the A regime, while in the Ω regime, at each time step, "a" is decreased by the following simple relationship: $a = a/d$. When "a" falls below a threshold ($a_{\min}$), the cells lose their ability to transfer to the A regime. Both $a_{\min}$ and d are system-specific parameters estimated based on experimental data.

The probability of a cell transferring between Ω and A regime is dictated by two main variables: the "a" level of the cell and the number of cells in the destination regime. This probability is calculated via the following relationships. For a cell that is in the Ω regime, the probability of transfer to the A regime, α, is:

$$\alpha = \frac{a}{a0\max} * \left[\frac{1}{aa + ba * \exp(ca * ma)} + da\right] \tag{1}$$

where ma is the number of cells in the A regime (the remaining parameters are specific to the system and are determined via parameter estimation). If a cell is in the A regime, the probability of transfer to the Ω regime, ω, is:

$$\text{if } a == 0\, \omega \leftarrow 1$$

else

$$\omega = \frac{amin}{a} * \left[\frac{1}{aw + bw * \exp(cw * \text{moip})} + dw\right] \tag{2}$$

where moip is the number of cells that are jointly in the Ω regime and the proliferation phase.

Once the transfer probabilities are calculated, a random number (uniform distribution) is generated, and if the random number is less than the transfer probability, then the cell transfers from one regime to another. If not, the cell will remain in its current phase. A final stipulation is that a cell must be in the G1 phase of the cell cycle to transfer from the Ω regime to the A regime.

2.1.2 Proliferation, Cell Death, and Differentiation Rules

An individual cell in the Ω regime is allowed to proliferate only after it loses the capability to pass into A regime, having crossed the "a" threshold value. Consistent with that, proliferation is allowed only in the Ω regime, for an amount of time which is cell dependent (in the range $t_{p\min} - t_{p\max}$), after which the cell enters a senescent stage and will not proliferate. Each cell is randomly assigned a maximum life span (in the range $l_{\min} - l_{\max}$), exceeding which it will die. While the cells age in the Ω regime, they neither proliferate nor age in the A regime.

Cellular differentiation is governed by the "lineage propensity" value, x, representing a cell's likelihood to differentiate into a particular lineage. Only the Ω regime allows an increase in lineage propensity. While updating the propensity of differentiation to a particular lineage, all the possible lineages are competing, and any can be updated, with the one with higher propensity having a higher probability of being selected. For each cell that is in the Ω regime and in the G1 phase, x for the chosen lineage k is updated in the following way:

$$x_k = x_k * \left(1 + b_{\text{prog}} * x_k + n_{\text{prog},\,k}\right) \tag{3}$$

where b_{prog} and $n_{\text{prog},k}$ are system-specific parameters estimated from the data. The propensities of the other lineages are modified proportionally so that $\Sigma x_k = 1$.

Once a cell's propensity for a specific lineage exceeds a threshold level (x_{com}, identical for all lineages), the cell is considered committed to that particular lineage and will retain its differentiated phenotype. If this threshold has not been exceeded and if the cell is chosen to be transferred to the A regime, the propensity values will converge to an average value. The model is therefore able to track specific germ layer populations, and through this the percent of the population positive for Sox17 (visceral and definitive endoderm marker) and CXCR4 (definitive endoderm and mesendoderm marker) [32, 33, 35] can be extracted.

2.1.3 Mechanism of hESC Differentiation

The current work is focused on the mechanistic investigation of the dynamics of hESC induction into endoderm. Using the platform of the stochastic population-based model, we investigated several alternate mechanisms and analyzed them for agreement with experimental data. Three characteristics of the differentiation process were chosen to be investigated: the presence/absence of an intermediate germ layer, mesendoderm, which subsequently gives rise to mesoderm and endoderm [36]; the presence/absence of CXCR4 in mesoderm; and whether proliferation of a specific differentiated cell phenotype is favored over others. Combination of the aforementioned attributes results in 12 alternate mechanisms. Each of these mechanisms was incorporated into the model and analyzed for agreement with experimental data. It is expected that the most likely mechanism will best describe the experimental dynamics of the stem cell system.

The incorporation of mesendoderm involved a two-stage differentiation scheme. In the first stage, hESC are able to differentiate into either mesendoderm or visceral endoderm. Once cells are committed to the mesendoderm lineage, several of their attributes are re-initialized, such as their "a" value and lineage propensities. The mesendodermal cells can then further differentiate into endoderm or mesoderm. Differences in the proliferation potential of different phenotypes were incorporated by considering three scenarios: proliferation of hESC and mesendoderm; proliferation of hESC, mesendoderm, and endoderm; and proliferation of all phenotypes.

In the next step, the developed stochastic model was used to test the proposed mechanisms for agreement with experimental data. The mathematical model involves multiple parameters which require detailed analysis before the model can be used for prediction. The parameters can be grouped into two categories: (a) simulation parameters which affect the convergence behavior of the simulation and (b) model parameters which affect specific model output for the converged simulation. Two parameters were identified to be simulation parameters: initial cell population and number of stochastic runs. Figure 3 illustrates the overall flowchart of the modeling framework.

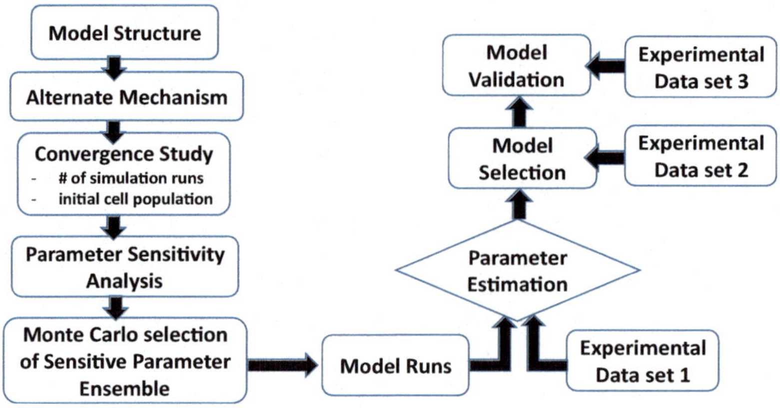

Fig. 3 Flowchart of the developed stochastic model. Diagram of the major steps of the overall modeling process

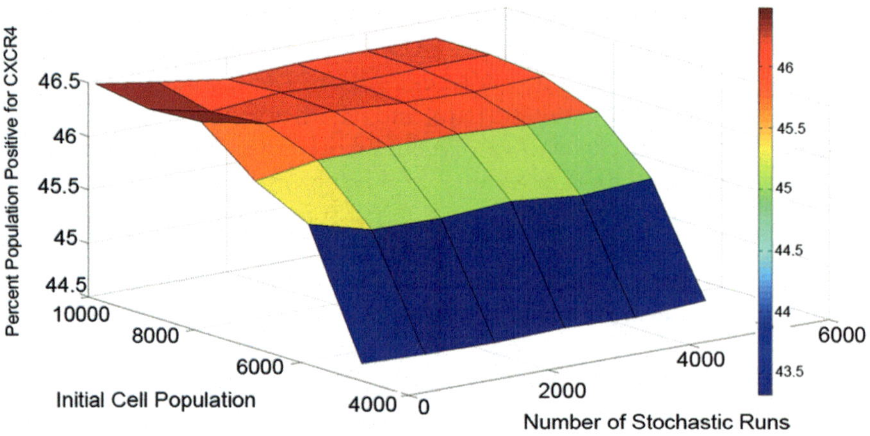

Fig. 4 Convergence study of simulated cell population over various initial cell populations and total stochastic runs. Output is percent of the simulated population positive for CXCR4, averaged over all stochastic runs at Day 5

2.2 Convergence Study

As with any stochastic model, the number of model runs necessary to obtain a converged solution needs to be determined. The run time of each model run also depends on the chosen size of initial cell population, which also affects the solution over a certain range. A two-dimensional convergence study was thus undertaken, wherein the effects of both stochastic run number and initial cell population on model output were determined. The convergence test allows determination of the minimum number of stochastic runs and initial cell population beyond which the model output does not significantly change. Figure 4 illustrates results from a two-dimensional convergence study for CXCR4-positive population output.

Overall it was observed that the simulation results were relatively insensitive to the number of stochastic runs, which converged beyond 2000 runs. However, the initial cell population size has a dominant effect on the output and convergence of the simulation.

Following this analysis an initial cell population of 9000 and total stochastic runs of 4000 were used for subsequent simulations.

2.3 Sensitivity Analysis

Sensitivity analysis was performed to determine the relative importance of parameters in affecting the outputs of cellular growth, death, and lineage commitment. Because this model is probabilistic in nature, traditional ways of determining local sensitivity, e.g., partial derivative of output with respect to an input, cannot be employed. A stochastic analysis was therefore chosen, using differences in output histograms of nominal and perturbed parameters, S, to estimate parameter sensitivity [37] as:

$$S = HD = \sum_{i=1}^{k} \left| \frac{\sum_{j=1}^{|x|} \mathrm{X}(x_j, I_i)}{|x|} - \frac{\sum_{j=1}^{|y|} \mathrm{X}(y_j, I_i)}{|y|} \right| \quad (4)$$

where x_j and y_j are the individual elements in the nominal and perturbed histograms, respectively; I_i represents the range of each bin i in the nominal histogram; X is a counting variable which takes on a value of 1 if x_j/y_j is within the interval I_i; $|x|$ and $|y|$ are cardinalities of data sets x (nominal) and y (perturbed), respectively; and k is the number of bins, which was determined by calculating the appropriate bin size of the nominal output histogram by the Freedman-Diaconis rule [38]:

$$\text{Bin size} = 2\mathrm{IQR}(P)n^{\frac{-1}{3}} \quad (5)$$

where IQR(P) is the interquartile range of a sample population P and n is the number of observations in P. The number of bins was different for each output and ranged from 37 to 51.

For sensitivity analysis, each parameter was perturbed by 10% while keeping the rest of the parameters at the nominal value. For each bin in the nominal output histogram, the difference between the percentage of total nominal histogram elements residing in that bin and the percentage of the total perturbed histogram elements residing in that bin is calculated. The sum of the absolute value of this difference over all of the bins is the histogram distance (HD).

Figure 5a illustrates the model parameter sensitivity to the output of cellular growth, as concluded from the shift in histogram distance (inset). A clear jump in the sensitivity is observed, with a large difference between parameters with low and high sensitivity.

While Fig. 5a represents the parameter sensitivity to model output of cellular growth, a similar analysis was also performed for all of the other model outputs (cell death, population positive for Sox17 and positive for CXCR4). Overall it was observed that even though the magnitude of sensitivity differs between outputs, the highly sensitive parameters were mostly conserved across

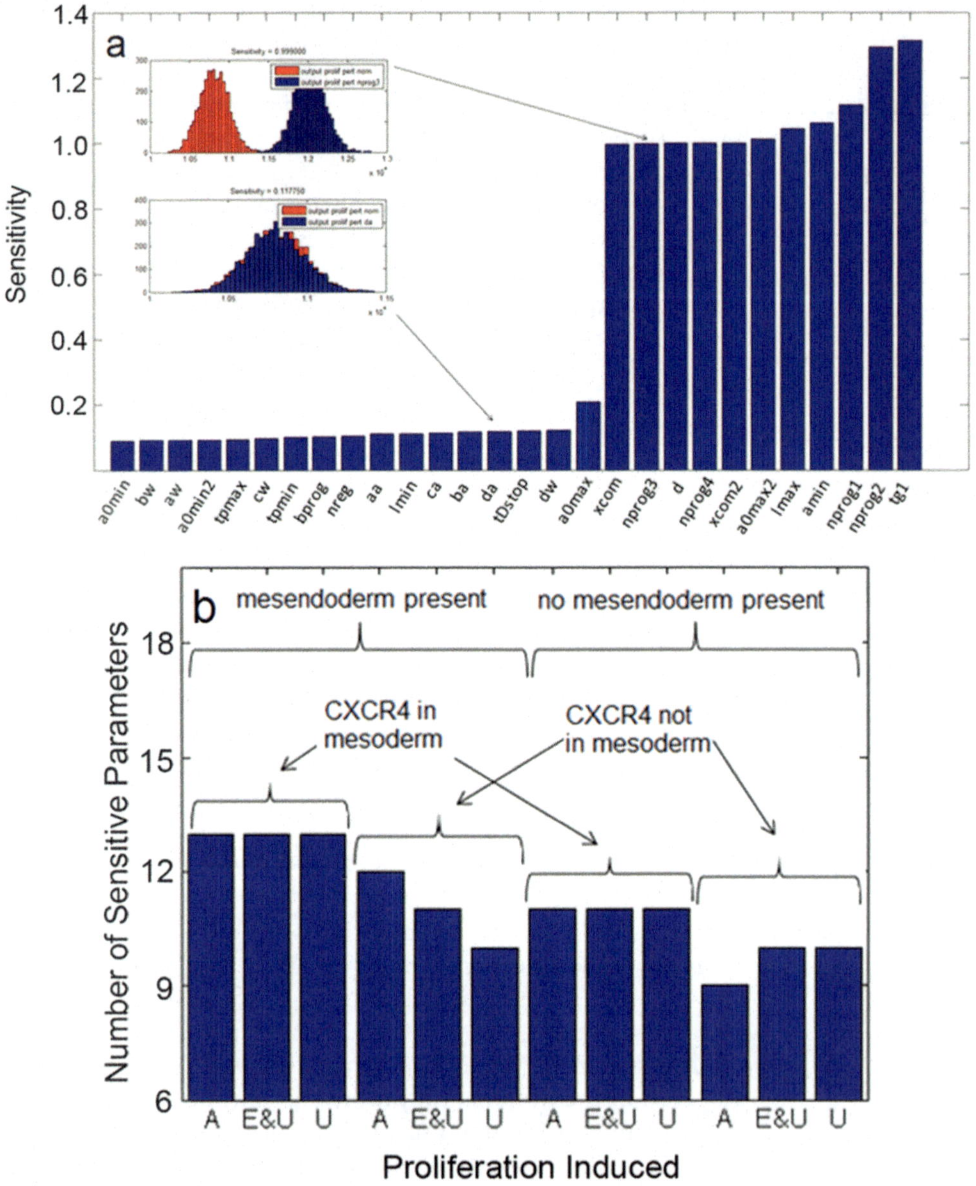

Fig. 5 Sensitivity analysis of population-based model. (**a**) Cellular growth sensitivity to each of the parameters, perturbed by 10%. Parameter definitions listed in Table 1 (inset). Comparison of cellular growth output histogram from nominal (red) and perturbed (blue) parameters. (**b**) Number of sensitive parameters determined for each mechanistic model. Proliferation induced: A, all phenotypes; E&U, endoderm and uncommitted (hESC and mesendoderm); U, uncommitted only

outputs. Furthermore, in the present study, we are investigating multiple competing mechanisms, which require modification of the model formulation. Since the effect of such modifications on the parameter sensitivity is not intuitively obvious, similar analysis was repeated for each of the 12 proposed mechanisms. Figure 5b summarizes the number of sensitive parameters for each of the mechanisms. From the analysis, eight classes of parameters were consistently observed to have the highest sensitivity:

- a_{min}: "a" value threshold beyond which a cell enters the proliferation phase.
- a_{0max}: the initial cell population is randomly assigned an "a" value, with an upper limit of a_{0max}.
- x_{com}: lineage propensity threshold beyond which a cell is committed to a particular lineage.
- d: factor with which "a" decreases.
- t_{g1}: time cell stays in G1 phase of cell cycle. Only in this phase can a cell differentiate and transfer to the A regime from the Ω regime.
- l_{max}: upper value of range of cell population's life span.
- $nprog_i$: factor in determining magnitude of propensity updates for each lineage i.
- aa: parameter in determining the probability of a cell transferring from the Ω regime to the A regime.

2.4 Ensemble Parameter Estimation

Having determined the sensitive parameters for each of the mechanisms, the next step is to determine the optimum parameter values which will result in best agreement with experimental data. We realize that a single parameter combination may not be adequate in describing the experimental data; instead, there exists a parameter hyperspace adequately satisfying the data. In literature, biological samples are typically defined as being "sloppy" [39] with a broad ensemble of parameters satisfying the error constraints (*see* **Note 2**). Hence, an ensemble parameter estimation was performed by randomly generating initial guesses from the hyperspace of the sensitive parameters. The model was simulated with 10,000 random parameter samples; for each of these simulations, the least square estimate is determined between experimental data and model output. These simulations were run for each mechanism and condition under investigation, and parameter ensembles were generated by considering only those parameter sets for which the least square error was within the specified error constraints. Following the methodology described above, we evaluated the model predictions obtained from the different mechanisms and compared them with experimental data to determine the most plausible mechanism.

The model is formulated to capture the dynamics of cellular growth, death, and differentiation, the output of differentiation being of most interest. Hence, the model parameters were optimized with respect to differentiation dynamics, while growth kinetics and the dynamics of cell death were used for verification and for further comparison between mechanisms (described below). Table 1 includes the description of model parameters along with representative parameter estimates. A projection of the simulated

Table 1
Description of the parameters used in the model, along with representative parameter values estimated for both the conditions

Parameter	Definition	Optimized values for Activin	Optimized values for FGF2/BMP4
a0max	Maximum value of "a" in initial cell population, first differentiation stage	0.3	0.3
a0max2	Maximum value of "a" in initial cell population, second differentiation stage	0.263	0.161
a0min	Minimum value of "a" in initial cell population, first differentiation stage	0.0001	0.0001
a0min2	Minimum value of "a" in initial cell population, second differentiation stage	0.0001	0.0001
aa	Used when determining the probability of a cell transferring to the alpha regime	2.55	2.07
amin	Threshold on "a" below which a cell is able to proliferate	0.0001	0.0001
aw	Used when determining the probability of a cell transferring to the omega regime	2.0	2.0
ba	Used when determining the probability of a cell transferring to the alpha regime	0.007	0.007
bprog	Used in updating propensity in omega regime	0.0	0.0
bw	Used when determining the probability of a cell transferring to the omega regime	0.05	0.05
ca	Used when determining the probability of a cell transferring to the alpha regime	0.01	0.01
cw	Used when determining the probability of a cell transferring to the omega regime	0.1	0.1
d	Factor by which "a" decreases in omega regime	1.366	1.324
da	Used when determining the probability of a cell transferring to the alpha regime	0.001	0.001
dw	Used when determining the probability of a cell transferring to the omega regime	0.01	0.01
lmax	Maximum bound on cell life span	194	221.8
lmin	Minimum bound on cell life span	0	0
nprog1[a]	Used in determining magnitude of propensity update in omega regime for lineage 1	0.1629	0.0879
nprog2[a]	Used in determining magnitude of propensity update in omega regime for lineage 2	0.0808	0.0611

(continued)

Table 1
(continued)

Parameter	Definition	Optimized values for Activin	Optimized values for FGF2/BMP4
nprog3[a]	Used in determining magnitude of propensity update in omega regime for lineage 3	0.0256	0.0278
nprog4[a]	Used in determining magnitude of propensity update in omega regime for lineage 4	0.04	0.0457
nreg	Used in updating propensity in alpha regime	0.1	0.1
tDstop	Time beyond which a cell enters into a senescent stage and will not die (counted from start of cell's life)	120	120
tg1	Time a cell stays in the g1 phase of the cell cycle	8.74	14.6
tpmax	Upper bound of time beyond which a cell enters into a senescent stage and will not proliferate (counted from start of cell's life)	120	120
tpmin	Lower bound of time beyond which a cell enters into a senescent stage and will not proliferate (counted from start of cell's life)	120	120
xcom	Threshold level of propensity beyond which a cell is considered committed, first differentiation stage	0.866	0.767
xcom2	Threshold level of propensity beyond which a cell is considered committed, second differentiation stage	0.6096	0.7908

[a]For the mechanisms which include mesendoderm: lineage 1-mesendoderm, lineage 2-visceral endoderm, lineage 3-definitive endoderm, lineage 4-mesoderm. For the mechanisms which exclude mesendoderm: lineage 1-definitive endoderm, lineage 2-mesoderm, lineage 3, visceral endoderm

error onto a two-dimensional parameter space (for the mechanism which incorporates mesendoderm and promotes proliferation of both uncommitted and endodermal cells without CXCR4 being expressed in mesoderm ("Mechanism B")) is shown in Fig. 6a. Although it was initially thought that a trend might be observed between the error and values of the parameter ensemble, Fig. 6a shows that there is a lack of any correlation between multiple parameters and associated errors (shown for parameter "*d*"; further analysis of all parameter combinations yielded similar results). Figure 6b illustrates the minimum ensemble error for each of the proposed mechanisms simulated under the two endoderm induction conditions, the error being evaluated according to least square formulation (with respect to differentiation dynamics).

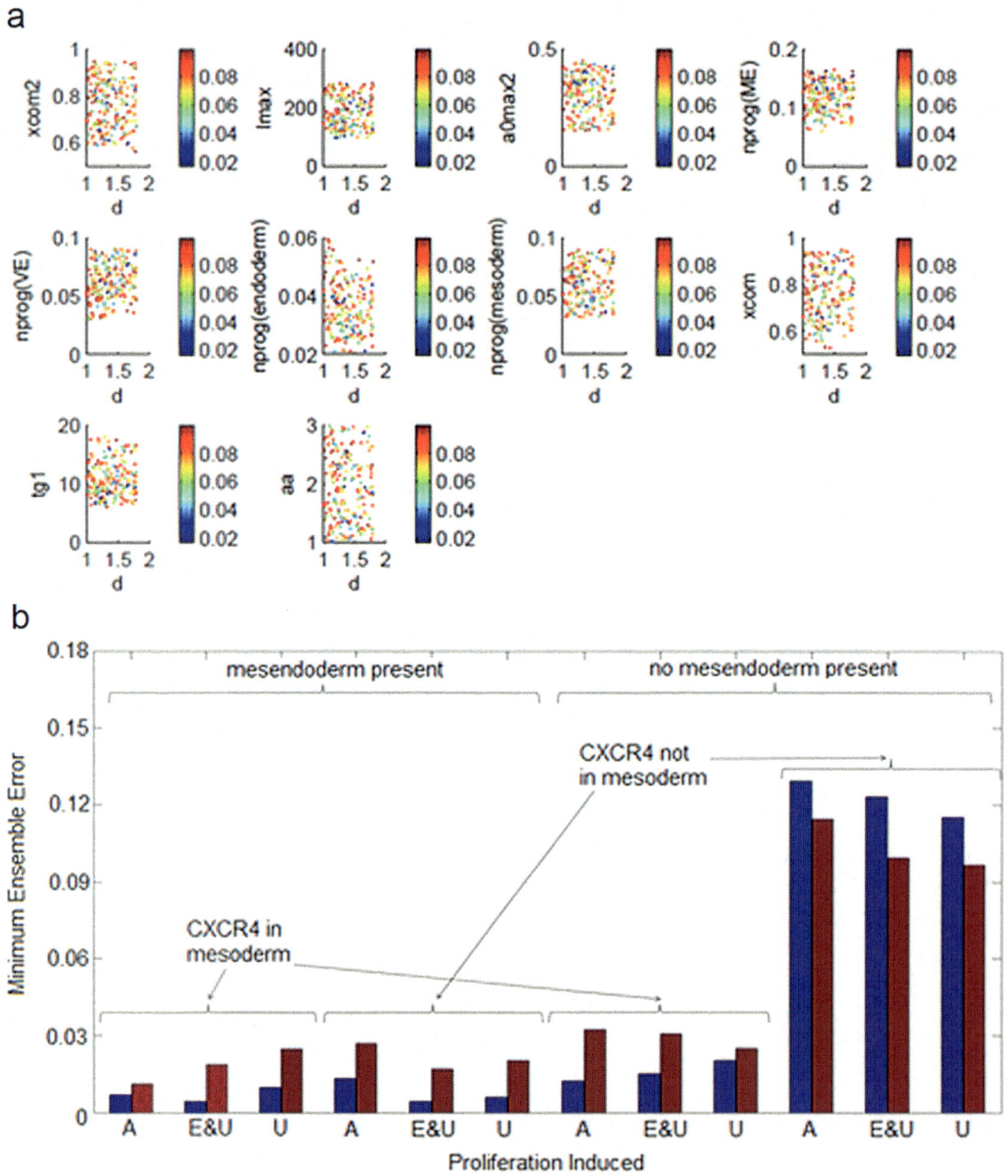

Fig. 6 Ensemble parameter estimation and model errors. (**a**) Parameter values for Mechanism B ensemble yielding errors of less than 0.025. Each parameter is compared to the most sensitive parameter, "*d*." Color bar denotes the ensemble error for that particular parameter value. (**b**) Minimum ensemble error generated for each mechanistic model. Proliferation induced: A, all phenotypes; E&U, endoderm and uncommitted (hESC and mesendoderm); U, uncommitted only. Blue, Condition A; red, Condition B

2.5 Mechanism Evaluation

2.5.1 Endoderm Induction by Activin A

In comparing the alternate simulated mechanisms and their associated error, as shown in Fig. 6b, it stands out that the absence of mesendoderm in general resulted in large errors. In some cases, the absence of mesendoderm resulted in an order of magnitude higher error than their counterpart models including mesendoderm (*see* **Note 3**).

Considering the Activin A only condition (Condition A), the most accurate mechanisms include those which incorporate mesendoderm and promote proliferation of both uncommitted and endoderm germ layer both with (henceforth denoted "Mechanism A") and without (henceforth denoted "Mechanism B") CXCR4 in mesoderm. However, Fig. 6b denotes the accuracy of the model in predicting differentiation dynamics only. Hence, in the next step, we further evaluated the performance of the two prospective mechanisms in predicting the growth kinetics and the dynamics of cell death. Figure 7 illustrates the ensemble simulation of all the model outputs and their comparison with experimental data. While both mechanisms had excellent performance in predicting Sox17 and CXCR4 dynamics (Fig. 7c, d, g, h), they differed significantly in predicting growth kinetics and cell death dynamics (Fig. 7a, b, e, f). Figure 7 clearly illustrates that Mechanism B performs better in describing both growth kinetics and cell death dynamics compared to Mechanism A. Hence, Mechanism B was chosen to be the most likely mechanism for Condition A.

2.5.2 Endoderm Induction by Activin A Supplemented by Growth Factors

Condition B (Activin A supplemented with FGF2 and BMP4) proved more difficult to describe via the investigated mechanisms, mainly because CXCR4 dynamics exhibits a faster and more prominent drop as compared to Activin A only condition. As shown in Fig. 6b, the two mechanisms which give lowest error for Condition B are the ones which incorporate mesendoderm, have CXCR4 present in mesoderm, and promote proliferation of all phenotypes (henceforth referred to as "Mechanism C") and the previously described Mechanism B. The simulated dynamics of these two mechanisms with experimental data of Condition B are shown in Fig. 8. Mechanism C, as shown in Fig. 8a–d, yields good agreement with experimental differentiation behavior, mainly due to the incorporation of CXCR4 in mesoderm; however, the simulated cell growth and death trends of this mechanism are quite different than the experimental trends. The mechanism with the next to lowest differentiation error, Mechanism B, exhibits agreement with experimental dynamics for all outputs, so it was chosen as the most likely mechanism for Condition B as well.

It is therefore reasonable to conclude that during endoderm induction with the conditions described above, undifferentiated stem cells first differentiate into a mesendoderm germ layer with subsequent differentiation to endoderm and mesoderm, the latter not expressing CXCR4 (*see* **Note 4**). Furthermore, the induction

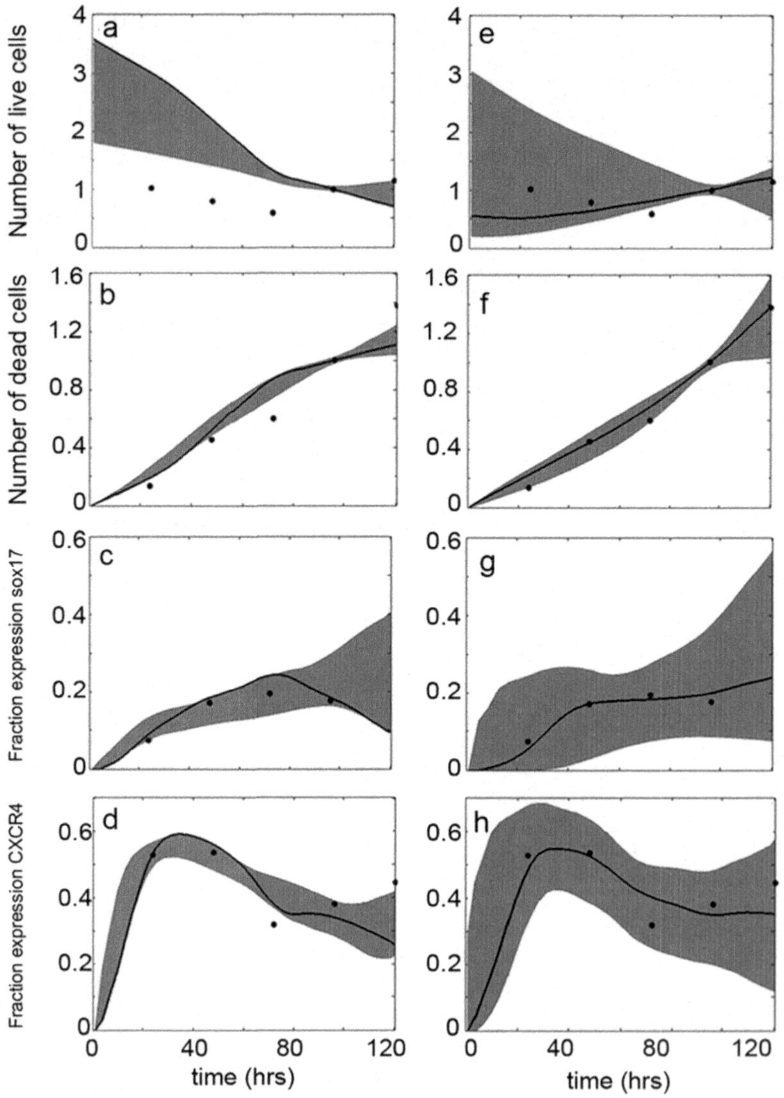

Fig. 7 Simulated output dynamics compared to experimental data (Condition A) Grey band denotes the ensemble of simulations having an error less than the threshold, with the single solid black curve showing the best fit. Black circles represent the experimental data points. (**a**–**d**): growth kinetics, cell death, fraction of population positive for Sox17 and CXCR4 dynamics, respectively, of Mechanism A; error threshold of 0.05. (**e**–**h**) Growth kinetics, cell death, fraction of population positive for Sox17 and CXCR4 dynamics, respectively, of Mechanism B; error threshold of 0.025. Cellular growth and death normalized to Day 4

condition seems to promote proliferation of both pluripotent and endoderm-like cells. The optimized parameters of this mechanism are shown in Table 2, with definitions of parameters in Table 1.

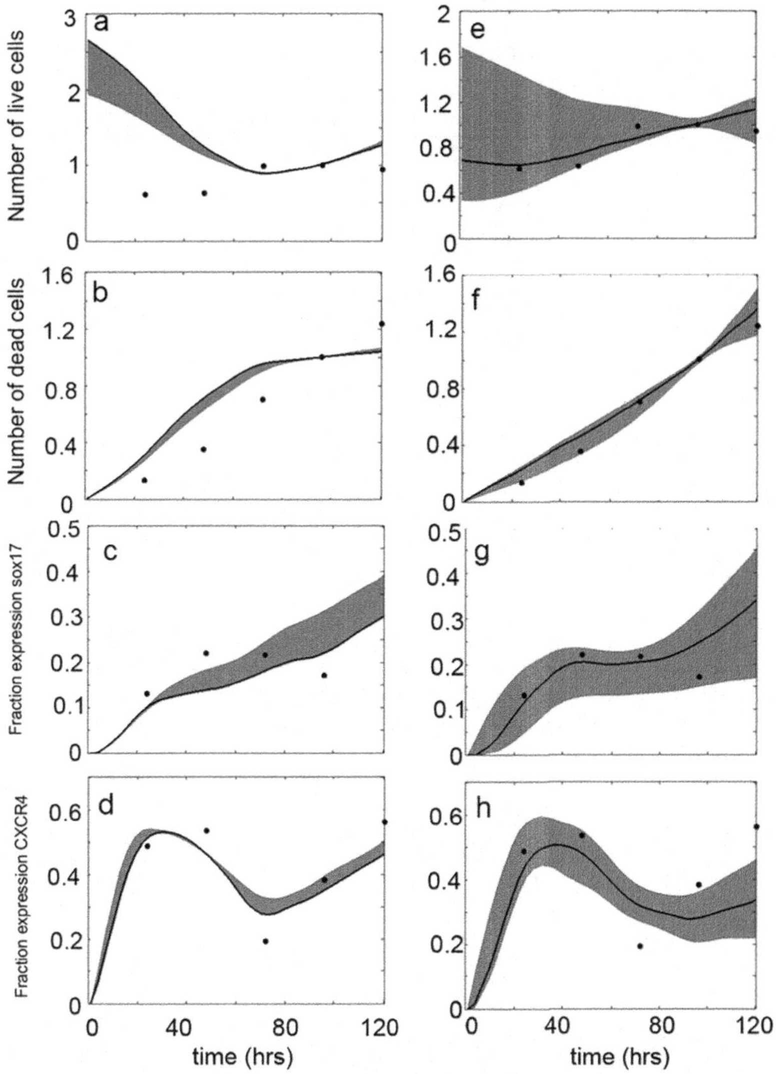

Fig. 8 Simulated output dynamics compared to experimental data (Condition B). Grey band denotes the ensemble of simulations having an error less than the threshold, with the single solid black curve showing the best fit. Black circles represent the experimental data points. (**a–d**): growth kinetics, cell death, fraction of population positive for Sox17 and CXCR4 dynamics, respectively, of Mechanism C; error threshold of 0.1. (**e–h**) Growth kinetics, cell death, fraction of population positive for Sox17 and CXCR4 dynamics, respectively, of Mechanism B; error threshold of 0.025. Cellular growth and death normalized to Day 4

2.6 Model Validation

The power of mathematical models lies in their predictive capacity. The predictive capacity of our proposed model was thus tested by simulating the population dynamics of cell types for which no a priori data was used in constructing the model. The chosen populations were that of undifferentiated cells and mesendoderm cells. The simulated profile of the undifferentiated cells (Fig. 9a–b)

Table 2
Comparison of the best fit parameter set between Conditions A and B

Parameter	Condition A	Condition B
a0max2	0.263	0.161
xcom	0.866	0.767
xcom2	0.61	0.79
d	1.37	1.32
tgl	8.74	14.6
lmax	194	222
nprog(ME)	0.16	0.0879
nprog(endoderm)	0.026	0.0278
nprog(mesoderm)	0.04	0.0457
nprog(VE)	0.081	0.0611
aa	2.55	2.07

Only the sensitive parameters are listed. A "2" after the parameter denotes the parameter for the second stage of differentiation (mesendoderm to mesoderm and endoderm) as opposed to the first stage (hESC to mesendoderm and visceral endoderm). *ME*, mesendoderm; *VE*, visceral endoderm

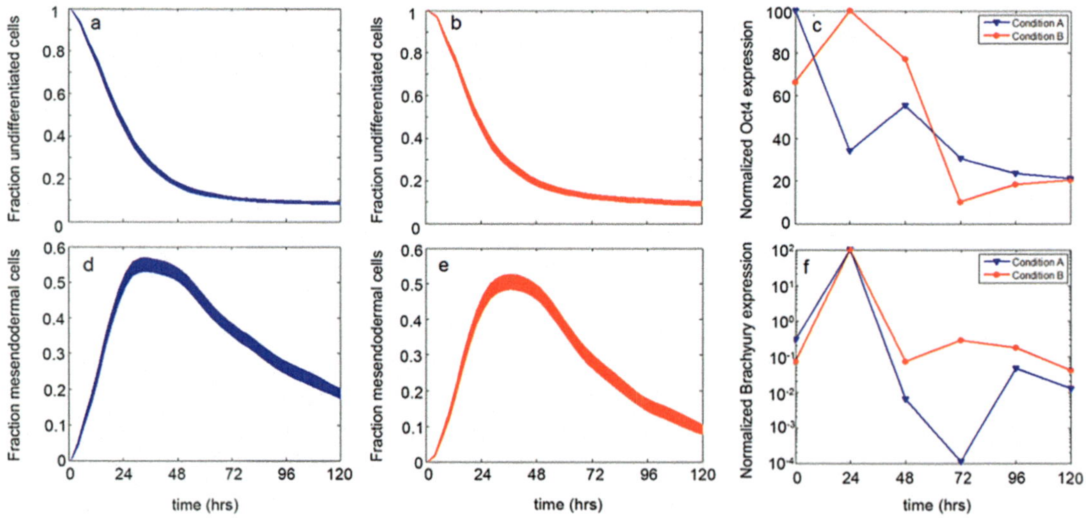

Fig. 9 Validation of model with experimental gene expression data. Simulated dynamics of the undifferentiated (**a** (Condition A), **b** (Condition B)) and mesendoderm (**d** (Condition A), **e** (Condition B)) phenotypes were compared to experimental data of their respective genes, measured by qPCR (markers, experimental measurements; lines, linear connections between data points): Oct4 (Undifferentiated, **c**) and Brachyury (Mesendoderm, **f**). The simulated dynamics bands represent 4000 stochastic simulations using the optimized parameters of Mechanism B. mRNA levels were measured with time using qPCR. Data was first normalized to the housekeeping gene Gapdh and then to undifferentiated cells. Fold change levels, determined by the $2^{-\Delta\Delta Ct}$ method, were then normalized to the maximum level for each respective gene (data reported as percent of maximum fold change)

shows an exponential decay to a final value of 10% of the cellular population. This final value was reached in approximately 3 days. The mesendoderm cell population was predicted to display more interesting dynamics, with a transient increase in cell population over the first day, followed by a decreasing trend over the next few days (Fig. 9d–e).

These model predictions were next verified by conducting further experiments to analyze the dynamics of undifferentiated cells by Oct4 gene expression and that of mesendoderm cells by Brachyury expression. To quantify mRNA levels, harvested cells were lysed, and mRNA was extracted and purified using a NucleoSpin RNA II kit. The RNA quantity and quality were measured using a SmartSpec™ Plus spectrophotometer, after which reverse transcription was performed with the ImProm-II Reverse Transcriptase System. cDNA levels of Gapdh, Oct4, and Brachyury were measured with quantitative polymerase chain reaction (qPCR) using an Mx3005P system and Brilliant SYBR Green qPCR Master Mix.

While the comparison of population dynamics with mRNA levels is not exact, under the assumption of efficient translation, the overall dynamics can be compared, but not the numerical values. Figure 9 illustrates the comparison of experimental data with model prediction, which are found to have excellent agreement given that the model was generated with no information of these specific cellular dynamics. Oct4 levels exhibit a decay to a final value of around 20% of the maximum, at a time which correlates with simulated predictions (3 days). Brachyury levels showed a similar non-monotonic trend as was predicted by the model. It reached a maximum around 24 h, following which it gradually decayed over time.

3 Notes

1. Definitive endoderm was induced in hESC through two different pathways: the addition of Activin A and Activin A supplemented with FGF2 and BMP4. The lineages to which hESC can differentiate are definitive endoderm, visceral endoderm, and mesoderm. Depending on the specific mechanism of the model, hESC can first give rise to visceral endoderm and mesendoderm, with the latter differentiating into definitive endoderm and mesoderm. In the current model, the ectoderm germ layer has been omitted. From previous literature [40], hESC induced toward endoderm show low expression of ectoderm markers (Sox1). Commitment levels to ectoderm would be low, and therefore, adding the additional ectoderm lineage would not enhance the model.

2. It is important to note that the nonlinearity observed in the differentiation dynamics contributed significantly toward identification of a robust mechanism. Sloppiness of biological parameters is well reported [39] with ranges of values being large and sensitivities between parameters varying considerably; this can make robust mechanism identification challenging. Quite interestingly, the observed dynamics of the presented study could only be explained by a single specific mechanism. Even a rigorous search of the parameter hyperspace did not yield an alternate potential mechanism. Regarding the nonlinearity of CXCR4 expression, two possible explanations are as follows: (1) mesendoderm, expressing CXCR4, further differentiates to phenotypes which might not express the surface protein; and (2) the cellular environment might promote a higher rate of death of a certain cell phenotype which expresses CXCR4. These dynamics, along with those of Sox17, proliferation, and cell death, led us to investigate a total of 12 possible mechanisms. The majority of the mechanisms investigated was unable to capture the temporal behavior of these outputs and therefore was discarded. The only mechanism which is able to accurately explain the experimental dynamics is one which does not have mesoderm expressing CXCR4, incorporates mesendoderm, and promotes proliferation of hESC and the mesendoderm and endoderm germ layers. This proposed mechanism is shown in Fig. 10.
3. One of the purposes of the present study was to investigate several aspects of differentiation which have faced conflicting reports in the past and to offer further insight using a mathematical analysis. One of these features is the presence of surface receptor, CXCR4. McGrath et al. [33] and Yusuf et al. [41] have reported that embryonic mesoderm expresses CXCR4 in vivo, depending on the stage of embryo development,

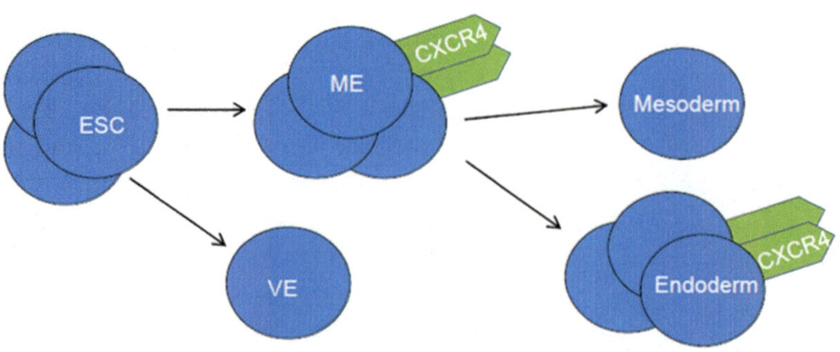

Fig. 10 Proposed differentiation scheme of hESC during endoderm induction as generated by the population-based model. Shown are the presence of mesendoderm, the lack of CXCR4 in mesoderm, and selective phenotype proliferation. *ME*, mesendoderm; *VE*, visceral endoderm

whereas Takenaga et al. [7] report using CXCR4 as a definitive endoderm marker with other markers used for mesoderm. Our results indicate that although both possibilities give low error with respect to Sox17 and CXCR4 population dynamics (depending on which phenotype proliferation is induced), only when CXCR4 is absent in mesoderm do we obtain qualitative agreement in the cellular growth and death temporal behavior. Furthermore, the majority of studies which follow embryo development in vivo or differentiation of ESC in vitro (e.g., [42–44]) include the mesendoderm as an intermediate phenotype arising from the differentiation of ESC which subsequently differentiates to endoderm and mesoderm rather than considering the latter two phenotypes differentiating directly from ESC. The model developed in the current study indeed comes to the same conclusion: the mesendoderm germ layer needs to be considered in order to accurately describe experimental dynamics.

4. The endoderm induction of hESC was conducted under two different conditions with the objective of investigating mechanistic differences between these two pathways. Quite interestingly, both conditions could be explained by the same, single mechanism, while the rejected mechanisms failed to describe the dynamics even after a thorough search of the parameter space. However, there were significant differences in optimum parameter values. One prominent difference between the two conditions was their differentiation potential after being committed to the mesendoderm germ layer. "a0max2" is lower for Condition B, indicating that mesendodermal cells will more quickly reach the pro-differentiation and proliferation regimes. This is also evident from the higher level of "d," although this is for both stages of differentiation. Also, cell commitment for Condition B can be considered expedited when considering the lower value of "xcom2," which is the propensity threshold beyond which a mesendodermal cell is considered committed to either endoderm or mesoderm. Therefore, Activin A supplemented with FGF2 and BMP4 drives differentiation toward endoderm/mesoderm to a higher degree than Activin A alone.

References

1. Thomson JA et al (1998) Embryonic stem cell lines derived from human blastocysts. Science 282(5391):1145–1147
2. Reubinoff BE et al (2000) Embryonic stem cell lines from human blastocysts: somatic differentiation in vitro. Nat Biotech 18(4):399–404
3. Grapin-Botton A (2008) Endoderm specification. In: Girard L (ed) Stem book. H.S.C.I, Cambridge
4. Wells JM, Melton DA (1999) Vertebrate endoderm development. Annu Rev Cell Dev Biol 15 (1):393–410
5. Grapin-Botton A, Melton DA (2000) Endoderm development: from patterning to organogenesis. Trends Genet 16(3):124–130
6. de Caestecker M (2004) The transforming growth factor-β superfamily of receptors. Cytokine Growth Factor Rev 15(1):1–11

7. Takenaga M, Fukumoto M, Hori Y (2007) Regulated Nodal signaling promotes differentiation of the definitive endoderm and mesoderm from ES cells. J Cell Sci 120:13
8. Kubo A, Shinozaki K, Shannon JM et al (2004) Development of definitive endoderm from embryonic stem cells in culture. Development 131:12
9. D'Amour KA et al (2005) Efficient differentiation of human embryonic stem cells to definitive endoderm. Nat Biotech 23 (12):1534–1541
10. Yasunaga M et al (2005) Induction and monitoring of definitive and visceral endoderm differentiation of mouse ES cells. Nat Biotech 23 (12):1542–1550
11. Davidson EH et al (2002) A genomic regulatory network for development. Science 295 (5560):1669–1678
12. Banerjee I et al (2010) An integer programming formulation to identify the sparse network architecture governing differentiation of embryonic stem cells. Bioinformatics 26 (10):1332–1339
13. Chambers I et al (2003) Functional expression cloning of nanog, a pluripotency sustaining factor in embryonic stem cells. Cell 113 (5):643–655
14. Mitsui K et al (2003) The homeoprotein nanog is required for maintenance of pluripotency in mouse epiblast and ES cells. Cell 113 (5):631–642
15. Boyer LA et al (2005) Core transcriptional regulatory circuitry in human embryonic stem cells. Cell 122(6):947–956
16. Loh Y-H et al (2006) The Oct4 and Nanog transcription network regulates pluripotency in mouse embryonic stem cells. Nat Genet 38 (4):431–440
17. Chickarmane V et al (2006) Transcriptional dynamics of the embryonic stem cell switch. PLoS Comput Biol 2(9):e123
18. Chickarmane V, Peterson C (2008) A computational model for understanding stem cell, trophectoderm and endoderm lineage determination. PLoS One 3(10):e3478
19. MacArthur BD, Please CP, Oreffo ROC (2008) Stochasticity and the molecular mechanisms of induced pluripotency. PLoS One 3(8):e3086
20. Cahan P, Li H, Morris SA, Lummertz da Rocha E, Daley GQ, Collins JJ (2014) CellNet: network biology applied to stem cell engineering. Cell:903–915
21. Prudhomme W et al (2004) Multivariate proteomic analysis of murine embryonic stem cell self-renewal versus differentiation signaling. Proc Natl Acad Sci U S A 101(9):2900–2905
22. Woolf PJ et al (2005) Bayesian analysis of signaling networks governing embryonic stem cell fate decisions. Bioinformatics 21(6):741–753
23. Mahdavi A et al (2007) Sensitivity analysis of intracellular signaling pathway kinetics predicts targets for stem cell fate control. PLoS Comput Biol 3(7):e130
24. Viswanathan S et al (2002) Ligand/receptor signaling threshold (LIST) model accounts for gp130-mediated embryonic stem cell self-renewal responses to LIF and HIL-6. Stem Cells 20(2):119–138
25. Viswanathan S, Zandstra P (2003) Towards predictive models of stem cell fate. Cytotechnology 41(2):75–92
26. Prudhomme WA, Duggar KH, Lauffenburger DA (2004) Cell population dynamics model for deconvolution of murine embryonic stem cell self-renewal and differentiation responses to cytokines and extracellular matrix. Biotechnol Bioeng 88(3):264–272
27. Pisu M et al (2008) A simulation model for stem cells differentiation into specialized cells of non-connective tissues. Comput Biol Chem 32(5):338–344
28. Henson MA (2003) Dynamic modeling of microbial cell populations. Curr Opin Biotechnol 14(5):460–467
29. Glauche I et al (2007) Lineage specification of hematopoietic stem cells: mathematical modeling and biological implications. Stem Cells 25 (7):1791–1799
30. Phillips BW et al (2007) Directed differentiation of human embryonic stem cells into the pancreatic endocrine lineage. Stem Cells Dev 16(4):18
31. Touboul T et al (2010) Generation of functional hepatocytes from human embryonic stem cells under chemically defined conditions that recapitulate liver development. Hepatology 51(5):1754–1765
32. Kanai-Azuma M et al (2002) Depletion of definitive gut endoderm in Sox17-null mice. Development 129:13
33. McGrath KE et al (1999) Embryonic expression and function of the chemokine SDF-1 and its receptor, CXCR4. Dev Biol 213 (2):442–456
34. Roeder I, Loeffler M (2002) A novel dynamic model of hematopoietic stem cell organization based on the concept of within-tissue plasticity. Exp Hematol 30(8):853–861
35. Nelson TJ et al (2008) CXCR4+/FLK-1+ biomarkers select a cardiopoietic lineage from

embryonic stem cells. Stem Cells 26 (6):1464–1473
36. Rodaway A, Patient R (2001) Mesendoderm: an ancient germ layer? Cell 105(2):169–172
37. Degasperi A, Gilmore S (2008) Sensitivity analysis of stochastic models of bistable biochemical reactions. In: Bernardo M, Degano P, Zavattaro G (eds) Formal methods for computational systems biology. Springer, Berlin/Heidelberg, pp 1–20
38. Freedman D, Diaconis P (1981) On the histogram as a density estimator:L2 theory. Probab Theory Relat Fields 57(4):453–476
39. Gutenkunst RN et al (2007) Universally sloppy parameter sensitivities in systems biology models. PLoS Comput Biol 3(10):e189
40. McLean AB et al (2007) Activin A efficiently specifies definitive endoderm from human embryonic stem cells only when phosphatidylinositol 3-kinase signaling is suppressed. Stem Cells 25(1):29–38
41. Yusuf F et al (2005) Expression of chemokine receptor CXCR4 during chick embryo development. Anat Embryol 210(1):35–41
42. Tada S et al (2005) Characterization of mesendoderm: a diverging point of the definitive endoderm and mesoderm in embryonic stem cell differentiation culture. Development 132 (19):4363–4374
43. Loose M, Patient R (2004) A genetic regulatory network for Xenopus mesendoderm formation. Dev Biol 271(2):467–478
44. Chng Z et al (2010) SIP1 mediates cell-fate decisions between neuroectoderm and mesendoderm in human pluripotent stem cells. Cell Stem Cell 6(1):59–70

Chapter 4

Automated Formal Reasoning to Uncover Molecular Programs of Self-Renewal

Sara-Jane Dunn

Abstract

The Reasoning Engine for Interaction Networks (RE:IN) is a tool that was developed initially for the study of pluripotency in mouse embryonic stem cells. A set of critical factors that regulate the pluripotent state had been identified experimentally, but it was not known how these genes interacted to stabilize self-renewal or commit the cell to differentiation. The methodology encapsulated in RE:IN enabled the exploration of a space of possible network interaction models, allowing for uncertainty in whether individual interactions exist between the pluripotency factors. This concept of an "abstract" network was combined with automated reasoning that allows the user to eliminate models that are inconsistent with experimental observations. The tool generalizes beyond the study of stem cell decision-making, allowing for the study of interaction networks more broadly across biology.

Key words Automated reasoning, Gene regulatory networks, Boolean networks, Satisfiability modulo theories, Biological programs, Biological computation

1 Introduction

Cellular decision-making occurs as the output of biological computation: information processing at the biochemical level, where chemical, mechanical, and electrical cues are transduced via the interactions of molecular components to produce an output, which may be the decision to divide, to differentiate, or to die. Understanding biological computation requires us to deduce the biological programs that lead to these outputs—the sets of functional components that are interconnected and regulate each other according to "rules" that collectively confer to the system the capacity to process input stimuli to output a biological function reliably and robustly. This is a nontrivial problem, as the information processing carried out by cells is demonstrably slow and noisy, with massively parallel operations, which motivates the need for mathematical abstractions that allow us to make sense of biological

Patrick Cahan (ed.), *Computational Stem Cell Biology: Methods and Protocols*, Methods in Molecular Biology, vol. 1975, https://doi.org/10.1007/978-1-4939-9224-9_4,

data, explore current hypotheses, and crucially generate new, testable hypotheses.

Stem cell decision-making during development is just one example of biological computation, in which cells receive and respond to cues that direct differentiation toward the germline or somatic lineages. The naive pluripotent state uniquely exhibited by embryonic stem cells (ESCs) had been considered to be controlled by a vast network of genetic interactions [1–5], and experimental investigations yielded insight into a core set of transcription factors (TFs), which were individually necessary or sufficient for the maintenance of the pluripotent state. While there was evidence of cross talk between these factors, it remained unclear how the cells interpreted environmental signals and how the TFs interacted to either sustain self-renewal or commit to differentiation. A challenge, therefore, was to derive the transcriptional program that could explain whether ESCs will remain in the self-renewing state or begin to differentiate under different conditions. This required an understanding of the dynamic interactions between the core pluripotency factors and how these interactions resolved to determine cell state over time.

Dynamic, mathematical models of genetic interactions have been proposed for a number of biological scenarios, from the cell cycle [6] to Wnt signalling [7] to Eric Davidson's seminal work on the development of sea urchin as a model organism [8]. Both continuous and discrete mathematical abstractions have been considered: from Boolean networks, first proposed by Kauffman [9] and Thomas [10], to ordinary differential equations, Bayesian networks, and stochastic simulation. An excellent review of these different approaches is that of Le Novère (2016) [11]. In this chapter, we will consider the domain of logical models known as Boolean networks (BNs, Fig. 1a). In this modelling scheme, network components are considered to have two possible states: ON or OFF. These states can also be thought of as high/low in the context of gene expression levels, present/absent, or active/inactive. Boolean functions at each node of the network define how that component updates its state in response to the logical combination of its regulators: for example, in Fig. 1a, component B will be active at the next timestep if either A or C is active at the current timestep. Timesteps are an abstract concept, which are not necessarily linked to real time, and updates can be either synchronous (all components evaluated at each step) or asynchronous (individual components are chosen randomly to be updated at each step). This logical formalism is useful in that it allows us easily to study the dynamic behavior of the system without the need for detailed biochemical descriptions, which require hard-to-measure kinetic parameters.

The first challenge in developing a model of an interaction network—regardless in the first instance of the formalism with which it will be explored—is to identify the set of components of

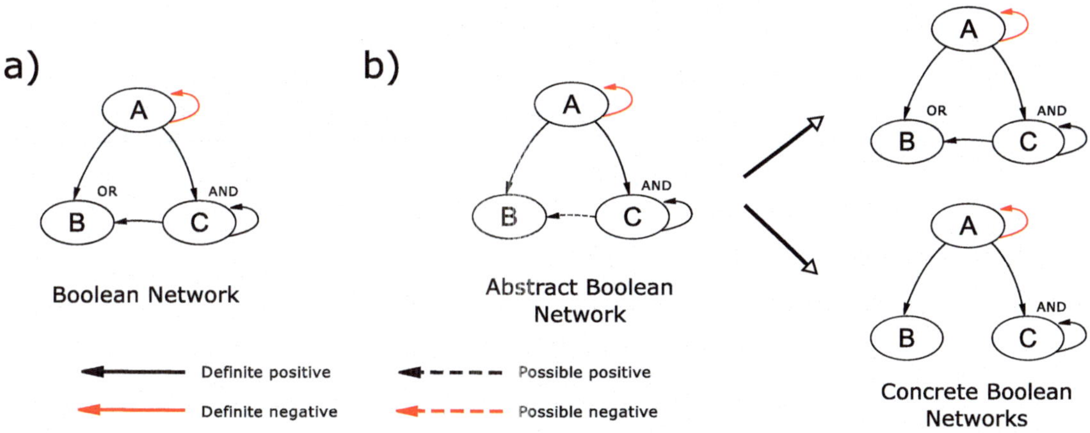

Fig. 1 Relationship between Boolean networks, Abstract Boolean networks, and concrete Boolean networks. (**a**) A Boolean network, which has a defined set of positive (black) and negative (red) interactions, as well as Boolean update functions for each component. (**b**) An Abstract Boolean network, which encodes one possible positive interaction from C to B. This ABN implicitly defines two concrete BNs, one where the possible interaction is present and one where it is absent

interest and activating or inhibitory interactions between them. Selection of the set of components is typically driven by functional studies, and is restricted to those found to be critical for the system under investigation. Interactions are often inferred from knockout studies or by identifying strong correlation between pairs of components across several experiments. However, a challenge arises at this early stage of model construction: it is often difficult to determine conclusively that an interaction exists between two factors. This could be due to uncertainty in the experimental data—correlation effects may be weak, one dataset may be inconsistent with another, or the data may be difficult to reproduce. Under any such scenario, making the case for a single network topology may prove problematic. A second, related challenge is that it is possible for alternative models to produce the same dynamic behavior—for example, two unique networks may stabilize in the same state given a specific initial condition (though take a different trajectory).

These two challenges present a problem to the modeller: do we simply make a case for one model and explore its behavior? Or do we try to consider a set of potential models and seek to eliminate those that are inconsistent with experiments as new data comes to light, or if they fail to predict untested behavior? The latter approach is preferable, as it allows a space of possible models to be explored, but it presents a third challenge in how to explore the space of possible models if the number of models is large. This has implications for the analysis strategy used to investigate network behavior. The most common network analysis approaches are either simulation—exploring the long-time behavior of the network from a given initial state—or exhaustive state-space exploration, which characterizes all paths from each initial state. Either approach becomes computationally infeasible as the set of networks grows too large to explore in reasonable time.

The methodology discussed in this chapter proposes one analysis strategy to address these challenges. The fundamental basis of the method is the application of program synthesis, employing techniques from formal verification. This is the branch of computer science that deals with computer-generated mathematical proofs, which in the context of computer software is typically used to verify the correctness of a software program. Engineered programs are designed according to a set of specifications, which serve as a blueprint for what that software should do. Program synthesis is the automatic construction of a program consistent with a set of specifications. Informally, we translate this approach to biology by using experimental observations as specifications that must be met by the biological program governing that behavior. This technique is encapsulated in a tool called the Reasoning Engine for Interaction Networks, or RE:IN.

Employing this approach allows us to construct an initial set of possible Boolean network models and, from this set, synthesize only those that are consistent with experimental observations. As we will go on to describe, this constrained set of models can subsequently be used to make predictions of untested behavior, by encoding hypotheses as test specifications and verifying whether they are satisfiable or not. In this way, we achieve a key goal in computational biology: the generation of models of interacting biochemical components that both explain existing experimental observations and are predictive of untested behavior.

2 Materials

The methodology described in this chapter, which is encoded in the software tool RE:IN, has been made freely available to run in the browser as an HTML5 application. It can be accessed directly via rein.cloudapp.net or indirectly via the project page: research.microsoft.com/rein. The project webpage provides example test files, a tutorial for using the tool and its syntax, and answers to frequently asked questions.

3 Methods

Throughout this section, the methodology will be discussed in the context of the network governing naive pluripotency in mouse ESCs, which is built up progressively as an example application. First, we will introduce the modelling formalism that is used to explore possible models of biological interaction networks, before describing how these models are analyzed and used to form predictions of untested behavior. We also show how the approach guides model refinement in light of new experimental data or in response to incorrect predictions.

3.1 Introducing the Network Modelling Formalism

3.1.1 Boolean Networks

Boolean networks are a type of logical model, which enable the dynamic behavior of a graph of interacting components to be investigated qualitatively. Accordingly, network components are considered to exist in one of the two discrete states, ON or OFF, and are updated according to Boolean functions that are evaluated based on the state of the target component's regulators. For example, the BN shown in Fig. 1a implements the following update rules for the three components, A, B, and C, where A', B', and C' refer to the active component state at the next timestep:

$$\begin{aligned} A' &= \neg A \\ B' &= A \vee C \\ C' &= A \wedge C \end{aligned} \tag{1}$$

Here, for example, B will be active at the next timestep (B') if either A or C is active at the current timestep, as it is positively regulated by both under the OR logic.

The dynamics of a BN evolve according to the update rules from some initial state, either following a synchronous or asynchronous update scheme: under the former, all components update their state at each step, and under the latter, components do so individually and in random order. The state space of transitions can be evaluated to identify reachable states, oscillatory states, or cycles, and to determine whether fixed points exist.

3.1.2 Abstract Boolean Networks

BNs can be formulated and investigated when the complete network topology is known. However, if there is uncertainty in one or more interactions between a pair of network components, or if the modeller wishes to explore the dynamics of competing models, it is useful to employ the concept of an Abstract Boolean Network (ABN) [12]. This formalism allows interactions to be marked as either definite, if significant experimental evidence supports the interaction, or possible, if there is uncertainty (Fig. 1b). Marking one interaction as possible implicitly defines two alternative network topologies: one in which the interaction is present and one in which it is absent. As such the number of networks scales exponentially: if n interactions are marked as possible, there are 2^n alternative network topologies.

3.1.3 Logical Update Functions

Given that the topology is well defined for a single BN, it is possible to construct an update function for each component that depends on its specific set of regulators. For example, the set of functions defined by Eq. (1) are consistent with the network topology defined in Fig. 1a but would not be consistent with both concrete network topologies defined by the ABN in Fig. 1b. This is because, in the ABN, B is not always dependent on C.

To navigate this challenge, we instead employ a set of update functions that are defined according to whether some, all, or none of the target's activators/repressors are active (Fig. 2a) and which collectively capture alternative regulatory mechanisms. We refer to

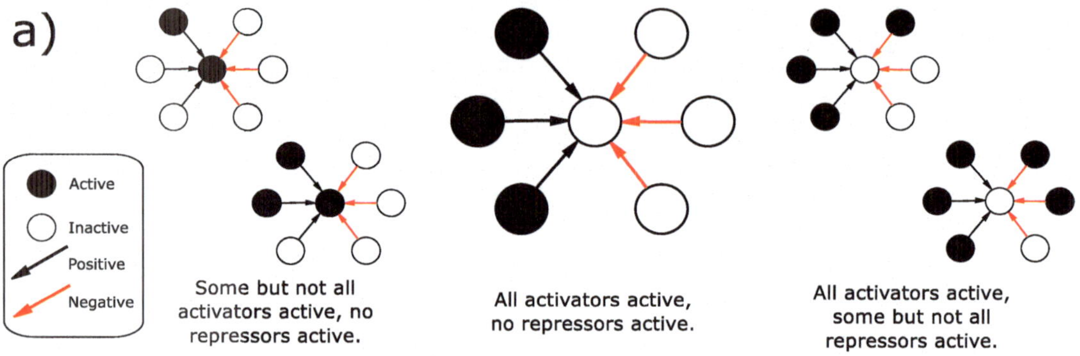

b)

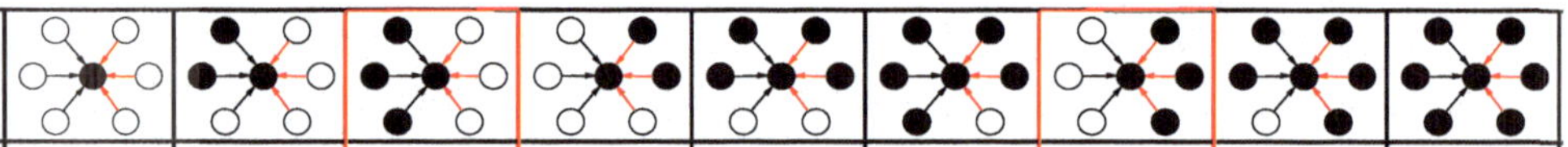

Regulation Condition									
0			✓						
1		✓	✓						
2			✓			✓			
3		✓	✓			✓			
4			✓			✓			✓
5		✓	✓			✓			✓
6		✓	✓		✓	✓			
7		✓	✓		✓	✓			✓
8		✓	✓		✓	✓		✓	✓
9	✓	✓	✓						
10	✓	✓	✓			✓			
11	✓	✓	✓		✓	✓			
12	✓	✓	✓	✓	✓	✓			
13	✓	✓	✓			✓			✓
14	✓	✓	✓		✓	✓			✓
15	✓	✓	✓	✓	✓	✓			✓
16	✓	✓	✓		✓	✓		✓	✓
17	✓	✓	✓	✓	✓	✓		✓	✓

✓: A condition under which the target gene is active

Fig. 2 (**a**) Illustrated is a target gene with three activators and three repressors. We can construct general update functions by considering whether some, all, or none of the target's activators and repressors are active, as depicted. (**b**) We define a set of 18 unique regulation conditions that are composed of possible scenarios under which the target will be active. The red columns highlight two key assumptions: targets with all activators active and no repressors active will be active at the next timestep; conversely targets with no activators active and all repressors active will not be active at the next timestep this set of functions as the set of regulation conditions, and there are 18 possible scenarios that they cover. For example, it might be the case that to be active at the next step, a target requires only one of its activators to be active, if none of its repressors are active. Assuming monotonicity, if that same target had *all* activators active but no repressors active, then it will also be active at the next step. This example is encapsulated by regulation condition 1 (Fig. 2b). In contrast, regulation condition 0 covers the case in which the target will be active only if all its activators are active and none of its repressors are active.

In addition to the assumption of monotonicity, the definition of these regulation conditions assumes that all regulators have equal significance. Furthermore, we assume that a target with all its activators active and none of its repressors active will be active at the next step, and conversely a target that has none of its activators active but all its repressors active will not be active at the next step. These two cases correspond to the highlighted columns of the table in Fig. 2b. This ensures that the target responds to the state of its regulators and will not remain permanently active or inactive.

In the same manner in which interactions can be defined as "possible," the set of regulation conditions defines possible update functions for each component in the ABN. With no prior information of the regulation logic governing a given target, all 18 of the regulation conditions can be assigned as possible update functions. Therefore, an ABN with c components and n possible interactions encodes $2^n \times 18^c$ concrete BNs, where each BN has a unique instantiation of interactions and a single regulation condition is assigned to each target. If prior information is known about the target based on experimental evidence, a subset of these regulation conditions in the ABN can be assigned.

Two additional rules are consistent with the abstract network topology and the requirement of monotonicity: the instant and delayed "threshold rule" [13], in which the balance of activators and repressors determines whether the component is activated or repressed at the next timestep. The delayed threshold rule specifies that if a target node is active at time t and the total input to the target is zero, then it will be degraded at time $t = t + 1$. The user can opt to use only these threshold rules, or in addition to the 18 regulation conditions, in the RE:IN tool (*see* **Note 1**).

3.2 Defining an Abstract Boolean Network

3.2.1 Components and Interactions

The methodology discussed in this chapter is applicable for the investigation of dynamic interaction networks for a given set of biological components, rather than to identify the critical components from the data at the outset. As such, the starting point in defining an ABN for investigation must be a set of functionally validated components—genes, proteins, noncoding RNAs, metabolites, signalling molecules, etc.—which have been found experimentally to have a substantial effect on the process under study when over- or under-activated. We discuss in a later section how the methodology can be employed to explore the need for additional components.

The next step is to identify potential interactions between the set of components. Interactions have direction, can be positive (activatory) or negative (inhibitory), and are either definite or possible. Interactions may represent the direct binding of a TF (source) to the promoter of a downstream gene (target), or a posttranscriptional modification of the gene's product, but may also represent indirect effects when a secondary regulatory effect has been captured by the data. In this manner, we assume that interactions are functional, not necessarily direct. Interactions can be mined directly from data, either by searching for significant correlation in expression for pairs of genes, from the literature (experimental conclusions as well as previously studied models), or using databases such as the Search Tool for the Retrieval of Interacting Genes/Proteins (STRING) [14]. The example below uses Pearson's correlation as a metric, which measures the extent and direction of linear correlation between two sets of values. Correlation analyses are readily available as R packages or MATLAB functions. If multiple sets of evidence support a given interaction, it can be assumed to be definite.

Together the set of components, possible and definite interactions, and regulation conditions comprise the ABN (Fig. 3a).

Example: Maintenance of the naive pluripotent state in mouse ESCs can be achieved in culture conditions that comprise the cytokine leukemia inhibitory factor (LIF) and two selective inhibitors (2i) of glycogen synthase kinase 3 (CH) and mitogen-activated protein kinase kinase (PD). ESCs can be stably cultured in any two of these conditions (2i + LIF, as well as 2i, LIF + CH, and LIF + PD). Direct targets of these input signals had previously been identified [15–20], *as well as a set of TFs that are either indispensable for self-renewal or support this state if overexpressed (Fig. 3b). In spite of this information, understanding how these TFs interacted to stabilize the pluripotency network was not understood.*

We analyzed gene expression for 12 TFs implicated in ESC maintenance in each combination of LIF, CH, and PD [21] *and measured the Pearson correlation between each pair of genes across the set of experiments (Fig. 3c). Pearson's correlation has a maximum value of* $+1$*, for perfect positive correlation, and a minimum of* -1 *for perfect negative correlation. We ranked all the pairs of genes according to their Pearson correlation coefficient. By setting a correlation threshold, we could select a set of possible interactions from this ranked list by taking only those pairs with coefficients above the threshold. We used this to generate the ABN (Fig. 3d). Different ABNs are defined by different Pearson thresholds—the choice of threshold is explained below, once we introduce the experimental constraints.*

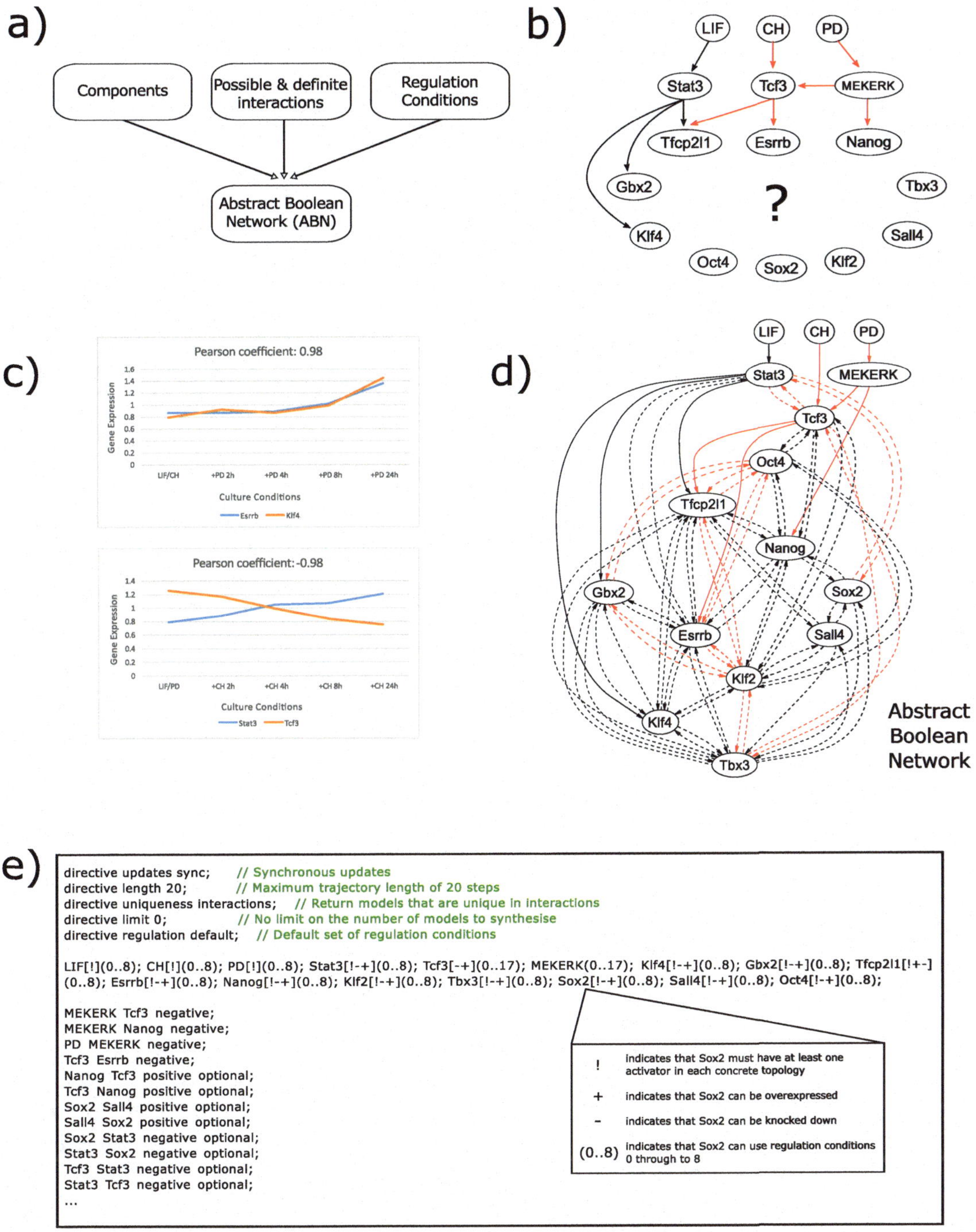

Fig. 3 (**a**) An ABN is composed of a set of components, a set of possible and definite interactions, and a set of regulation conditions assigned to each component. (**b**) A set of TFs known to be critical for maintenance of the naive pluripotent state in mouse ESCs, which is sustained in the presence of three signals: LIF, CH, and PD. How these TFs interact with each other was unknown. (**c**) Pearson's correlation provides a measure of the linear correlation between two variables. We measured the extent of correlation in gene expression between gene pairs across several experiments. Strong positive correlation (e.g., between Esrrb and Klf4) is indicative

As will become clear throughout the discussion of this problem, a good rule of thumb in this approach is to minimize the number of possible interactions in the initial ABN, which minimizes the number of concrete BNs it defines, which has implications for the number of predictions that can be made of untested behavior.

3.2.2 Encoding an ABN Using the RE:IN Syntax

Users can directly encode the components, regulation conditions, and interactions for their model via the RE:IN tool, which provides an easy-to-navigate interface for adding each model element. These can also be saved directly from the tool. An alternative, if preferred, is to upload a text file that describes the ABN in the RE:IN syntax, as described here.

Each component must first be defined, along with the set of regulation conditions that it can use. For example:

```
Oct4(0..8); Sox2(0,2,4); Klf4(2..5);
```

This indicates that there are three components: Oct4, which can use regulation conditions 0–8; Sox2, which can use regulation conditions 0, 2, or 4; and Klf4, which can use regulation conditions 2–5.

Interactions are listed according to.

```
Source target positive/negative [optional];
```

The last argument is only used if the user wishes to define a specific interaction as optional. For example:

```
PD MEKERK negative;
Tcf3 Esrrb negative;
Nanog Tcf3 positive optional;
Tcf3 Nanog positive optional;
```

Here four interactions are defined, two of which are optional, i.e., possible.

Example: The encoding of the ABN defined to study pluripotency is shown in Fig. 3e. Here, we assumed that the components required at least one activator to be active, and therefore restricted the set of possible regulation conditions to {0, … , 8}. *The two exceptions to this were Tcf3 and MEKERK, which do not have any activators and therefore used the unrestricted set* {0, … , 17}. *Please see* ***Note 1*** *for a full explanation of the directives.*

Fig. 3 (continued) of a possible positive interaction between these components, while strong negative correlation (e.g., between Stat3 and Tcf3) is indicative of a possible negative interaction. (**d**) An ABN defined by a Pearson's correlation threshold of 0.792. (**e**) An example of the RE:IN syntax used to encode this pluripotency ABN

3.3 Experimental Constraints

An ABN captures what is hypothesized about the set of interactions and critical components that regulate a specific system or cell decision. This set of assumptions should be tested against observations of cell behavior to refine our understanding of the system. The goal is to test whether any of the concrete BNs defined by the ABN are consistent with experimental observations and, if so, to reveal whether some (or all) of the possible interactions are required (or not) or indeed whether additional interactions should be included. This will guide us toward the most explanatory and predictive model of the system. Here we describe how we translate experimental observations into expressions that describe the corresponding expected trajectories of a BN.

Consider, for example, the observation of a change in the expression level of each network component from some starting state upon loss of the input signals to the system. For a given BN to be a potentially explanatory model of the system, it should follow a trajectory from the observed initial state to the observed final state if the signals are switched off. Indeed, if time course measurements are available, the BN should also pass through the expected states along the trajectory before reaching the final state. Extrapolating to the problem that we have set up, we would like to constrain the set of BNs defined by the ABN to only those that are consistent with the expected trajectory and thereby eliminate models that are inconsistent with the data (*see* **Note 2**).

To compare model behavior with experimental data, we construct formal representations of the observations, where a single experiment corresponds to a single execution of the system. Each experimental constraint contains four terms:

- $e \in E$, the experiment label, where E is the complete set of experiments.
- $n \in 0 \ldots K$, the timestep, where K is the maximum trajectory length.
- $c \in C$, the component of the ABN, where C is the complete set of components.
- $v \in \mathbb{B}$, the observed (discretized) state of component c.

For example, consider an experiment in which component c was initially observed to be inactive but later observed to be active. Both observations describe the same execution of the system and are therefore denoted by the same experiment label, e. If we allow 20 steps for this transition, we would construct the expression $(e, 0, c, \bot) \wedge (e, 19, c, \top)$, which requires the existence of a trajectory that recapitulates this observation. The use of experiment labels allows us to represent observations for different components within the same experiment, as well as to represent different experiments, for example, when the system is initialized in different states.

Expressions related to different labels, e.g., e_1 and e_2, are not necessarily mutually exclusive, and the same trajectory might satisfy both.

Additional terms are used to encode experimental observations related to genetic perturbations or stable states. The terms *KO* (e, c, v) and *FE*(e, c, v) are used to define knockout and forced expression perturbations, which are assigned to a specific experiment and component, but do not depend on time. That is, these perturbations modify the dynamics of the system along the trajectory such that component c is always active (*FE*) or inactive (*KO*) when $v = \top$, regardless of the regulation conditions for c or the state of each of its regulators (note that the regulation conditions are applied as before when $v = \bot$, i.e., if the *K*0 or *FE* is set to false). *See* **Notes 3** and **4**, which highlight common pitfalls in implementing these perturbations.

The constraint *Fixpoint*(e, n) is used to indicate that the trajectory satisfying all constraints labelled by e must reach a fixed point at step n. In other words, the only possible transition from the state reached at timestep n is a self-loop. As appropriate, the terms (e, n, c, v), *KO*(e, c, v), *FE*(e, c, v), and *Fixpoint*(e, n) are combined into logical expressions using the operators $\{\wedge, \vee, \Rightarrow, \Leftrightarrow, \neg\}$, to allow us to formalize different experimental observations.

Given that we are working within a Boolean framework, the constraints must describe each component as either ON/OFF, which can also be thought of as active/inactive or high/low in the context of gene expression. The user can select their own method of discretization. Clustering methods such as k-means can aid in this step [22]. Note that it is not necessary to define the state of all components at each step. If the state of a particular component was not measured during an experiment, this can be omitted from the constraints and the solver will assign values to the component to ensure that the constraint is satisfied. This can be useful to predict the value of an unknown component under specific experimental conditions.

Lastly, timesteps do need to be specified in a constraint, but the values do not need to match precisely with physical time. For example, if the user wishes to encode that a stable state is reached "eventually," selecting a large final timestep will also permit trajectories in which this state is reached at an earlier step. Note that the maximum trajectory length to be considered is a parameter to be set by the user (Fig. 3e). Functionality exists within RE:IN to identify the recurrence diameter for the set of models. Informally, this reveals the length of the longest trajectory possible between any initial and any final state for a given network. In principle, it provides an over-approximation of the longest trajectories that need to be considered to reproduce any experiment.

3.3.1 Encoding Constraints Using the RE:IN Syntax

As for defining the ABN, RE:IN has a syntax that is used to encode experimental constraints. A simple example to illustrate this is a constraint that defines the initial state of the set of components and then the state after nine steps:

```
#ExperimentLabel[0] |= $InitialState;
#ExperimentLabel[9] |= $FinalState;
fixpoint(#ExperimentLabel[9]);
```

This example makes use of the "fixpoint" keyword to encode that the final state should be stable. This trajectory is defined to be ten steps long, but alternative trajectory lengths can be used if appropriate, and intermediate states can also be defined. We write constraints succinctly as above by making use of predicates, which allow us to define the complete state of a set of components A, B, and C:

$InitialState :=
{
A = 1 and
B = 0 and
C = 1
};
	$FinalState :=
	{
	A = 1 and
	B = 1 and
	C = 1
	};

Should the user wish only to encode the state of one component at a specific timestep, it can be written as:

```
#ExperimentLabel[9].A = 0;
```

We can also make use of the keywords 'and', 'or', 'not' to encode expected behavior, such as the following:

```
#ExperimentLabel[0] |= $InitialState;
not(#ExperimentLabel[6] |= $FinalState);
#ExperimentLabel[7] |= $FinalState or
#ExperimentLabel[8] |= $FinalState or
#ExperimentLabel[9] |= $FinalState;
```

This encoding requires that the final state be reached at **step** 7 or **step 8** or **step 9** and is not reached at **step 6**. It is possible to compose expressions to encode constraints such that one state must be reached before another, without defining the exact step numbers. This requires the user to explicitly encode that if `$SecondState` is reached at timestep k, then `$ThirdState` is not reached at timestep $(k - 1)$, $(k - 2)$, ... 1, but can be reached at steps $(k + 1)$, $(k + 2)$, ... , n, where n is the maximum trajectory length. An expression of this type would be written thus:

```
(#ExperimentLabel[1] |= $SecondState and
(#ExperimentLabel[2] |= $ThirdState or
#ExperimentLabel[3] |= $ThirdState or
#ExperimentLabel[4] |= $ThirdState)) or

(#ExperimentLabel[2] |= $SecondState and
not(#Experiment[1] |= $ThirdState) and
(#ExperimentLabel[3] |= $ThirdState or
#ExperimentLabel[4] |= $ThirdState or
#ExperimentLabel[5] |= $ThirdState)) or
...;
```

Should the user wish to encode a knockout or overexpression in an experiment, this must be set at the initial timestep. RE:IN recognizes whether a component is knocked out with `KO()`and whether a gene is overexpressed by `FE()`. Therefore, `KO(A) = 1` corresponds to A being knocked out, while `KO(B) = 0` means that B is not. Likewise, `FE(C) = 1` means that C is overexpressed. If you require this, when you define a component, check the "KO" or "FE" box appropriately in the interaction editor of the RE:IN tool. In the RE:IN syntax, where the set of components are first defined, you must immediately append `[-]` for KO and/or `[+]` for FE just before the round brackets that define the allowed regulation conditions.

For example,

```
A[-](0..8); B(0..8); C(0..8);
```

defines the set of components, while

```
#ExperimentLabel[0] |= $InitialStateWithAKO;
#ExperimentLabel[0] |= $GeneAKnockedOut;
#ExperimentLabel[9] |= $FinalState;
fixpoint(#ExperimentLabel[9]);
```

```
$InitialStateWithAKO :=
{
 A = 0 and
 B = 0 and
 C = 1
};
```

```
$GeneAKnockedOut :=
 {
 KO(A) = 1;
 };
```

encodes the perturbation in the experiment. Here we have ensured that the initial state is in agreement with the knockout (*see* **Note 3**). It is important to ensure that in other experiments where A is not knocked out, this is explicitly set (*see* **Note 4**):

```
#SecondExperiment[0] |=
  $InitialState;
```

```
$NoKnockOuts
  :=
 {
```

(continued)

a)

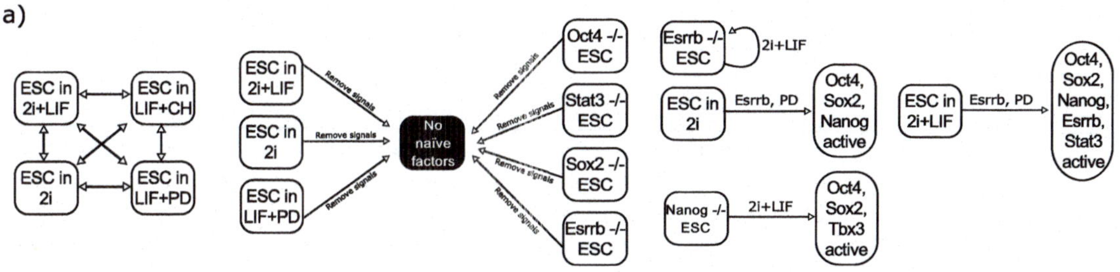

b)

Experiments

```
#ExperimentOne[0] |= $TwoiPlusLif;
#ExperimentOne[0] |= $TwoiCultureConditions;
#ExperimentOne[0] |= $NoKnockouts;
#ExperimentOne[18] |= $Twoi;
fixpoint(#ExperimentOne[18]);

#ExperimentTwo[0] |= $TwoiWithEsrrbKO;
#ExperimentTwo[0] |= $TwoiCultureConditions;
#ExperimentTwo[0] |= $EsrrbKnockout;
#ExperimentTwo[18] |= $NoNaiveFactors;
fixpoint(#ExperimentTwo[18]);
```

Predicates

```
$TwoiPlusLif:=
{
 MEKERK = 0 and
 Oct4=1 and
 Sox2=1 and
 Nanog=1 and
 Esrrb=1 and
 Klf2=1 and
 Tfcp2l1=1 and
 Klf4=1 and
 Gbx2=1 and
 Tbx3=1 and
 Tcf3=0 and
 Sall4=1 and
 Stat3=1
};

$Twoi:=
{
 MEKERK = 0 and
 Oct4=1 and
 Sox2=1 and
 Nanog=1 and
 Esrrb=1 and
 Klf2=1 and
 Tfcp2l1=1 and
 Klf4=0 and
 Gbx2=0 and
 Tbx3=1 and
 Tcf3=0 and
 Sall4=1 and
 Stat3=0
};

$TwoiCultureConditions :=
{
 LIF = 0 and
 CH = 1 and
 PD = 1
};

$NoKnockouts:=
{
 KO(Esrrb)= 0
};

$EsrrbKnockout:=
{
  KO(Esrrb) = 1
};
```

Fig. 4 (**a**) A graphical depiction of the experimental constraints that correspond to known behavior of mouse ESCs under different experimental scenarios. For example, (left) ESCs can be switched from one culture media to another without loss of pluripotency, and the gene expression pattern changes accordingly. (**b**) An example of how two experiments are encoded

```
#SecondExperiment[0] |=          KO(A) = 0;
$NoKnockOuts;                    };
#SecondExperiment[9] |=
$FinalState;
fixpoint(#SecondExperiment[9]);
```

Examples of ABNs and constraints that have been explored in RE:IN are available at research.microsoft.com/rein.

Example: We curated a set of 23 experimental observations of the behavior of mouse ESCs in vitro, *which are summarized in Fig. 4a* [21]. *We defined 12 constraints to capture the gene expression change for each component as you switch from 2i + LIF to 2i, from LIF + CH to LIF + PD,* etc. *A further three constraints corresponded to removing all signals starting from each of 2i + LIF, 2i, and LIF + PD. The remaining eight constraints corresponded to knockout and forced expression experiments* [15, 16].

We discretized the experimental data for each constraint so that the expression level of each gene was high (1) or low (0). A threshold of 0.55 was selected for discretization, above which normalized gene expression was deemed to be high. As a guide, we sought a threshold that would generate distinct discrete gene expression patterns in each of 2i + LIF, 2i, LIF + CH, and LIF + PD culture conditions, such that it would reflect the dynamic nature of the system. In fact, this *threshold is one among several valid thresholds (0.55 < x < 0.7), and we found the approach was robust within this range. An initial and final discrete state was defined for each experiment. Where a complete gene expression pattern was known, this was defined. However, in those cases where only a partial pattern was published, only those genes with known expression had a defined final state, and we did not constrain the remaining genes in these cases.*

The trajectory length was fixed at 20 steps, which was sufficient to allow convergence to the expected final state. Furthermore, we required that the final state be a fixed point when a complete gene expression pattern was defined, given that under these culture conditions the stem cells are held in a homogeneous state [23].

Two examples of how these constraints were encoded for analysis in RE:IN are shown in Fig. 4b. The first experiment corresponds to the change in gene expression pattern for ESCs cultured initially in 2i + LIF and then switched to just 2i. The second experiment corresponds to the loss of naïve factor expression if you knock out Esrrb under 2i culture conditions. In each case, we exploit the syntax to write each experiment succinctly through the use of predicates, which define the state of a set of network components.

3.4 Satisfiability Analysis

Thus far we have defined an ABN and a set of constraints that correspond to experimental observations of system behavior that we wish to explain. The next step is to determine whether any of the concrete BNs from the ABN are consistent with the constraints. We introduce the concept of a constrained Abstract Boolean Network (cABN) as the formal representation of the ABN together with the constraints.

Two common routes to explore the behavior of a BN are to use simulation and state-space exploration. Under the simulation strategy, the modeller seeds the network with an initial state and allows it to evolve that state over a large number of steps according to the interactions and update rules. This can be used to examine whether the system stabilizes or demonstrates cyclic behavior. Exhaustive state-space exploration maps out every transition between all reachable states. Both strategies are easily implemented for one or a handful of models but quickly become impractical to explore a large number of models. In the case of the pluripotency network, the ABN (Fig. 3d) defines 10^{35} different BNs. Even if we could simulate each network in just 1 s and ran 1000 network simulations in parallel, and if we assume that a lifetime is roughly 80 years, it would still take 10^{23} lifetimes (and a considerable cost in compute!) to simulate each and every model.

We implement an alternative approach that utilizes methods from the fields of formal verification and program synthesis, where we seek instead to prove whether the set of expected network trajectories defined by the experimental observations can be met by a given model from the ABN. To achieve this, we set up a *satisfiability problem*: for a given logical formula, does there exist an instantiation of variables for which the formula evaluates to TRUE? A simple example of this is the following. Consider the formula

$$A \Rightarrow (B \vee C). \tag{2}$$

This reads that if A is TRUE, then B or C is TRUE (A implies B or C). For this trivial example, it is easy to identify those assignments for A, B, and C that will satisfy the formula:

$$\{A, B, C\}, \{A, \neg B, C\}, \{A, B, \neg C\}, \{\neg A, \neg B, \neg C\}. \tag{3}$$

While this example is from classical Boolean satisfiability, to tackle the challenge posed by the ABN and experimental constraints, we employ satisfiability modulo theories (SMT) as it extends the decision problem to include additional background theories, such as bit vectors. The problem that we wish to solve is the following: given a cABN, which consists of a set of network components, interactions (possible and definite), regulation conditions, and experimental constraints, find one concrete BN that is consistent with all experimental observations. A solution would be an instantiation of interactions and regulation conditions such that there is a trajectory of the transition system that satisfies each of the experimental constraints. We use bit vectors to encode the individual elements of the cABN. For example, a bit vector of the possible interactions encodes the Boolean choice variable representing whether each interaction is selected or not. For a complete mathematical explanation of the encoding, the reader is referred to Yordanov and Dunn et al. (2016) [12]. RE:IN uses the SMT solver Z3 [24] to obtain a solution for the encoded problem.

3.4.1 Unsatisfiable Constraints

By encoding the ABN and set of experimental constraints into RE:IN, the first analysis step is to use the tool to confirm whether any concrete BNs are consistent with the constraints (Fig. 5a). If RE:IN determines that the constraints are unsatisfiable, it means that none of the concrete BNs implicitly defined by the ABN are capable of explaining the experimental observations. This is a strong result, which reveals that it is necessary to revisit the set of components and interactions that comprise the ABN—should you include additional components or additional possible interactions? Are you confident about any definite interactions that you have included? If you have a ranked set of possible interactions, then one unbiased approach to revise the ABN is to reduce the threshold for including interactions as possible until the constraints can be satisfied. Note that we assume here that the set of constraints has been encoded correctly.

3.4.2 Satisfiable Constraints

If RE:IN can synthesize even just one BN consistent with experimental observations, then this proves that the constraints are satisfiable, and your analysis can proceed. RE:IN does allow consistent BNs to be enumerated. However, unless you are considering a

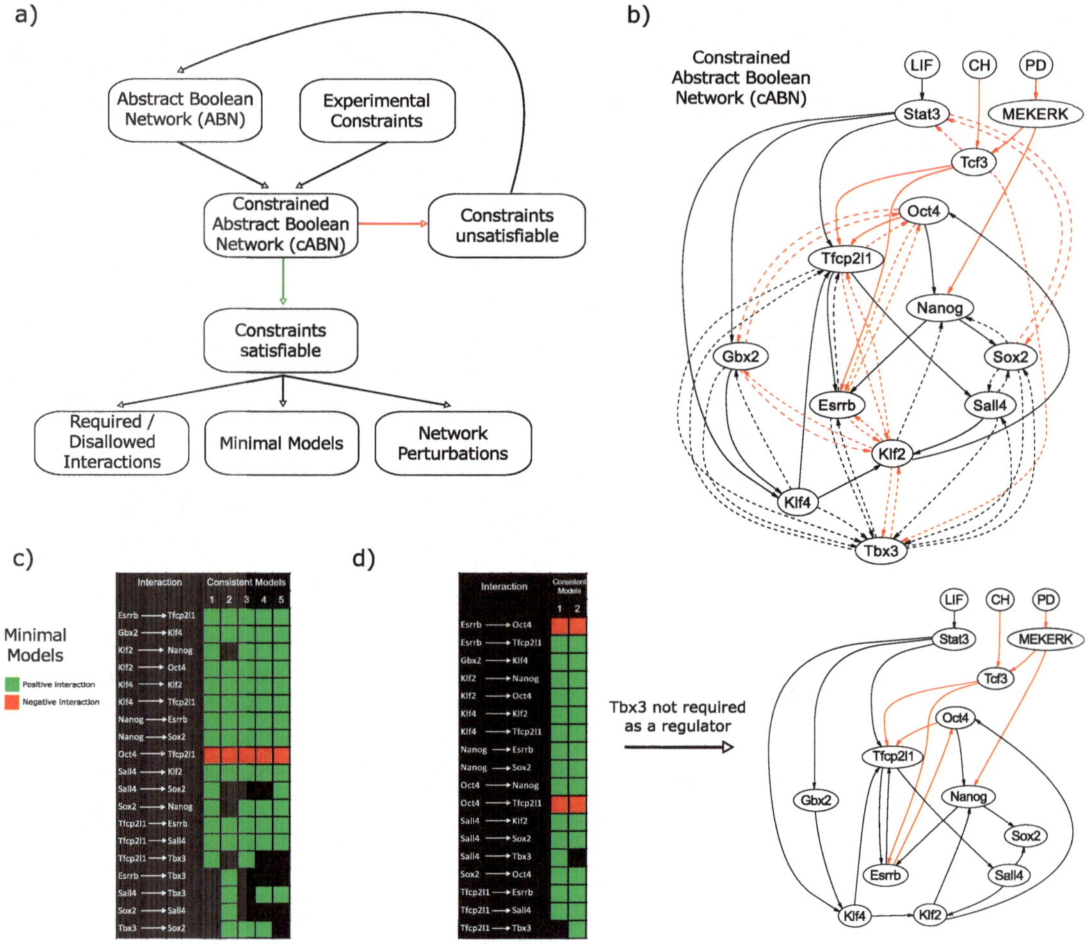

Fig. 5 (**a**) Together, the ABN and the set of experimental constraints combine to give the constrained ABN. If the constraints are not satisfiable, the user should revise the ABN. If they are satisfiable, required/disallowed interactions can be identified, minimal models identified, and predictions generated such as the response to genetic perturbations. (**b**) The cABN that is consistent with the set of 23 experimental constraints concerning pluripotency (Fig. 4). Here, required and disallowed interactions have been incorporated to simplify the representation of the set of consistent models (compare with the initial ABN shown in Fig. 3d). (**c**) The set of five minimal networks. (**d**) The set of two minimal networks derived from the refined pluripotency cABN, after the incorrect prediction has been corrected. These two models differ only in which TF activates Tbx3. Given that Tbx3 is not required as a regulator, it can be removed from the system, and the two minimal networks collapse onto the same topology

relatively small number of possible interactions, and accordingly there is a small number of concrete BNs, it is prudent not to take this approach as there may be a very large number of results that will take a long time to identify. Rather, we can use RE:IN to draw conclusions about the complete set of consistent models by setting up queries and testing satisfiability only once as described below.

3.5 Analyzing the cABN

3.5.1 Required and Disallowed Interactions

We would like to learn as much as we can about the concrete BNs that are consistent with the experimental observations. In situations where there are too many concrete models to enumerate, we can use RE:IN to identify whether any of the possible interactions are present in every concrete BN—such interactions are therefore required to satisfy the constraints. Similarly, we can identify those possible interactions that are never present in any concrete BN consistent with the constraints—such interactions, if enforced as definite, would prevent the constraints from being satisfied and are therefore disallowed.

RE:IN has an explicit function that identifies required and disallowed interactions, and this analysis can be run once it has been confirmed that the constraints are satisfiable. The algorithm simply iterates over each possible interaction and tests whether the constraints are satisfiable if it is removed completely or if it is fixed as definite. If removing the interaction makes the constraints unsatisfiable, then it must be present in all consistent models. If fixing the interaction as definite makes the constraints unsatisfiable, then it must never be present in any consistent model. If the outgoing interactions from a specific component are all found to be disallowed, then this component can be removed from the analysis as it is not required to act as a regulator.

3.5.2 Minimal Models

It can often be informative to understand whether a rich set of behaviors can be explained by a relatively simple BN. We therefore enable users to identify one type of "minimal" network: those that have the fewest possible interactions instantiated. These networks are easy to examine, and doing so can reveal components and interactions that are essential for the biological process. It is also a useful tool to explore how additional components might need to connect into a pre-defined network to explain the set of experimental observations, as described later.

Example: Having defined a set of experimental constraints corresponding to the behavior of ESCs under tested conditions (Fig. 4a), we sought to identify an ABN that could allow us to satisfy these observations. Furthermore, we sought to identify the ABN with the fewest possible interactions, in order to limit the number of alternative models and to prioritize interactions according to the measure of correlation. We sought the maximum Pearson correlation threshold that would define a set of models that could satisfy the constraints. We found that the interactions with a Pearson correlation coefficient of 0.792 or above were sufficient to satisfy the complete set of 23 experimental constraints (Fig. 3d).

Next, we explored whether the cABN could be characterized by identifying required and disallowed interactions. RE:IN analysis showed that 11 interactions were present in all consistent BNs. That is, we can conclude that without any one of these interactions, it would not be possible to satisfy the experimental constraints. Furthermore, 26 interactions were found to be disallowed. The updated network diagram for this cABN is shown in Fig. 5b, which can be compared with Fig. 3d to show how this analysis has revealed insight into the structure of the possible networks that govern pluripotency. A large number of possible models have been eliminated, and the theoretical maximum number of possible models has reduced from approximately 10^{35} *to* 10^{24}.

We also sought the simplest networks that satisfied the constraints. We found 5 minimal networks, each with 16 possible interactions instantiated in addition to the definite interactions encoded in the ABN. The alternative minimal models are summarized in Fig. 5c.

3.6 Testing Hypotheses and Formulating Predictions

As illustrated in the pluripotency example, while applying constraints on the ABN does reduce the number of possible models that explain experimental observations, in realistic scenarios (such as exploring the pluripotency network), the number of consistent models can remain large. Simulation or state-space exploration remains intractable to explore the constrained set, and it would be hard to justify selecting one model over another at this point. This underscores the need for an approach to derive predictions based on the entire set of models that are each individually consistent with available knowledge, until one or more can be ruled out. This is achievable by encoding hypotheses as "test" constraints and examining satisfiability. A key advantage of our approach is that, in this manner, it does not require us to prioritize individual networks for analysis, and we only formulate a prediction when the complete set of consistent models are in agreement.

To formulate a prediction from our model set, we encode a new hypothesis as an additional constraint and test whether it is satisfied by the cABN. Note that we keep the original set of constraints so that predictions are based only on the consistent models. Crucially, we must also test the null hypothesis separately—the converse of the constraint. The prediction holds across all consistent BNs if the initial hypothesis is satisfiable, while the null hypothesis is unsatisfiable. If both the hypothesis and its null are satisfiable, then it means that some networks support the hypothesis and others support the null hypothesis. Without further information, we cannot conclude that one subset of the models is correct while the remainder are not, and therefore in such instances we do not make a prediction. That said, useful insights are identified even when no prediction can be generated for a given query, as this suggests a discriminating biological experiment to refine the set of models further.

The fact that predictions can be made only when all concrete models agree reveals how the size of the cABN relates to its predictive capacity and motivates the need to limit the number of possible interactions. Increasing the number of possible interactions increases the number of possible BNs that can potentially produce different dynamic behaviors. This increases the likelihood of scenarios in which a hypothesis and its null are both satisfiable, preventing predictions from being made.

The simplest hypothesis to test is whether a particular state is reachable under some conditions (such as a change in signals, a gene knockout, or forced expression). In this case, the hypothesis constraint specifies that under the perturbation the defined gene expression state is reachable. The null constraint specifies instead that the gene expression state is unreachable, making use of the 'not' keyword. However, richer enquiries can also be encoded, such as whether states are reached after a specific number of steps or that a specific order of events must occur along a particular trajectory. Accordingly, encoding hypotheses follows the same format as encoding constraints.

Example: Having identified a set of concrete BNs consistent with the 23 experimental constraints we encoded, we next sought to test whether this set of models could predict the response of ESCs to genetic knockouts. We tested all possible single and double knockouts, examining whether Oct4 and Sox2 could remain ON. As one example, we tested the effect of Esrrb and Gbx2 double knockout in ESCs cultured in 2i + LIF. The constraint laid out below was added to the set of 23 to test this hypothesis:

```
#TestDoubleKnockout[0] |= $TwoiPlusLifWithEsrrbGbx2KO;
#TestDoubleKnockout[0] |= $TwoiPlusLifCultureConditions;
#TestDoubleKnockout[0] |= $EsrrbAndGbx2KO;
#TestDoubleKnockout[18] |= $Oct4AndSox2On;
#TestDoubleKnockout[19] |= $Oct4AndSox2On;
```

Here, the predicates are defined as:

```
// Initial gene expression
  state
$TwoiPlusLifWithEsrrbGbx2KO
:=
{
MEKERK = 0 and
Oct4=1 and
Sox2=1 and
Nanog=1 and
Esrrb=0 and
Klf2=0 and
Tfcp2l1=1 and
Klf4=1 and
Gbx2=1 and
Tbx3=1 and
Tcf3=0 and
Sall4=1 and
Stat3=1
};
$Oct4AndSox2On :=
{
Oct4 = 1 and
Sox2 = 1
};
```

```
// Culture conditions
$
TwoiPlusLifCultureConditions:=
{
LIF = 1 and
CH = 1 and
PD = 1
};
// Specifying knockouts
$EsrrbAndGbx2KO :=
{
KO(Esrrb) = 1 and
KO(Klf2) = 1
};
```

RE:IN finds that the updated set of 24 constraints are all satisfiable. However, this result only allows us to determine that at least one model can satisfy the new set of constraints. Therefore, we also encoded the null hypothesis instead of the above constraint to test it separately:

```
#TestNull[0] |= $TwoiPlusLifWithEsrrbGbx2KO;
#TestNull[0] |= $TwoiPlusLifCultureConditions;
#TestNull[0] |= $EsrrbAndGbx2KO;
not(#TestNull[18] |= $Oct4AndSox2On);
not(#TestNull[19] |= $Oct4AndSox2On);
```

RE:IN finds that the set of 24 constraints that contain the null hypothesis are not satisfiable. Given that the hypothesis is satisfiable, and the null is unsatisfiable, we can determine that all models must have Oct4 and Sox2 ON under this double knockout. Therefore, the set of models predict that ESCs will remain pluripotent under Esrrb/Gbx2 double knockout.

3.7 Refining the Set of Models

Once a set of predictions have been formulated, these can be tested experimentally. New information gleaned from these experiments can be used to refine the cABN and, accordingly, our understanding of the system. Where the cABN has correctly predicted previously untested behavior, we gain support for the set of models currently defined, and they can be used to understand why this behavior arises. Incorrect model predictions do not mean that we need simply to throw away the models, but rather that we have learned that the assumptions that the current cABN were based upon are insufficient and the current understanding of the system is incomplete. In such scenarios, we need to consider the possibility of additional possible interactions or remove strong assumptions such as interactions that have been set as definite.

Consider the scenario that the cABN was used to formulate a prediction that was subsequently invalidated by experiment. Given the approach we implement to generate predictions, *all* concrete models will be inconsistent with this new experimental observation. This observation can be included as an additional constraint, and the process of deriving a set of models consistent with the updated constraints should be followed again to refine the cABN, so that the models can explain the correct behavior (Fig. 5a).

Example: In 2i + LIF conditions, 11 knockout predictions were tested experimentally by siRNA transfection followed by clonal assay. These experiments confirmed all but one of the predictions: Tbx3/Klf2 double knockdown was found experimentally not to induce differentiation, but this was incorrectly predicted by the cABN. Given that all models incorrectly predicted the effect of this knockdown tells us that if this new experimental observation is added as a permanent constraint, it will not be satisfiable by the current set of models. We therefore must revisit the ABN to identify whether additional possible interactions can be included so that the expanded set of constraints can be met (Fig. 5a).

Given that we started with a ranked list of possible interactions ordered by their Pearson correlation coefficient, we examined lowering the Pearson threshold. A threshold of 0.647 defined an ABN that allowed the 24 constraints to be satisfied, while a threshold of 0.648 was not sufficient. However, the ABN defined by a threshold of 0.647 could not generate any predictions when the knockout tests were run—this was due to the higher number of models defined by this threshold, increasing the chance that some models exhibit different behaviors under the same initial conditions, preventing predictions from being made. The single difference between the 0.647 threshold and the 0.648 threshold was one interaction: a possible positive interaction between Oct4 and Sox2. We found that if only this bidirectional interaction was included in the original ABN defined by the threshold 0.792 then the 24 constraints were satisfiable, and the set of predictions were preserved.

Subsequently, we searched for the minimal models that satisfied the updated set of constraints and found that there were only 2, each with 17 of the possible interactions instantiated. While both models did not require Tbx3 to behave as a regulator, they differed only in which gene activated Tbx3. Therefore, Tbx3 was found to be dispensable in each of these networks, and they collapse onto the same topology (Fig. 5d).

3.8 Exploring Additional Components

The pluripotency example illustrates how the RE:IN can be used to explore a set of possible networks, by first constraining the set of models against known experimental behaviors and subsequently using the constrained set to formulate predictions of untested behavior. If it is not possible to satisfy the constraints with a given ABN, it can be extended by incorporating additional possible interactions as described above. However, it may also be necessary or of interest to explore the need for additional components. The following scenario illustrates how RE:IN can facilitate the search for new components.

A single BN with a concrete network topology (Fig. 6a) is found to be inconsistent with experimental constraints using RE: IN, indicating that the network must be refined. The user explores whether all of the interactions in the BN are required by setting them as possible not definite (Fig. 6b) but finds that this set of models are all inconsistent with the constraints. The user next explores whether additional interactions might be required, by creating an ABN that contains all possible positive and negative interactions between the components (Fig. 6c). This extended ABN is also found to be inconsistent with the constraints. This

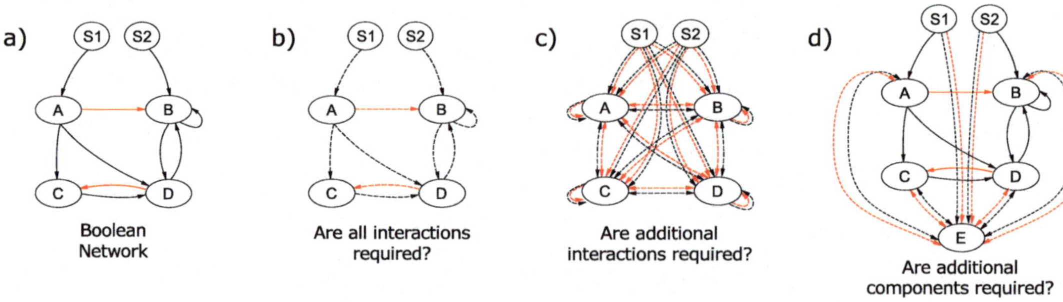

Fig. 6 (**a**) A concrete BN proposed for a given system, which is found to be inconsistent with experimental constraints. (**b**) The user creates an ABN by setting the hypothesized interactions as possible, not definite. (**c**) The user creates an ABN that includes all possible positive and negative interactions to explore whether BNs with more interactions than in (**a**) and (**b**) satisfy the constraints. (**d**) The user creates a "dummy" component E and connects it to the remaining components with all possible positive and negative interactions

last step demonstrates that no network exists between the set of components that satisfies the observed system behavior, and therefore additional components must be required.

To identify candidate components, the user adds a "dummy" component, *E*, and connects it to the input signals and remaining components via positive and negative possible interactions. In this way, flexibility is introduced such that the new component can interact with the others, but it is not required to do so. If this ABN is sufficient to satisfy the experimental constraints, then the user should search for required and disallowed interactions—this will reveal whether any of the possible interactions are necessary to satisfy the constraints. Searching for minimal models can also help to identify the smallest number of connections that the dummy node has to make to the other components. This information can be used to identify which critical gene should be added to the set of components, as RE:IN has exposed the connections that it must make.

4 Notes

This chapter has described the RE:IN methodology, invoking the pluripotency network as an example to demonstrate its utility. Here we include some useful notes for those wishing to implement the approach for their own studies. FAQ are also provided on the RE: IN website, which are updated as necessary over time.

1. Several directives need to be set to run a RE:IN analysis, which can be modulated individually via the "Options" tab in the tool itself or included at the top of the text file that includes the definition of the ABN. The directives are:

Solution limit: How many concrete BNs to enumerate. Setting to 0 tells the solver to enumerate all possible BNs that are consistent with the constraints.

Interactions limit: The maximum number of possible interactions to allow in any consistent BN. Setting to 0 tells the solver not to limit the number of interactions. Note that this excludes definite interactions that were defined in the ABN.

Regulation conditions: There are three possible settings: "default" corresponds to the set of 18 regulation conditions plus the 2 threshold rules; "No threshold rules" corresponds to just regulation conditions 0 ... 18; and "Legacy" refers to the 15 regulation conditions that were used by Dunn et al. [21], which were subsequently revised by Yordanov et al. [12].

Updates: These can be set as synchronous or asynchronous.

Unique solutions: This setting instructs RE:IN on how to distinguish BNs that are consistent with the constraints. This is relevant because two concrete BNs can differ both in their topology and in the regulation condition assigned to each component. You can enumerate concrete BNs that are unique in topology ("Interactions only"), unique in either topology or regulation conditions assigned to each component ("Interactions and regulation conditions"), or also unique in the trajectories required to satisfy the different experiments ("Interactions, regulation conditions, and experiments").

Experiment length: The maximum trajectory length for each experimental constraint.

If the user wishes to upload a text file containing the definition of the ABN, the directives should appear as follows at the top of the file:

```
directive updates sync;
directive length 20;
directive uniqueness interactions;
directive limit 0;
directive regulation default;
```

2. RE:IN allows the user to execute network trajectories via both synchronous and asynchronous updates. Under the asynchronous update scheme, setting a constraint only ensures that there exists at least one concrete network and path for which that constraint is satisfiable. That is, under the same conditions it is possible that the genes could be updated in a different order in the same model and get to an altogether different

state. For this reason, it is difficult to formulate predictions of behavior under an asynchronous scheme, and additional assumptions would be required that restrict the update scheme in some sense (e.g., to prevent the same gene from being updated at each and every timestep in isolation).

3. If you have one constraint in which a component is knocked down, then you must ensure that the component is defined to be OFF in any other states defined for that experiment. If, for example, the component is set to be active in the initial state, and the knockout is set to TRUE, the constraint will be unsatisfiable. Similarly, if a gene is set to be overexpressed, then it must be ON in all states of that experiment.
4. If there are any experiments in which a gene is knocked down (or overexpressed), then it must be explicitly set not to be knocked down (overexpressed) in the remaining experiments. This is because the solver may exploit the possibility of knocking down the gene in order to satisfy the constraint.

References

1. Chen X, Xu H, Yuan P et al (2008) Integration of external signaling pathways with the core transcriptional network in embryonic stem cells. Cell 133:1106–1117. https://doi.org/10.1016/j.cell.2008.04.043
2. Lu R, Markowetz F, Unwin RD et al (2009) Systems-level dynamic analyses of fate change in murine embryonic stem cells. Nature 462:358–362. https://doi.org/10.1038/nature08575
3. van den Berg DLC, Snoek T, Mullin NP et al (2010) An Oct4-centered protein interaction network in embryonic stem cells. Cell Stem Cell 6:369–381. https://doi.org/10.1016/j.stem.2010.02.014
4. Som A, Harder C, Greber B et al (2010) The PluriNetWork: an electronic representation of the network underlying pluripotency in mouse, and its applications. PLoS One 5:e15165. https://doi.org/10.1371/journal.pone.0015165
5. Yeo J-C, Ng H-H (2013) The transcriptional regulation of pluripotency. Cell Res 23:20–32. https://doi.org/10.1038/cr.2012.172
6. Tyson JJ, Chen KC, Novak B (2003) Sniffers, buzzers, toggles and blinkers: dynamics of regulatory and signaling pathways in the cell. Curr Opin Cell Biol 15:221–231. https://doi.org/10.1016/S0955-0674(03)00017-6
7. Kühl M, Kracher B, Groß A, Kestler HA (2014) Mathematical models of Wnt signaling pathways. In: Wnt Signal. Dev. Dis. John Wiley & Sons, Inc, Hoboken, NJ, USA, pp 153–160
8. Davidson EH (2010) Emerging properties of animal gene regulatory networks. Nature 468:911–920. https://doi.org/10.1038/nature09645
9. S a K (1969) Metabolic stability and epigenesis in randomly constructed genetic nets. J Theor Biol 22:437–467
10. Thomas R (1973) Boolean formalization of genetic control circuits. J Theor Biol 42:563–585
11. Le Novère N (2015) Quantitative and logic modelling of molecular and gene networks. Nat Rev Genet 16:146–158. https://doi.org/10.1038/nrg3885
12. Yordanov B, Dunn S-J, Kugler H et al (2016) A method to identify and analyze biological programs through automated reasoning. npj Syst Biol Appl 2:-16010. https://doi.org/10.1038/npjsba.2016.10
13. Li F, Long T, Lu Y et al (2004) The yeast cell-cycle network is robustly designed. PNAS 101:4781–4786. https://doi.org/10.1073/pnas.0305937101
14. Szklarczyk D, Morris JH, Cook H et al (2017) The STRING database in 2017: quality-controlled protein–protein association networks, made broadly accessible. Nucleic Acids Res 45:D362–D368. https://doi.org/10.1093/nar/gkw937

15. Martello G, Bertone P, Smith AG (2013) Identification of the missing pluripotency mediator downstream of leukaemia inhibitory factor. EMBO J:1–14. https://doi.org/10.1038/emboj.2013.177
16. Martello G, Sugimoto T, Diamanti E et al (2012) Esrrb is a pivotal target of the Gsk3/Tcf3 axis regulating embryonic stem cell self-renewal. Cell Stem Cell 11:491–504. https://doi.org/10.1016/j.stem.2012.06.008
17. Hall J, Guo G, Wray J et al (2009) Oct4 and LIF/Stat3 additively induce Krüppel factors to sustain embryonic stem cell self-renewal. Cell Stem Cell 5:597–609. https://doi.org/10.1016/j.stem.2009.11.003
18. Silva J, Nichols J, Theunissen TW et al (2009) Nanog is the gateway to the pluripotent ground state. Cell 138:722–737. https://doi.org/10.1016/j.cell.2009.07.039
19. Tai C-I, Ying Q-L (2013) Gbx2, a LIF/Stat3 target, promotes reprogramming to and retention of the pluripotent ground state. J Cell Sci 126:1093–1098. https://doi.org/10.1242/jcs.118273
20. Lanner F, Lee KL, Sohl M et al (2010) Heparan sulfation-dependent fibroblast growth factor signaling maintains embryonic stem cells primed for differentiation in a heterogeneous state. Stem Cells 28:191–200. https://doi.org/10.1002/stem.265
21. Dunn S-J, Martello G, Yordanov B et al (2014) Defining an essential transcription factor program for naive pluripotency. Science (80-) 344:1156–1160. https://doi.org/10.1126/science.1248882
22. Jain AK (2010) Data clustering: 50 years beyond K-means. Pattern Recogn Lett 31:651–666. https://doi.org/10.1016/j.patrec.2009.09.011
23. Ying Q-L, Wray J, Nichols J et al (2008) The ground state of embryonic stem cell self-renewal. Nature 453:519–523. https://doi.org/10.1038/nature06968
24. De Moura L, Bjørner N (2008) Z3: An efficient SMT solver. In: Tools algorithms constr. anal. syst. Springer, Berlin Heidelberg, pp 337–340

Chapter 5

Mathematical Modelling of Clonal Stem Cell Dynamics

Philip Greulich

Abstract

Studying cell fate dynamics is complicated by the fact that direct in vivo observation of individual cell fate outcomes is usually not possible and only multicellular data of cell clones can be obtained. In this situation, experimental data alone is not sufficient to validate biological models because the hypotheses and the data cannot be directly compared and thus standard statistical tests cannot be leveraged. On the other hand, mathematical modelling can bridge the scales between a hypothesis and measured data via quantitative predictions from a mathematical model. Here, we describe how to implement the rules behind a hypothesis (cell fate outcomes) one-to-one as a stochastic model, how to evaluate such a rule-based model mathematically via analytical calculation or stochastic simulations of the model's Master equation, and to predict the outcomes of clonal statistics for respective hypotheses. We also illustrate two approaches to compare these predictions directly with the clonal data to assess the models.

Key words Stem cell fate, Stem cell heterogeneity, Clonal dynamics, Master equation

1 Introduction

In recent decades, rapid advances in experimental techniques have led to astonishing insights into fundamental biological processes. Nowadays, it is possible to study the effect of essentially any gene on an organism's physiology, through sophisticated genetic manipulation techniques, e.g., CRISPR/Cas9 [1–3]. Yet, while these tools allow to directly obtain insights into biological molecular mechanisms, it is still challenging to identify the time course and outcomes of dynamic physiological processes in living cells and tissue. A detailed understanding of cellular processes on the tissue level is further confounded by functional properties of cells depending very sensitively on their environment. Thus, insights from in vitro studies of cell dynamics are not always reproducible in vivo. For example, the proliferation and differentiation of stem cells, in development, tumors, and renewing tissues, depend crucially on their in vivo cellular environment (the stem cell "niche" [4]). In living tissues, however, the time course of biological processes is more difficult to observe. Ideally, one would like to observe

Patrick Cahan (ed.), *Computational Stem Cell Biology: Methods and Protocols*, Methods in Molecular Biology, vol. 1975, https://doi.org/10.1007/978-1-4939-9224-9_5,

a given cell or cell lineage in a tissue over time, and monitor divisions and differentiation simultaneously with gene expression, to identify the cell (differentiation) state. In practice, however, this is often impeded by the lack of available reliable molecular markers and inaccessibility of tissue for in vivo live observation. Although advances in intravital live imaging of cell lineages have been made via multiphoton microscopy, to observe cell dynamics in mammalian tissues [5–7], surgical removal of tissue is still required in most cases. By looking at those removed tissue samples, only static "snapshots" of the observed process, e.g., clonal distributions, are possible [8–11], not their dynamic time course. Thus, experimental data often do not represent a complete picture of the underlying process under study, and it is for that reason that the experimental data alone is often not sufficient to make conclusions about cellular dynamics.

Parallel to experimental advances, the aid of powerful computational techniques nowadays allows to handle and analyze the plethora of experimental data available by high-throughput experimental techniques as used in genomics and transcriptomics [12]. However, while bioinformatics provides a powerful machinery for processing data, it cannot answer hypothesis-based research questions if the data does not reflect a direct measurement of the process of interest. Traditional statistical inference and hypothesis testing, based on standardized statistical tests (e.g., Student's t-test), only apply if there is a one-to-one correspondence between the quantity addressed by the hypothesis and the experimental data; nonetheless (as discussed below) it is in many cases not possible or too expensive to obtain such direct measurements. This chapter will describe how in such cases mathematical modelling can solve problems and answer biological research questions, if there is no one-to-one correspondence between hypothesis and experimental data, by mathematically transforming research hypotheses into a form which eventually can be directly compared with the data.

Mathematical modelling is the (simplified) mathematical description of assumptions (e.g., a hypothesis) based on which a mathematical evaluation produces quantitative predictions. Model predictions can be obtained by an analytical formula or a numerical algorithm, such as Monte Carlo simulations. Since most numerical algorithms are ideally implemented by a computer, we would often talk of "computational modelling" in this case, although the basic principles of how models are developed and tested are essentially of mathematical nature. By implementing biological hypotheses as mathematical or computational models, the experimental scenario can be simulated in silico under the hypothesis assumptions, producing virtual pseudo-data. These results can then be compared with the experimental data, and depending how well this matches the experimental data, hypotheses can be accepted or rejected.

The use of mathematical modelling approaches for hypothesis validation is not just a development of recent years; in fact, some of the greatest biological discoveries of the mid-twentieth century have only been possible by modelling approaches similar to those discussed here. While in more recent times discoveries have been largely driven by the advance of experimental techniques and computational power, harvesting the richness of "big data" and utilizing large-scale complex computational models to do "in silico" biology, some researchers have rediscovered the use of more simplistic models as powerful tools for validating hypotheses [9, 13, 14].

An elucidating historical example of how mathematical modelling can aid advances in basic biology is the discovery that mutations in bacteria occur in the absence of selective pressure, by Max Delbrück and Salvador Luria in 1943 [15]. At that time, mutations could only be observed by their phenotypic effect, which often involved placing cells into a selective environment. It was therefore not possible to directly test experimentally whether mutations in bacteria occur only in response to selective pressure or also in absence of the latter, in a random fashion. Nonetheless, Luria and Delbrück approached this problem by (1) measuring the distribution of phage-resistant bacterial colonies after exposure to phages that kill nonresistant bacteria and (2) developing a mathematical model for either complementary hypothesis (pre-selective mutations vs. selection-induced mutations). Then they obtained mathematical predictions for the statistics of resistant colonies and in particular about the relation of the mean and variance of the distribution of resistant mutants. Thereby they could show that predictions from the hypothesis of selection-induced mutants were not consistent with the measured statistical quantities, while the hypothesis of pre-selection mutations was. This prooved the random nature of mutations which was eventually awarded with the Noble Prize for Physiology and Medicine in 1969.

Another example, and a defining discovery of the twentieth century, was the revelation of the DNA structure [16–18]. It might not be very common to view the discovery of the DNA structure as an accomplishment of mathematical modelling, yet also in this case the problem was that experimentally it was not possible to directly "see" the molecular structure of molecules via pure experimental means. The experimental data was given in the form of diffraction patterns from molecular DNA crystals radiated with X-rays [18], not a direct image of the molecular structure. Effectively, it was the *mathematically predicted* X-ray diffraction patterns, following from Bragg's and von Laue's crystal diffraction laws, which were compared with those X-ray patterns to find the correct structure. Although it was possible to narrow down the possible molecular configurations of the DNA structure to a few options by the knowledge of chemical interactions, some ambiguity in chemically possible DNA structures remained. For example, based on physicochemical considerations, Linus Pauling first suggested a triple helix structure for DNA [19], but this suggestion lacked an experimental confirmation. While a direct image of DNA was not able to obtain, a mathematical framework which predicts diffraction patterns from molecular structures existed, in form of the physical laws of diffraction discovered by the Bragg family [20] and von Laue [21]. Applying this mathematical framework for an assumed DNA structure predicts X-ray diffraction patterns (based on Bessel functions J_n) which can then be compared with the experimentally obtained ones.

By such a mathematical analysis, it turned out that Pauling's triple helix—i.e., the predicted diffraction patterns predicted from it—did not match those observed by Rosalind Franklin et al. [18]. Instead, the double helix structure proposed by Watson and Crick [16] did match Franklin's data, as confirmed by Wilkins et al. [17]. Thereby, through the interplay of chemical modelling, X-ray refraction experiments, and mathematical modelling, the DNA structure had been revealed.

From the present standpoint, most biologists would probably not see this discovery as a result of mathematical modelling. Mathematical/computational modelling is nowadays perceived as being more complex, with predictions following from extensive computer simulations of models including a plethora of elements and processes of a system. In fact, the mathematical formula used to predict X-ray diffraction patterns by Wilkins et al., i.e., the physical diffraction law, was rather simplistic and had already been known decades before the discovery of the DNA structure. It was just the application of this law, in form of a mathematical formula, which essentially led to the discovery. In this chapter, I will describe how approaches essentially similar to the historical examples above, based on models of limited complexity, can widen our opportunities to gain insight into biological processes. We will see how mathematical modelling, which deliberately *neglects* certain aspects of the dynamics, can be used to drive biological discoveries, which will be exemplified by the study of in vivo cell fate dynamics through the interplay of mathematical modelling and genetic cell lineage tracing of cell clones. The challenge in this endeavor is twofold: (1) the research questions/hypotheses are about the behavior of individual cells, while the data is multicellular (clones) and (2) the data is available only as static "snapshots" and only yields the statistics of clonal distributions, but the tracing of individual clones is not possible (we limit ourselves to situations in which this is the case; in certain tissue types, intravital clonal tracing is indeed possible [5–7], yet information about a cell's state is then limited). The search for "dynamics" of cell fate choice is not directly experimentally accessible, and thus, standardized hypothesis testing cannot be applied on this data. Nonetheless, the clonal data is effectively an outcome of cell fate dynamics, so that a mathematical description of the latter allows to make predictions on clonal statistics. Thus, biological hypotheses can be validated by translating them into (simple) mathematical models and comparing model predictions with experimental data, following the philosophy of the scientific method [22].

2 Methods

2.1 General Modelling Approach

The general pathway to answer biological research questions via mathematical modelling is along the following steps. First a scientific hypothesis is formulated, which is then "translated" into a mathematical form, i.e., a mathematical model. This model is then evaluated, analytically or numerically, to yield quantitative predictions which are then compared with the experimental data. To do this comparison, one requires a way to "score" the model's goodness and assign it a "certainty" value, which will be described in detail later in Subheading 2.5.2. If the certainty is below a given threshold, hypotheses can be rejected; otherwise, one may choose a "most certain" model as the one which most likely describes the real dynamics. In principle this follows the lines of classical hypothesis testing, only that we consider situations in which the hypothesis cannot be directly tested on the data, and the mathematical model is needed to make quantitative predictions in a form that can be compared with the data. How such a model is designed and evaluated, and how the predictions are then compared with the data to validate the hypothesis, will be subject of the following sections, which is aligned with the strategy applied by previous studies to reveal the cell fate choices upon cell division [9–11, 23–27].

2.2 Clonal Statistics and Experimental In Vivo Lineage Tracing

To exemplify the power of this modelling approach, we will, despite a general wide applicability of this method, focus on modelling cell fate by comparison with clonal (cell lineage) data. The idea is that the observation of a cell's progeny—a clone—contains some information about the cell fate choices taken by its constituent cells, in terms of whether a cell divides, dies, or differentiates.

A way to obtain clonal data from in vivo tissues is to use genetic labelling enabled in transgenic mouse lines via a Cre-Lox recombination system [28, 29]. As shown in Fig. 1, left, these animals carry a gene for a Cre recombinase together with a gene for a green fluorescent protein (GFP) whose expression is normally blocked by a stop cassette. The stop cassette is surrounded by LoxP sites. Cre is normally not expressed but it can be activated by tamoxifen. If tamoxifen is administered, Cre protein is expressed and cuts out the LoxP sites, thereby removing the stop cassette, so that the GFP is expressed. This labels the cell fluorescently, and since the removal of the stop cassette is irreversible, the labelling is inherited to all the

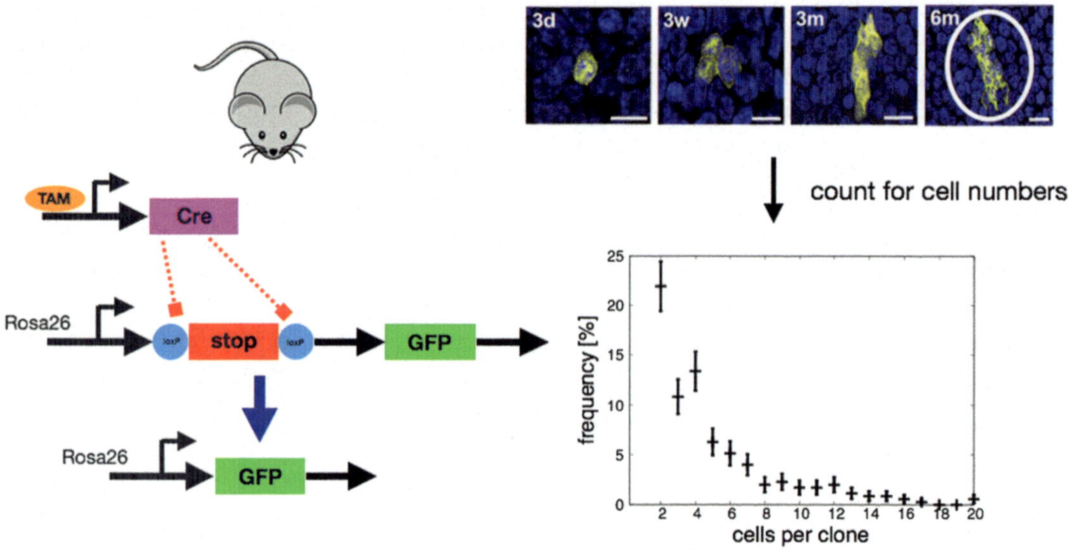

Fig. 1 Illustration of cell lineage tracing experiments and clonal distributions. (Left) Illustration of the genetic construct for genetic lineage tracing (*see* text). (Top right) Imaging of genetically labelled clones in mouse epidermis (from [9]). By labelling sporadic single cells with an inheritable genetic label (a Cre-Lox-mediated removal of a stop cassette), the progeny of a single cell—a clone—retains the label (mouse image from freeclipartimage.com). Individual clones can be separated, if initially labelled cells are sparse enough. (Bottom right) The Frequency distribution of measured clone sizes (clone size distributions, in terms of cell number per clone) is obtained after surgical removal of tissue (here: from mouse esophageal epithelium [10]) and scoring clones via fluorescent microscopy. The clone size distribution can be compared to model predictions

progeny of the initially labelled cell—a clone. If the administered dose of tamoxifen is very low, Cre is activated only sporadically, so that initially induced cells are far from each other. In solid, cohesive tissues where the cell's progeny remains locally confined, clusters of cells form which are, with a high probability, monoclonal (i.e., originating from one cell each) and can thus be identified as clones (Fig. 1, top right). After surgical removal of the tissue and analysis via fluorescent imaging, clones can be counted for their size and scored to obtain a frequency distribution of clone sizes and the clone size distribution [8] (*see* Fig. 1, bottom right). This data can then be used for comparison with mathematical models [27].

2.3 One Hypothesis-One Model

To test a hypothesis, it first needs to be properly formulated. This apparently trivial task often requires more thought than one would expect, since it is crucial that hypotheses are formulated in a mathematically precise way, so that they can be translated into mathematical models. Furthermore, it is important that a hypothesis is not too complex; otherwise the corresponding model would contain too many parameters, making it impossible to validate it, due to "over-fitting." As an example, we consider the cell fate choice of stem cells in a healthy, homeostatic (stationary) tissue. In principle, cells can undergo a plethora of choices; they can divide and differentiate; their progeny might again be stem cells, or progenitor cells which lose proliferative potential at each division, or maturing cells of different types. Furthermore, the fate behavior of stem cells may be determined by other cells in their vicinity, through biochemical signalling or mechanical interactions. Nonetheless, integrating all these options into one model would require a plethora of parameters and render model validation impossible, as the large number of parameters reduces the inference power due to "over-fitting." Instead, we take a more minimalistic modelling approach, by coarse graining the model as far as it just represents only one particular (atomistic) research hypothesis. For example, a long-standing paradigm about fate behavior in homeostatic epithelial tissues was that cell fate outcomes of stem cells are always asymmetric [30], i.e., that upon division of a stem cell, exactly one daughter cell remains a stem cell, while the other commits to differentiation. The latter can potentially progress via a progenitor state in which it still may divide, yet only a finite amount of times before its progeny terminally differentiates. While there had been evidence for this scenario, and it would provide a simple way of maintaining stationary stem cell pool (required in a healthy tissue), only from 2007 onward, a combined strategy of cell lineage tracing and mathematical modelling could unanimously test this hypothesis in various tissues [9–11, 23], as described in the following.

To formulate this hypothesis mathematically, we distinguish only between two (meta-)types of cells: *renewing* cells (stem cells, denoted by "S") which maintain their full proliferative potential and continue to divide (despite potential intermittent pausing) and *committed, nonrenewing* cells, viz., those that may divide, but progressively lose their potential and eventually differentiate, denoted by "D." The latter would include all cells that are committed to differentiation, into whatever cell type, or to die, and include non-stem cell committed progenitor cells, as well as mature cells. Furthermore, only considering the "fate" of the cell, we only distinguish whether after division a cell commits or not, irrespective of when this occurs exactly. While this distinction appears rather crude, it is sufficient to address the question whether divisions are asymmetric, symmetric, or of mixed nature. In this picture, only three outcomes after division are possible, without specifying a hypothesis: two S-cells, two D-cells, or one S-cell and one D-cell. Furthermore, we would assume that D-cells disappear after some time. Visually, such outcomes can be illustrated as in Fig. 2, top.

The hypothesis that only asymmetric divisions occur corresponds to the invariant outcomes with one S-cell and one D-cell as shown in Fig. 2, left. The alternative hypothesis would be that there is a non-zero probability of symmetric outcomes (two S-cells or two D-cells), as in the general model, called mixed cell fate outcomes (Fig. 2, right). Furthermore, the condition of

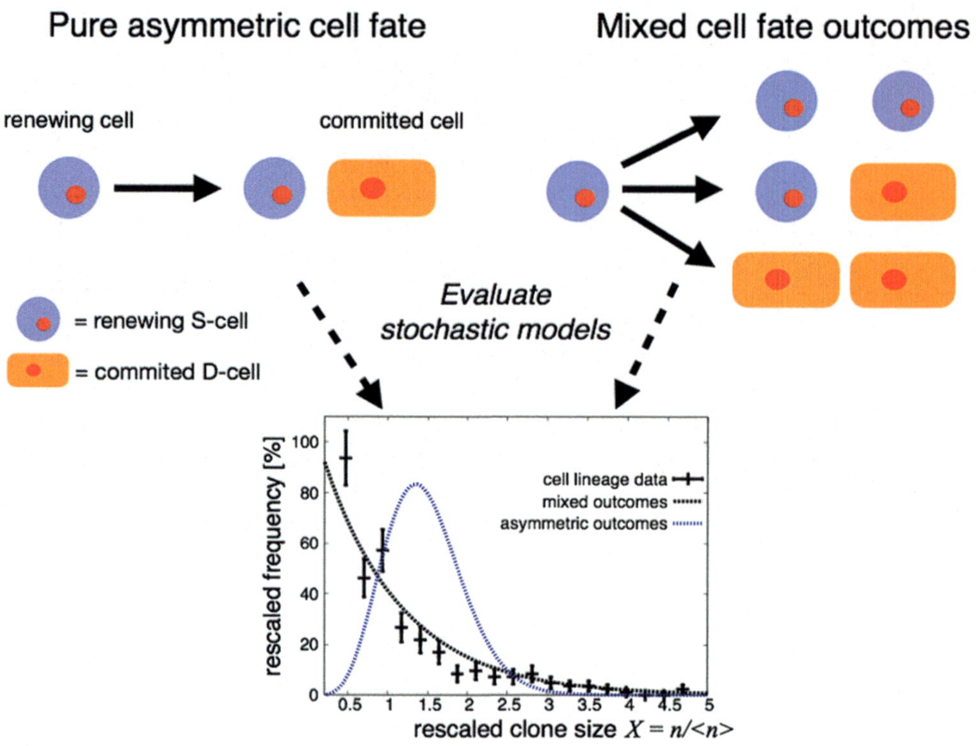

Fig. 2 Generic models of cell fate choices in homeostatic renewing tissue. In homeostasis, defined by a stationary distribution of cell types and constant mean cell number, after each division on average one cell commits to differentiation (D, orange noncircular cells) and one retains stemness (S, circular cells). This can occur via two ways: (Left) invariant asymmetric division, in which each division leads to exactly one S-cell and one D-cell, or (Right) mixed cell fate outcomes by which stem cells sometimes duplicate and sometimes produce two D-cells, yet at equal rates, so that the mean cell number stays constant. These hypotheses can be implemented as stochastic models. Predictions for clone size distributions are shown in the bottom figure (dashed blue, invariant asymmetric model; dashed black, mixed cell fate outcomes model) together with clonal data from cell lineage tracing in mouse esophageal epithelium [10]. Comparison of model predictions and data leads to rejection of the pure asymmetric cell fate outcomes model. The mixed cell fate model, however, predicts an exponential distribution of clone sizes [40] which matches the experimental data

homeostasis requires that at some point a D-cell and all its progeny are removed from a tissue and that the ratios of S + S and D + D outcomes are the same.

Crucially, those hypotheses can be directly translated into a stochastic model, if it is assumed that the transitions occur stochastically (an assumption that will be justified in the following section). We assume that cell divisions occur at a constant stochastic rate λ and that cell fate choices are taken with probability p_{SS}, p_{SD}, and p_{DD}, respectively. Finally, D-cells can terminally differentiate with rate γ after which they are not counted anymore in the experiments. Such a process is a *Markov process* which is mathematically described by the Master equation [31]. How this equation is formulated and evaluated will be described in the following section. Notably, given our coarse distinction between cell types, this simple model is the most generic one.

It might be remarked that the here presented model does not include any aspects of the interaction with other cells or nuances in cell differentiation states; it not even distinguishes if D-cells would divide or not. One could raise the issue and ask "How can you learn something about reality if you don't include all these important aspects?". The point about this modelling approach is not to deem these other processes irrelevant, yet, after all, they are not part of the stated hypothesis and therefore including these processes would be irrelevant for testing this particular hypothesis. If the hypothesis were stated in a more detailed manner, for example, "Cell fate commitment occurs at cell division," then the model would need to include details about the commitment state (*see* our discussion in Subheading 2.5.1). However, it is advisable to proceed in small steps by formulating hypotheses preferably in an atomistic manner, so that the corresponding models do not include too many processes and parameters that would lead to issues that are discussed later on.

2.4 Model Evaluation

In general, dynamics of a system, which are according to some hypothesis on a biological process, can be expressed in two ways: *deterministic* dynamics are described as a dynamical system in the form of differential equations or as recursive difference equations [32]. The other modelling type, which will be considered here, are stochastic dynamics, which are described by stochastic processes (in discrete or continuous time) [31]. The latter approach is favourable to address uncertainties in individual outcomes, and if the studied hypothesis considers mixed outcomes for a given state, such as in Fig. 2, right. Furthermore, stochastic processes allow the description of complex systems by treating unknown external cues as random noise. A stochastic model can be formulated mathematically by the *Master equation* [31] (*see* **Note 1**). The Master equation describes the time evolution of the probability of a random state. In general, a stochastic process can be described by the collection of all its possible states i and all transitions between any states i and j ($i, j = 1, 2, \ldots$), determined by the stochastic rates ω_{ij}, for the transitions from state j to i. With this terminology, the Master equation is a set of coupled differential equations for the probabilities p_i of each state [31]:

$$\frac{\mathrm{d}}{\mathrm{d}t} p_i(t) = \sum_{i=1}^{\infty} \left(\omega_{ij} p_j(t) - \omega_{ji} p_i(t) \right)$$

This equation is based on the "flux" of probability; whenever there is a non-zero transition rate ω_{ij} from j to i, the probability "flows" from j to i. The Master equation sums over all terms for state i to gain probability via influx ($\omega_{ij} p_j$) from other states j and to lose probability due to outflux ($\omega_{ji}p_i$) from state i. The solution of the Master equation then gives the probability of each state over time, by which in principle all statistical properties of the model can be determined.

To solve the Master equation, or at least to obtain reasonable approximations, various numerical and analytical techniques exist. Some analytical methods are based on a spectral decomposition of the matrix $A_{ij} = \omega_{ij} - \delta_{ij} \Sigma_j \omega_{ji}$ and others on the use of *generating functions* [31]. In most cases, however, an exact analytical solution is not possible to obtain, yet asymptotically for large times or system sizes, the Master equation can often be approximated by a simpler partial differential equation which often can be solved. On the other hand, numerical techniques can be used to obtain accurate model predictions. One very powerful method to solve the Master equation numerically is the *Gillespie algorithm* [33], also called *kinetic Monte Carlo* simulation. By this algorithm, the stochastic process is directly simulated, using (pseudo-)random numbers to determine which state transitions occur and at which time points. Through repeated simulations, the frequency distribution of outcomes yields an estimate for the probability distribution $p_i(t)$, which becomes increasingly exact for large simulation ensembles. For details on these methods, I refer to the extensive literature on this topic (e.g., [33, 34]).

For our general model of cell fate outcomes above (Fig. 2), the Master equation for the probability of finding a clone with s S-cells and d D-cells reads

$$\frac{\mathrm{d}}{\mathrm{d}t} p_{s,d}(t) = \lambda\Big(r\, p_{s+1,d-2} + r\, p_{s-1,d} - (1-2r) p_{s,d-1}\Big) + \gamma\, p_{s,d+1} - (\gamma + \lambda) p_{s,d} \tag{1}$$

where $r = p_{SS} = p_{DD}$ and thus for pure asymmetric outcomes, $r = 0$. This comprises all possible stochastic events of the model (here the state $i = (s, d)$ is a vector). In fact, this Master equation can be solved exactly for its generating function $G(k, l, t) := \sum_{s,d=0}^{\infty} k^s l^d p_{s,d}(t)$ [35], yet no closed form for $p_{s,d}(t)$ is available. From a practical perspective, however, it is often more elusive to obtain $p_{s,d}$ from numerical simulations or to obtain an asymptotic solution for large times (*see* Subheading 2.5.1).

Importantly, if as initial condition an individual S-cell is chosen (corresponding to $p_{s,d}(t = 0) = \delta_{s,1}\delta_{d,0}$), the solution represents the stochastic time evolution of clones with s S-cells and d D-cells. Taking the sum of both yields then the clone size $n = s + d$, and thus $p_n(t) = \sum_{s|d=n-s} p_{s,d}(t)$ represents the probability distribution of clone sizes, an approximation for the frequency distribution of clone sizes, the *clone size distribution (CSD)*, which can be directly compared with the experimental data. Together with a scoring function which assigns a "goodness" to this comparison, we can assign a certainty to each model and thereby to each hypothesis.

In the following, two approaches for the modelling and hypothesis validation will be presented: the first one is based on analytical insights, which extracts robust, universal features of the model and which is not sensitive on the details of the models. This approach cannot distinguish nuanced aspects of models, but it provides a high distinctive power for hypotheses that differ structurally, such as the two presented in Fig. 2. The second method is able to distinguish more details of the model but also bears the risk of over-fitting in case that too many details of the model are included. The latter is based on Bayesian analysis and provides means to assign a quantitative certainty to a model/hypothesis, in view of the data, given certain assumption on the model structure.

2.5 Hypothesis Validation by Comparison of Model Results with Data

2.5.1 Approach 1: Universality and Robustness of Stochastic Models as Tools for Hypothesis Validation

In the previous section, we represented the hypotheses, Fig. 2, by Markov processes with constant stochastic transition rates, for which transition times are exponentially distributed [34]. One may raise the question whether such a simplification is justified. In reality, cell division times are not distributed exponentially, as the Markovian assumption implies, but often there is a lag time during which cells cannot divide, and a certain degree of synchronicity between cell divisions [36]. However, The Markov approximation can be justified by the concept of "universality": different processes may have asymptotically the same statistical distributions in the long run, if the results are expressed in terms of effective, rescaled variables (e.g., rescaled by the mean or standard deviation of the distribution) [37, 38]. This means that details of a model, such as the distribution of state transition times (e.g., cell division times), do not matter if one considers long-term behavior. Therefore, we can use Markovian dynamics, with exponentially distributed times, instead of the a priori unknown distribution of cell division times.

The probably most well-known example for universality in statistics is the central limit theorem, which applies to stochastic processes that correspond to adding independent random variables. If the latter have finite mean and variance, the frequency distribution of the final outcomes x_i converges to a normal distribution, irrespective of any details of the distribution of individual random numbers. If one rescales a distribution obtained as such by its mean $\langle x \rangle$ and standard deviation σ, the distribution of *rescaled* outcomes $X_i = (x_i - \langle x \rangle)/\sigma$ is always a normal distribution with mean zero and unit variance. Thus, with this rescaling, all distributions of added random variables "collapse" on the same curve. Note that it does not matter for model predictions what is the distribution of individual variables. In Fig. 3 we show the example of three different stochastic processes, each involving the addition of random

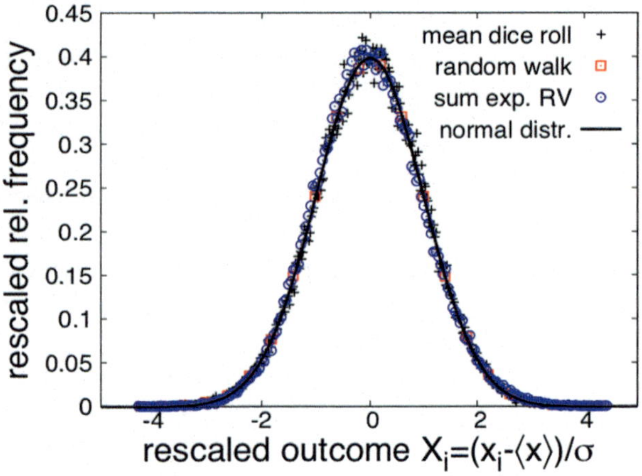

Fig. 3 Illustration of central limit theorem as an example for the universality principle. The frequency distribution of outcomes of three different stochastic processes, simulated on a computer by a Gillespie algorithm [33]. Shown are the outcomes in terms of the rescaled variable $X = (x - <x>)/\sigma$, where x are the raw outcomes, $<x>$ the mean of all outcomes, and σ their standard deviation. The processes are (+) the mean value of N dice rolls, (squares) a symmetric random walk with N step length $\pm$ 0.5, and (circles) the sum of N exponentially distributed random numbers. The black line is a normal distribution with mean zero and unit standard deviation. $N = 10{,}000$

numbers, yet with different distributions of individual random variables, whose outcomes are all distributed according to a standard normal distribution in the variable X_i.

Similar universal principles exist for different types of stochastic processes, and in these cases, the distribution of individual transitions does not matter (they are asymptotically equivalent [39]). Therefore, modelling cell divisions as a Markov process is usually justified, assuming that the distribution of cell division times is at least slightly stochastic.

For modelling purposes, the feature of universality has advantages and disadvantages. On the one hand, the principle of scaling allows to express model predictions as a function of a single, rescaled variable, without any parameters. Thus, by comparison of rescaled data and model predictions, one can usually immediately see, without any data fitting, whether the model (and thus the underlying hypothesis) is compliant with the data or not. In the case of our two hypotheses from the last section (Fig. 2), an analytical analysis of the Master equation in the limit of large times shows that the two hypotheses are in different universality classes: while a model of pure asymmetric divisions (Fig. 2, left) leads to a Poisson-like distribution (which becomes a Normal distribution, asymptotically for large mean values), mixed (stochastic) cell fate outcomes lead to an exponential distribution of clone sizes [9, 40]. The two distributions look entirely different when plotted as a function of the rescaled variable $X = n/\langle n \rangle$ (n = clone size, $\langle n \rangle$ its mean) and can be clearly distinguished, as shown in Fig. 2, bottom, together with measured clonal distributions from cell lineage experiments in adult mouse esophageal epithelium [10]. It becomes clear that the hypothesis of invariant asymmetric division does not comply with the data from mouse esophagus; in fact, it looks qualitatively different, as the model prediction displays a peak, while the experimental distribution is without peak. This hypothesis can therefore be rejected. On the other hand, the hypothesis of mixed cell fate outcomes matches the clonal data very well, without fitting. Given that the two hypotheses are complementary (if only renewing vs. committed cells are distinguished), the second hypothesis must, by exclusion, be correct. The same exponential distribution can be observed for cell lineage data in many other mammalian tissues, such as mouse intestine [11], germ line [23], and epidermis [9].

However, does this fully define the cell fate dynamics in these tissues? What about signalling cues from other cells? Is cell fate really intrinsically stochastic, or just variable, and controlled by a random environment? And are cell fate choices in fact taken at cell division or may there be changes in commitment during interphase? All these questions are actually not addressed in this model, since they were not considered by the stated hypothesis. In fact, cell fate may or may not be controlled by other cells, and the "stochasticity" may just express the randomness in the cellular environment, rather than cell-intrinsic randomness. After all, the modelling has only shown that the data complies with the universality class of mixed (stochastic) cell fate, which includes occasional gain and loss of stem (S) cells, but it does not specify how this occurs. It can only be stated that the cell fate dynamics do not comply with the long-upheld hypothesis of invariant asymmetric cell fate [30].

In the discussed model, the actual state of a cell (e.g., the gene expression state) is not considered. Now, let us consider a model in which intermediate cell states are distinguished, as shown in Fig. 4. In that model, cell divisions itself are asymmetric: upon division, daughter cells are in two different states, a naïve stem cell and a cell that is primed toward differentiation yet not committed. Furthermore, assume that a primed cell has the option to terminally commit to differentiation or to return into the naïve state. For a cell in such a primed state, the eventual cell fate is not decided yet. On the other hand, naïve stem cells would have the option to spontaneously switch into the primed state, independently of cell division, but could not differentiate directly. Such a mechanism, displayed in Fig. 4, right, has been suggested for embryonic stem cells [41] and in adult mouse germ line [7] (in the latter case in the form of multinuclei syncytia). The alternative process, which is commonly assumed to happen in adult tissue, is that cells decide their cell

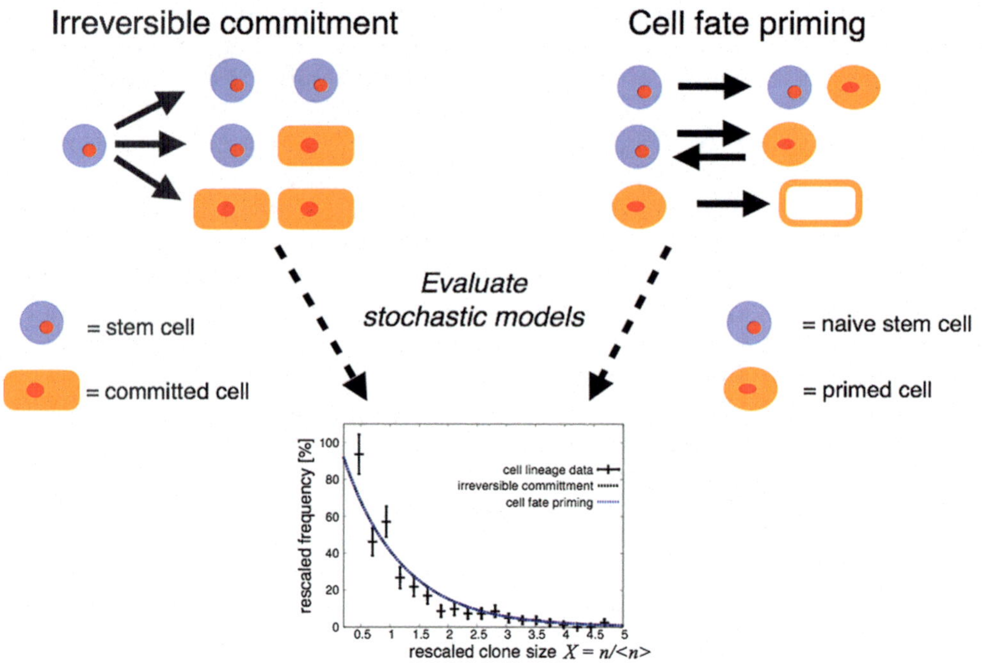

Fig. 4 Illustration of a model with mixed fate outcomes and irreversible commitment of daughter cells versus a model with asymmetric divisions but the possibility of reversible switching between cell states which are primed for either differentiation or renewal. Both models predict the same asymptotic rescaled clone size distribution—an exponential distribution—as shown in the bottom of the figure. There, the black dashed line is the prediction from the irreversible committment model and the blue dashed line from the cell fate priming model, the pluses with error bars are data points from mouse esophageal epithelium [10]. X is the clone size, rescaled by the mean clone size $<n>$, while the frequency is rescaled such that the probability distribution is normalised

fate irreversibly at cell division, when it is fully decided whether a cell remains a renewing cell or commits to differentiation. Both these models are illustrated in Fig. 4. What would be the predictions of both these a model in terms of clonal statistics? In fact, it was shown mathematically that for both models, the predicted clone size distribution is an exponential distribution [42]. This outcome is also shown in Fig. 4, bottom. This means that both models fall within the same universality class, having the same asymptotic distribution. Since the asymptotic distribution is already reached after few cell divisions, these two hypotheses cannot be distinguished based just on the clonal data. Additional information is required to answer the question whether reversible state transitions, including a cell state primed for differentiation, do occur in adult tissue.

At first glance, this result, that invariant asymmetric cell divisions can also result in an exponential clone size distribution, seems to contradict our previous statement that only mixed fate outcomes can achieve that. However, one needs to notice that while the cell states after division are asymmetric, the switching between naïve and primed states, with eventually different outcomes long term, eventually corresponds to mixed cell fate outcomes. This example shows that whether two models are analogue depends very much on the level of detail of a hypothesis: the two here-considered models may correspond to the same hypothesis on a coarse-grained level which only distinguishes between eventual fate outcomes, but they are different when one explicitly considers cell states and the time points at which transitions between the latter occur. In the former case, our modelling approach, based on distinguishing universality classes, is successful to distinguish and validate hypotheses, while on the more detailed level, such distinction is not possible. Thus, this approach, relying on the universal features of stochastic processes, is very powerful to distinguish qualitative structural differences in models, but it is less suited to distinguish more nuanced details.

Note that there is still one property which may have a large influence on the model outcome and thus the universality class: the spatial dimension. If we assume that the cells are embedded in a d-dimensional tissue with other cells (*see* **Note 2**) and that new cells can only be born if neighboring cells are removed, the model outcome might differ. In particular, it has been shown that for lower than three dimensions, the predicted clone size distributions are different if cell fate choices are taken cell-intrinsically or cell-extrinsically (although in two dimensions, the difference is marginal and is difficult to distinguish given the noisy data) [13].

2.5.2 Approach 2: Bayesian Modelling

If one wishes to distinguish details of models and to determine quantitative parameters, we need a quantitative way to assess a model's goodness in view of the data. Here we present how the Bayesian theorem and optimization by ***simulated annealing*** can provide effective means to assign a certainty value to a model and its parameters and to find the optimal set of parameters to fit the data. Furthermore, this method is able to combine the information of two or more experiments to test models, as it is not always possible to measure all relevant quantities simultaneously in the same assay.

The Bayesian theorem assigns a probability, or, better, a "certainty," to a model and a particular parameter configuration, following the laws of probability. The certainty of a set of parameter values θ in view of the data D can be quantified by the "Bayesian posterior probability,"

$$P(\theta|D) = N\ P(D|\ \theta)\ P(\theta) \tag{2}$$

where $P(D|\ \theta) =: L(\theta)$ is the likelihood of θ, i.e., the probability that the model with parameter values θ reproduces exactly the data D, and the *prior* $P(\theta)$ is the a priori certainty of the model parameters, without taking into account D. The factor $N = \left(\sum_{\theta'} P(D|\theta')P(\theta')\right)^{-1}$ is a normalization factor to ensure that the total probability is unity [43]. The prior $P(\theta)$ is an estimate of the parameter certainty due to information from other sources, e.g., measurements from other experiments. The possibility to include this external information provides a very powerful tool to combine the information from different experimental assays (which may have an entirely different data structure) if not all information can be measured by one assay. The prior could, for example, be the Bayesian posterior from the Bayesian analysis of modelling another experiment. If the prior has not been explicitly measured via Bayesian analysis, it is normally chosen as the distribution with the maximum entropy for given information. For example, if a previous measurement resulted in measuring the mean value and variance of some quantity, the maximum entropy prior to choose is a normal distribution with precisely this mean and variance [44, 45]. Without further information one can use non-informative priors such as the uniform distribution, conjugate priors, or "Jeffreys prior" [46]. If the data is rich enough, the Bayesian certainty is usually dominated by the likelihood distribution, and thus the choice of the (uninformative) prior is less relevant.

In the case of stochastic models, the likelihood can be determined from the predicted probability distribution of outcomes. For clonal dynamics, since any two outcomes (clone sizes) are uncorrelated, the observed *absolute* frequency distribution of clone sizes n, f_n follows a multinomial distribution (independent count statistics). As discussed above, the model evaluations yield an estimate for the probabilities p_n that a given clone size n is attained. Note that in general, n can be a vector of numbers, if clones would be distinguished by different cell types, such as S- and D-cells in the models discussed above.

Then, the model likelihood, i.e., the probability that the model with predicted relative frequencies p_n would exactly reproduce the measured absolute frequencies f_n, is a multinomial distribution [10, 24, 25]

$$L(\theta) = P(D, \theta) = \frac{(\sum_n f_n)!}{\prod_n f_n!} \prod_n p_n(\theta)^{f_n} \tag{3}$$

Together with the prior information $P(\theta)$, the evaluation of the likelihood provides an estimate for the certainty of the model and the parameters via Eq. 2. In principle, by evaluating $P(\theta| D)$ analytically, or numerically for a closed-meshed set of parameter values, one can obtain the Bayesian posterior distribution which not only gives the "most certain" set of parameter values but also the range of reasonably certain parameter values (e.g., in terms of the standard deviation of the Bayesian posterior distribution) [24, 25].

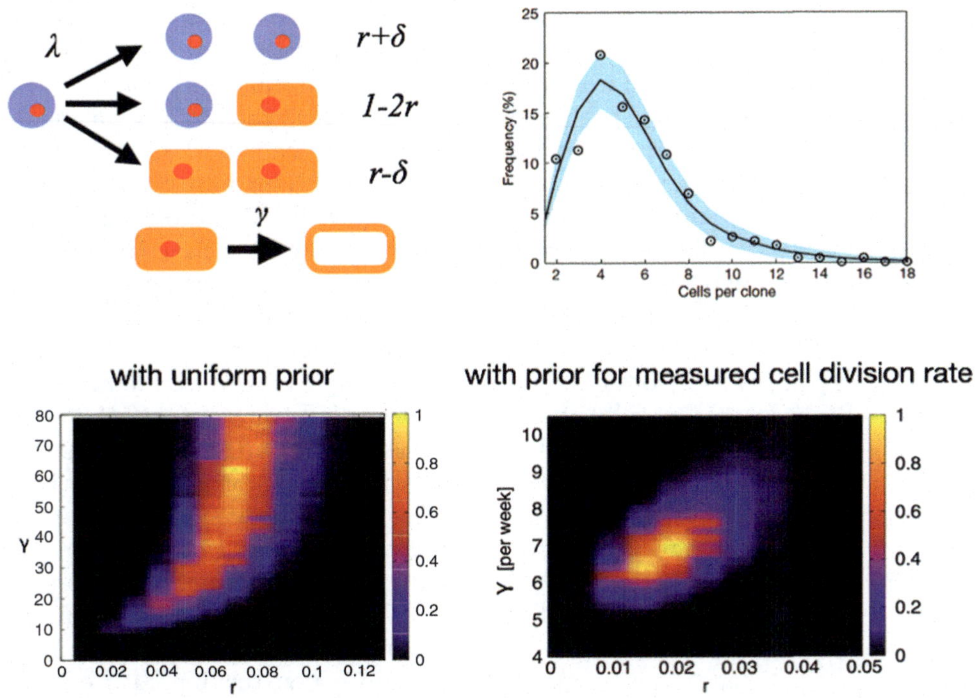

Fig. 5 Bayesian inference of cell fate parameters in esophageal tumors. Top left: the model is structurally the same as mixed outcomes for normal tissue (Fig. 2) but with an imbalance of cell fate, δ, reflected by a higher probability of stem cell gain than loss, resulting in a predominant growth of cell numbers in the tumor. Top right: comparison of model (black line) and clonal data (circles) after Bayesian fitting (selection of parameter set with highest Bayesian posterior), taken from [25]. The shaded area denotes the 95% confidence interval. Bottom (taken from [25]): the Bayesian posterior distribution projected on the parameter plane spanned by the terminal differentiation rate γ, and the ratio of symmetric divisions, r. The "hotter" the color, the higher the Bayesian posterior (interpreted as the model certainty). (Bottom left) with a non-informative, flat prior $P(\theta)$, (Bottom right) with a Gaussian prior $P(\lambda) = \exp(-(\lambda - \lambda_0)^2/2\sigma^2)(2\pi\sigma^2)^{-1/2}$ where λ_0 and σ denote mean and standard deviation of a direct measurement of the cell division rate via another experiment (*see* text)

In order to distinguish details of the model, such as particular parameter values, we need to test the model at "short times," i.e., before reaching the asymptotic scaling distribution by which individual parameters cannot be distinguished. To illustrate how Bayesian inference together with stochastic modelling on short-term cell lineage data can lead to quantitative insights on cell fate dynamics, we consider a cell lineage tracing study of esophageal epithelium [25]. In this study, transgenic mice were treated with a mutagen to induce esophageal tumors, and the kinase inhibitor sorafenib was given to accelerate tumor growth. Clones were traced and scored for their cell number after 10 days, in the same way as described in Subheading 2.5.1, and a model analogue to Fig. 2 was used, which is displayed in Fig. 5, top left. As a slight modification to the previously used homeostatic model, here unequal rates of progenitor cell gain and loss were considered, as tumors are growing (S + S with probability $r + \delta$ while D + D with probability r-δ, where

r quantifies the ratio of stem cell gain/loss events and δ denotes the fate bias. γ denotes the rate of cell death/shedding). Without any further information on the system, a uniform prior $P(\theta) = const.$ was used. Evaluation of the model via stochastic simulations predicted the expected frequencies (probabilities) p_n for clones with size n, which were then compared with the measured frequencies f_n to give the likelihood, according to Eq. 3. Doing this for a close-meshed set of parameters gives, via Eq. 2, a Bayesian distribution, which is displayed in Fig. 5, bottom left. This figure displays the Bayesian certainty, in dependence of two chosen parameters, the symmetric cell fate ratio r and the terminal differentiation rate γ, with "hotter" colors corresponding to higher certainties.

Crucially, in Fig. 5, bottom left, one observes a "ridge" of similarly high values of Bayesian certainties; a unique parameter set with a highest certainty cannot be clearly identified. It has to be noted that the Bayesian landscape itself is stochastic and rugged, since the predictions p_n are coming from stochastic simulations; therefore, even if a maximum would be found, it could possibly not be distinguished from noise. Thus, it appears that based just on the clonal data alone, no unanimous decision on the maximum certainty can be made. The reason is that the approach of a scaling distribution may emerge already after a few cell divisions and the distribution thus may effectively only depend on the combined scaling parameter $\lambda r/\gamma$ (the rate of growth of the mean clone size). Thus, changing the parameter r accordingly with γ leads to the same likelihood, resulting in a line of equal probability in the Bayesian landscape. Possibly, the determination of at least one parameter would narrow down the set of parameters sufficiently to get a unanimous result. For that reason another experiment was performed in that study [25] to determine the cell division rate more accurately, based on measuring the retention of an H2B-GFP molecular label. The information of this experiment, however, cannot be incorporated into the same likelihood function as the clonal data, Eq. 3. Nonetheless, the option to integrate an informative prior probability into the Bayesian analysis allows to study another experiment separately and using this information to shape the prior distribution $P(\theta)$.

To measure the cell division rate, Frede et al. [25] measured the dilution of H2B-GFP protein after stopping its expression. H2B, a DNA-related histone, is highly stable; hence after stopping its expression, the number of proteins is conserved. Thus, upon each cell division, the H2B-GFP concentration halves, and the number of undergone cell division can be determined as the logarithm (to base 2) of its fluorescence level. Thereby the cell division rate was measured, giving the mean and variance, which was then used to construct an informative Bayesian prior for the clonal analysis. The distribution with the maximum entropy for fixed mean and variance is a normal distribution with the latter [45]. Taking this prior for the Bayesian analysis, Eq. 2, one obtains a narrower Bayesian posterior distribution, as shown in Fig. 5, bottom right.

This demonstrates how, by gathering data from different sources, Bayesian inference can be used to validate hypotheses on cell fate dynamics and accurately determine cell fate parameters.

In principle, the same procedure can be followed for other biological data and respective models. In particular, whenever data is given in the form of frequency distributions of independent events, one can use the same formula, Eq. 3, to determine the model likelihood.

2.5.3 Simulated Annealing

Reproducing the whole Bayesian posterior distribution by simulating a closed-meshed grid of parameter value sets is computationally very expensive, in particular in the case of many parameters. Normally, however, one is only interested in the most likely set of parameter values and its proximity. To find just the maximum Bayesian posterior, one can use a variety of optimization algorithms. Crucially, however, the chosen algorithm needs to assure that it does not merely find a local maximum, since the Bayesian posterior distribution is highly rugged with a large number of local maxima, due to the stochastic nature of the Monte Carlo simulations. Thus, one should avoid greedy optimization algorithms such as a simple gradient ascent. Here, I wish to describe a "simulated annealing" algorithm [47]. This algorithm is based on random variation of parameters and then selecting preferably the outcome with higher Bayesian posterior. However, this iterative algorithm accepts, with a lower probability, also parameter sets with a lower Bayesian posterior, i.e., at each iteration step a parameter θ^* with lower certainty compared to the initial parameter θ is chosen, with Prob(accept smaller certainty θ^*) $=$ $\exp\left(-(P(\theta| D) - P(\theta^*| D))/T\right)$, where the "temperature" T parametrizes this probability (this choice of parametrization follows the convention of statistical physics and thermodynamics, in which high temperatures are associated with large noise and high probability to escape local optima). The simulated annealing optimization algorithm follows the rules:

1. Start with a random parameter set θ and temperature $= T_0$.
2. Compute the Bayesian posterior $P(\theta| D)$ of parameters θ by comparing with data set D via Eqs. (2) and (3).
3. Variate parameters by a small random vector $d\theta$.
4. Compute $P(\theta + d\theta| D)$.
5. Choose new parameters $\theta + d\theta$ with probability $\min\left(1, \exp\left(-\frac{P(\theta|D)-P(\theta+d\theta|D)}{T}\right)\right)$ (i.e., it is always accepted if $P(\theta + d\theta| D) > P(\theta| D)$, but sometimes also a smaller certainty $P(\theta + d\theta| D) < P(\theta| D)$ is accepted, to escape potential local maxima).

6. Reduce T by $T \rightarrow T\alpha$, $\alpha < 1$.
7. Stop algorithm if $T < \varepsilon$ (defined as stopping condition a priori), otherwise continue by returning to **step 3**.

This algorithm has a resemblance to the random walk of a physical particle in a potential given by an inverted Bayesian certainty, subject to Brownian motion with temperature T, which is slowly cooled down ("annealed"), a process that is often used in engineering to reach a global optimum. Here, the initial temperature T_0, the annealing step α, and the step width $d\theta$ are free meta-parameters. It cannot be said a priori which choice of these meta-parameters is optimal, but any $T_0 > 0$ will lead more likely to the global maximum of $P(\theta| D)$ than a greedy algorithm, although reaching the global maximum cannot be assured. The correct choice of the annealing parameters is a delicate art and goes beyond the scope of this article; we therefore refer to Ref. [48] for a detailed discussion. Nonetheless, by repeating this algorithm many times, one obtains an ensemble of results close to the maximum, which then can be averaged. If the algorithm is stopped at a higher level of ε, the remaining fluctuations will also give some information about the width of the distribution (around the maximum) and thus the standard error of the estimated parameters.

3 Notes

1. Alternative descriptions by stochastic differential equations or continuous space Fokker-Planck equations exist but are not described here.
2. Although all tissues are three-dimensional, the environment of stem cells, by which their fate is regulated, can have effectively lower dimensions. For example, epithelial basal cells are within a two-dimensional cell sheet, while intestinal stem cells form a "ring" in the crypt base, which is effectively one-dimensional.

4 Conclusions

In many circumstances experimental data is not sufficient to validate biological hypotheses, since hypotheses and the data may not be directly comparable, for example because the data and the hypothesis represent different scales in size and time. In that case, standard statistical tests cannot decide if a given hypothesis is compliant with the data. Nonetheless, mathematical modelling may then be a valuable tool for hypothesis validation, since it is able to "bridge" the scales between a hypothesis and measured data via quantitative predictions from a mathematical model which represents the hypothesis in a one-to-one way. In particular, for studying cell fate dynamics, often a direct in vivo observation of individual cell fate outcomes is not possible, and only multicellular statistical data of cell clones can be obtained. By implementing the rules behind a hypothesis (cell fate outcomes) one-to-one as a stochastic model, a mathematical evaluation of such a rule-based model, via analytical evaluation or stochastic simulations of the model's Master equation, can predict the outcomes of clonal statistics for respective hypotheses. These predictions can then be directly compared with the clonal data, and the certainty of the hypothesis underlying this model can be assessed by how well the predictions match the data. We described two approaches that use stochastic modelling. The first is based on the concept of universality, which states that different stochastic processes that only differ in certain details will after some time (e.g., after a few cell divisions) depend only on few effective, "rescaled" parameters, and converge to the same distribution. This means on the one hand that nuanced details of a model cannot be distinguished by this approach, but on the other hand, if two models (i.e., the underlying hypotheses) are structurally different, they may predict qualitatively different distributions that can be very clearly distinguished. In that case, if the data is properly rescaled, a correct hypothesis would yield a prediction that matches the data without requiring any fitting, while an incorrect one often leads to a qualitatively very different distribution. This non-fit approach is therefore very powerful to distinguish hypotheses if they are in different universality classes. However, if two hypotheses fall within the same universality class and produce the same long-term predictions, one needs (1) a method to distinguish parameters in a more detailed way, and (2) provide the opportunity to combine information from different experimental assays to specify some parameters. We have shown that Bayesian analysis provides both of these: it provides a way to assign a quantitative certainty value to a model and its parameters, which through optimization of the Bayesian posterior function allows to fit a model and determine the best matching model and parameter values. Furthermore, via the Bayesian prior, this approach allows to include the information from other experiments. In particular, the latter property of Bayesian analysis provides a powerful method to test nuanced differences in hypotheses, by measuring different quantities in independent assays and thereby accumulating a variety of information required to distinguish the hypotheses in question.

References

1. Jinek M, Chylinski K, Fonfara I, Hauer M, Doudna JA, Charpentier E (2012) A programmable dual-RNA – guided DNA endonuclease in adaptive bacterial immunity. Science 337:816–822
2. Yang L, Esvelt KM, Aach J et al (2013) RNA-guided human genome engineering via Cas9. Science 339:823–827
3. Cong L, Ran FA, Cox D et al (2013) Multiplex genome engineering using CRISPR/Cas systems. Science 339:819–823
4. Scadden DT (2006) The stem-cell niche as an entity of action. Nature 441(7097):1075
5. Rompolas P, Mesa KR, Kawaguchi K et al (2016) Spatiotemporal coordination of stem cell commitment during epidermal homeostasis. Science 7012:1–9
6. Ritsma L, Ellenbroek SIJ, Zomer A et al (2014) Intestinal crypt homeostasis revealed at single-stem-cell level by in vivo live imaging. Nature 507:362–365
7. Hara K, Nakagawa T, Enomoto H et al (2014) Mouse spermatogenic stem cells continually interconvert between equipotent singly isolated and syncytial states. Cell Stem Cell 14:658–672
8. Kretzschmar K, Watt FM (2012) Lineage tracing. Cell 148:33–45
9. Clayton E, Doupé DP, Klein AM, Winton DJ, Simons BD, Jones PH (2007) A single type of progenitor cell maintains normal epidermis. Nature 446:185
10. Doupé DP, Alcolea MP, Roshan A et al (2012) A single progenitor population switches behavior to maintain and repair esophageal epithelium. Science 337:1091
11. Lopez-Garcia C, Klein AM, Simons BD, Winton DJ (2010) Intestinal stem cell replacement follows a pattern of neutral drift. Science 330:822
12. Mount DW (2004) Bioinformatics: sequence and genome analysis. Cold Spring Harbor, NY: Cold Spring Harbor Laboratory Press
13. Klein AM, Simons BD (2011) Universal patterns of stem cell fate in cycling adult tissues. Development 138:3103
14. Angel A, Song J, Dean C, Howard M (2011) A polycomb-based switch underlying quantitative epigenetic memory. Nature 476:105–109
15. Luria S, Delbrück M (1943) Mutations of bacteria from virus sensitivity to virus resistance. Genetics 28s:491–511
16. Watson JD, Crick FHC (1953) Molecular structure of nucleic acids. Nature 171:737–738
17. Wilkins MHF, Stokes AR, Wilson HR (1953) Molecular structure of desoxypentose nucleic acids. Nature 171:738–740
18. Franklin RE, Gosling RG (1953) Molecular configuration in sodium thymonucleate. Nature 171:740–741
19. Pauling L, Corey RB (1953) A proposed structure for nucleic acids. Proc Natl Acad Sci 39:84–97
20. Bragg WH, Bragg WL (1913) The reflection of x-rays by crystals. Proc Roy Soc Lond 88:428–438
21. von Laue M (1913) Kritische Bemerkungen zu den Deutungen der Photogramme von Friedrich und Knipping. Phys Z 14:421–423
22. Popper K (1959) The logic of scientific discovery, vol 12. Hutchinson & Co.
23. Klein AM, Nakagawa T, Ichikawa R, Yoshida S, Simons BD (2010) Mouse germ line stem cells undergo rapid and stochastic turnover. Cell Stem Cell 7:214
24. Alcolea MP, Greulich P, Wabik A, Frede J, Simons BD, Jones PH (2014) Differentiation imbalance in single oesophageal progenitor cells causes clonal immortalization and field change. Nat Cell Biol 16:615
25. Frede J, Greulich P, Nagy T, Simons BD, Jones PH (2016) A single dividing cell population with imbalanced fate drives oesophageal tumour growth. Nat Cell Biol 18:967
26. Mascré G, Dekoninck S, Drogat B et al (2012) Distinct contribution of stem and progenitor cells to epidermal maintenance. Nature 489:257–262
27. Blanpain C, Simons BD (2013) Unravelling stem cell dynamics by lineage tracing. Nat Rev Mol Cell Biol 14:489–502
28. Soriano P (1999) Generalized lacZ expression with the ROSA26 Cre reporter strain. Nat Genet 21:70
29. Sauer B (1998) Inducible gene targeting in mice using the Cre/lox system. Methods 14:381
30. Potten CS (1974) The epidermal proliferative unit: the possible role of the central basal cell. Cell Tissue Kinet 7:77
31. van Kampen NG (2003) Stochastic processes in physics and chemistry. North-Holland Personal Library

32. Strogatz SH (1994) Nonlinear dynamics and chaos: with applications to physics, biology, chemistry, and engineering. Addison Wesley
33. Gillespie DT (1977) Exact stochastic simulation of coupled chemical reactions. J Phys Chem 81:2340
34. Doob JL (1990) Stochastic processes. Wiley-Interscience
35. Antal T, Krapivsky PL (2010) Exact solution of a two-type branching process: clone size distribution in cell division kinetics. J Stat Mech Theory Exp 2010:P07028
36. John PCL (ed) (1981) The cell cycle. Cambridge University Press
37. Feigenbaum MJ (1983) Universal behavior in nonlinear systems. Phys D Nonlinear Phenom 7:16–39
38. Barndorff-Nilsen OE, Cox DR (1989) Asymptotic techniques for use in statistics. Springer
39. Epps TW (2013) Probability and statistical theory for applied researchers. World Scientific Publishing Company
40. Klein AM, Doupé DP, Jones PH, Simons BD (2007) Kinetics of cell division in epidermal maintenance. Phys Rev E 76:021910
41. Chambers I, Silva J, Colby D et al (2007) Nanog safeguards pluripotency and mediates germline development. Nature 450:1230–1234
42. Greulich P, Simons BD (2016) Dynamic heterogeneity as a strategy of stem cell self-renewal. Proc Natl Acad Sci 113:7509
43. Box GEP, Tiao GC (1973) Bayesian inference in statistical analysis. John Wiley and Sons, New York
44. Jaynes ET (1957) Information theory and statistical mechanics. Phys Rev 106:620–630
45. Sivia D, Skilling J (1996) Data analysis: a bayesian tutorial. Claredon Press, Oxford
46. Jeffreys H (1946) An invariant form for the prior probability in estimation problems. Proc R Soc Lond A Math Phys Sci 186:453–461
47. Kirkpatrick S, Gelatt JR (1983) Optimization by simulated annealing. Science 220:671
48. Press WH, Teukolsky SA, Vetterling WT, Flannery BP (2007) Simulated annelaing methods. In: Numerical recipes: the art of scientific computing, 3rd edn. Cambridge University Press, New York

Chapter 6

Computational Tools for Quantifying Concordance in Single-Cell Fate

J. A. Cornwell and R. E. Nordon

Abstract

Cells are dynamic biological systems that interact with each other and their surrounding environment. Understanding how cell extrinsic and intrinsic factors control cell fate is fundamental to many biological experiments. However, due to transcriptional heterogeneity or microenvironmental fluctuations, cell fates appear to be random. Individual cells within well-defined subpopulations vary with respect to their proliferative potential, survival, and lineage potency. Therefore, methods to quantify fate outcomes for heterogeneous populations that consider both the stochastic and deterministic features of single-cell dynamics are required to develop accurate models of cell growth and differentiation. To study random versus deterministic cell behavior, one requires a probabilistic modelling approach to estimate cumulative incidence functions relating the probability of a cell's fate to its lifetime and to model the deterministic effect of cell environment and inheritance, i.e., nature versus nurture. We have applied competing risks statistics, a branch of survival statistics, to quantify cell fate concordance from cell lifetime data. Competing risks modelling of cell fate concordance provides an unbiased, robust statistical modelling approach to model cell growth and differentiation by estimating the effect of cell extrinsic and heritable factors on the cause-specific cumulative incidence function.

Key words Single-cell fate mapping, Concordance, Competing risks, Clonal analysis

1 Introduction

1.1 Stochastic and Deterministic Features of Single-Cell Dynamics

In order to study the dynamics of cell function, one requires a means to continuously observe single cells over days to weeks in culture. Time-lapse imaging and single-cell tracking are powerful technologies that enable one to record the entire life history of individual cells over extended time periods.

Time-lapse imaging and single-cell tracking generate cell pedigree data, which contain a record of cell fate outcomes (division, death, self-renewal, differentiation), motility, and kinship relationships. Used in combination with fluorescent protein reporters or other viable cell stains, cell tracking can be used to quantify the molecular dynamics of single cells over time and correlate gene expression with cell function. Importantly, continuous observation

Patrick Cahan (ed.), *Computational Stem Cell Biology: Methods and Protocols*, Methods in Molecular Biology, vol. 1975, https://doi.org/10.1007/978-1-4939-9224-9_6,

of single cells and their progeny allows one to study biological processes that span multiple generations of a cell lineage, e.g., lineage commitment or cellular reprogramming.

Importantly, single-cell dynamics show both deterministic and stochastic features of cell growth. Cells within a clonal population tend to have correlated fate outcomes, whereas fate outcomes for unrelated cells are independent. Logically, the underlying mechanism is related to ancestral memory, where equal proportions of protein, mRNA, miRNA, and other fate determinants are received by daughter cells at mitosis. Similarly, daughter cells receive DNA with high fidelity, with possible vertical transmission of epigenetic marks (DNA methylation, histone modification) that determine which genes can be read. Therefore, one may assume that concordant fate outcomes for related cells is a product of heritable fate determinants that are transmitted vertically through a pedigree. Importantly, correlation structures should be considered when modelling population dynamics as they have a strong influence on population dynamics. These findings have motivated the development of robust quantitative tools for quantifying concordant fate outcomes. The following section discusses contemporary methods for quantifying cell growth dynamics and the association in cell fate.

1.2 Quantitative Measures of Cell Fate Concordance

There are a number of statistics that quantify the association between two cell fate outcomes. The Pearson's correlation coefficient statistic allows one to quantify the temporal association in fate for individual fate outcomes [1]. However, Pearson's correlation test excludes discordant fate outcomes which bias results toward symmetric outcomes. Yule's Q measures the ratio of discordant to concordant outcomes but excludes right-censored outcomes [2]. In a typical cell-tracking experiment, there are usually many cells that are right-censored (cell fate was not observed during the observation period) [3]. Therefore, Yule's Q handles discordant fate outcomes but does not quantify temporal association in fate, while the intraclass correlation coefficient (ICC) statistic does.

The ICC statistic quantifies the temporal association in fate outcomes using permutation tests. For example, a bootstrapped Monte Carlo permutation test can be used to quantify the similarity in time to fate outcome by comparing unrelated cells and related cells in simulated data sets [4]. However, such an approach does not consider censored outcomes and is therefore prone to selection bias.

A general mathematical framework for describing the complex relationships between individual cell fates, heterogeneity, inheritance, and population level dynamics has been described using multi-type branching processes [5]. Cell heterogeneity, or stochastic behavior, was simulated using empirical or parametric probability distributions. Parametric distributions described random behavior using a family of probability distribution, e.g., exponential distributions. However, a more pragmatic approach would make use of empirical (nonparametric) probability distributions, which are based on the frequency of events. Nonparametric models are appropriate when nothing is known about the physical or chemical laws that govern system dynamics, for example, the Kaplan-Meier estimator [6] is commonly applied to the statistical analysis of survival data where there is a single fate such as death. Here we are looking at a more complicated situation where cells have multiple mutually exclusive fates, called competing risks.

These limitations are overcome by a branch of survival statistics, known as competing risks statistics, to quantify concordance in cell fate [3]. CR statistics were originally developed to quantify patient lifetime data which contain both censored and competing outcomes. By analogy, CR concordance statistics have utility in quantifying concordant fate outcomes for related cells in the presence of censored and competing fate outcomes [3]. Therefore, in the following sections, we demonstrate the application of CR statistics to quantify association in cell fate in a heterogeneous population with censored and competing fate outcomes.

1.3 Chapter Outline

This chapter will describe the core concepts underpinning CR statistics and discuss the motivation for its application to cell-tracking data. This is followed by a description of the process of constructing nonparametric and semi-parametric CR models to quantify the incidence of cell fate outcomes and concordance in cell fate. Therefore, the outline of this chapter is as follows:

Section 2: Materials including R libraries and cell lifetime datasets

Section 3: Introduction to CR statistics

Section 4: Worked example of CR statistics applied to cell lifetime data

Section 5: Notes and example R code

Readers interested in detailed mathematical concepts that are outside the scope of this chapter should review the work of Scheike and Zhang, who developed flexible methods to construct CR regression models for clinical patient data [7]. Additionally, the work of Scheike and Sun on concordance probability statistics to model heritable influences on lifetime events in monozygotic and dizygotic twins provides a comprehensive statistical overview of CR analysis [8].

2 Materials

The methods presented in this chapter are applied to cell lifetime data as generated by single-cell tracking experiments. The materials required include:

1. Cell lifetime dataset.
2. Statistical programming package R.
3. R libraries.

2.1 Cell Lifetime Data

Cell lifetime data is recorded from time-lapse films by single-cell tracking software. There are a number of tools available with different levels of functionality. It is assumed that cell-tracking data is saved in a specified tab-delimited format, as described below.

2.1.1 Downloading and Installing Libraries

1. Install *mets* (*version 0.1–13*), *cmprsk* (*version 2.2–7*), *survival* (*version 2.37–7*), and *timereg* (*version 1.7.0*) packages.
2. Initiate libraries in. R script file using library (*packagename*).

2.1.2 Importing Cell lifetime Data

1. Open R Studio.
2. Set working directory to folder containing data and code.
3. Import data into R (*see* **Note 1**).

2.1.3 Subset Data, Prepare Data for CR Analysis

Cell lifetime data generated from cell-tracking experiments should be loaded as a data frame containing the fields shown in Table 1.

Table 1
Example cell-tracking dataset and cell-tracking features

Cell tracking feature	Comments
CloneID	An integer value that is unique to cells belonging to the same pedigree
CellID	An integer value that specifies the location of an individual cell within a pedigree
Generation number	An integer value that specifies the how many prior divisions tracked before the current cell
Fate outcome	An integer value indicating the number of divisions preceding the birth of the cell. The founding cells of a pedigree have a generation number = 0
Lifetime	A cell's lifetime is defined as the time elapsed between the birth of the cell and the end of the cell's track. A cell's track ends when one of the fate outcomes above is realized. A continuous variable (usually expressed in minutes, hours, or days)
Treatment condition/ covariate	Cells tracked under different treatment conditions are grouped using a categorical variable that defines distinct treatment groups
Kinship cluster	Kinship cluster variables are used to identify pairs of related cells within a pedigree. For each kinship relationship (sisters, mother-daughters, first cousins, second cousins, etc.), a unique set of kinship cluster variables exist, such that each pair of related cells has the same cluster variable

3 Methods

3.1 Competing Risks Statistics

Competing risks (CR) are defined as mutually exclusive event outcomes called risks for time-to-event data [9, 10]. Originally defined in the context of clinical medicine, CR was as an advancement to the KM estimator and Cox regression models [11] for estimating the efficacy of treatment conditions on patient survival where there were multiple risks (e.g., death from disease or treatment). Importantly, while the KM estimator includes right-censored outcomes, it overestimates the probability of the event of interest in the presence of a competing event.

Competing risks statistics are used to model the failure time caused by one of several causes [12]. In such a scenario, the distribution of event times for the population sample is estimated. One may, in addition, estimate the effect of treatments or other groupings, called covariates, on the distribution of event times for each risk. The cumulative incidence function is used to estimate the probability of an event resulting from a specific cause (e.g., cause = 1), before and including time t. Expressed in mathematical form the cumulative incidence function is defined as

$$F_1(t) = \Pr(T \leq t, \epsilon = 1)$$

where T is a random variable for the event time and ϵ indicates the event cause (risk). The cause-specific probability density function is also defined $f_i(t) = \frac{\mathrm{d}F_i(t)}{\mathrm{d}t}$.

The other quantities of interest are the cause-specific hazard, which is the instantaneous failure rate at time t given the subject is "at risk," i.e., the failure time $T \geq t$:

$$\lambda_1(t) = \lim_{\Delta t \to 0} \frac{\Pr(T \in [t, t + \Delta t], \epsilon = 1 \mid T \geq t)}{\Delta t}$$

The integral of the hazard is called the cumulative hazard:

$$\Lambda_i(t) = \int_0^t \lambda_i(s)\mathrm{d}s$$

The cause-specific hazard is related to the cause-specific probability distribution:

$$f_1(t) = \lim_{\Delta t \to 0} \frac{\Pr(T \in [t, t + \Delta t], \epsilon = 1 \mid T \geq t)}{\Delta t} \Pr(T \geq t) = \lambda_1(t)S(s-)$$

$$F_i(t) = \int_0^t \lambda_1(t)S(s-)\mathrm{d}s = \int_0^t S(s-)d\Lambda_i(s)$$

where $S(s) = \Pr(T > s)$ is the survival probability from all causes, and $S(s-) = \Pr(T \geq s)$.

The subtle distinction between $S(s)$ and $S(s-)$ is more relevant for empirical (nonparametric) survivor functions. They are discontinuous, decreasing step functions, with right continuous with left limits. Therefore, the left limit denoted by $S(s-)$ is the proportion of subjects (or cells) that are at risk just before time s.

One can model the effect of covariates on the cause-specific hazard using the proportional hazards semi-parametric model [13]

$$\lambda_i(t; \boldsymbol{Z}) = \lambda_i^0(t)\exp(\beta_i \boldsymbol{Z})$$

where $\boldsymbol{Z}$ is the covariate vector. The baseline cause-specific hazard function $\lambda_i^0(t)$ is a nonparametric estimate, while cause-specific parameter vector β_i models time-invariant constant effects, using the proportional hazards model $\{\exp(\beta_i \boldsymbol{Z})\}$.

However, the effect of a covariate on the cause-specific hazard may be very different to its effect on the cause-specific cumulative hazard [10]. Indeed, investigators are mostly interested in determining the effect of a covariate on the cause-specific cumulative incidence function. Fine and Gray developed a proportional hazards model for the cause-specific cumulative incidence function [9]. A link function $g(\cdot)$ is chosen to transform $F_i(t)$ to satisfy linearity

$$g\{F_i(t; \boldsymbol{Z})\} = h_i^0(t) + \beta_i \boldsymbol{Z}$$

such that when two transformed cumulative incidence functions with covariates $\boldsymbol{Z}_1$ and $\boldsymbol{Z}_2$ are subtracted from one another, i.e., $g\{F_i(t; \boldsymbol{Z}_1)\} - g\{F_i(t; \boldsymbol{Z}_2)\} = \beta_i \boldsymbol{Z}_1 - \beta_i \boldsymbol{Z}_2$, the baseline function $h_i^0(t)$ should vanish. This means that the covariates $\boldsymbol{Z}$ have a constant time-invariant effect. The inverse function $g^{-1}(\cdot)$ when applied to $h_i^0(t)$ gives the baseline cumulative incidence function for $\boldsymbol{Z} = \boldsymbol{0}$. The Fine and Gray paper suggested using the proportional hazards model $g[F_i(t; \boldsymbol{Z})] = log\{-log(1 - F_i(t; \boldsymbol{Z}))\}$. The Fine and Gray model will not fit the data if there are time-varying covariate effects, i.e., β_i vary over time.

Scheike developed a comprehensive semi-parametric modelling approach addressing time-varying covariate effects [7]. A goodness-of-fit test for time invariance was implemented to evaluate link functions and select the optimal combination of parametric and nonparametric covariates:

$$g\{F_i(t; \boldsymbol{X}, \boldsymbol{Z})\} = \boldsymbol{\eta}_i(t)\boldsymbol{X} + \beta_i \boldsymbol{Z}$$

where $\boldsymbol{X}$ and $\boldsymbol{Z}$ are covariate vectors describing the nonparametric and parametric components of the model, respectively. The $\boldsymbol{X}$ covariates account for time-varying effects, while the $\boldsymbol{Z}$ covariates describe constant effects. The paper suggests a method for selection of a model, starting with the least parsimonious, nonparametric model (no $\boldsymbol{Z}$ covariates). If the jth component of $\boldsymbol{\eta}_i(t)$, $\eta_{ij}(t)$, is time invariant, one can substitute the nonparametric covariate X_j with a parametric covariate Z_j. This is repeated iteratively till all components of $\boldsymbol{\eta}_i(t)$ are time variant. This results in a model which has the fewest estimated coefficients. This flexible modelling approach is implemented by the R packages *comp.risk* and *timereg*.

3.2 The Concordance Function

The concordance function is a time-dependent function used to estimate whether the cause-specific cumulative incidence for individuals within a related group called a cluster (e.g., twins) are correlated. Statistical test of significance relies on comparing the concordance function for members within a cluster to randomly selected individuals. One may also wish to know whether the temporal association of a risk within the cluster is influenced by a covariate, for example, to estimate the cumulative incidence of breast cancer (BRCA1 mutation) while accounting for competing mortality as well as within-family correlation [14].

The concordance function for cause 1 (e.g., cancer) is defined as a bivariate probability distribution

$$\mathsf{C}(t) = P_{1,1}(t,t) = P(T_1 \leq t, \epsilon_1 = 1, T_2 \leq t, \epsilon_2 = 1)$$

where T_{ki} and C_k are the event time and the right censoring time for the ith individual ($i = 1, 2$) in the kth twin pair. The events are $\epsilon_{ki} \in \{1, \ldots, J\}$ and for notational convenience the cluster is fixed such that $T_1 = T_{1k}$.

The dependence in fate may be easily inferred from analysis of the lifetime concordance, which is given by,

$$P_{1,1}(\infty,\infty) + P_{1,2}(\infty,\infty) + P_{2,1}(\infty,\infty) + P_{2,2}(\infty,\infty) = 1$$

Estimation of the concordance function given a set of covariates X is achieved by the following:

$$\mathsf{C}(t; X) = P_{1,1}(t, t; X) = P(T_1 \leq t, \epsilon_1 = 1, T_2 \leq t, \epsilon_2 = 1 | X)$$

and a specific semi-parametric regression structure,

$$\mathsf{C}(t; X) = h\big(\Lambda(t), X^T\beta, t\big)$$

where h is a known link function such as the logit link

$$h(x, y, t) = \left(\frac{e^{(x+y)}}{1 + e^{(x+y)}}\right)$$

and $\Lambda(\mathrm{t})$ is an increasing baseline function and β is a set of regression coefficients.

3.2.1 Tests and Casewise Concordance

The concordance function described above allows determination of the casewise concordance. Put simply, the casewise concordance is the probability that a twin developed cancer ($\epsilon_1 = 1$) before time t given that the other twin had cancer ($\epsilon_2 = 1$) before time t:

$$\mathsf{C}_p(t; X) = \mathsf{C}(t; X)/F_1(t; X)$$

In the case where there is no censoring present within the dataset, then the above estimator can be computed by

$$n_{11}(t) \Big/ \left(n_{11}(t) + \frac{1}{2} n_d(t) \right)$$

where $n_{11}(t)$is the number of concordant twin pairs at age t, that is, twin pairs where both have cancer, and $n_d(t)$ is the number of discordant twin pairs at age t (pairs where exactly one twin in a twin pair has cancer). However, the above estimator becomes severely biased when applied to censored competing risks data [12].

3.3 The Cross-Odds Ratio

Scheike et al. proposed a cross-odds ratio (COR) function to measure the association of cause-specific event times within a cluster [15]. The COR was used to determine if a cell's fate depends on the fate of a related cell. Therefore, the COR can provide evidence that a cell's fate is heritable and to determine if sibling cell fates are symmetric. The cross-odds ratio is used to judge whether fates are concordant (symmetric) or discordant (asymmetric).

The odds of event A is defined as the probability of event A divided by the probability of its complement:

$$\text{ODDS}[A] = \frac{\Pr[A]}{1 - \Pr[A]}$$

The conditional odds of event A given B is defined

$$\text{ODDS}[A|B] = \frac{\Pr[A|B]}{1 - \Pr[A|B]}$$

And the cross-odds ratio is

$$\text{XODDS}[A, B] = \frac{\text{ODDS}[A|B]}{\text{ODDS}[A]}$$

In this context, condition A is where cell i belonging to cluster k would "fail" from cause e_i before time t $\left[A =^{\text{def}} \{T_{ki} \leq t,\} \cap \{\varepsilon_{ki} = e_i\}\right]$ where T_{ki} and ε_{ki} are random variables denoting the event time and cause, respectively. Likewise, condition B is a failure of cell j from cause e_j before time t $\left[A =^{\text{def}} \{T_{kj} \leq t,\} \cap \{\varepsilon_{kj} = e_j\}\right]$.

If events within the cluster are independent, then ODDS $[A] = \text{ODDS}[A|\ B]$ and $\text{XODDS}[A,\ B] = 1$. If events are concordant, then the occurrence of event B increases the odds of event A, so $\text{XODDS}[A,\ B] > 1$. If cells are sibling pairs, then synchronous events have a COR that is significantly greater than 1. If events are discordant, occurrence of event B decreases the odds of event A, so $0 < \text{XODDS}[A,\ B] < 1$. A cross-odds ratio that is significantly less than 1 indicates asymmetric fates that could not occur by chance alone.

In this chapter the degree of concordance of cells within a cluster is shown by graphing the cumulative distributions $\Pr[A]$ and $\Pr[A|\ B]$ (probandwise concordance). If $\Pr[A|\ B]$ are within the confidence interval of $\Pr[A]$, then one has less confidence that cause-specific events within a cluster are dependent. Thus one needs to compare the conditional distribution $\Pr[A|\ B]$ with the unconditional distribution $\Pr[A]$ to test whether symmetric or asymmetric events occur by chance alone.

4 Example

4.1 Background

In this section we demonstrate how CR concordance may be used to quantify the rate of symmetric and asymmetric divisions within a heterogeneous population of self-renewing cardiac stem cells. Computational tools to construct CR models have been implemented in the statistical programming language R. We draw upon existing lifetime data that is publicly available in an online repository (https://github.com/Jamcor/crpaper) describing the response of cardiac stem cells to individual cytokines and cytokine combinations in factorial design experiments [3].

Self-renewal is a mitotic event in which at least one daughter inherits the stem/progenitor cell traits of its mother, which then persists throughout the daughters' lifetime. At the time a cell divides, it is impossible to determine if it has undergone a self-renewing division, since the definition of self-renewal requires that its daughter cell has maintained its traits throughout its lifetime. Therefore, it's only possible to study self-renewal by single-cell pedigree analysis.

In this context we are interested in applying CR concordance to quantify the probability of cell self-renewal under different cytokine treatments. To understand how cardiac stem cells self-renew in serum-free medium (SFM), the concordance in fate outcomes for sister cells was compared. The CR models presented below were developed in order to address this question. The analysis presented below was used to show that the dominant mode of cardiac stem cell division was symmetric amplifying divisions, rather than asymmetric divisions.

Briefly, the experimental design is as follows. Cardiac colony-forming unit-fibroblast was isolated from adult mouse hearts and cultured for 96 h in serum-free medium (SFM) with either no cytokines or in the presence of 20 ng/mL platelet-derived growth factor, 2 ng/mL basic fibroblast growth factor. A total of 733 cells were tracked of which 327 divided (44.6%), 57 died (7.7%), and 349 were right-censored (47.7%). A more detailed description of the experimental design may be found in [3].

4.2 Method

Constructing CR models in R is an iterative process, requiring some trial and error to find the most suitable form for the model. This section describes a general workflow for building CR models, provides basic R syntax, and includes methods for visualizing data. It is assumed that data has already been loaded to the R workspace and subset as described in Subheading 2.1.3. Firstly, the process is described in general terms and then a detailed example including R code is included.

4.2.1 Constructing Semi-parametric CR Regression Models

CR models are represented using Wilkinson-Rogers notation, as shown below:

$$Y \sim 1 + C_1X_1 + C_2X_2 + C_3X_3 + C_4X_4 + C_5(X_1 : X_2)$$

where Y is the dependent (response) variable, X is a set of covariates ($X_1 \ldots X_n$), and C is a set of coefficients ($C_1 \ldots C_n$) to be estimated by the model. Interactions between covariates are separated by a semicolon (e.g., the interaction of X_1 and X_2 is denoted X_1:X_2). Once a model has been constructed that contains the covariates of interest, the following steps are followed:

1. Select a link function:
   (a) Additive, proportional, or Fine-Gray link functions are available in *comp.risk.*
2. Test covariates for time-invariant effects (Kolmogorov-Smirnov) test (*see* **Note 2**):
   (b) Time-variant covariates are modelled with nonparametric coefficients.
   (c) Time-invariant parameters are modelled with parametric coefficients.
3. Repeat **steps 1** and **2** until the number of time-invariant parameters has been maximized.
4. Estimate parameters for time-invariant effects and nonparametric coefficients (*see* **Note 3**).

4.2.2 A Semi-parametric CR Model that Quantifies the Effects of Cytokines on Cardiac Stem Cell Division and Death

Here, we present example CR models used to quantify division and death outcomes in cardiac stem cells treated with basic fibroblast growth factor (bFGF) and platelet-derived growth factor (PDGF). We draw upon existing cell lifetime data available from Cornwell et al. [3] to model the effect of covariates on the cumulative incidence function (CIF). We start by developing nonparametric CR models with division and death as specific outcomes of interest to test for time-invariant effects (*see* **Note 2**: *constructing a non-parametric CR model to test covariates for time-invariant effects*). The nonparametric models tested include:

$$\text{CIF} \quad \sim 1 + (\text{PDGF}) * (\text{FGF}) \tag{1}$$

$$\text{CIF} \quad \sim 1 + \text{const}(\text{PDGF}) * \text{const}(\text{FGF}) \tag{2}$$

Based on the results shown in Table 2, we can see that the additive link function maximized the number of time-invariant

Table 2
Results from the tests for nonparametric terms for a range of different link functions applied to a nonparametric CR regression model for division and death outcomes

	Additive link function	
Covariate	**Kolmogorov-Smirnov test**	**p value (H_0 = constant effect)**
CR model for division		
Intercept	0.292	0.27
PDGF	0.173	0.92
FGF	0.248	0.55
PDGF:FGF	0.338	0.68
Proportional link function		
Intercept	2.12	0.01
PDGF	1.65	0.28
FGF	1.16	0.33
PDGF:FGF	2.34	0.12
Fine-Gray link function		
Intercept	0.77	0.03
PDGF	0.547	0.33
FGF	0.407	0.38
PDGF:FGF	0.282	0.82
CR model for death		
Intercept	0.0553	0.36
PDGF	0.1460	0.31
FGF	0.0752	0.75
PDGF:FGF	0.199	0.22
Proportional link function		
Intercept	2.58	0.01
PDGF	0.908	0.39
FGF	1.34	0.23
PDGF:FGF	1.82	0.14
Fine-Gray link function		
Intercept	0.0758	0.33
PDGF	0.14	0.25
FGF	0.273	0.07
PDGF:FGF	0.421	0.04

Table 3
Results for parametric terms for their effect on division and death outcomes

Covariate	Additive link function			
	Coefficient	SE	*z*-score	*p* value
CR model for division				
PDGF	0.097	0.045	2.183	0.029
FGF	0.1	0.05	2.884	0.004
PDGF:FGF	0.014	0.067	0.214	0.83
CR model for death				
PDGF	0.038	0.018	2.099	0.036
FGF	0.057	0.022	2.635	0.008
PDGF:FGF	−0.07	0.028	−2.522	0.012

effects and thus is the most appropriate selection for use in a semi-parametric CR model. We construct a semi-parametric model with an additive link function to investigate division and death outcomes by adding const() wrappers to each covariate, as shown in Eq. 2 (*see* **Note 3**: *testing covariates for their effect on the cumulative incidence function*).

After the model has been evaluated, one is able to view a summary that contains the result of statistical tests for the effect of each covariate (*see* **Note 3**: *testing covariates for their effect on the cumulative incidence function*). The results from Eq. 3 are summarized in Table 3. These results provide strong evidence that PDGF and FGF both have positive effects on the cumulative incidence of division. FGF and PDGF alone had a positive effect on death, while the combination of PDGF:FGF alone had a positive effect on survival.

We can use the same approach to quantify the effect of any set of covariates and their interactions. In the following section, we describe methods to simulate and visualize the impact of covariates on the cumulative incidence function.

4.3 Simulating and Visualizing Cumulative Incidence Functions and Testing for Goodness of Fit

Semi-parametric CR models, such as Eq. 3, will allow us to simulate how covariates will influence the cumulative incidence of a cause-specific event. This is achieved in practice by using the *predict* function (*see* **Note 4**: *simulating a cumulative incidence function given a set of covariates*).

The output from the predict function contains estimates of the cumulative incidence function that can be plotted, as shown in Fig. 1 (*see* **Note 5**: *plotting a cumulative incidence function*). To demonstrate this we simulate CIFs which show the incidence of division without cytokines (control), with individual cytokines (PDGF and FGF), and with cytokine interactions (PDGF:FGF). Similarly, we simulate CIFs for death under the same condition, shown in the bottom panel of Fig. 1a.

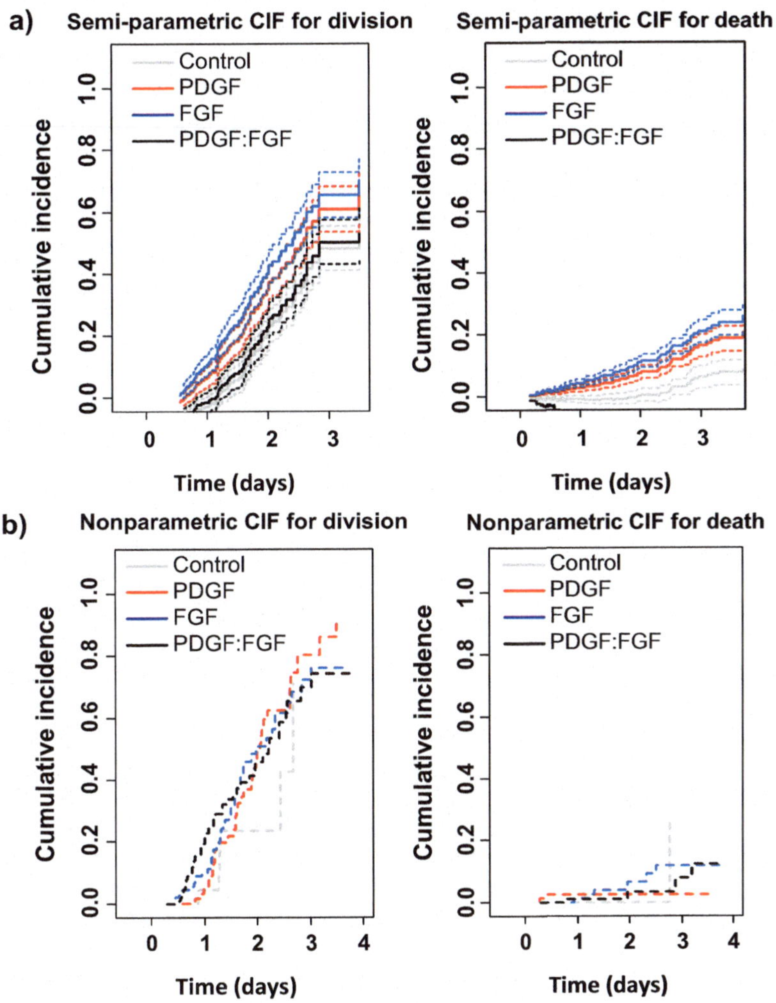

Fig. 1 Estimates of the cumulative incidence function. (**a**) Semi-parametric CIFs showing the incidence of division and death in cultures of cardiac stem cells left untreated (gray), treated with PDGF (red), treated with FGF (blue), and treated with both PDGF and FGF (black). Dotted lines represent 95% confidence intervals (CI). (**b**) Nonparametric CIFs showing the incidence of division and death in cultures of cardiac stem cells left untreated (gray), treated with PDGF (red), treated with FGF (blue), and treated with both PDGF and FGF (black)

To test for goodness of fit, we compare the nonparametric estimate of the cumulative incidence function with the semi-parametric CIF [16]. Nonparametric CIFs can be estimated using the *cuminc* function (*see* **Note 6**: *empirical estimation of the cumulative incidence function*). As shown in Fig. 1b, we find a good agreement between our model and our experimental observations.

Now that we have demonstrated that the CRR model in Eq. 3 appropriately fits our data, in the following section, we will extend the model so that we can use it to study concordance in cell fate and how covariates may impact cell fate concordance.

4.4 Application of CR Models to Quantifying Concordance in Cell Fate

We have covered how to construct CRR models that quantify the effect of covariates on competing and possibly censored fate outcomes, such as division and death; we are now prepared to extend our analysis to quantify the influence of covariates on the association in fate outcomes in cellular kin. As above, we will start with a general overview of the process, highlighting the key steps and making note of potential pitfalls. Once we have described the general details of the process, we will provide an example of how to quantify concordance using CR.

Measures of association in cell fate outcomes include Pearson's correlation, ICC, and Yule's Q test. For the cardiac stem cell lifetime data analyzed in this chapter, we can show the association in division and death times for sibling cells using Pearson's correlation (Fig. 2a). However, this test only included paired observations, i.e., when both cells have known fate outcomes and known lifetimes. Out of a total of 273 sibling pairs, 84 pairs both divided (30.7%); 10 pairs both died (3.7%); 179 pairs were right-censored or had discordant outcomes (65.5%). Therefore, this analysis excludes a significant amount of data. Similarly, while Pearson's correlation estimates no association for mother-daughter pairs (Fig. 2a), a large number of discordant and right-censored outcomes are excluded.

The probabilistic association in cell fate may be obtained by modelling the association in cell fate for sibling cells as a bivariate probability distribution, known as a concordance probability distribution. As we highlighted in Subheading 3, this approach has the advantage that the concordance function can be estimated from competing and right-censored observations. Another significant advantage is that we can do regression on the concordance function to quantify the effect of covariates on the association in cell fate (*see* Subheading 4.5).

The process for developing CR concordance models follows the same process as describe above except an additional variable is added to the CR model that identifies kinship clusters (*see* **Note 7**: *constructing a CR concordance model*). In the cellular context, a cluster is defined as any pair of cells that share a common ancestor and are separated by the same familial distance (i.e., are in the same generation). Clusters include siblings, mothers, and daughters, as well as cousins (1st, 2nd, 3rd, etc.).

At first we will demonstrate how such a model can be applied to quantify concordance in division outcomes for sibling cells. We start the process by introducing a clustering variable to Eq. 2 that uniquely identifies sibling cell pairs, as shown in Eq. 3 below.

$$\begin{aligned}\text{CIF} \quad &\sim 1 + \text{const(PDGF)}*\text{const(FGF)} \\ &+ \text{cluster(ClusterID)}\end{aligned} \tag{3}$$

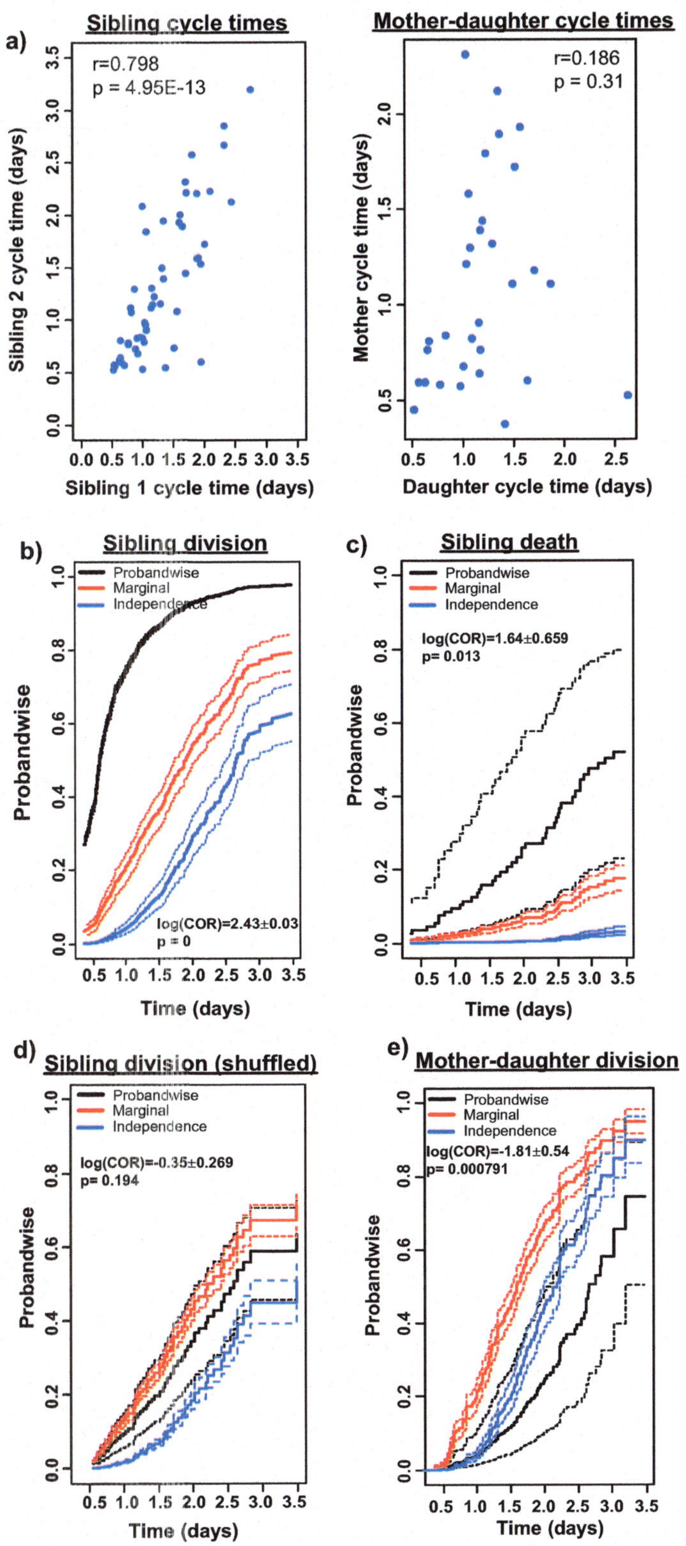

Fig. 2 Influence of covariates on the association in fate outcomes in cellular kin. (**a**) Bivariate dot plots showing the association between sibling cell cycle times

This clustering variable allows us to fit a parametric model for the log-cross-odds ratio that will describe the predictive effect of one sibling undergoing division given that its sibling has already divided (*see* Subheading 3.3 and **Note 8**: *calculating the cross-odds ratio*). Put simply, the COR provides a measure of how the timely occurrence of division in one cell influences the odds of division for its sibling [15]. This approach also allows us to estimate the concordance function over time (*see* **Note 9**: *plotting probandwise concordance*). Therefore, we can calculate the COR and estimate the concordance function for different kinship clusters as a means to study possible associations in cell fate outcomes for related cells.

The results from this model showing concordance in sibling division and death are shown in Fig. 2b, c. In all panels the marginal probability of division is shown in red; the unconditional probability (squared marginal) distribution is shown in blue and represents independence; and the probandwise concordance probability is shown in black and represents the probability that a cell will divide given that its sibling has already divided. One can see that the concordance probability of division and death for siblings is substantially higher than the marginal distribution, indicating strong concordance.

The *log(COR)* for sibling cell division was 2.43 ± 0.03 ($p < 0.001$), indicating that sibling cells have strongly concordant outcomes for division. By changing the event of interest in our CR model (*see* **Note 3**: *testing covariates for their effect on the cumulative incidence function*), we compared concordance in death outcomes and found that death outcomes for sibling cells are also concordant with a $\log(\text{COR}) = 1.64 \pm 0.659$ ($p = 0.013$). These results clearly demonstrate that the molecular machinery involved in preparing a cell for division and death is common between siblings since the degree of concordance we observe would not be observed by chance alone. This result is also in agreement with Fig. 2a.

Fig. 2 (continued) (left panel) and mother-daughter cycle times (right panel). r = Pearson's correlation coefficient. (**b**) Probandwise (left panel) probability for division in sibling cells. Probandwise concordance probability (black), marginal distribution of division (red), and independence (blue). (**c**) Probandwise (left panel) probability for death in sibling cells. Probandwise concordance probability (black), marginal distribution of death (red), and independence (blue). (**d**) Probandwise (left panel) probability for division in randomly assigned sibling cells. Probandwise concordance probability (black), marginal distribution of division (red), and independence (blue). (**e**) Probandwise (left panel) probability for division in mother-daughter cells. Probandwise concordance probability (black), marginal distribution of division (red), and independence (blue). Dotted lines represent 95% confidence intervals (CI)

We will take a moment to demonstrate this point by quantifying the COR and concordance probability for a simulated data set in which sibling pairs have been randomly assigned. After randomly assigning cells to a sibling cluster, we estimated the COR and probandwise concordance. Not surprisingly, randomization means that division outcomes were not significantly concordant (Fig. 2d, $\log(\text{COR}) = -0.35 \pm 0.269$, $p = 0.194$).

Next, we will consider concordance of mother-daughter pairs. While sibling cell cycle times are highly correlated, mother-daughter cycle times can be directly correlated, inversely correlated, or independent. The Pearson's correlation analysis in Fig. 2a suggests that there is no association in division outcomes for mother-daughter pairs.

To model mother-daughter concordance we substitute the clustering variable in Eq. 3 with a new clustering variable that identifies mother-daughter pairs (*see* **Note** 7: *constructing a competing risks concordance model*). Because each mother can be counted twice (one for each mother-daughter pair), it is necessary to remove duplicate measurements of mother cycle times. Here, it is also important to make a change to the symmetry condition of the *cor.cif*, since symmetry is not satisfied for mother-daughter pairs. Siblings are considered to be symmetric since interchanging lifetime data for sibling 1 and sibling 2 would not result in a different COR. In contrast, mother-daughter cycle times cannot be interchanged and result in the same COR.

The probandwise concordance mother-daughter pairs are shown in Fig. 2e. We can clearly see that the probandwise concordance is much lower than the squared marginal (black versus blue line), which represents perfect independence. Accordingly, we calculated a $\log(\text{COR}) = -1.81 \pm 0.54$ ($p = 0.00079$). A significantly negative COR indicates that mother-daughter division outcomes are discordant, which was not detected by the Pearson's correlation coefficient because Pearson's correlation only included paired observations and thus excluded right-censored outcomes.

So far we have demonstrated how to quantify concordance in a specific fate outcome for different kinship clusters. We have focused on sibling clusters and mother-daughter clusters, though the same approach can be used for other clusters including cousins, aunt-niece, etc. [3]. In the following section, we demonstrate how the competing risks regression models we have developed can be used to quantify how intrinsic and extrinsic factors influence cell fate concordance. A more detailed discussion on the mathematical framework underpinning the cross-odds ratio, which is outside the scope of this chapter, can be found in Scheike et al. [15].

4.5 Regression on Covariates Influencing the Cumulative Incidence and Concordance Function

Contemporary approaches for quantifying concordance in cell fate do not allow statistical testing for differences between treatment groups or to model the effect of covariates on concordance in cell fate. In this section we demonstrate how to quantify the effect of covariates on concordance by regression modelling of the cross-odds ratio that we introduced in Subheading 4.3.

To demonstrate regression on covariates that influence the concordance function, we work through an example where we quantify the effect of *Pdgfra*-GFP expression on concordance in cCFU-F division. GFP marks self-renewing stem/progenitor cCFU-F and loss of GFP expression indicates the onset of differentiation [3]. Thus, it is likely that kindred cells with differential expression of GFP will be more different in their cell cycle dynamics than kindred cells pairs that both express GFP. Reflecting on the biological context, we would interpret this result to mean that concordance is lost upon differentiation (indicated by loss of GFP expression). Therefore, the hypothesis we are testing is that concordance in division will be higher for kinship pairs with common GFP expression. We test this hypothesis for siblings and mother-daughter pairs.

First, it is necessary to establish a threshold for GFP expression in order to classify cells as GFP^{+} or GFP^{-}. We define our threshold as the mean fluorescent intensity measured at the end of a cell's lifetime below which cells are not responsive to treatment with PDGF. This is a functional classification for cells that express PDGFRα and respond to treatment with PDGF [3]. The reason that GFP expression is measured at the end of each cell's lifetime is because some cells that were born GFP^{+} transitioned to a GFP^{-} state [3].

After establishing a threshold, we classified kinship pairs based on the GFP status and included this classification as a covariate in our regression model. Regression on GFP expression will allow us to make a statement regarding the effect of GFP expression on the degree of concordance for kindred cells.

We define three types of kinship pairs:

1. **GFP^{+} twins:** kinship pairs (sibling pairs or mother-daughter pairs) where both cells in the pair are GFP^{+}.
2. **GFP^{-} twins:** kinship pairs (sibling pairs or mother-daughter pairs) where both cells in the pair are GFP^{-}.
3. **Mixed pairs:** kinship pairs where one cell in the pair is GFP^{-} and the other cell in the pair is GFP^{+}. **Note:** For mother-daughter pairs, the mother was always GFP^{+} and the daughter was always GFP^{-}, since GFP^{-} mothers did not give rise to any GFP^{+} daughters.

Next, we specify a regression design for the COR parameters that includes a term for GFP status (*see* **Note 10:** *how to construct a* *design matrix*). Using the *cor.cif* we fit a regression model that includes GFP status as a categorical variable (*see* **Note 11:** *regression on covariates influencing concordance*). We can then estimate the probandwise concordance that results from our regression model using the *concordance* function (*see* **Note 9**: *plotting probandwise concordance*).

Effect of covariates on sibling cell concordance. The cross-odds ratio and results of statistical tests (*see* **Note 8**: *calculating the cross-odds ratio*) are summarized in Table 4 for each covariate tested. The probandwise concordance probability distributions showing the effect of covariates on sibling cell concordance are shown in Fig. 3.

We find that GFP status had a significant effect on sibling division concordance ($\log(\mathrm{COR}) = 1.24 \pm 0.03, p = 0$). Inspection of Fig. 3a shows that GFP^{++} twins had very strong concordance ($\log(\mathrm{COR}) = 2.79 \pm 0.03, p = 0$) in division outcomes, while GFP^{-} twins showed no association ($\log(\mathrm{COR}) = -0.17 \pm 0.67, p = 0.806$). As seen in the last panel of Fig. 3, mixed pairs trend toward discordant (independent) outcomes ($\log(\mathrm{COR}) = -0.369 \pm 0.45, p = 0.417$). We attribute the influence of GFP status on kinship correlations to the relationship between GFP status and cell cycle; i.e., GFP^{+} cells have a greater incidence of division than GFP^{-} cells.

Table 4
Results from regression on covariates influencing concordance and the cross-odds ratio

	Siblings					
Covariate	Link function	Log (coefficient)	SE	z-score	p value	Cross-odds ratio
Effect of covariates on sibling and mother-daughter division concordance						
No covariate (baseline concordance)	Additive	1.23	0.521	2.37	0.018	3.43
PDGF	Additive	0.259	1.05	0.246	0.81	1.3
FGF	Additive	−0.368	0.933	−0.395	0.69	0.69
Generation #	Additive	−1.20	0.485	−2.47	0.014	0.302
GFP status	Additive	1.24	0.0328	37.8	0	3.46
Shuffled GFP status	Additive	0.397	0.449	0.886	0.376	1.49
Shuffled generation #	Additive	−0.07	0.18	−0.388	0.698	0.932
Clone ID	Additive	−0.03	0.0692	−0.435	0.663	0.97
Mother-daughter pairs						
No covariate (baseline concordance)	Additive	−1.81	0.54	−3.36	0.000774	0.163
PDGF	Additive	0.54	1.25	0.431	0.666	1.72
FGF	Additive	0.112	1.24	0.0899	0.928	1.12
Generation #	Additive	0.317	0.792	0.4	0.689	1.37
GFP status (GFP^{-+} rel. to GFP^{++})	Additive	−3.09	1.4	−2.2	0.0276	0.0453
Shuffled GFP status	Additive	1.33	0.969	1.38	0.169	3.79
Shuffled generation #	Additive	0.826	0.871	0.948	0.343	2.28

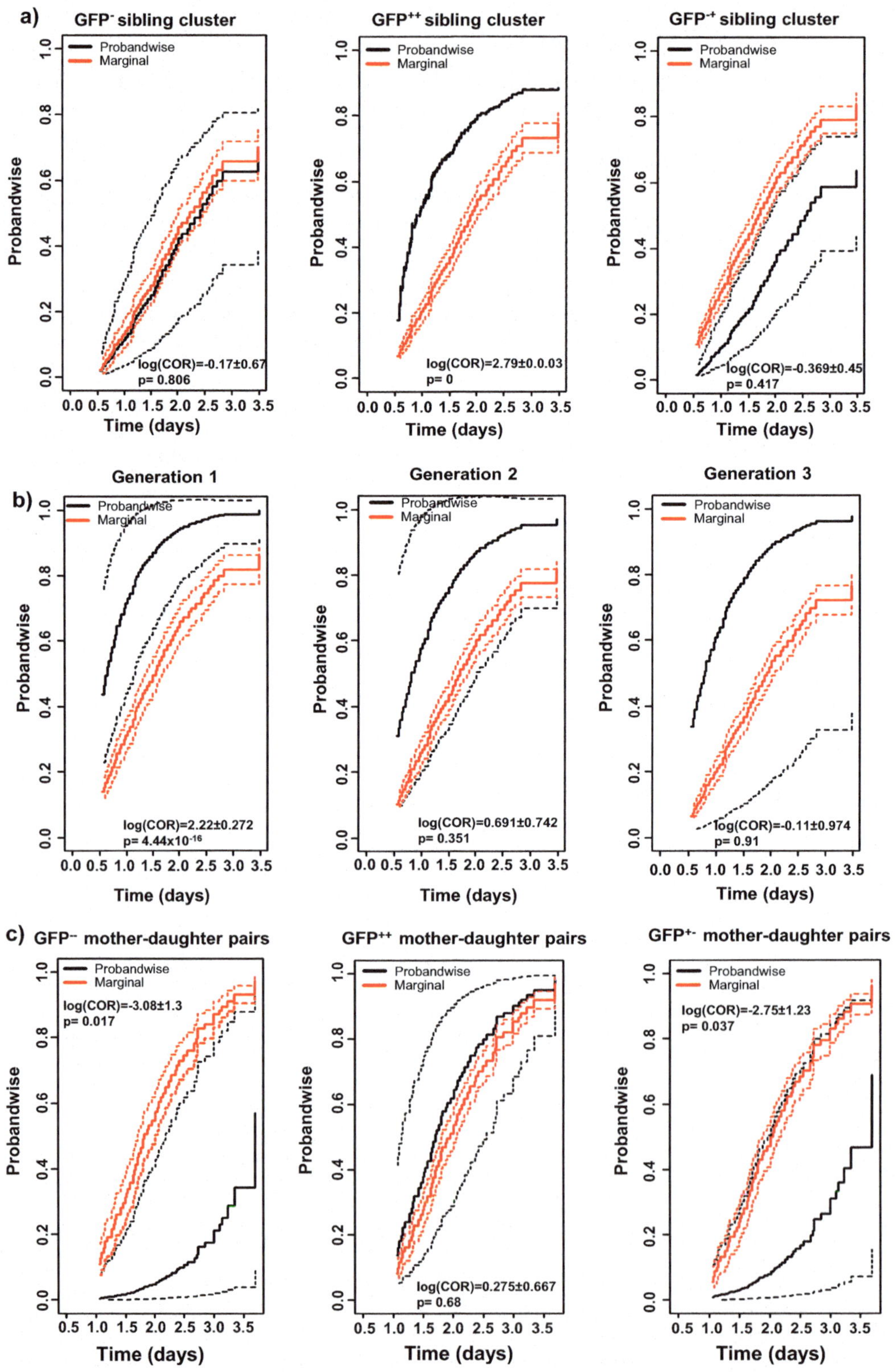

Fig. 3 Effect of covariates on sibling cell concordance. Estimated probandwise concordance functions showing concordance in division outcomes for sibling cells classified as GFP^{+} twins, GFP^{-} twins, or mixed pairs. (**a**) Simulated probandwise concordance function for GFP^{-}sibling twins (left panel), GFP^{++} sibling twins (middle

We then used this same approach to screen the effect of other extrinsic and intrinsic factors on concordance and the COR. We test the influence of (1) cytokine treatment, (2) generation number (all cells pooled), and (3) clone ID. The results from statistical tests for each of these covariates are summarized in Table 4. FGF, PDGF, and clone ID had no significant effects on sibling division concordance. However, generation number had a negative effect on sibling concordance (log(coefficient) $= -1.2 \pm 0.485$, $p = 0.014$), as can be seen from the probandwise concordance shown in Fig. 3b.

Effect of covariates on mother-daughter concordance. Importantly, the discordance that we measured for GFP^{-+} sibling cells may help us to understand the source of discordance for the mother-daughter pairs we observed above (*see* **Note 4**, Fig. 2d). This is because if a GFP^{+} mother gives rise to one GFP^{-} daughter and one GFP^{+} daughter, then we may expect to observe concordant and discordant outcomes for GFP^{+} and GFP^{-} mother-daughter pairs, respectively. Taking the same approach we used for sibling cells, to test this hypothesis, we will estimate the probandwise concordance and COR for GFP^{--}, GFP^{++}, and GFP^{-+} mother-daughter pairs.

If mother and daughter have concordant division times, it means that the daughter cell has been instructed by her mother to copy her division time. On the other hand, discordant mother-daughter division times indicate that mother has instructed daughter to lengthen (or shorten) her cycle time. Therefore, the concordance model detects determinism between mother and daughter. As can be seen from Fig. 3c, the probandwise concordance for GFP^{+-} mother-daughter pairs was substantially lower than the marginal distribution indicating strong discordance (log (COR) $= -2.75 \pm 1.23$, $p = 0.037$), as were GFP^{--} pairs (log (COR) $= -3.08 \pm 1.3$, $p = 0017$). This is not surprising, because differentiation from GFP^{+} to GFP^{-} cell types is associated with longer division times. Thus, changes in cycle length were deterministic and not random. However, the probandwise concordance of GFP^{++} mother-daughter pairs was similar to the marginal distribution (log(COR) $= 0.275 \pm 0.667$, $p = 0.68$). The null hypothesis, mother does not instruct daughter's division time, was not rejected, so mother-daughter division times were neither concordant nor discordant.

Fig. 3 (continued) panel), and GFP^{-+} sibling twins (right panel). (**b**) Simulated probandwise concordance function showing sibling cell concordance in generation 1 (left panel), generation 2 (middle panel), and generation 3 (right panel). (**c**) Simulated probandwise concordance function for GFP^{--} mother-daughter pairs (left panel), GFP^{++}mother-daughter pairs (middle panel), and GFP^{-+} mother-daughter pairs (right panel). In all panels black lines show the probandwise concordance, and red lines show the marginal distribution of division. Dotted lines indicate 95% confidence intervals (CI)

A similar conclusion can be drawn when the concordance function for mother-daughter pairs is regressed with respect to GFP status (Table 4). The regression coefficient was -3.09 ± 1.4 ($p = 0.03$), the negative coefficient indicating that a loss of GFP expression triggers a discordant division time in the daughter cell. To summarize, concordance analysis allows one to identify whether a cell's fate is determined by its mother, sister, or other familial relation. Furthermore, concordance regression analysis can identify cell intrinsic or extrinsic factors that alter the differentiation program at the single-cell level.

5 Notes

1. Loading and Preparing Cell Lifetime Data.

   To load data into R from a tab-delimited file, the read.delim function is used. The R code to load cell lifetime data is as follows:

```
Data <- read.delim("cCFU data.txt")
```

Variable	Data type	Comment
Data	*Data frame*	*Contains cell lifetime, cell fate outcomes, and covariates of interest*
Data $Event_Time	*Decimal*	*The time elapsed between birth of a cell and the end of its lifetime*
Data$Cause	*Integer*	A unique identifier for each possible outcomes, i.e., 1 = division, 0 = right-censored, 2 = death
Model	*Character/ string*	*The selected link function (FG, additive, prop). Determined by minimizing the number of time-invariant variables*
causeS	*Categorical*	A code that specifies the cause-specific event of interest, i.e. 1 = division, 0 = right-censored, 2 = death
ModelOutput	*comp.risk object*	Output of the CR model containing model parameters and statistical tests for the effects of covariates on the event of interest

2. Constructing a Nonparametric CR Model to Test Covariates for Time-Invariant Effects.

   To test covariates for time-invariant effects, one must first construct a nonparametric competing risks model using the comp.risk function. Each covariate in the model is included without a const() wrapper. The R code to test covariates for time-invariant effects is as follows:

```
ModelOutput <- comp.risk(Surv(Event_Time,Cause>1) ~ 1 + FGF*PDGF, data=Data,
cause=Data$Cause, causeS=1, model="additive")
summary(ModelOutput)
```

Note: To model different events of interest (e.g., division and death), the *causeS* argument of *comp.risk* must be changed accordingly (e.g., causeS = 1 for division and causeS = 2 for death). By default *comp.risk* will expect that the code for censored events is cause = 0, if not otherwise specified.

3. Testing Covariates for Their Effect on the Cumulative Incidence Function.

   To test time-invariant covariates for their effect on the event of interest, a const() wrapper is included around each term. The R code to test covariates for their effect on the event of interest is as follows:

```
ModelOutput <- comp.risk(Surv(Event_Time,Cause>1) ~ 1 const(FGF)*const(PDGF),
data=Data, cause=Data$Cause, causeS=1, model="additive")
summary(ModelOutput)
```

   Note: To model different events of interest (e.g., division and death), the *causeS* argument of *comp.risk* must be changed accordingly (e.g., causeS = 1 for division and causeS = 2 for death). By default *comp.risk* will expect that the code for censored events is cause = 0, if not otherwise specified.

4. Simulating a Cumulative Incidence Function Given a Set of Covariates.

   To predict the cumulative incidence function given a set of covariate values, the predict function is applied to the semi-parametric competing risks model (ModelOutput) from **Note 3**. The R code for predicting a CIF for the covariates from **Note 2** is as follows:

```
CIF<-predict(ModelOutput,X=1,Z=c(0,0),uniform=TRUE) # CIF without PDGF and FGF
CIF<-predict(ModelOutput,X=1,Z=c(1,1),uniform=TRUE) # CIF with PDGF and FGF
CIF<-predict(ModelOutput,X=1,Z=c(0,1),uniform=TRUE) # CIF with PDGF and without FGF
```

   where X and Z are values for the nonparametric and parametric covariates in the model. If there are no covariates, then Z is not necessary. The three examples above show how one would simulate the CIF for various conditions with and without FGF or PDGF.

5. Plotting a Cumulative Incidence Function (CIF).

   The output of the predict function above outputs a cumulative incidence function that may be plotted by accessing CIF $time and CIF$P1. The R code to plot a CIF including 95% confidence intervals is as follows:

```
plot(CIF$time, CIF$P1, type="s",xlab="Time to division (days)", ylab="Cumulative
incidence", ylim=c(0,1), xlim=c(0,3.5), lty=1, lwd=1, col="grey", main="Division")
lines(CIF$time, CIF$P1+CIF$se.P1, type="s",lty=1, lwd=1, col="grey")
lines(CIF$time, CIF$P1-CIF$se.P1, type="s",lty=1, lwd=1, col="grey")
```

6. Empirical Estimation of the Cumulative Incidence Function.

    An empirical estimate of the cumulative incidence function may be obtained using the cuminc function for models that do not satisfy the condition of time invariance. The R code to estimate and plot a CIF using the cuminc function is as follows:

```
EmpiricalCIF <- cuminc(ftime=Data$Event_Time, fstatus=Data$Cause,
group=Data$Group, cencode="Censored")
plot(EmpiricalCIF$time, EmpiricalCIF$est, type="s", xlab="Time to division
(hours)",
ylab="Cumulative incidence", ylim=c(0,1), xlim=c(0,4), lty=2, lwd=2,
col="green", main="Division")
```

7. Constructing a Competing Risks Concordance Model.

    To model "within pair dependence" over time and quantify the effect of covariates on the marginal risk and dependence structure, a clustering variable is introduced into the model. The R code to estimate the conditional and unconditional probability distributions for a kinship cluster is as follows:

```
ConcordanceModel <- comp.risk (Surv(Event_Time,Cause>1) ~ 1 const(FGF)+
const(PDGF)+cluster(ClusterID), data=Data, cause=Data$Cause, causeS=1,
model="additive")
Summary (ModelOutput)
Note: here the ClusterID argument is a unique identifier for kinship
clusters. Clusters may be any pair of related cells. Or in the case where
we shuffle data the ClusterID will is assigned to cell pairs randomly.
```

8. Calculating the Cross-Odds Ratio.

    To determine the cross-odds ratio (COR) for the CR model from **Note 7**, the *cor.cif* and *concordance* functions are used. The R code to calculate the COR for the probandwise concordance is as follows:

```
CrossOddsRatio <- cor.cif (ConcordanceModel, data=Data, cause1=1,
cause2=1, sym=0 or 1)
Concordance <- concordance (CrossOddsRatio, cif1=ConcordanceModel$P1,
cif2=ConcordanceModel$P1)
```

    It should be noted here that the *sym* = 1 for daughter pairs and *sym* = 0 for mother-daughter pairs. The concordance functions can then be plotted as described in **Note 9**.

9. Plotting Probandwise Concordance.

   To plot the probandwise concordance, the output from the *concordance* function is used. The R code to plot the conditional and unconditional cumulative incidence for the model (shown in **Note 8**) is as follows:

```
plot(ConcordanceModel$time, Concordance$intercept$probandwise[,1],type="s",
xlim=c(0,1), ylim=c(0,1), lwd=2, xlab="Time (days)", ylab="Probability",
main="Concordance ouput")
lines(ConcordanceModel$time, Concordance$intercept$probandwise[,2], type="s",
lty=2)
lines(ConcordanceModel$time, Concordance$intercept$probandwise[,3], type="s",
lty=2)
lines(ConcordanceModel$time, Concordance$P1, type="s", lty=2)
```

10. Constructing a Regression Design Matrix.

    To fit a parametric model to the concordance function, a regression design matrix must be specified. To construct a model design matrix for a SelectedCovariate the model.matrix function (stats package) is used. The R code to construct a design matrix as discussed is as follows:

```
DesignMatrix<-model.matrix(~-1+factor(SelectedCovariate)) #
DesignMatrix<-model.matrix(~+1+factor(SelectedCovariate))
DesignMatrix<-model.matrix(~1+const(SelectedCovariate))
```

11. Testing for the Effect of Covariates on the Concordance Function.

    To test for the effect of covariates on the concordance function, the design matrix (*see* **Note 10**) and the ConcordanceModel (*see* **Note 7**) are input into the *cor.cif* function. The R code to test for the effect of covariates on the concordance function is as follows:

```
CrossOddsRatio<-cor.cif(ConcordanceModel, data=Data, the-
ta.des=DesignMatrix, sym=0 or 1)
```

    Note: *sym* = 1 for daughter pairs and *sym* = 0 for mother-daughter pairs. The concordance functions can then be plotted as described in **Note 9**. The above code is similar to **Note 8** then with the addition of *theta.des* which is used to specify the regression design structure for the cross-odds ratio parameters.

References

1. Dowling MR, Kan A, Heinzel S, Zhou JHS, Marchingo JM, Wellard CJ, Markham JF, Hodgkin PD (2014) Stretched cell cycle model for proliferating lymphocytes. Proc Natl Acad Sci 111(17):6377–6382
2. Duffy KR, Wellard CJ, Markham JF, Zhou JHS, Holmberg R, Hawkins ED, Hasbold J, Dowling MR, Hodgkin PD (2012) Activation-induced B cell fates are selected by intracellular

stochastic competition. Science 335 (6066):338
3. Cornwell JA, Hallett RM, der Mauer SA, Motazedian A, Schroeder T, Draper JS, Harvey RP, Nordon RE (2016) Quantifying intrinsic and extrinsic control of single-cell fates in cancer and stem/progenitor cell pedigrees with competing risks analysis. Sci Rep 6:27100. https://doi.org/10.1038/srep27100
4. Scherf N, Franke K, Glauche I, Kurth I, Bornhäuser M, Werner C, Pompe T, Roeder I (2012) On the symmetry of siblings: automated single-cell tracking to quantify the behavior of hematopoietic stem cells in a biomimetic setup. Exp Hematol 40(2):119–130.e119
5. Nordon RE, Ko KH, Odell R, Schroeder T (2011) Multi-type branching models to describe cell differentiation programs. J Theor Biol 277(1):7–18. https://doi.org/10.1016/j.jtbi.2011.02.006
6. Kaplan EL, Meier P (1958) Nonparametric-estimation from incomplete observations. J Am Stat Assoc 53(282):457–481. https://doi.org/10.2307/2281868
7. Scheike TH, Zhang MJ (2008) Flexible competing risks regression modeling and goodness-of-fit. Lifetime Data Anal 14 (4):464–483. https://doi.org/10.1007/s10985-008-9094-0
8. Scheike TH, Sun YQ (2012) On cross-odds ratio for multivariate competing risks data. Biostatistics 13(4):680–694. https://doi.org/10.1093/biostatistics/kxs017
9. Fine JP, Gray RJ (1999) A proportional hazards model for the subdistribution of a competing risk. J Am Stat Assoc 94 (446):496–509. https://doi.org/10.2307/2670170
10. Gray RJ (1988) A class of K-sample tests for comparing the cumulative incidence of a competing risk. Ann Stat 16(3):1141–1154. https://doi.org/10.1214/aos/1176350951
11. Cox DR (1972) Regression models and life-tables. J R Stat Soc B 34(2):187–18+
12. Scheike TH, Holst KK, Hjelmborg JB (2014) Estimating twin concordance for bivariate competing risks twin data. Stat Med 33 (7):1193–1204. https://doi.org/10.1002/sim.6016
13. Larson MG (1984) Covariate analysis of competing-risks data with log-linear models. Biometrics 40(2):459–469. https://doi.org/10.2307/2531398
14. Chen BE, Kramer JL, Greene MH, Rosenberg PS (2008) Competing risks analysis of correlated failure time data. Biometrics 64 (1):172–179. https://doi.org/10.1111/j.1541-0420.2007.00868.x
15. Scheike TH, Sun Y (2012) On cross-odds ratio for multivariate competing risks data. Biostatistics (Oxford, England) 13(4):680–694. https://doi.org/10.1093/biostatistics/kxs017
16. Scheike TH, Zhang M-J (2011) Analyzing competing risk data using the R timereg package. J Stat Softw 38(2):i02

Chapter 7

Quantitative Modelling of the Waddington Epigenetic Landscape

Atefeh Taherian Fard and Mark A. Ragan

Abstract

C.H. Waddington introduced the *epigenetic landscape* as a metaphor to represent cellular decision-making during development. Like a population of balls rolling down a rough hillside, developing cells follow specific trajectories (valleys) and eventually come to rest in one or another low-energy state that represents a mature cell type. Waddington depicted the topography of this landscape as determined by interactions among gene products, thereby connecting genotype to phenotype. In modern terms, each point on the landscape represents a state of the underlying genetic regulatory network, which in turn is described by a gene expression profile. In this chapter we demonstrate how the mathematical formalism of Hopfield networks can be used to model this epigenetic landscape. Hopfield networks are auto-associative artificial neural networks; input patterns are stored as attractors of the network and can be recalled from noisy or incomplete inputs. The resulting models capture the temporal dynamics of a gene regulatory network, yielding quantitative insight into cellular development and phenotype.

Key words Waddington epigenetic landscape, Gene regulatory network, Hopfield network, Attractor, Energy landscape, Quantitative model

1 Introduction

1.1 Waddington's Epigenetic Landscape

In 1942 [1] the British developmental biologist Conrad Waddington coined the term *epigenetics* to describe the causal interactions between genes and their products that result in the appearance of cellular phenotype. He later [2] depicted these interactions as giving rise to an *epigenetic landscape* across which cellular developmental trajectories unfold, much as a population of marbles would roll down a furrowed hillside, committing to ever-deeper valleys until they come to rest in one or another stable state (mature phenotypes). The overall slope of the landscape, its detailed topography, and the course of cell movement are controlled by complex interactions among the underlying genes. Through this metaphor, Waddington made the point that it is not single genes, but rather a complex network of their interactions, which control cellular differentiation.

Patrick Cahan (ed.), *Computational Stem Cell Biology: Methods and Protocols*, Methods in Molecular Biology, vol. 1975, https://doi.org/10.1007/978-1-4939-9224-9_7, © Springer Science+Business Media, LLC, part of Springer Nature 2019

1.2 Gene Regulatory Networks and the Epigenetic Landscape

A gene regulatory network (GRN) orchestrates the action of genes across the genome, generating genome-wide expression patterns (profiles) that in turn determine cellular phenotype. At a given time t, each profile corresponds to a state S of a network. Thus $S(t)$ is defined by expression levels x of each gene $i \in 1, \ldots, n$ in the network, $S(t) = [x_1(t), x_2(t), \ldots, x_n(t)]$. The network dynamic is determined by the changes of S over time. The set of possible changes of $S(t)$ is the *state space*, each point of which corresponds to a particular gene expression pattern [3].

In a two-gene system consisting of x_1 and x_2 that regulate each other's activity, state space is two-dimensional, i.e., a plane. The combined changes in x_1 and x_2 result in movement of $S(t)$ in state space. This movement of $S(t)$ through state space (its trajectory) is dictated by the underlying interaction network and continues until $S(t)$ reaches a stable (equilibrium) state, called the *attractor state*. A system with multiple (potential) stable states corresponds to a state space with multiple attractors [3].

In Waddington's heuristic model, at the entry point, cells start rolling down the hillside. As they become increasingly committed to a valley (lineage), their transcriptional program becomes established (increasingly predominating over stochastic or environmental noise) and enables the maintenance of a developmental pathway (canalization). In the absence of perturbation, a trajectory down the hillside follows a series of valleys and would be expected to end in a well (a stable phenotypic state) that corresponds to a specific cell fate. These stable states are hence attractors on the landscape. Changes in the underlying expression profile, for example, triggered by regulatory signals such as cytokines, may allow a cell to transition from one trajectory to another, leading to a switch between cell types [4–6].

Each point in state space is assigned an "elevation" (a quasi-potential energy), the magnitude of which is inversely related to the probability that it is occupied by a cell. Attractors are stable (more probable) phenotypic states relatively likely to be occupied by cells and thus characterized by a low energy, whereas unstable (less probable) states are less likely to be occupied and are characterized by a relatively higher energy. In the context of a biological system, cells seek to occupy the most probable (most stable) accessible state. Low and high elevations correspond to wells and ridges in Waddington's landscape, thus providing a quantitative measure for the landscape.

Large-scale gene regulatory networks present high-dimensional state spaces. As it is impractical to depict more than three dimensions (as a 2D projection), typically the x- and y-axes are chosen to correspond to the two features that best describe that cell state, and the z-axis represents quasi-potential energy of the network.

1.3 Computational Models of the GRN

For decades, researchers have proposed different methods to represent the complexity in biological systems and thereby address key fundamental questions in the regulation of cellular development and cell-fate determination. Computational models of large-scale GRNs facilitate cooperation between experimental and computational scientists, providing insight into the mechanisms of development and disease, reducing the need for costly wet-lab experiments, and revealing new aspects of biological processes.

Depending on the system being modelled, different entities have to be considered. These include the mathematical formulation (to describe the relationship among components), appropriate parameter values, and initial conditions that collectively represent, reproduce, and enable predictions of a biological system [7]. Quantitative or continuous modelling is the most common approach to provide precise predictions of gene regulatory networks, signalling, and metabolism. Such models are based on a bottom-up or knowledge-based approach for which prior information on the network components and their relationship is required. Considering the number of components and the complexity of interactions underlying processes such as cellular development, a common limitation on quantitative models is the lack of knowledge of parameter values such as rate or equilibrium constants of the interactions associated with a system. Alternatively, top-down or data-driven approaches aim to infer biological networks from experimental datasets. The input for such mathematical approaches is usually gene expression data, although other types of molecular information including methylation state and copy-number variation can be utilized.

Computational models inspired by Waddington's heuristic model of the epigenetic landscape provide a framework to discuss and understand the mechanism underlying cellular differentiation, development, and disease, enabling us to capture the dynamics of the GRN and map genotype to the overall behavior of a biological system. The most commonly used methods for modelling the GRN-based landscape include Boolean networks [8–10], differential equations [11, 12], and neural networks [13–17].

Pujadas and Feinberg (2012) described an epigenetic landscape in which the epigenome regulates the effect of noise in cellular plasticity and development and thereby mediates canalization during cell-fate determination. Over the course of development, as cells respond to differentiation factors, the stochastic noise decreases, enabling cells to commit to a specific lineage. In the context of cancer, such a model helps us to understand the variability of DNA methylation and gene expression in cancer cells. Mitra et al. (2014) incorporated time delays into a single-gene model, providing a framework to understand feedback circuits in the context of a simple gene regulatory network in development. In the absence of delays, the model has two attractors, representing the pluripotent and the differentiated state of the cell. Introducing delays into the system creates a new region on the landscape with sustained oscillations between the two attractors. They suggested that such oscillatory states might correspond to the unique intermediate states observed in reprogramming [18].

1.4 Limitations of the Current Models

Continuous and logical models are the two most common frameworks used for modelling biological systems. Each of these frameworks is suited for particular questions and situations, and each provides different insights into a biological process. Continuous models (dynamic systems) typically use ordinary differential equations to describe a biological process with rates of change in the concentration of gene products, whereas logical models commonly use sets of Boolean rules on binarised gene expression data in which genes are either active (ON) or inactive (OFF). However, these two types of models are not without limitations. Continuous models can capture the dynamics of a system, but the lack of information on reaction kinetics, interaction functions and related parameters, and their high computational cost (time and memory) hampers their application to many situations [19, 20]. While logical networks avoid many of these drawbacks, it may be asked whether they are sufficiently representative of real biology, as much information is lost by binarising the gene expression values [21]. Moreover, both approaches require prior or partial knowledge of the modelled system [20]. Unsupervised and nonparametric approaches may be better able to address some of these limitations.

Another limitation relates to the choice of input data. Simulated data are extensively used to test hypotheses in silico. Predictions arising from such data need to be tested in vitro and in vivo [19]. Last but not the least, most systems biology simulations tend to target relatively small subnetworks of interacting molecules [19]. Examples include simulation of MAPK signal transductions pathways [22], a two-gene model of lysis-lysogeny switching in bacteriophage lambda [23], and a quantitative model of the Waddington epigenetic landscape based on a feedback circuit between two genes [11]. This limitation could be avoided by utilizing large-scale gene expression empirical data but necessitates applying an efficient dimensionality reduction method so that the model reflects the global view of system dynamics rather than only a small circuit.

1.5 Landscape Models

1.5.1 Feature Selection

Machine-learning models can learn by example. The accuracy of the learned function depends on properties of the training set, which needs to contain enough features to accurately predict the output. Accuracy will suffer if the feature number is too great (the "curse of dimensionality") or too small (because the information is insufficient to describe the problem). Transcriptomic data are a typical example of high-dimensional data, in which each data point (biological sample) can have thousands of variables (genes). One approach to deal with high-dimensional data is to project the data into a smaller number of variables (features) while preserving the data structure as much as possible.

In feature selection [24], the features that contribute most to the prediction variable are automatically identified and retained for further analysis. In the context of gene expression profiles, feature selection techniques select a subset of genes (or probes) out of several tens of thousands that best describe a specific cell state without altering the original representation of the dataset. Two techniques are often used for dimensionality reduction [25–27]:

Feature subset selection removes non-relevant or redundant features. These include *filter methods* that apply a statistical measure to assign a score to each feature and rank them by the score to keep or remove them from the dataset; *wrapper methods* that prepare different combinations of features and use a predictive model to evaluate and compare different combinations and assign a score based on model accuracy; and *embedded methods* that combine the feature-selection step with model construction, such that the features that best contribute to the accuracy of the model are selected while the model is being created.

Feature extraction reduces dimensionality by creating new variables as combinations of others. These include *linear methods* that transform data from high-dimensional into a lower-dimensional linear subspace and *nonlinear methods* that find a nonlinear relationship of the features in a low-dimensional space which is then mapped into higher dimensions.

1.5.2 Network Robustness

Huang [28] explained perturbations of the Waddington landscape as a "wind gust blowing onto Waddington's descending marble" that can cause cells to deviate from their trajectory and, if large enough, traverse higher-energy (less probable) topography into a different valley or phenotypic state. Kauffman first proposed that cancer attractors are preexisting stable states in the GRN landscape [29], such that the tumor cell phenotype is already inherent in a healthy genome, although it would not normally be revealed. Mutations can lower the barriers on this landscape, thus opening access to the unused disease attractor. Although some mutations have no effect on a specific phenotype, others (e.g., in specific codons or genes, or series or combinations thereof) may trigger a cascade of events that eventually results in malfunctioning of the GRN and the appearance of a disease phenotype.

1.5.3 Landscape Visualization

As noted above, in three-dimensional visualizations of the Waddington landscape, the x- and y-axes typically correspond to a quantitative measure of two key features representing cell state, and the z-axis corresponds to a potential or energy. For instance,

Bhattacharya et al. [11] computed the epigenetic landscape of a two-gene regulatory network in which the *x*- and *y*-axes correspond to the expression of the two genes and the *z*-axis to a quasi-potential describing their interaction. For a GRN with more than two genes, dimensionality reduction techniques can be applied to retain the global information that best describes the GRN (above). There are many supervised and unsupervised dimensionality reduction techniques including principal component analysis (PCA), Isomap (an extension of Kernel PCA), maximum variance unfolding, Bayesian Gaussian process latent variable modeling (Bayesian GPLVM), and others for analyzing linear or nonlinear structures in the data [16, 30, 31]. Guo and Zheng [15] applied a Gaussian process dynamical model (GPDM), a nonlinear dimensionality reduction method, to project the high-dimensional time-series data onto a two-dimensional latent space as the *x,y*-plane of the landscape. GPDM takes into account the temporal dynamics of the data and is able to project in both directions between the low-dimensional latent space and the high-dimensional observation space. Similar to previous models [13, 14, 16, 17], the *z*-axis in this landscape corresponds to a quasi-potential or energy calculated described by a Lyapunov function, a family of monotonically decreasing functions that guarantee convergence to the local minima of the energy landscape. After each iteration, the energy associated with each pattern is updated until samples converge to their low-energy (attractor) state.

1.6 The Hopfield Network

Hopfield networks (HNs), introduced by John Hopfield in 1982 [32], are auto-associative artificial neural networks. Input patterns are stored as attractors of the network, and patterns can be recalled from noisy or incomplete inputs [32–34]. An HN is a fully connected graph composed of nodes $i \in 1, \ldots, n$ and weighted undirected edges w_{ij} between nodes i and j. Nodes in an HN usually have binary states, either $x_i \in \{0, +1\}$ or $\{-1, +1\}$. Network dynamics are achieved by recursive synchronous or asynchronous updating. At each iteration, the state of node i is updated to:

$$x_i^{(t+1)} = \text{sign}\left(\sum_j^n x_i^{(t)} w_{ij}\right)$$

where $x_i^{(t)}$ is the state of node i at time t and the sign function is defined as:

$$\text{sign}(x) = \begin{cases} +1 & \text{if } x \geq 0 \\ -1 & \text{otherwise} \end{cases}$$

The HN is guaranteed to converge to a (local) attractor state, with monotonically decreasing energy:

$$E(X) = -\frac{1}{2}\sum_{i,j}^{n} x_i w_{ij} x_j = -\frac{1}{2}X^T W X$$

where $X^0 = \left(X_1^0, \ldots, X_n^0\right)^T$ are the initial states in the state space. The network state $X^{(t)}$ will converge to the local minimum of the energy function as $t \to \infty$. Maetschke and Ragan interpreted this convergence as the system "relaxing from [a] higher energy state to a minimum energy state" [16].

The patterns $P_{k \in \{1, \ldots, m\}}$ are stored as attractors of the network in the weight matrix W, typically using Hebbian learning [35]:

$$W = \frac{1}{m}\sum_{k}^{m} P_{(k)} P_{(k)}^T$$

Weights are computed as the sum over the outer product of the patterns. Alternatively, any similarity matrix that describes the relation between nodes can be utilized as W, such as the covariance matrix.

Ideally, attractors in an HN act as attractors not only for the original patterns stored in the network but also for patterns similar to those. In short, patterns are stored as attractors of the network and can be retrieved from noisy or incomplete inputs. In the context of a GRN, considering each gene expression pattern as an input, samples with similar gene expression profiles are expected to converge to similar attractors [36]. The capacity of an HN using Hebbian learning rule is approximately 0.138 N uncorrelated patterns in a network of N nodes [37, 38].

1.7 Hopfield Network-Based Models of GRNs

Koulakov and Lazebnik (2012) modelled fusion between different cell types as attractors of an HN, concluding that fusion allows the cells to reach an attractor state that would otherwise be inaccessible [39]. Lang et al. (2014) used a Hopfield network to model a Waddington landscape and explained partially reprogrammed cells as hybrids that co-express genes from multiple cell fates [14]. Maetschke and Ragan (2014) applied Hopfield networks on static gene expression data and demonstrated that cancer subtypes can be characterized as attractors [16]. Guo and Zheng (2017) used a continuous Hopfield network to develop a method to map the trajectories of cell-state transition based on time-course single-cell transcriptional data to a Waddington landscape. Apart from normal development, this model has been shown to be useful in discovering potential cancer therapeutic strategies by combining computational modelling with experimental techniques [15].

1.8 Mapping Biological Cell States to the Hopfield Energy Landscape

Attractor landscape models provide a quantitative approach to understand the dynamics of GRN in biological systems. Nevertheless, both real biological GRNs and computational models of attractor landscapes employ their own concepts of states, trajectories, and attractors. In the context of a biological system, a cell state corresponds to a specific configuration of GRN at a given time; trajectories correspond to the course of developmental or other biological event; and attractors are stable phenotypic states. By contrast, in the HN formalism *state* refers to the value of $X^{(t)}$ where t is the update cycle of the HN; trajectories are constructed as a result of continuous updating of the nodes in HN, represented as a descent in a virtual landscape described by (for example) a Lyapunov function, and attractors are the local minima of this function.

It is important to note that in the Hopfield formalism, energy does not relate to any physical form of energy but instead is a scalar value associated with each state of a cell at a given time. Since local stable states, i.e., attractors, can be identified as local minimum of this function, an attractor can be distinguished from other (attractors) states by comparing their energy values. Moreover, in the proposed HN model, energy does not reflect the likelihood of a state space; given that the network is constructed from covariance of all pairs of genes in the network, the energy measures the extent of this covariance. Strong covariation gives rise to a relatively low (i.e., favorable) energy, whereas looser covariation gives a high energy [13, 17]. Differential regulation of GRN during the course of development results in changes in the patterns of covariation, reflected as a trajectory in the proposed Hopfield landscape model.

The use of the HN formalism need not be limited to normal cellular development. For example, looking at disease as perturbations of the network [40], and considering disease as a pre-existing attractor in the landscape [41], Taherian and Ragan constructed an HN from disease-progression data and tracked trajectories of cells as they progress from a normal to a cancerous state. Moreover, they measured the size of attractor by comparing the energy of attractors before and after perturbation [17]. Such a model could be further extended by employing targeted perturbations to subsets of genes in the GRN, to assess their contribution to the overall stability of an attractor state.

2 Data and Software

2.1 Input Data

Any quantitative measure, continuous or discrete, which describes the activity of elements in a GRN can be used as input to construct and visualize the landscape. Data can be generated from experiments on DNA, RNA, or protein, reflecting the overall state of a cell. Such data are accessible from a wide variety of databases, some general and others specific to species, cell lines, or diseases. Freely available online databases include Gene Expression Omnibus datasets, a high-throughput microarray and next-generation sequence database [42]; HIstome, a database of human histones [43]; and miRBase, a microRNA database [44]. One can use a combination of data types to construct the landscape, providing a more accurate and comprehensive model of the GRN. For instance, a landscape constructed from gene expression and histone modification data provides information on the dynamics of GRN and epigenetic events that potentially result in cellular identity and differentiation [14].

2.2 Software

A computational implementation of HN in Python is freely available at http://bioinformatics.org.au/tools/hclust/ [16]. Current computational models of GRNs typically lack a user-friendly interface. To address this problem, Guo et al. (2017) have developed NetLand, a user-friendly tool that facilitates model construction using HNs, dynamical simulation, and landscape plotting [45].

3 Methods

Quantifying the Waddington epigenetic landscape using the mathematical formalism of HN (Fig. 1):

Step 1: Data preprocessing—normalization and feature selection.

Step 2: Training phase—construct the Hopfield network from the preprocessed data.

Step 3: Recall phase—an iterative process that relaxes samples to their attractor state (*see* **Note 1** for an alternative approach).

Step 4: Visualize the Hopfield energy landscape in 3D.

3.1 Algorithm

Input: Gene expression data $G \in \mathbb{R}^{p \times m}$ where p is the number of samples and m is number of genes (or probes).
Output: Hopfield energy landscape in three dimensions (3D), displaying the trajectories of cells converging toward their respective attractors

1. Normalize the gene expression matrix G to $\mu = 0$ and $\sigma = 1$ (z-score normalization).
2. $D_{p \times n}$ = feature selection on G by selecting genes with the highest variance, where

   p is the number of samples and n is number of feature selected genes or probes.
3. Construct the HN (training phase)

   Initialize a zero matrix $W_{n \times n}$:

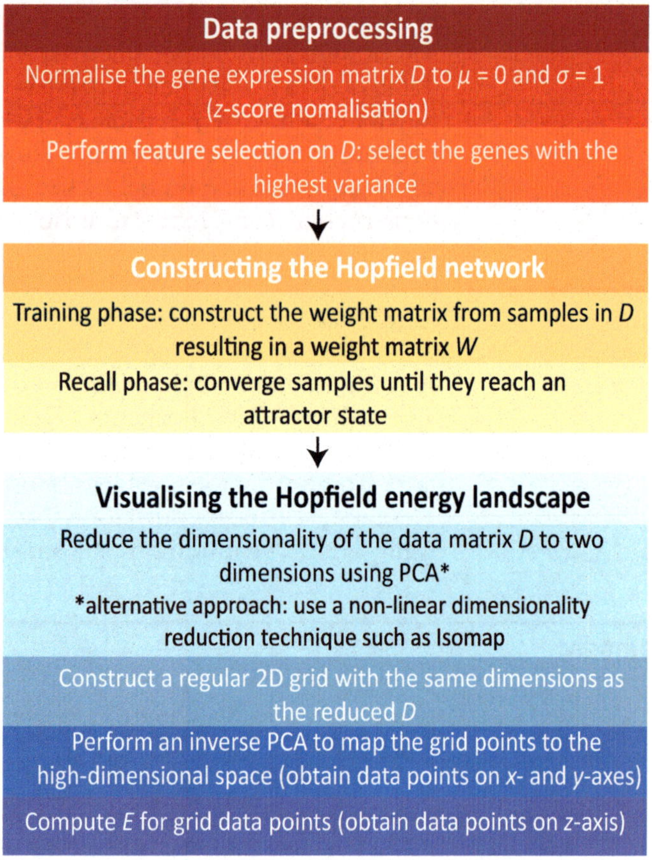

Fig. 1 Workflow, from data preprocessing to visualizing a 3D model of the Hopfield energy landscape. *PCA* principal component analysis

For example, in D:

$W = W + \text{sum (sample} \otimes \text{sample)}$

return $\frac{W}{P}$

Alternative approach: W = Pearson's correlation coefficient (D)

4. Generate the trajectories (an iterative process, convergence of samples to their respective attractors).
   For iterate = 1, 2, . . ., attratorState do.
   Get trajectory for each p in $D_{p\ .\ n}$: $p\ .\ W$
   Return convergence of samples in high- and lower-dimensional data space.
5. Visualize the energy landscape in 3D.
6. Reduce dimensionality of D:
   $D_{p\times 2} = \text{PCA}\ (D_{p\times n}, \text{number of components} = 2)$

7. Define a regular 2D grid $G_{k \times 2}$ with k points in the same range as $D_{p \times 2}$
8. Perform inverse PCA to map the grid points to the high-dimensional space:
   $G_{k\times n} = G_{k\times 2}T^{-1}$
9. Compute the energy for grid $E(G_{k \times n})$ and sample data $E(G_{p \times n})$ in high dimensions.
10. Interpolate the grid energy values over the grid data to render the surface of the landscape. The x- and y-axis are the first and the second principal components, and the z-axis is the energy.

This visualization method is not limited to HN but can be applied to render a 3D surface from a dataset with points of interest and an energy function [16]. Please refer to *see* **Note 2** for visualization source code in the Python programming language.

3.2 Network Perturbation

In a computational model, robustness can be tested by introducing noise to the system. In our model, to evaluate the robustness of the Hopfield energy landscape, we [13] perturbed the network by randomly changing the expression values of a randomly selected subset (5, 10, 20, 50, or 90%) of genes. This was followed by constructing the HN from the perturbed data and computing E for the network after each round of perturbation. Next we compared the resulting energy value with that of a completely random network of the same size ($\Delta E = E_{\text{random}} - E_{\text{perturbed}}$). A random network represents an unstable state of the cell and provides a base of comparison with the unperturbed state of the system. To construct these random networks, values were randomly selected from a uniform distribution within the range of the corresponding original gene expression data. Since all values in each such network are generated randomly, patterns of coactivity between network elements are expected to be nonexistent (or at least extremely rare) and without biological meaning. Each set of 100 random networks provided the base against which we compared the original and perturbed states of the corresponding system. This process was repeated 100 times for each dataset and each proportion of perturbed genes [13].

3.3 Limitations of the Model

In a standard HN, the weight matrix W is computed using a Hebbian learning rule [32]. Introduced by Donald Hebb in 1949 [35], this refers to a local incremental learning rule such that the update of weights between nodes depends only on the information available on either side of the nodes (*local*) and the learning process can memorize new patterns without referring to the previous learned patterns (*incremental*) [46].

In Hebbian learning, the weight matrix results from an unsupervised learning process in which the algorithm learns different unlabelled patterns to set the edge weights between nodes; this eventually allows the network to converge to the stored patterns (attractors). The Hebbian learning rule adjusts the weights between nodes in such a way that the output echoes its similarity with the input. The limitation associated with this method is that the input data have to be properly normalized or else the weights can increase in value and strengthen after each iteration, potentially leading to creation of sparse attractors in the network. Maetschke and Ragan (2014) evaluated high-capacity learning rules as an alternative learning algorithm that might allow storage of *n* linearly independent patterns but found that their complexity resulted in prohibitively long learning time compared to Hebbian learning. They also showed that using a similarity matrix rather than a sample matrix as input improves the accuracy of clustering. They compared the performance of HNs with clustering algorithms including DBSCAN [47], k-means [48], and WARD [49] and found that HNs perform well (if not better) than other unsupervised clustering algorithms and that using a similarity matrix as input consistently improved performance [16].

4 Notes

1. An alternative approach to identify the trajectories of cell development

   As explained above, HN formalism attractors are local minima of the energy landscape, constructed from an artificial updating of the network until the samples (patterns) reach their attractor state. Thus, in the present context, attractors do not necessarily correspond to what Waddington envisioned as biological phenotypes. In order to map real developmental trajectories to the Hopfield energy landscape, Taherian Fard et al. [13] showed that the energy landscape could be constructed from energy values of samples in their original state, such that each attractor corresponds to a biological phenotype. In this proposed landscape, cells start out in a stable phenotypic state (e.g., a pluripotent phenotype) associated with a low energy. As cells move along a developmental trajectory, they pass through one or more unstable transient states corresponding to higher-energy (i.e., less favorable) states and eventually reach another stable low-energy state, e.g., a specialized phenotype, in the Hopfield landscape. As a result, unlike in the Waddington metaphor, the inferred trajectories in this representation do not run (primarily) downhill into a low-energy attractor at one edge of a state-space dimension; rather, they find low-energy states at both ends of every trajectory [13].

2. Hopfield energy landscape visualization

The following Python code describes the 3D visualization of the Hopfield energy landscape, namely, transforming data from high- to low-dimensional space, creating a grid, and interpolating high-dimensional data points onto a 2D grid. Please note that this code has been modified to fit the scope of this chapter. The original source code is available at [16].

```
def create_mesh_points(model2D, W, res):
    """Create high-dimensional mesh points
    W -- weight matrix of Hopfield network
    res -- resolution of mesh
    """
    w = max(W.min(),W.max())*2.0
    x = y = linspace(-w, +w, res)
    X, Y = meshgrid(x, y)
    points2d = array([[x, y] for x, y in zip(ravel(X), ravel(Y))])
    return model2D.inverse_transform(points2d)
def plot_landscape(samples, W, trajectories, res=50):
    """Creates data points for x-, y- and z-axis
    samples - data matrix with samples in rows and genes in columns
    W -- weight matrix of Hopfield network
    trajectories -- relaxation of samples from initial state to
    their respective attractor state
    """
    from sklearn.decomposition import PCA
    model2D = PCA(n_components=2)
    model2D.fit(samples)
    mesh = create_mesh_points(model2D, W, res)
    points = vstack((mesh, trajectories))
    energies = compute_hopfield_energy(W, points)
    points = model2D.transform(points)
    w = max(points.min(),points.max())*1.5
    x,y = mgrid[-w:w:complex(0,res),-w:w:complex(0,res)]
    from scipy.interpolate import griddata
    z = griddata(points, energies, (x,y), method='nearest')
    return x, y, z
```

Acknowledgments

We thank Dr Stefan Maetschke for valuable comments and the Australian Research Council for funding (grant no. DP11010 3384 to MAR).

References

1. Waddington CH (1942) Canalization of development and the inheritance of acquired characters. Nature 150:563–565. https://doi.org/10.1038/150563a0
2. Waddington CH (1957) The strategy of the genes. Allens & Unwin, London
3. Huang S, Ernberg I, Kauffman S (2009) Cancer attractors: a systems view of tumors from a gene network dynamics and developmental perspective. Semin Cell Dev Biol 20(7):869–876. https://doi.org/10.1016/j.semcdb.2009.07.003
4. Huang S (2009) Reprogramming cell fates: reconciling rarity with robustness. BioEssays 31(5):546–560. https://doi.org/10.1002/bies.200800189
5. Huang S, Eichler G, Bar-Yam Y, Ingber DE (2005) Cell fates as high-dimensional attractor states of a complex gene regulatory network. Phys Rev Lett 94(12):128701
6. Moris N, Pina C, Arias AM (2016) Transition states and cell fate decisions in epigenetic landscapes. Nat Rev Genet 17(11):693–703. https://doi.org/10.1038/nrg.2016.98
7. Le Novere N (2015) Quantitative and logic modelling of molecular and gene networks. Nat Rev Genet 16(3):146–158. https://doi.org/10.1038/nrg3885
8. Alvarez-Errico D, Vento-Tormo R, Sieweke M, Ballestar E (2015) Epigenetic control of myeloid cell differentiation, identity and function. Nat Rev Immunol 15(1):7–17. https://doi.org/10.1038/nri3777
9. Davila-Velderrain J, Martínez-García JC, Alvarez-Buylla ER (2015) Modeling the epigenetic attractors landscape: towards a post-genomic mechanistic understanding of development. Front Genet 6:160. https://doi.org/10.3389/fgene.2015.00160
10. Davila-Velderrain J, Villarreal C, Alvarez-Buylla E (2015) Reshaping the epigenetic landscape during early flower development: induction of attractor transitions by relative differences in gene decay rates. BMC Syst Biol 9(1):20
11. Bhattacharya S, Zhang Q, Andersen ME (2011) A deterministic map of Waddington's epigenetic landscape for cell fate specification. BMC Syst Biol 5:85
12. Li C, Wang J (2013) Quantifying cell fate decisions for differentiation and reprogramming of a human stem cell network: landscape and biological paths. PLoS Comput Biol 9(8):e1003165. https://doi.org/10.1371/journal.pcbi.1003165
13. Taherian Fard A, Srihari S, Mar JC, Ragan MA (2016) Not just a colourful metaphor: modelling the landscape of cellular development using Hopfield networks. Npj Systems Biol Applicat 2:16001. https://doi.org/10.1038/npjsba.2016.1
14. Lang A, Li H, Collins J, Mehta P (2014) Epigenetic landscapes explain partially reprogrammed cells and identify key reprogramming genes. PLoS Comput Biol 10(8):e1003734. https://doi.org/10.1371/journal.pcbi.1003734
15. Guo J, Zheng J (2017) HopLand: single-cell pseudotime recovery using continuous Hopfield network-based modeling of Waddington's epigenetic landscape. Bioinformatics 33(14):i102–i109. https://doi.org/10.1093/bioinformatics/btx232
16. Maetschke SR, Ragan MA (2014) Characterizing cancer subtypes as attractors of Hopfield networks. Bioinformatics 30(9):1273–1279. https://doi.org/10.1093/bioinformatics/btt773
17. Taherian Fard A, Ragan MA (2017) Modeling the attractor landscape of disease progression: a network-based approach. Front Genet 8:48. https://doi.org/10.3389/fgene.2017.00048
18. Mitra MK, Taylor PR, Hutchison CJ, McLeish TCB, Chakrabarti B (2014) Delayed self-regulation and time-dependent chemical drive leads to novel states in epigenetic landscapes. J R Soc Interface 11(100):20140706. https://doi.org/10.1098/rsif.2014.0706
19. Kitano H (2002) Computational systems biology. Nature 420(6912):206–210
20. Esfahani M, Yoon B-J, Dougherty E (2011) Probabilistic reconstruction of the tumor progression process in gene regulatory networks in the presence of uncertainty. BMC Bioinf 12 (Suppl 10):S9
21. Shmulevich I, Dougherty ER, Wei Z (2002) From Boolean to probabilistic Boolean networks as models of genetic regulatory networks. Proc IEEE 90(11):1778–1792. https://doi.org/10.1109/jproc.2002.804686
22. Schoeberl B, Eichler-Jonsson C, Gilles ED, Muller G (2002) Computational modeling of the dynamics of the MAP kinase cascade activated by surface and internalized EGF receptors. Nat Biotech 20(4):370–375
23. Zhu XM, Yin L, Hood L, Ao P (2004) Robustness, stability and efficiency of phage λ genetic switch: dynamical structure analysis. J Bioinforma Comput Biol 02(04):785–817. https://doi.org/10.1142/S0219720004000946
24. Blum AL, Langley P (1997) Selection of relevant features and examples in machine learning. Artif Intell 97(1):245–271

25. Guyon I, Elisseeff A (2003) An introduction to variable and feature selection. J Mach Learn Res 3:1532–4435
26. Saeys Y, Inza I, Larrañaga P (2007) A review of feature selection techniques in bioinformatics. Bioinformatics 23(19):2507–2517. https://doi.org/10.1093/bioinformatics/btm344
27. Hira ZM, Gillies DF (2015) A review of feature selection and feature extraction methods applied on microarray data. Adv Bioinforma 2015:13. https://doi.org/10.1155/2015/198363
28. Huang S (2012) The molecular and mathematical basis of Waddington's epigenetic landscape: a framework for post-Darwinian biology? BioEssays 34(2):149–157. https://doi.org/10.1002/bies.201100031
29. Kauffman S (1971) Differentiation of malignant to benign cells. J Theor Biol 31(3):429–451
30. Bartenhagen C, Klein H-U, Ruckert C, Jiang X, Dugas M (2010) Comparative study of unsupervised dimension reduction techniques for the visualization of microarray gene expression data. BMC Bioinf 11:567–567. https://doi.org/10.1186/1471-2105-11-567
31. Zwiessele M, Lawrence ND (2016) Topslam: Waddington landscape recovery for single cell experiments. bioR xiv:057778
32. Hopfield JJ (1982) Neural networks and physical systems with emergent collective computational abilities. Proc Natl Acad Sci U S A 79 (8):2554–2558
33. Amit DJ, Gutfreund H, Sompolinsky H (1985) Spin-glass models of neural networks. Phys Rev A 32(2):1007–1018
34. Amit DJ (1989) Modeling brain function: the world of attractor neural networks. Cambridge University Press, New York
35. Hebb DO (1949) The organization of behavior: a neuropsychological theory. Wiley, New York
36. Brouwer RK (1993) Using the Hopfield neural network as a classifier by storing class representatives (1993) In: Proceedings of Canadian Conference on Electrical and Computer Engineering, 14–17 Sep 1993. pp 337–340, vol. 331 https://doi.org/10.1109/ccece.1993.332325
37. Clift F, Martinez TR (2001) Improved Hopfield networks by training with noisy data (2001) In: Proceedings of International Joint Conference on Neural Networks, pp 1138–1143. vol. 1132. https://doi.org/10.1109/ijcnn.2001.939521
38. Alemi A, Baldassi C, Brunel N, Zecchina R (2015) A three-threshold learning rule approaches the maximal capacity of recurrent neural networks. PLoS Comput Biol 11(8): e1004439. https://doi.org/10.1371/journal.pcbi.1004439
39. Koulakov AA, Lazebnik Y (2012) The problem of colliding networks and its relation to cancer. Biophys J 103(9):2011–2020
40. del Sol A, Balling R, Hood L, Galas D (2010) Diseases as network perturbations. Curr Opin Biotechnol 21(4):566–571
41. Huang S (2011) On the intrinsic inevitability of cancer: from foetal to fatal attraction. Semin Cancer Biol 21:183–199. https://doi.org/10.1016/j.semcancer.2011.05.003
42. Barrett T, Wilhite SE, Ledoux P, Evangelista C, Kim IF, Tomashevsky M, Marshall KA, Phillippy KH, Sherman PM, Holko M, Yefanov A, Lee H, Zhang N, Robertson CL, Serova N, Davis S, Soboleva A (2013) NCBI GEO: archive for functional genomics data sets-update. Nucl Acids Res 41(D1):D991–D995. https://doi.org/10.1093/nar/gks1193
43. Khare SP, Habib F, Sharma R, Gadewal N, Gupta S, Galande S (2012) HIstome-a relational knowledgebase of human histone proteins and histone modifying enzymes. Nucl Acids Res 40(DI):D337–D342. https://doi.org/10.1093/nar/gkr1125
44. Kozomara A, Griffiths-Jones S (2014) miRBase: annotating high confidence microRNAs using deep sequencing data. Nucl Acids Res 42 (D1):D68–D73. https://doi.org/10.1093/nar/gkt1181
45. Guo J, Lin F, Zhang X, Tanavde V, Zheng J (2017) NetLand: quantitative modeling and visualization of Waddington's epigenetic landscape using probabilistic potential. Bioinformatics 33(10):1583–1585. https://doi.org/10.1093/bioinformatics/btx022
46. Storkey AJ, Valabregue R (1999) The basins of attraction of a new Hopfield learning rule. Neural Netw 12(6):869–876
47. Ester M, Kriegel H-P, Sander J, Xu X (1996) A density-based algorithm for discovering clusters in large spatial databases with noise. AAAI Press, Palo Alto, pp 226–231
48. MacQueen J (1967) Some methods for classification and analysis of multivariate observations. In: Proceedings of the fifth Berkeley symposium on mathematical statistics and probability, Statistics, vol 1. University of California Press, Berkeley, CA, pp 281–297
49. Ward JH (1963) Hierarchical grouping to optimize an objective function. J Am Stat Assoc 58 (301):236–244. https://doi.org/10.1080/01621459.1963.10500845

Chapter 8

Modeling Gene Networks to Understand Multistability in Stem Cells

David Menn and Xiao Wang

Abstract

Stem cells are unique in their ability to differentiate into diverse phenotypes capable of displaying radically different, yet stable, gene expression profiles. Understanding this multistable behavior is key to rationally influencing stem cell differentiation for both research and therapeutic purposes. To this end, mathematical paradigms have been adopted to simulate and explain the dynamics of complex gene networks. In this chapter, we introduce strategies for building deterministic and stochastic mathematical models of gene expression and demonstrate how analysis of these models can benefit our understanding of complex observed behaviors. Developing a mathematical understanding of biological processes is of utmost importance in understanding and controlling stem cell behavior.

Key words Modeling, Differential equations, Stochasticity, Parameterization, Network topology, Synthetic biology

1 Introduction

Multistability is a mathematical property describing a system capable of having two or more mutually exclusive states. These dynamical states could provide stem cells a portfolio of decision options that allow cells carrying a homogeneous genetic code to express phenotypic variety [1]. Through differential expression of genes in response to environmental factors, cells lock into various expression patterns, upregulating some genes while downregulating others. Network topology shapes the ways that these genes' expression influences one another, either reinforcing or inhibiting each other or themselves. These gene interactions, usually in the form of transcriptional regulation, can be modeled mathematically, and the information gained from the models can be used to more thoroughly understand the behaviors and limits of the network [2, 3]. The utility of modeling for understanding complex gene networks has been demonstrated in a variety of contexts: understanding the link between external stimuli and gene expression [4],

Patrick Cahan (ed.), *Computational Stem Cell Biology: Methods and Protocols*, Methods in Molecular Biology, vol. 1975,
https://doi.org/10.1007/978-1-4939-9224-9_8,

exploration of gene topologies capable of specific behavior [5, 6], and development of a functional multistable synthetic gene networks [7–9].

In this chapter, we cover the basics of constructing mathematical models using ordinary differential equations (ODEs), discuss ways to adapt them to simulate noisy biological environments, and discuss implementation for two common applications.

2 Materials

The work we discuss is purely in silico and requires a personal computer, preferably one with a fast processor, and a programming environment. We prefer MATLAB because of its usability and host of built-in functions. MATLAB also can be used with the MATCONT add-on, which adds additional analysis options for solving systems of ODEs. With a knowledgeable user, however, any programming environment with continuation software can be appropriate. Additionally, for some computationally intensive processes, access to a distributed computing center is necessary. This will require some knowledge of parallelization that is beyond the scope of this chapter.

3 Methods

3.1 Applications

Before getting into details on how to build and run gene network models, we will discuss two main applications of mathematical modeling to stem cell behavior. While these are obviously not the only contexts in which modeling can be leveraged, in some ways they are fundamental for understanding stem cell behavior. The defining feature of stem cells is their ability to differentiate into various lineages which hold a diverse range of protein expression patterns, so we will focus here on two applications centered on understanding multistability.

In the first, the topology of the network is known, and it is a matter of exploring the range of parameters and conditions under which the system behaves in a desired manner. This will be referred to as parameterizing a network. A detailed example of this type of network analysis can be found in the work of Wu et al. [7]. In the case of stem cells, this means exploring the conditions under which the network exhibits multistability and/or will preferentially differentiate to one state over another. In the second application, the topology itself is unknown or not fully known, so one must interrogate a set of potential networks to discover those that could exhibit multistability. We refer to this as exploring network topologies. One can look at the work of Faucon et al. for a good example of this type of analysis [5]. There is a lot of commonality in the methods employed in both applications, so we will look at both simultaneously.

As an important note before beginning, biological networks are highly interconnected and complex systems, making them difficult to study directly due to the great number of confounding factors. Because of this, simplification methods are often used to reduce network complexity. Please *see* **Note 1** for a more thorough discussion of this topic.

3.1.1 Formulating a System of ODEs

Gene network models typically begin as a system of ordinary differential equations (ODEs), each equation describing a single subcellular population: a specific protein, mRNA, metabolite, etc. While each individual member of these populations may exhibit unique, semi-random behavior, the population at large will behave relatively deterministically. Because of this, equations have been developed to represent the aggregate behavior of molecular populations. Two very important and interrelated equations in this respect are the Michaelis-Menten equation which describes enzymatic activity [10] and the Hill equation which captures the nonlinear behavior of multi-protein binding dynamics [11]. We do not cover derivations of these equations here, but both are worth being familiar with, as they are the most common formulations used extensively in the biomolecular modeling literature [4–8, 12]. With that in mind and being cognizant of the need for simplification discussed in **Note 1**, we will now explore how to formulate ODE models.

In a system of ODEs, several equations are constructed, each representing the fundamental populations that are being studied. Figure 1 shows several gene topologies and the related equations that describe their behavior. One may notice that nearly all equations will have two parts: a production term and a degradation term. In a transcriptional regulation pathway, interacting nodes will either activate or repress the production of themselves or another population. Both interactions have basic forms, based on Hill equation kinetics:

$$\frac{\mathrm{d}x}{\mathrm{d}t} = a*\frac{y^n}{k^n + y^n} \tag{1}$$

$$\frac{\mathrm{d}x}{\mathrm{d}t} = a*\frac{k^n}{k^n + y^n} \tag{2}$$

Here, the change in our population (x) has a maximum expression rate (a) which is modified by the abundance of activators or repressors (y). In activation (Eq. 1), the production rate of x asymptotically approaches a as y increases. In repression (Eq. 2), production of x asymptotically falls from a to 0 with increasing y. A rate constant (k), sometimes referred to as the "half-max" value, serves as a sort of center point that helps determine the scale of *y* needed to influence *x*'s expression. Finally, the Hill coefficient (n) determines the linearity of *x*'s response to *y*, with the steady-state expression of *x* vs. *y* becoming increasingly sigmoidal as n increases. A third type of production term—constitutive production—may be included in some circumstances:

$$\frac{dx}{dt} = a \tag{3}$$

However, unless there is some sort of dynamic interaction with other nodes in the network, once degradation is factored in (see below), constitutively expressed components will end up expressing at a steady state. Therefore, this steady-state expression can be substituted for the ODE, lowering the number of equations that must be solved and reducing the computational requirements for running the system of ODEs. Frequently, a constant production term like this will be used in conjunction with a repression term (Eq. 2) to represent leaky repression. In this case, the node produces a small amount of product even when fully repressed.

A

$$\frac{dx}{dt} = a_1 + b_1\frac{k_1^{n_1}}{k_1^{n_1} + y^{n_1}} - d_1 * x$$

$$\frac{dy}{dt} = a_2 + b_2\frac{k_2^{n_2}}{k_2^{n_2} + x^{n_2}} - d_2 * y$$

B

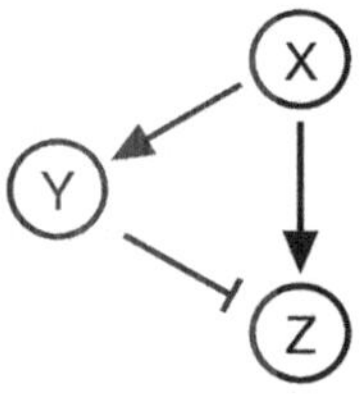

$$\frac{dx}{dt} = a_1 - d_1 * x$$

$$\frac{dy}{dt} = a_2\frac{x^{n_2}}{k_2^{n_2} + x^{n_2}} - d_2 * y$$

$$\frac{dz}{dt} = a_3\frac{x^{n_2}}{k_2^{n_2} + x^{n_2}} + b_3\frac{k_3^{n_3}}{k_3^{n_3} + y^{n_3}} - d_3 * z$$

C

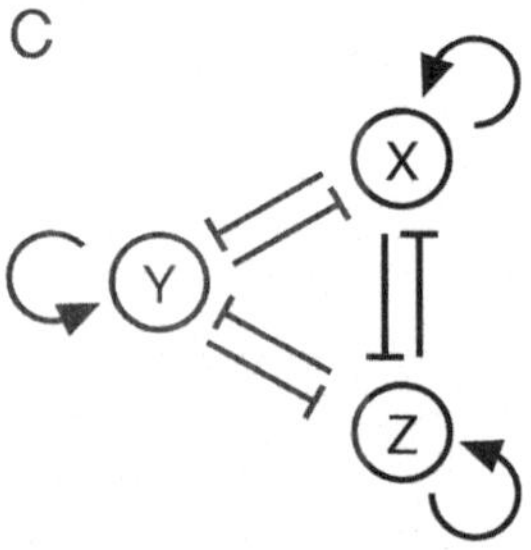

$$\frac{dx}{dt} = a_1\frac{x^{n_4}}{k_4^{n_4} + x^{n_4}} + b_1\frac{k_2^{n_2}}{k_2^{n_2} + y^{n_2}} + c_1\frac{k_3^{n_3}}{k_3^{n_3} + z^{n_3}} - d_1 * x$$

$$\frac{dy}{dt} = a_2\frac{k_1^{n_1}}{k_1^{n_1} + x^{n_1}} + b_2\frac{y^{n_5}}{k_5^{n_5} + y^{n_5}} + c_2\frac{k_3^{n_3}}{k_3^{n_3} + z^{n_3}} - d_2 * y$$

$$\frac{dz}{dt} = a_3\frac{k_1^{n_1}}{k_1^{n_1} + x^{n_1}} + b_3\frac{k_2^{n_2}}{k_2^{n_2} + y^{n_2}} + c_3\frac{z^{n_6}}{k_6^{n_6} + z^{n_6}} - d_3 * z$$

Fig. 1 Network diagrams and potential equations to describe them. Arrow links indicate activation, while barred links indicate repression. (**a**) Mutual inhibition toggle switch. (**b**) Incoherent feed-forward loop. (**c**) Fully connected triad with autoactivation and full mutual inhibition

Populations will also degrade with time, and there are two primary forms of that as well: constant and enzymatic. Typically, constant degradation is sufficient, but networks which produce many transcripts may see those transcripts competing for degradation machinery, in which case the enzymatic model may be more appropriate.

$$\frac{dx}{dt} = -d*x \tag{4}$$

$$\frac{dx}{dt} = -d*\frac{x}{x+z} \tag{5}$$

In the case of constant degradation (Eq. 4), the degradation rate (d) is simply a fraction of the amount of x present in the system. In the enzymatic case (Eq. 5), however, the total number other things (typically proteins) being degraded in the system (z) imposes an upper limit on the rate at which x can be degraded.

3.1.2 Alternate Equation Forms

From these basic forms, equations can be expanded as needed to integrate relevant information into the model. Leaky repressible promoters, for example, may include a low level of constitutive expression and a repression production term [7]. An activator that works only in the presence of a drug may be modified by a term to indicate the fraction of functional activator at various dosages of drug [4]. As a guideline, however, it is best to construct the simplest model that still captures the behavior or interactions that you care about. Simple models can always be expanded.

This becomes slightly less straightforward once one has multiple regulators affecting the same gene. If a gene is both activated and repressed, it may not be obvious whether to combine the production terms additively or to make some sort of hybrid equation. Figure 2 shows two formulations of the same network. In practice, the differences in outputs between various formulations of the same network are often insignificant. Because modeling is an abstraction of the incredibly complex observed cellular behavior, there is no single accepted "correct" way to model a gene network. For parameterization purposes, provided that the ODEs produce the experimentally observed behavior, they are likely okay. In the case of network exploration, it is often a good idea to experiment with alternate equation formulations on a subset of your networks, to ensure that the observed behavior is not simply an artifact of the model's form.

3.1.3 ODE Formulation for Network Exploration

For parameterization of a network, once the equation form is determined, model creation is finished. However, for exploring network topologies, additional steps are required. Because of the combinatorial nature of this type of work, automation is a necessity. First, you need to generate a list of all possible networks within the

$$\frac{dx}{dt} = a_1 - d_1 * x$$

$$\frac{dy}{dt} = a_1 \frac{k_1^{n_1}}{k_1^{n_1} + x^{n_1}} + b_1 \frac{y^{n_2}}{k_2^{n_2} + y^{n_2}} - d_2 * y$$

$$\uparrow\downarrow$$

$$\frac{dy}{dt} = a_1 \frac{y^{n_1}}{k_1^{n_1} + y^{n_1} + x^{n_2}} - d_2 * y$$

Fig. 2 The ODE describing *y*'s expression can be constructed in several different ways. The combined effects of repression by *x* and autoactivation could be modeled as either the sum of two production terms (top) or as a single term comprising both processes (bottom)

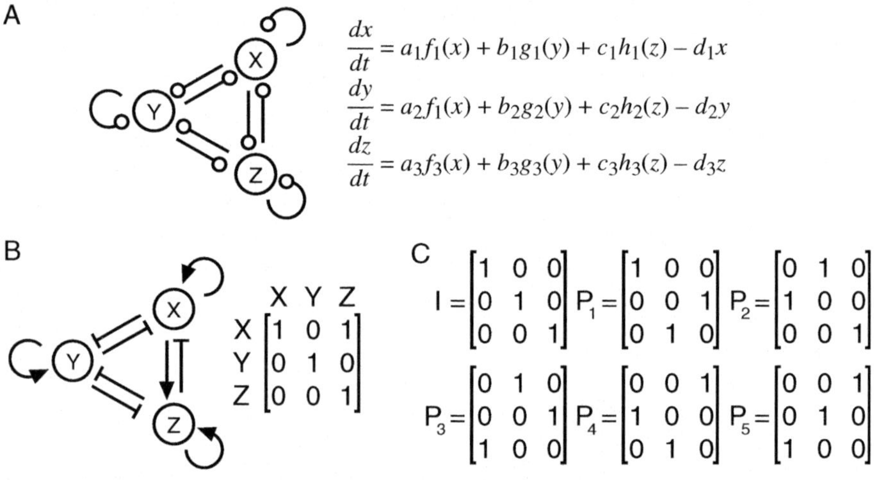

$$\frac{dx}{dt} = a_1 f_1(x) + b_1 g_1(y) + c_1 h_1(z) - d_1 x$$
$$\frac{dy}{dt} = a_2 f_1(x) + b_2 g_2(y) + c_2 h_2(z) - d_2 y$$
$$\frac{dz}{dt} = a_3 f_3(x) + b_3 g_3(y) + c_3 h_3(z) - d_3 z$$

	X	Y	Z
X	1	0	1
Y	0	1	0
Z	0	0	1

$$I = \begin{bmatrix} 1 & 0 & 0 \\ 0 & 1 & 0 \\ 0 & 0 & 1 \end{bmatrix} P_1 = \begin{bmatrix} 1 & 0 & 0 \\ 0 & 0 & 1 \\ 0 & 1 & 0 \end{bmatrix} P_2 = \begin{bmatrix} 0 & 1 & 0 \\ 1 & 0 & 0 \\ 0 & 0 & 1 \end{bmatrix}$$

$$P_3 = \begin{bmatrix} 0 & 1 & 0 \\ 0 & 0 & 1 \\ 1 & 0 & 0 \end{bmatrix} P_4 = \begin{bmatrix} 0 & 0 & 1 \\ 1 & 0 & 0 \\ 0 & 1 & 0 \end{bmatrix} P_5 = \begin{bmatrix} 0 & 0 & 1 \\ 0 & 1 & 0 \\ 1 & 0 & 0 \end{bmatrix}$$

Fig. 3 There are several stages to exploring network topologies. (**a**) A general form of the ODEs that can describe all possible networks is formulated. In the diagram, the circular ends of the regulatory links indicate that they could be either activating or repressing. (**b**) For a specific network, a connection matrix is made. A value of 0 indicates repression, while a 1 is activation. (**c**) Permutation matrices are used to eliminate redundant networks. Here, all permutation matrices (P_{1-5}) for the FCT topology are shown, with the identity matrix (I) for reference

bounds of whatever constraints you've imposed on the system. In the case of Faucon et al., for example, the constraint used was fully connected triads (FCTs) [5]. To decrease downstream computation, you should identify networks which are rotations or projections of one another. These networks will behave identically, so there is no need to solve more than one of them. Once redundant networks have been eliminated, you will be left with a set of all unique networks for further analysis.

To generate all possible networks, first formulate a generic form of your system of ODEs. For a system of fully connected triads, this might be that shown in Fig. 3a. The parameters a_{1-3}, b_{1-3}, c_{1-3}, and

d_{1-3} will not affect the form of the equations, but the functions f_{1-3}, g_{1-3}, and h_{1-3} will differ depending on whether they are activating (Eq. 1) or repressing (Eq. 2). If we assume that all nodes must activate or repress (they can't not interact), the system can be represented by a 3×3 connection matrix (Fig. 3b). Once this is realized, it is a trivial matter to automate the generation of all possible matrices. This approach can be modified easily to fit any topology with any number of interactions.

Examining the network in Fig. 3b, one can see that if the activation from x to z were switched to be from x to y, we are left with a network which functions identically to the original, with only the names of the nodes changed. To eliminate this sort of redundancy, one can utilize linear algebra transformations to check for equivalence. In the fully connected triads example, this involves generating a set of permutation matrices that account for all possible row and column switching (Fig. 3c). Two connection matrices (M_1 and M_2) are considered equivalent to one another if they satisfy the equation $M_1 = P_i{}^{*} M_2{}^{*} P_i'$ $(i = 1, 2, 3, 4, 5)$, where i represents the permutation matrices. If matrices are found to be equivalent, one can safely be dropped to reduce downstream computation. It is worth noting explicitly that, while the algorithm for generating all possible systems of ODEs is relatively straightforward when moving into higher-order and nonsymmetrical networks, this is not necessarily the case for permutation matrix generation and matrix elimination. Additionally, if nodes are intended to represent specific genes, rather than generic transcription factors, network elimination of this type may be entirely inappropriate. In short, when working with more complex networks, thorough knowledge of linear algebra is required to determine an appropriate method for eliminating redundancy.

3.2 Deterministic Analysis

ODEs return purely deterministic outputs; given the same initial conditions (ICs) and parameters, the output will be the same every time the equations are run. ODE models may be used to fit experimental data, providing additional information on the underlying regulatory framework. This may be performed separately from or in conjunction with bifurcation analysis, which reveals multistable behaviors inherent in stem-like cell types. This technique identifies ranges of parameters under which the system exhibits hysteresis: it can hold more than one steady-state expression level. The specific state the system holds is determined by the ICs provided.

3.2.1 Bifurcation Analysis

At its most basic, bifurcation analysis traces steady states in parameter space. The system of ODEs is run to steady-state expression, then one or more parameters are altered slightly, and the new steady-state behavior is compared to the previous levels. This is done iteratively, allowing the steady states to be determined for a range of parameter values. In addition to identifying stable steady

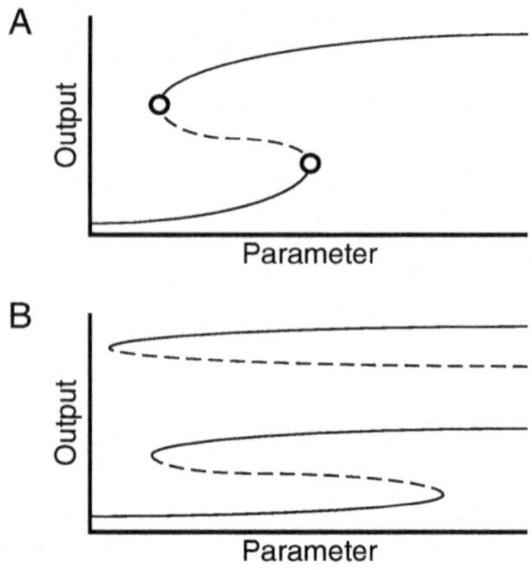

Fig. 4 Bifurcation analysis can reveal regions of multistability within parameter space. (**a**) S-shaped hysteresis curve with bifurcation points indicated with circles. These are the parameter values at which the system undergoes a state change. (**b**) For more complex bifurcation analyses, a wide range of ICs should be chosen to avoid missing relevant information. Here, the topmost SSS could be missed entirely if a very large IC is not tested

states, bifurcation analysis uncovers unstable states (which divide the space between stable attractors), a host of unique bifurcation points (specific values at which behavior changes), and more exotic behaviors (like oscillations). To perform bifurcation analysis, a continuation toolbox, like MATCONT, is required.

An important factor in running a bifurcation analysis is to set a wide enough range on the parameter under observation. Figure 4a shows the hysteresis curve of the simple mutual inhibition switch from Fig. 1a, made using the information from bifurcation analysis. This "S" shape is common, indicative of a region of multistability. Within this parameter range, the system can settle into either a high- or low-expressing state, depending on the ICs. In setting the range of parameters to search for this multistable behavior, it is necessary to set it wide enough to at least include both bifurcation points. Because the location of these points is not usually known prior to analysis, setting the limits beyond what you might consider a physiologically relevant range is suggested. This may include the use of negative parameter values which, while not even physiologically attainable, may be necessary for uncovering bifurcation points and thereby tracing a complete hysteresis curve.

In addition to parameter range selection, in more complex systems, it is good practice to perform bifurcation analyses from a variety of initial conditions. Because this technique essentially traces a single steady state as it changes with a given parameter, it cannot reveal additional steady states if they are discontinuous with the state being traced. As an example of this, see the hypothetical hysteresis plot in Fig. 4b. For relatively simple networks, this may be done empirically; however, as network complexity grows, more computationally rigorous methods may be required. These are covered in Subheading 3.3.3.

3.2.2 Parameter Fitting

If experimental data on a specific network is available, it can be used to determine which parameter values best fit the data. For this to be effective, a range of data points, either distributed in time (a time course) or across parameter space (a dose-response curve), is needed. Using a nonlinear curve fitting algorithm, such as MATLAB's *lsqcurvefit*, the model is run with an initially random parameter set and compared to the experimental data, generating a measurement of error from the observed values. The parameters are then iteratively altered, and the model rerun, improving the fit and thereby reducing error with each iteration, until it arrives at a local minimum of error. This algorithm is very powerful, but there are several ways its implementation can break down, which we will discuss.

One of the first decisions that must be made is which parameters to analyze. Running an algorithm like this on a large system of ODEs or one with many parameters can be computationally expensive and time consuming. It is therefore often beneficial to fix any parameters whose values can be determined with reasonable certainty. Protein degradation times, for example, are often closely linked to cell division and can, in many contexts, be approximated using the cellular doubling time. One must simply solve for the exponential decay constant (λ) in the exponential decay function $N(t) = N_0 e^{-\lambda t}$. Here were know the doubling time (t), and it is assumed that $N(t)$ is ½ of N_0, as the actual protein decay rate is minor in comparison to the dilution rate due to cell division. Similarly, some parameter values, such as maximum production rates, may be deducible from experimental data. In a simple repressor defined by Eqs. (2) and (4), for example, if the degradation rate is deduced as discussed above, in the absence of a repressor, $a = d * x$. If data are available giving a value for x under these conditions, a can be set with some confidence. Finally, parameters may be able to be inferred from prior work.

Once any parameters which can be inferred have been fixed, then the task at hand is setting bounds on those which remain unknown. If these bounds are set too narrowly, one risks excluding the best fit for the system. However, if the bounds are set too wide, the best fit may be missed as the algorithm converges to local minima far from the global best fit. Developing an empirical understanding of your system of ODEs is necessary for choosing proper bounds. This comes in part from understanding prior literature on similar cellular processes. Hill coefficients (n) in Eqs. 1 and 2 are typically <10, as high degrees of nonlinearity tend to make processes unstable. Relatedly, repression or activation coefficients (k) from Eqs. 1 and 2 are sometimes referred to as "half-max values" because they are related to the amount of activator or repressor required to transition expression to half its maximum value. A range of potential values may be deduced if additional information is known about the system. Once a good initial guess is determined, setting bounds around this guess is trivial. If the analysis is run and one or more parameter values consistently converge to the bounds that you have chosen, this is a good indicator that your bounds should be expanded.

Once all parameters have been set or bounded, the algorithm is run hundreds of times. On each run, bounded parameters' initial values are chosen randomly, and the algorithm finds a set of parameters whose error converges to a local minimum. Unlike discussed in Subheading 3.3.1, ODE ICs should remain fixed across all of these runs, corresponding to whatever ICs were used to generate the underlying data against which the model is being fit. Each run will result in a set of parameters and corresponding error. These can be ranked from lowest to highest error, and a subset of those with the lowest error can be further analyzed to glean additional information on levels and interrelations between different parameters within the network.

If any confusion remains on the difference between random initial conditions and initial parameters, *see* **Note 2**.

3.2.3 Network Topology Exploration

If you are exploring network topologies for multistable behavior, a more computationally expensive approach is required. For this application, access to a distributed computing cluster is necessary. As discussed in Subheading 3.2.3, you must first generate a minimal set of all possible networks that fit your predetermined criteria. Then, each network must be evaluated with a range of parameters and ICs to determine if it exhibits multistable behavior and, if so, how many unique states it can hold. Like in the previous section, this application explores a range of values for several parameters; however, in this case, preset values are applied in combination with a set of preset ICs across the full range of all possible networks. There is no randomness here.

This type of analysis is, by definition, not based on data; rather, it is a more theoretical exploration of topology behaviors. As such, the prior discussion of setting accurate parameter bounds and selecting physiologically relevant values is less relevant here than when parameterizing a known network. Here it is more important that parameter values make sense relative to one another, with the assumption that findings would scale to an actual biological system. This altered focus typically leads researchers to select easy to work with ranges for parameters, often utilizing powers of 10, such as 0.1, 0.3, 1, 3, and 10. Due to the highly combinatorial nature of network topology exploration, only a few values can be chosen for each set of parameters, and one should attempt to fix as many parameters as possible to reduce complexity. Faucon et al., for example, reduced their fully connected triad system to 9 parameters, 3 with 3 levels and 6 with 5 levels, but this still left 3^3 * $6^5 = 421{,}875$ parameter sets to check for every IC of every possible network [5].

Similar to how parameter space was divided, initial condition space must also be divided. The range of ICs must be large enough to encompass all possible expression levels and subdivided enough to have a high likelihood of finding all stable steady states. The form of Hill equations makes finding their maximum expression easy: it is simply the value of a in Eqs. 1 and 2. Maximum steady-state expression is therefore the sum of all production terms divided by the degradation term. Once this is known for each variable, the range of [0 max] can be subdivided into segments covering the whole of IC space. The number of segments will vary depending on the system being analyzed but should generally be much more than the total number of theoretical roots on the right-hand side of the ODE, which is numerically equivalent to the steady states of the ODE. The goal is to have enough coverage of IC space so that it is likely that at least one IC will converge to each stable steady state (SSS). Around 10 segments per variable are a good starting value, which can be adjusted depending on the system being analyzed.

Finally, for each network, for each parameter set, for each IC, right-hand side of the system of ODEs can be solved numerically. In MATLAB, there are several built-in algorithms, such as *fsolve*, for solving nonlinear equations like the ones we have constructed. After solving for all ICs within one parameter set of a given network, unique roots can be consolidated for further analysis. Unique roots can be entered into the Jacobian matrix and the eigenvalues calculated. Solutions with all negative eigenvalues are SSS.

3.3 Stochastic Analysis

While deterministic analysis is useful for identifying SSS and predicting average population behavior (in many circumstances), it does not recapitulate expression variance inherent in cellular systems. Some processes, like stability switching or random differentiation from a neutral state, cannot be exhibited by a simple ODE model. To explore these sorts of behavior, the model must be modified to introduce stochasticity to the genes' expression. This is frequently done in one of two ways: converting the system of ODEs to a Gillespie algorithm [13] or converting the ODEs into stochastic differential equations (SDEs) with a method like the Langevin equation [14].

In both implementations covered here, the level of noise is correlated with the size of the associated population. A population with a SSS level of 100 will exhibit much more relative variability than one with a SSS level of 10,000. For this reason, it may be necessary to rescale parameters to approximate a physiologically relevant level of gene expression variation. Many common data acquisition methods, such as flow cytometry and qPCR, give outputs which are correlated with, but not identical to, the size of intracellular protein or RNA populations. Fitting parameters to this data may give upscaled parameter values which may subsequently underestimate the role of noise on expression. Therefore, it may be helpful to modify your equations' production terms with a rescaling factor which allows you to easily up- or downscale the simulation without changing its qualitative behavior. This can be achieved by multiplying *a* and *k* terms in Eqs. 1 and 2 by a constant value. Adjusting the SSS value to be around 100 is a good starting point for achieving a reasonable amount of expression noise. Of course, this level can be adjusted either to explore the effects of noise or to simulate data in which the noisiness is known.

3.3.1 Gillespie Algorithms

The Gillespie algorithm breaks the simulation down into discrete events in which the populations being modeled experience some change: production or degradation of individual proteins or molecules, typically. Each event is assigned a probability based on its associated rate, relative to the rates of all other events in the network. As simulation time progresses, events occur one after another with variable time between each, dependent on the current state of the system. The benefit of this approach is that it fairly accurately simulates what is actually happening in the intracellular environment by converting all populations to discrete values. However, this algorithm tends to be relatively slow, particularly when running larger gene networks. Methods to improve simulation time have been developed (such as tau-leap methods), but these introduce approximations which may affect the accuracy of the simulation [13]. Here we will cover the basics of constructing a Gillespie algorithm, but know that there are many ways that this framework can be modified for different applications.

The algorithm is a simple four-step process: initialization, random step, update, and iterate. Initialization involves defining the system and setting ICs. Then, it will take random-sized steps through time, at each step selecting a single event to occur. These random events will be used to update the state of the network. The new state will then be used to generate a new random step and further update the network state, iterating between taking random steps and updating the network state. At each step, rates of change are calculated, and two random numbers are generated to determine which event happens and at what time, making this a relatively computationally intense process.

It is easy to convert a system of ODEs into a Gillespie algorithm. Each equation must be split into the component parts which cause a single change (event) to a population. In a simple transcriptional repressor network like that shown in Fig. 5a, the unique

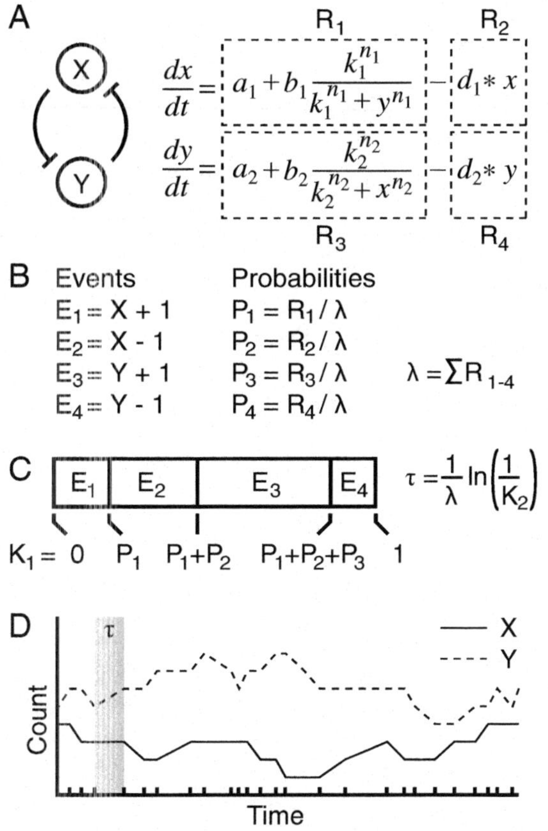

$$\frac{dx}{dt} = a_1 + b_1 \frac{k_1^{n_1}}{k_1^{n_1} + y^{n_1}} - d_1 * x$$

$$\frac{dy}{dt} = a_2 + b_2 \frac{k_2^{n_2}}{k_2^{n_2} + x^{n_2}} - d_2 * y$$

$$\tau = \frac{1}{\lambda} \ln\left(\frac{1}{K_2}\right)$$

Fig. 5 Implementation of the Gillespie algorithm. (**a**) The ODE system can be deconstructed into production and degradation terms. (**b**) Discrete events (E_{1-4}), with related rates (R_{1-4}), and probabilities (P_{1-4}) are defined. At each iteration of the algorithm, rates and probabilities are recalculated from the system's current state. (**c**) Random numbers (K_1 and K_2) are used to determine which event happens and the time since the previous event (τ). (**d**) Sample output of several iterations of the algorithm. Notice that at each τ only one variable changes and only every by ± 1

events which can occur are production of protein *x*, degradation of protein *x*, production of protein *y*, and degradation of protein *y*. The rates at which these events occur—and by extension the probabilities that they will happen at each time point—are obtained directly from deconstructing the ODEs into their production and degradation terms (Fig. 5b).

Based on their rates relative to one another, each event can be assigned a probability. An event's probability of occurring is $P_i = (R_i/\lambda)$, where R_i is the rate of the individual event and λ is the sum of all events' rates: $\lambda = \Sigma R_{1\text{-}n}$. When calculated for all events, $\Sigma P_{1\text{-}n} = 1$. At each iteration of the algorithm, a random number, K_1, chosen from the uniform distribution [0, 1] can be used to randomly select a single event to occur. The timing between events, usually referred to as τ, is governed by the exponential function: $\tau = (1/\lambda) * \ln(1/K_2)$, where λ is the sum of the rates of all events, as above, and K_2 is a random number from the uniform distribution [0, 1]. Figure 5c shows a visualization of how the random numbers, K_1 and K_2, are used to generate each time step, and Fig. 5d shows a sample output of the mutual inhibition toggle.

Production and degradation of intracellular populations is not the only thing that can be modeled with a Gillespie algorithm. Protein multimerization, for example, could be described as a series of discrete steps from monomers to dimers to tetramers, with reversals of each of these interactions and degradation events. For more complex interactions, it should be noted that a single event might see multiple populations changing in relation to one another. For example, in a dimerization event, two monomers are lost, and one dimer is gained.

3.3.2 Langevin Equations

A similar outcome can be obtained by modifying the ODEs into chemical Langevin equations, which introduces some randomness with each step forward in time in the ODE solver. Relative to the Gillespie algorithm, the use of a chemical Langevin equation is much faster to run; however, the mathematical derivation is much more complex and not as easy to grasp intuitively. To avoid abuse while implementing this method, a thorough investigation of primary literature is suggested [14]. For our purposes, at each iteration of the ODE, all chemical species are updated with the formula:

$$N(t+\Delta t) = N(t) + \Delta t A(N(t)) + \alpha * S\sqrt{\Delta t(F(N(t)) + B(N(t)))}z(t) \qquad (6)$$

The terms before α are standard for ODE solvers, but α and everything that follows have been added to introduce noise. Here, *N(t)* is the abundance of the population, *A* is the right-hand side of the ODE, and *F* and *B* are the forward (production) and backward (degradation) reaction terms in each equation. α is a scalar term for tuning the level of noise. *z(t)* and *S* are the standard normal variables and stoichiometric matrix, respectively, of the biochemical reactions. For a fuller description of these terms, as well as software tools to implement chemical Langevin equations, see work by Adalsteinsson, McMillen, and Elston [14].

3.3.3 Stochastic Analysis

With stochasticity integrated into the model, the methods for exploring the network to its fullest extent are expanded. As this is a relatively small field of research, however, there are few definitive methods for exploring stochasticity in biochemical models. Because stochastic simulations have noise, each run of the simulation will be different from every other run. For this reason, analysis of stochastic simulations can be time consuming, as the model must be run enough times to generate a characteristic picture of the effect that noise is having on expression. To account for low probability behavior, this can mean running the simulation tens, hundreds, or thousands of times.

Stochastic simulation can be used to explore the pull that different SSS exert on the system. While bifurcation analysis of an ODE model can separate state space into different attractor regions, stochastic simulations can quantify how likely those regions are to maintain stability in the face of noise. Faucon et al. used an approach like this to quantify the strength of different attractor states and measure the mean time to transition between states under variable noise conditions [5]. As one example, they show that increasing noise can cause some attractor nodes to transition from highly stable to nearly incapable of holding steady expression. Additionally, because expression noise can cause networks to transition between different SSS expression profiles, stochastic simulations can reveal preferred transition paths between SSS attractors. As computational power increases, allowing modeling of ever more complex networks, stochastic analysis of this sort can help reveal the mechanisms underlying complex differentiation pathways.

4 Notes

1. Reducing Network Complexity.

   Mathematical modeling of complex cellular behavior is still a relatively new discipline, limited by both our computational power and understanding of the network topologies underlying observed cellular behavior. For this reason, we often use abstracted or synthetic networks as stand-ins for the actual networks which regulate stem cells. For example, instead of trying to model and understand the full mechanism underlying pluripotency, the network of dozens of interconnected genes is frequently condensed down to its most basic form: the Oct4-Sox2-Nanog triad [5]. This sort of abstraction can reduce the system to one that is simple enough to study and understand while simultaneously being accurate enough to still represent the underlying regulatory complexities. Inevitably, some of the details are lost, and if one is not careful, you may end up eliminating the expected behavior from your model. Nonetheless, it is common practice to reduce networks to as few moving parts as possible.

   Alternatively, researchers can build gene networks from the ground up to decouple the behavior of interest from the rest of the cellular regulatory machinery. In these synthetic networks, fundamental interactions between genetic components can be more accurately observed and simulated via modeling, since they have been separated from many of the confounding factors which make natural networks so difficult to study. This is one of the fundamental differences between synthetic biology and more traditional disciplines attempting to apply mathematical methods to biology.

In addition to how the system is built and/or conceptualized, when one begins building models, it quickly becomes apparent that the model could be as complex as you could possibly make it. For a common process like transcription/translation, there are endless steps that could be added to the model: transcription factor interactions with the DNA, polymerase recognition of the promoter, assembly of the polymerase complex, transcriptional starting and pausing, incomplete transcription, termination dynamics, RNA capping and polyadenylation, nuclear export, transport through the cytoplasm to the ribosome, tRNA recruitment, and on and on. The list could be expanded nearly infinitely. Running mathematical models can be computationally expensive, and biology is endlessly complicated. If too many details are included, model run times can skyrocket, slowing down analysis considerably. It is therefore necessary to reduce model complexity while maintaining the fundamental behavior that you are trying to simulate.

2. Random Initial Conditions, Parameters, and Stochasticity.

To ensure that there is no confusion, we would like to talk briefly about the different types of randomness employed in this work. Random initial conditions refer to randomizing the starting values of the variables (each equation in the system of ODEs represents one variable). This may be used in some network exploration algorithms, but in most cases ICs are defined rationally based on experimental constraints or to cover all of potential IC space. Alternatively, parameters are all non-variable terms in the ODE. These can be randomized for curve fitting algorithms, which will then iteratively adjust the parameter values to arrive at a local best fit for a model against a set of experimental data. Finally, stochasticity refers to randomness in gene expression, introduced into the model either by converting it to a discreet Gillespie algorithm or through implementation of the Langevin equation. In both the Gillespie and Langevin cases, parameter values and initial conditions remain fixed, but the changes in variables' expression as the simulation runs are subject to some randomness.

References

1. Menn DJ, Wang X (2001) Stochastic and deterministic decision in cell fate. eLS. https://doi.org/10.1002/9780470015902.a0025319
2. Menn DJ, Su R-Q, Wang X (2017) Control of synthetic gene networks and its applications. Quant Biol 5:124–135. https://doi.org/10.1007/s40484-017-0106-5
3. Wang L-Z, Wu F, Flores K et al (2016) Build to understand: synthetic approaches to biology. Integr Biol 8:394–408. https://doi.org/10.1039/C5IB00252D
4. Ellis T, Wang X, Collins JJ (2009) Diversity-based, model-guided construction of synthetic gene networks with predicted functions. Nat Biotechnol 27:465–471. https://doi.org/10.1038/nbt.1536
5. Faucon PC, Pardee K, Kumar RM et al (2014) Gene networks of fully connected triads with complete auto-activation enable multistability and stepwise stochastic transitions. PLoS One 9:e102873. https://doi.org/10.1371/journal.pone.0102873
6. Ma W, Trusina A, El-Samad H et al (2009) Defining network topologies that can achieve biochemical adaptation. Cell 138:760–773. https://doi.org/10.1016/j.cell.2009.06.013
7. Wu M, Su R-Q, Li X et al (2013) Engineering of regulated stochastic cell fate determination. Proc Natl Acad Sci 110:10610–10615. https://doi.org/10.1073/pnas.1305423110
8. Gardner TS, Cantor CR, Collins JJ (2000) Construction of a genetic toggle switch in *Escherichia coli*. Nature 403:339–342
9. Wu F, Su R-Q, Lai Y-C, Wang X (2017) Engineering of a synthetic quadrastable gene network to approach Waddington landscape and cell fate determination. eLife 6:e23702. https://doi.org/10.7554/eLife.23702
10. Johnson KA, Goody RS (2011) The original michaelis constant: translation of the 1913 Michaelis–Menten paper. Biochemistry (Mosc) 50:8264–8269. https://doi.org/10.1021/bi201284u
11. Weiss JN (1997) The hill equation revisited: uses and misuses. FASEB J 11:835–841. https://doi.org/10.1096/fasebj.11.11.9285481
12. Elowitz MB, Leibler S (2000) A synthetic oscillatory network of transcriptional regulators. Nature 403:335–338. https://doi.org/10.1038/35002125
13. Li H, Cao Y, Petzold LR, Gillespie DT (2008) Algorithms and software for stochastic simulation of biochemical reacting systems. Biotechnol Prog 24:56–61. https://doi.org/10.1021/bp070255h
14. Adalsteinsson D, McMillen D, Elston TC (2004) Biochemical network stochastic simulator (BioNetS): software for stochastic modeling of biochemical networks. BMC Bioinformatics 5:24. https://doi.org/10.1186/1471-2105-5-24

Part II

Single Cell Approaches

Chapter 9

Trajectory Algorithms to Infer Stem Cell Fate Decisions

Edroaldo Lummertz da Rocha and Mohan Malleshaiah

Abstract

Single-cell trajectory analysis is an active research area in single-cell genomics aiming at developing sophisticated algorithms to reconstruct complex cell-state transition trajectories. Here, we present a step-by-step protocol to use CellRouter, a multifaceted single-cell analysis platform that integrates subpopulation identification, gene regulatory networks, and trajectory inference to precisely and flexibly reconstruct complex single-cell trajectories. Subpopulations are either user-defined or identified by a graph-clustering approach in which a k-nearest neighbor graph (kNN) is created from cell-to-cell distances in a low-dimensional embedding. Edges in this graph are weighted by network similarity metrics (e.g., Jaccard index) to robustly encode phenotypic relatedness, creating a representation of single-cell transcriptomes suitable for community detection algorithms to identify clusters of densely connected cells. This subpopulation structure represents a map of putative cell-state transitions. CellRouter implements a flow network algorithm to explore this map and reconstruct cell-state transitions in complex single-cell, multidimensional omics datasets. We describe a step-by-step application of CellRouter to hematopoietic stem and progenitor cell differentiation toward four major lineages—erythrocytes, megakaryocytes, monocytes, and granulocytes—to demonstrate key components of CellRouter for single-cell trajectory analysis.

Key words Single-cell genomics, Stem cell differentiation, Trajectory reconstruction, Cell fate transitions, Hematopoietic stem cells, Single-cell analysis, Computational biology, Systems biology

1 Introduction

Single-cell trajectory analysis aims to reconstruct the structure and dynamics of cell-state transitions [1–4]. This includes, for example, lineage diversification, when stem or progenitor cells differentiate to multiple lineages; lineage convergence, when distinct precursor cells give rise to a common cell type; and re-specification of cell identity by cellular reprogramming. Several algorithms have been developed to organize cells according to molecular divergence in unidimensional trajectories [5–7], when a stem or progenitor cell gives rise to only one cell type, and bifurcations, when two or more cell lineages are generated from precursor cells [8–14]. These algorithms were designed to reconstruct single-cell trajectories describing cell-state transitions by exploring asynchronous cellular

Patrick Cahan (ed.), *Computational Stem Cell Biology: Methods and Protocols*, Methods in Molecular Biology, vol. 1975, https://doi.org/10.1007/978-1-4939-9224-9_9,

behaviors, stochastic and regulated cell-to-cell variation, or the intrinsic heterogeneity of single-cell multidimensional omics profiles. Application of these algorithms to a variety of experimental systems has provided important insights on cell fate decision processes [8–10, 15–18].

The existence of complex subpopulations with diverse cell types or states within single-cell datasets impose several analytical challenges to identify cell-state transition trajectories. Although single-cell transcriptional profiles allow in-depth characterization of complex cellular ecosystems in health and disease, sophisticated analytical strategies are required. Here we describe a step-by-step protocol for using CellRouter [8]—a multifaceted single-cell analysis platform to reconstruct complex cell-state transition trajectories with high resolution and flexibility. As an example or a case study, we apply CellRouter to analyze gene expression dynamics during hematopoietic stem and progenitor cell differentiation. Moreover, it can be easily applied to analyze any single-cell transcriptomic or proteomic datasets. CellRouter explores the subpopulation structure of single-cell multidimensional data to precisely identify transitions between cell states. Moreover, CellRouter integrates gene expression dynamics reconstructed along cell-state transition trajectories with gene regulatory networks (GRNs) to identify putative regulators of these fate transitions. The analysis pipeline consists of six main steps (Fig. 1): Starting from single-cell transcriptomes (1), GRN reconstruction and determination of subpopulation structure are performed (2). Then, this subpopulation structure is explored to reconstruct multistate transition trajectories (3) and (4). Finally, identification of dynamically regulated genes and putative regulators as well as characterization of the waves of transcriptional regulation during cell-state transitions are performed (5) and (6).

2 Methods

Single-cell gene expression profiles are often represented as a two-dimensional matrix M. Typically, rows represent genes and columns represent individual cells. CellRouter explores the subpopulation structure of single-cell profiles to precisely identify cell-state transition trajectories in complex cellular ecosystems. To illustrate how to use CellRouter to study gene expression dynamics during lineage diversification, we will reanalyze a hematopoietic stem and progenitor cell dataset that includes differentiated lineages [19].

2.1 Required Software

To complete this protocol, Java and the R environment for statistical computing are required. We recommend R version $\geq$3.4 and Java version $\geq$1.8. RStudio is also recommended. Several R

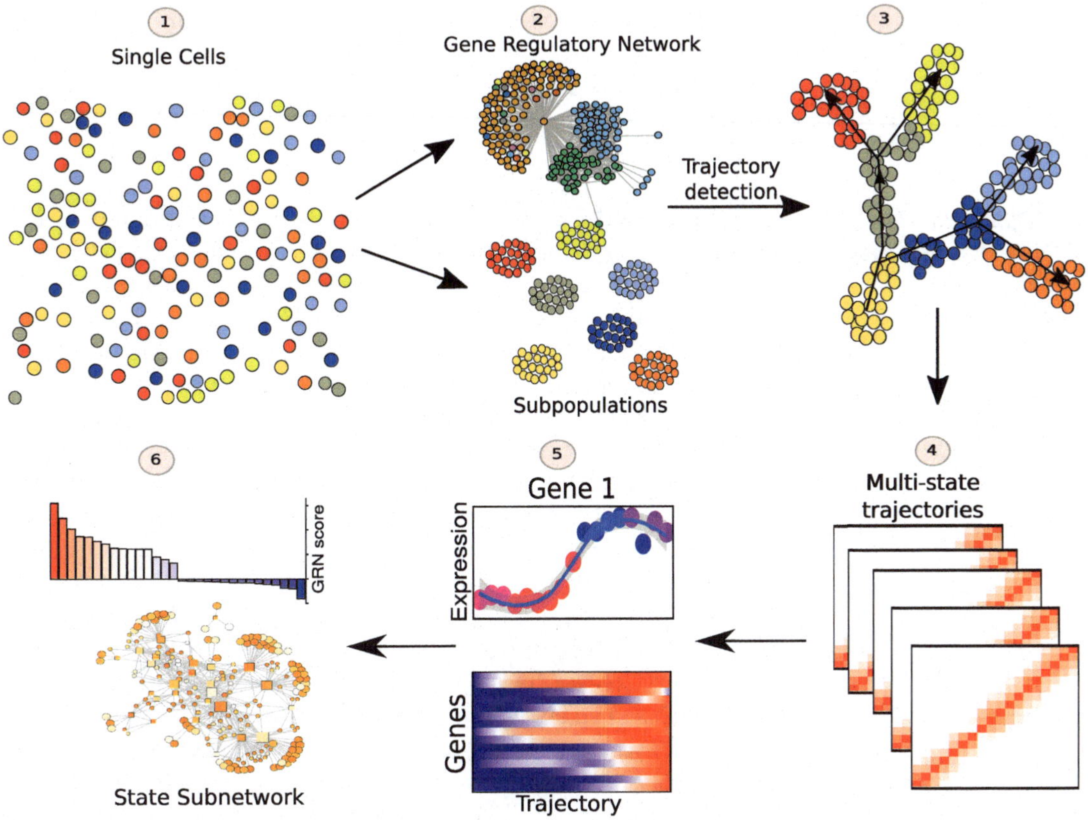

Fig. 1 Overview of CellRouter (*1*). Starting from single cells representing multiple cell states, including stable and in transition, a gene regulatory network was built to identify putative gene regulatory relationships, and subpopulations were identified by a combination of learning the structure of the data and community detection algorithms (*2*). High-resolution multistate trajectories are identified from any subpopulation to any other, including intermediate and mature cell types (*3*). Multistate trajectories illustrating the divergence of single-cell transcriptomes along each trajectory progression (*4*). Identification of genes implicated in the dynamic biological process under study, such as differentiation (*5*), and identification of regulators driving or mediating cell fate transitions at the gene and network level (*6*)

packages are required. These packages can be installed by typing in RStudio:

```
list.of.packages ←
c('reshape','reshape2','pheatmap','clusterProfiler','ReactomePA','plotrix',
'tsne','igraph','ggplot2','mclust','grid','scde','gplots','genefilter',
'Rtsne', 'DESeq2')
new.packages <- list.of.packages[!(list.of.packages %in%
installed.packages()[,"Package"])]
if(length(new.packages)) install.packages(new.packages,
repos=c("http://cran.rstudio.com/", "https://bioconductor.org/biocLite.R"))
```

CellRouter can be downloaded from GitHub at https://github.com/edroaldo/cellrouter, where processed datasets and fully reproducible examples are available. Processed data and files to complete this protocol are available inside the folder "mixed-states" of the CellRouter GitHub repository.

2.2 Application of CellRouter to Mouse Hematopoietic Differentiation

Here, we applied CellRouter to single-cell RNA-Seq data consisting of murine hematopoietic stem/progenitor cells (HSPCs; lin^-, $Sca1^+$,$c\text{-}Kit^+$ (LSKs)), common myeloid progenitors (CMPs), granulocyte-monocyte progenitors (GMPs), and $LK34^+$ cells (lin^-,$c\text{-}Kit^+$, $CD34^+$). Typically, a set of informative genes is identified prior to dimensionality reduction. These gene sets can be obtained by differential expression analysis between subpopulations or by identifying the most variable genes across single cells, as these gene sets potentially contribute to cell-state specific molecular programs (*see* **Note 1**). We used genes identified by iterative clustering as reported in the original study [19]. We performed dimensionality reduction using t-distributed stochastic neighbor embedding (t-SNE) [20] and annotated cell types based on the original publication [19] and subpopulation-specific gene expression signatures identified by CellRouter (Fig. 2a). Due to the stochastic nature of t-SNE analysis, usually we initialize t-SNE coordinates from multidimensional scaling (MDS) coordinates, which produces t-SNE plots that look very similar across multiple runs (*see* **Note 1**). The following example code loads the gene list used for t-SNE analysis from the processed data.

1. **Load the gene list containing genes to be used for dimensionality reduction, expression matrix, and sample metadata.**

```
guideGenes <- read.csv('ICGS-s5.txt', sep='\t', row.names = 1)
guideGenes <- guideGenes[-1,]
ndata <- get(load('Olsson_clean_dataset.R'))
samples <- get(load('results/samples.R'))
```

2. **Perform t-SNE analysis using the following:**

```
di <- dist(t(ndata[rownames(guideGenes),]))
m <- tsne(di, initial_config=cmdscale(di,k=2))
colnames(m) <- c('tSNE1', 'tSNE2')
```

Next, we predict gene regulatory networks from single-cell transcriptomes. Although this can be performed using all genes, we perform this analysis using a subset of potentially informative genes based on principal component analysis, which will also be used for trajectory analysis.

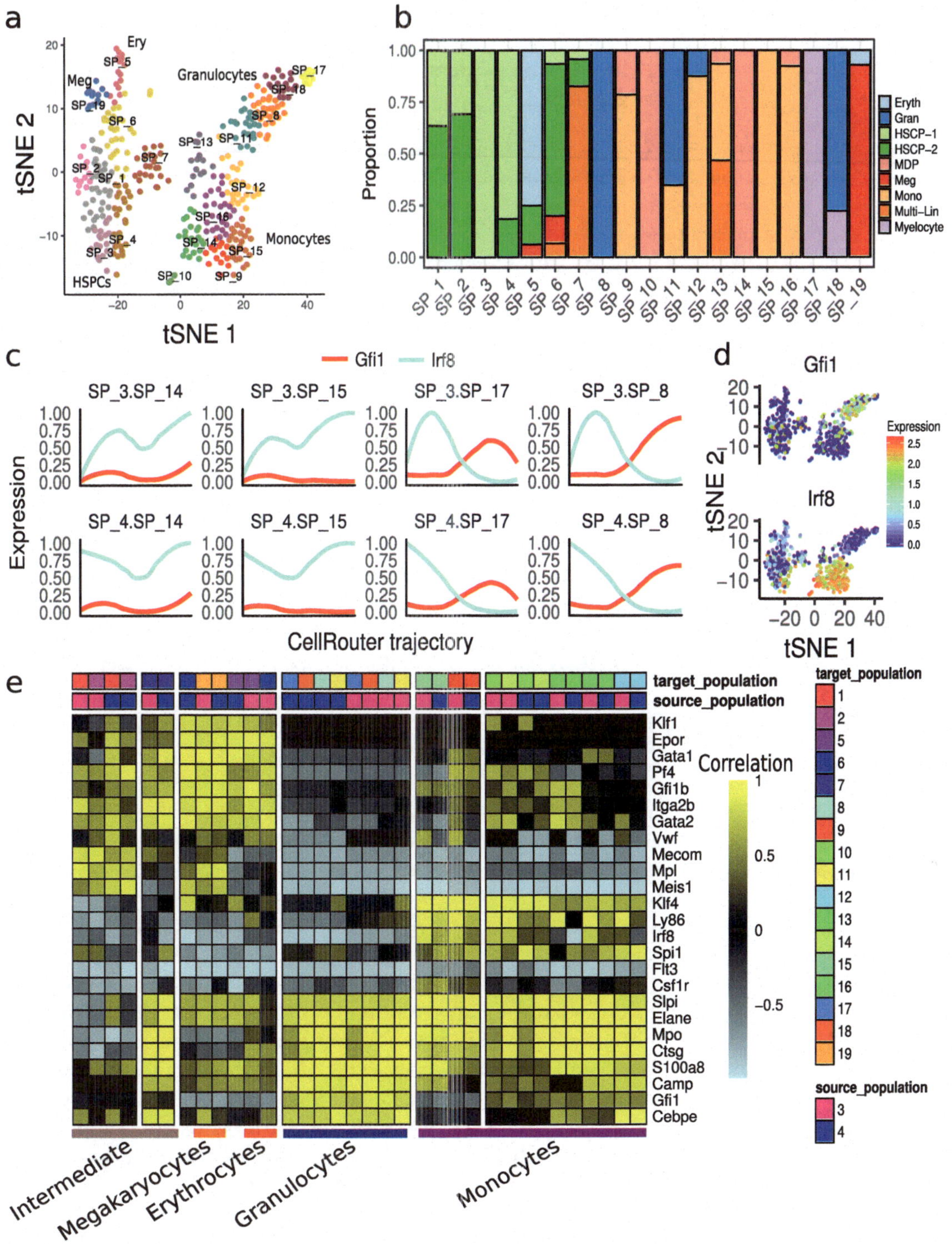

Fig. 2 Multi-lineage differentiation dynamics from HSPCs. (**a**) t-SNE analysis color coded by subpopulations identified using graph-based clustering. (**b**) Proportion of cells in each cluster previously annotated in the original publication as hematopoietic populations. (**c**) Expression trends of master regulators during differentiation from HSPCs to granulocytes and monocytes as well as myelocytes. (**d**) t-SNE analysis colored by expression of granulocyte (top panel) and monocyte (bottom panel) master regulators. (**e**) Transcriptional dynamics of selected transcriptional factors and lineage specifying genes

3. **Principal component analysis to select a subset of informative genes.**

```
pca <- prcomp(t(ndata), scale=TRUE, center=TRUE)
loadings <- pca$rotation
num_pc <- 5
quantile <- 0.975
genes2use <- unique(as.vector(unlist(apply(loadings[,1:num_pc], 2,
function(x){print(quantile(x, quantile)) names(x[which(abs(x) >=
quantile(x, quantile))])}))))
```

4. **Predict transcription factor → target associations.**

```
ggrn <- buildGRN('Mm', ndata, genes2use, 5, 'results/GRN.R')
```

Next, we create a CellRouter object that will store information to reconstruct cell-state transition trajectories, identify subpopulations and gene expression signatures, and identify markers of each cell type or state identified (*see* **Note 2**).

5. **Create CellRouter object.**

```
cellrouter <- CellRouter(expdata=ndata[genes2use,], annotations=colnames(ndata))
cellrouter@rdimension <- as.data.frame(m)
cellrouter <- findsubpopulations(cellrouter, 4, 'jaccard', 'results/kNN_network.
gml')
cellrouter <- diffexpr(cellrouter, column='population', pvalue = 0.05)
markers <- findmarkers(cellrouter)
write.csv(markers, file='results/gene_signatures.csv') #save markers
plotProportion(cellrouter, 'conditions', 'population', 5.5, 2.5, 'results/propor-
tion.pdf')
plotReducedDimension(cellrouter, 3.5, 3.5, filename='results/tSNE.pdf')
```

We found a refined subpopulation structure and two distinct and presumably bipotent subpopulations upstream of a potential lineage bifurcation, one in an intermediate position between HSPCs and monocytes/granulocytes and another one preceding the megakaryocyte/erythrocyte divergence (Fig. 2a). We then assessed the proportion of known cell types in each cluster and further annotated subpopulations (Fig. 2b). Subpopulation identification was based on graph-based clustering, where a k-nearest neighbor graph (kNN) is generated using Euclidean distances. This graph is transformed to encode network similarities (e.g., Jaccard index) to represent phenotypic similarity. The resulting graph is suitable for community detection algorithms, such as the Louvain method, to robustly identify communities of densely interconnected cells

[21, 22]. Moreover, subpopulations identified using alternative clustering strategies, such as hierarchical clustering, model-based clustering, or density-based clustering, can also be used as input to CellRouter.

CellRouter explores this subpopulation structure as a reference map of cell-state transitions to reconstruct trajectories between any cell states identified. Therefore, dimensionality reduction and subpopulation identification are important steps. Moreover, cell cycle may have a strong confounding effect and should be carefully considered, as well as other hidden sources of variation [23] (*see* **Note 3**). We selected subpopulations 3 and 4, which are enriched for HSPCs, as starting points for trajectory identification (Fig. 2b). This allows to study differentiation dynamics from subpopulations 3 and 4 toward all other subpopulations, regardless of branching structure or position within a branch. To reconstruct differentiation trajectories, use the commands described below.

6. **Create kNN graph.**

```
cellrouter <- createKNN(cellrouter, 15, 'jaccard',
'results/paths/kNN_network_trajectory.gml')
filename <- "results/paths/cell_edge_weighted_network.txt"
write.table(cellrouter@graph$edges, file=filename, sep='\t', row.names=FALSE,
col.names = FALSE, quote=FALSE) #input network
```

7. **Select starting subpopulations.**

```
sources <- c('SP_3', 'SP_4')
#remaining subpopulations are targets
targets <- setdiff(as.vector(cellrouter@sampTab$population), sources)
cellrouter <- findpaths(cellrouter, libdir, paste(getwd(), 'results/paths',
sep='/'),
method="graph")
```

Next, we have to preprocess the trajectories identified by the *findpaths* function.

8. **Preprocessing of trajectories.**

```
ranks <- c('path_cost', 'path_flow', 'rank', 'length')
cellrouter <- processtrajectories(cellrouter, genes2use, path.rank=ranks[3],
num.cells = 3, neighs = 1)
names <- unique(names(cellrouter@pathsinfo$distr))
cellrouter <- correlationpseudotime(cellrouter, type='spearman')
cellrouter <- topgenes(cellrouter, 0.85, 0.15)
cellrouter <- smoothdynamics(cellrouter, names)
cellrouter <- clusterGenesPseudotime(cellrouter, 4)
save(cellrouter, file='results/CellRouter_Mixed_States_Processed.R')
```

Transcriptional dynamics of lineage-specific transcription factors confirmed the anticorrelated expression dynamics of Irf8 and Gfi1 in GMPs and revealed a potential early lineage priming in HSPCs (subpopulation 4 expresses Irf8) toward monocyte differentiation (Fig. 2c, d). Gfi1 and Irf8 are important pro-differentiation factors for granulocytes and monocytes, respectively. These results suggest that progenitor cells primed for monocytes intrinsically have low expression levels of transition-specific regulators of other lineages. However, state transitions to granulocytes might require concurrent upregulation of Gfi1 and downregulation of Irf8. As differentiation progresses, Gfi1 is downregulated in myelocytes (subpopulation 17), the most distant cell state in the granulocyte branch, suggesting that it is no longer required in late differentiation. To perform these analyses, use the following commands.

9. **Select specific transitions.**

```
transitions <- c('SP_4.SP_8', 'SP_4.SP_17', 'SP_3.SP_8', 'SP_3.SP_17', 'SP_4.
SP_14', 'SP_4.SP_15', 'SP_3.SP_14', 'SP_3.SP_15')
plottrajectories(cellrouter, transitions, c('Gfi1', 'Irf8'), rescale=FALSE,
columns=4, width=6, height=2, filename='results/dynamics_Irf8_Gfi1.pdf')
plotDRExpression(cellrouter, c('Gfi1', 'Irf8'), TRUE, 1, 3, 3, 'results/
tSNE_Irf8_Gfi1.pdf')
```

We then investigated the transcriptional dynamics of lineage-specific transcription factors and marker genes (Fig. 2e). The plot shows Spearman's rank correlation of each gene along each differentiation trajectory. Consistently, Irf8 is positively correlated to cell fate transitions from HSPCs to monocytes and anticorrelated to transitions in the granulocyte branch. Gfi1 showed an opposite pattern. Similarly, erythroid-specific genes, such as Klf1, Epor, and Gata1, are highly correlated to cell fate transitions in the erythroid branch, while megakaryocyte genes, such as Vwf, are highly correlated to transitions in the megakaryocyte branch (Fig. 2e).

10. **Plot heatmap of lineage-specific genes.**

```
genelist <- c('Mecom', 'Flt3', 'Meis1', 'Gata2', 'Mpl', 'Vwf', 'Pf4', 'Itga2-
b','Epor', 'Klf1', 'Gata1', 'Gfi1b', 'Cebpa', 'Spi1', 'Ctsg', 'Mpo','Elane',
'Slpi', 'Ly86', 'Csf1r', 'Irf8', 'Klf4', 'Gfi1', 'Cebpe', 'S100a8', 'Camp',
'Mmp9')
df <- plotheatmapCategory(cellrouter, threshold=0.6, ccols = 6, genelist=gene-
list, width=9, height=5)
```

Next, we assessed whether dynamically regulated genes along each trajectory show branch-specific gene expression patterns. Specifically, we evaluated how genes involved in erythrocyte differentiation (transition from subpopulations 4 to 5) look like in the megakaryocyte trajectory (transition from subpopulations 4 to 19) (Fig. 3a) and vice versa (Fig. 3b), as well as how genes implicated in granulocyte differentiation (transition from subpopulations 4 to 8) look like in the monocyte trajectory (transition from subpopulations 4 to 15) (Fig. 3c) and vice versa (Fig. 3d).

11. **Branch-specific expression (erythroid/megakaryocyte branch).**

```
plotbranch(cellrouter, 'up','SP_4.SP_5', 'SP_4.SP_19', 1, width=3, height=4,
filename='results/SP_4.SP_5_branch_dynamics.png')
plotbranch(cellrouter, 'up','SP_4.SP_19', 'SP_4.SP_5', 1, width=3, height=4,
filename='results/SP_4.SP_19_branch_dynamics.png')
```

12. **Branch-specific expression (granulocyte/monocyte branch).**

```
plotbranch(cellrouter, 'up','SP_4.SP_8', 'SP_4.SP_15', 1, width=3, height=4,
filename='results/SP_4.SP_8_branch_dynamics.png')
plotbranch(cellrouter, 'up','SP_4.SP_15', 'SP_4.SP_8', 1, width=3, height=4,
filename='results/SP_4.SP_15_branch_dynamics.png')
```

Next, we used CellRouter to investigate two potential regulatory mechanisms during erythroid and granulocyte differentiation: (1) upregulation of transcription factors and their predicted target genes, to capture induction of lineage-specific transcriptional programs, and (2) downregulation of transcription factors leading to increased expression of target genes, to capture de-repression of lineage-specifying genes during differentiation. In both cases, target genes were upregulated during differentiation, but their predicted regulators could be upregulated (in red) or downregulated (in blue) (Fig. 3e, f).

13. **Predicted regulators of cell fate transitions.**

```
tfs <- find_tfs(species = 'Mm')
transitions <- c('SP_4.SP_5','SP_3.SP_5','SP_4.SP_19', 'SP_3.SP_19','SP_4.
SP_15',
'SP_3.SP_15', 'SP_4.SP_8','SP_3.SP_8')
x <- grnscores(cellrouter, tfs, transitions,direction='both', flip=TRUE, q.
up=0.85,
q.down=0.15,dir.targets='up', columns=4, width=8, height=7,
filename='results/GRN_score')
```

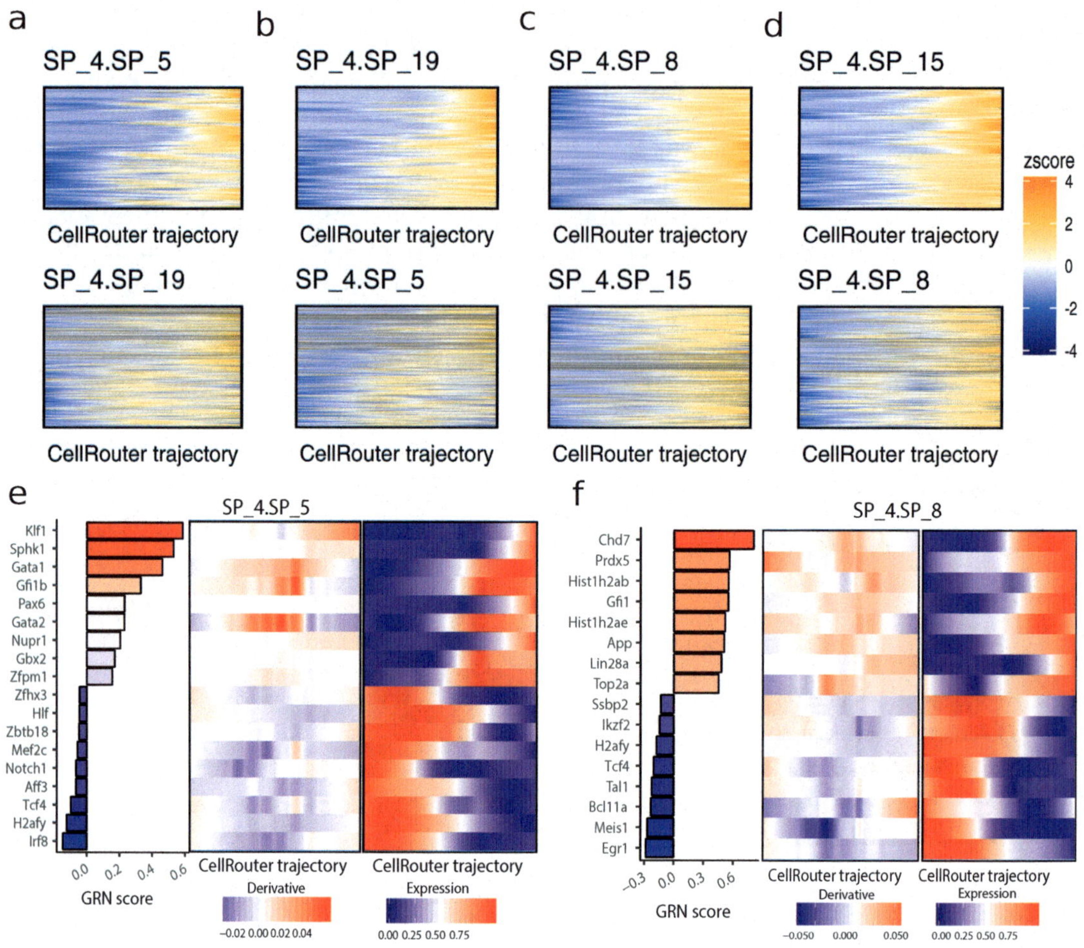

Fig. 3 Transition-specific gene expression dynamics. (**a**) Erythroid-specific gene expression kinetics (transition from subpopulation 4 to 5) in the erythroid/megakaryocyte branch. (**b**) Megakaryocyte-specific gene expression kinetics (transition from subpopulation 4 to 19) in the erythroid/megakaryocyte branch. (**c**) Granulocyte-specific gene expression kinetics (transition from subpopulation 4 to 8) in the granulocyte/monocyte branch. (**d**) Monocyte-specific gene expression kinetics (transition from subpopulation 4 to 15) in the granulocyte/monocyte branch. (**e**) Left panel: predicted regulators of transitions to the erythroid fate (transition from subpopulation 4 to 5). Middle panel: derivative of the gene expression dynamics of each regulator. Right panel: Kinetic profile of each regulator along the differentiation trajectory. (**f**) Left panel: predicted regulators of transitions to the granulocyte fate (transition from subpopulation 4 to 8). Middle panel: derivative of the gene expression dynamics of each regulator. Right panel: kinetic profile of each regulator along the differentiation trajectory

14. **Predicted regulators of erythroid differentiation.**

```
p <- transitions[1]
m2 <- plottr(cellrouter, p, x[[p]]$scores, cluster=FALSE,
2, 4.5, 4, paste('results/', p, 'GRN_Score_1.pdf',sep="))
```

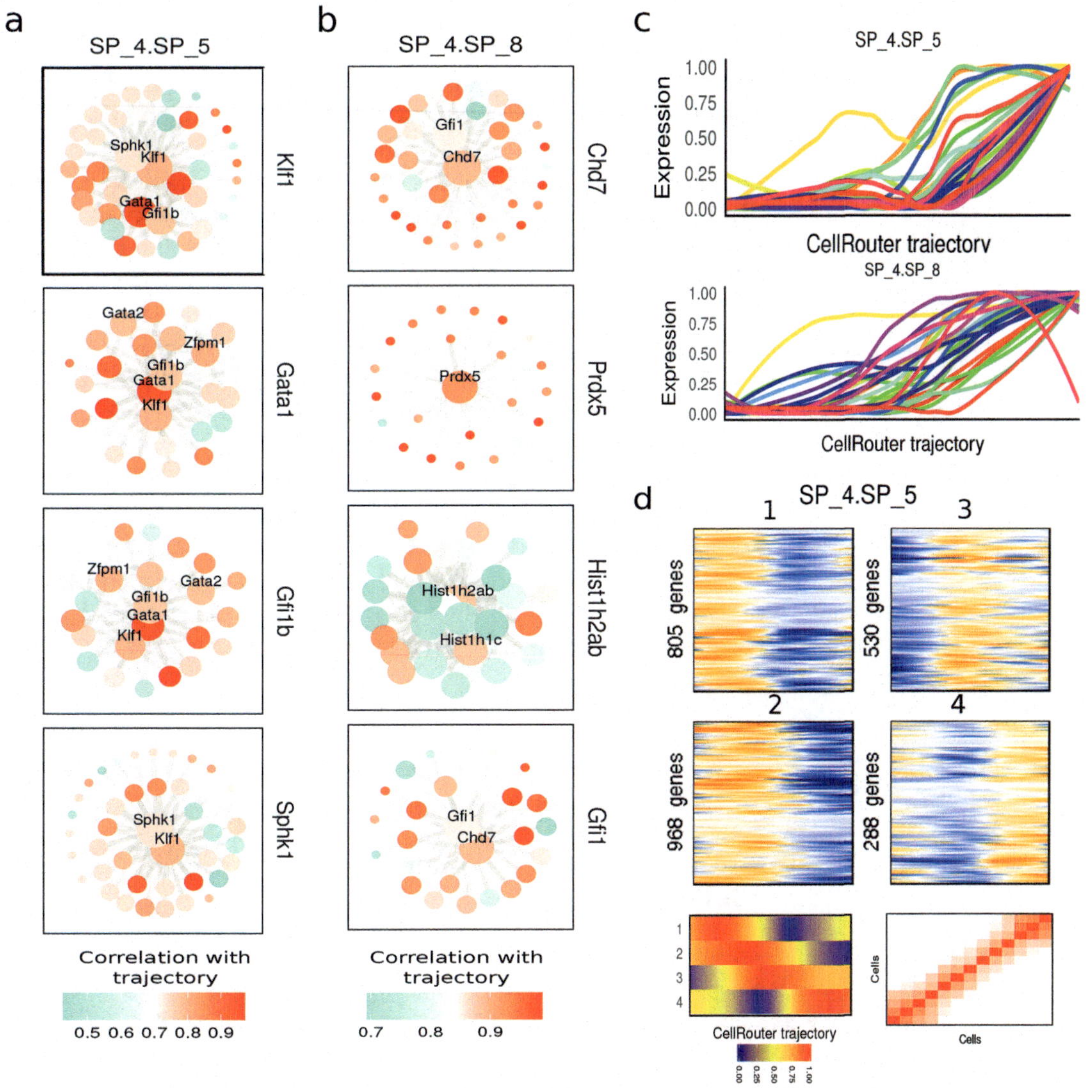

Fig. 4 Regulatory circuits and expression patterns. (**a**) Regulator-centered subnetworks of known top predicted regulators for transitions from HSPCs to erythrocytes (transition from subpopulation 4 to 5). (**b**) Regulator-centered subnetworks of top predicted regulators for transitions from HSPCs to granulocytes (transition from subpopulation 4 to 8). (**c**) Gene expression dynamics of the Klf1-centered network module (top panel) and Gfi1-centered network module (bottom panel). (**d**) Clustering of gene expression trends during erythrocyte differentiation from HSPCs into four transcriptional patterns, with a summary of the patterns in each cluster shown at the bottom left panel and Jaccard similarity matrix of the erythroid trajectory demonstrating the gradual divergence of single-cell transcriptomes

15. **Predicted regulators of granulocyte differentiation**

```
p <- transitions[7]
m2 <- plottr(cellrouter, p, x[[p]]$scores, cluster=FALSE,
2, 4.5, 4,paste('results/', p, 'GRN_Score2.pdf',sep=''))
```

Consistently, Klf1, Gata1, and Gfi1b were among the top genes driving the erythroid fate (transition from subpopulations 4 to 5). Consistent with its known biology, Gata2 was upregulated before Gata1 during erythroid differentiation, while Klf1 upregulation was observed in later stages (Fig. 3e). In the granulocyte branch, CellRouter rediscovered, in a fully unbiased manner, Gfi1, a critical gene controlling cell fate decisions between the granulocyte and monocyte lineages (Fig. 3f). Transcriptional regulators downregulated during HSPC differentiation include known hematopoietic genes, such as Hlf, Meis1, and Tal1 (Fig. 3e, f). Overall, these CellRouter results demonstrate not only an unbiased rediscovery of lineage-specific genes and potential new candidates but also their kinetic patterns along each cell fate transition trajectory.

Finally, CellRouter can also be applied to study potential cell circuits mediating these cell fate transitions. CellRouter allows the extraction of regulator-specific network modules from each trajectory, thus shedding light on potential regulatory interactions and their dynamics during differentiation (Fig. 4a, b).

Load additional libraries:

```
library('ggnetwork');library('GGally');library('geom-
net');library('network');library('sna')
```

16. **Regulator-centered subnetworks during erythroid differentiation.**

```
transition <- c('SP_4.SP_5')
x.ery <- grnscores(cellrouter, tfs, transition, direction='both', flip=TRUE, q.
up=0.85, q.down=0.15, dir.targets='up', columns=4, width=8, height=7,
filename='results/GRN_score_Ery')
regulators <- c('Klf1', 'Gata1', 'Gfi1b', 'Sphk1')
regulatornetwork(x.ery, regulators, 4, 1, 7, 6, 'results/regulator_network-
s_ery.pdf')
```

17. **Regulator-centered subnetworks during granulocyte differentiation.**

```
transition <- c('SP_4.SP_8')
x.gran <- grnscores(cellrouter, tfs, transition, direction='both', flip=TRUE,
q.up=0.85, q.down=0.15, dir.targets='up', columns=4, width=8, height=7,
filename='results/GRN_score_Gran')
regulators <- names(x.gran$SP_4.SP_8$scores)[1:4]
regulatornetwork(x.gran, regulators, 4, 1, 7, 6, 'results/regulator_networks_-
Gran.pdf')
```

To plot the gene expression dynamics of the Klf1 or Gfi1 network modules (Fig. 4c), use the commands in sections 18 and 19, respectively:

18. **Dynamics of Klf1 and its target genes during erythroid differentiation.**

```
plottrajectories(cellrouter, c('SP_4.SP_5'), x.ery$SP_4.
SP_5$targets$Klf1, rescale=TRUE, columns=4, width=6,
height=1.6, filename='results/dynamics_Klf1_ery.pdf')
```

19. **Dynamics of Gfi1 and its target genes during granulocyte differentiation.**

```
plottrajectories(cellrouter, c('SP_4.SP_8'), x.gran$SP_4.
SP_8$targets$Gfi1, rescale=TRUE, columns=4, width=6,
height=1.6, filename='results/dynamics_Gfi1_gran.pdf')
```

These results showed that CellRouter can identify cell fate transition-specific expression dynamics and regulatory factors in intermediate and terminal cell states. Finally, we clustered gene expression kinetic profiles into transcriptional clusters containing genes with similar trends during differentiation (Fig. 4d).

20. **Kinetic trends along erythroid differentiation.**

```
transition <- c('SP_4.SP_5')
plotClusterHeatmap(cellrouter, transition, 5, 1.5, 2,
'results/dynamics.pdf')
plotclusters(cellrouter, transition, 2, 400, 350,
'results/cluster_dynamics.png')
```

We clustered gene expression patterns along each trajectory into four transcriptional clusters. Clusters 1 and 2 showed distinct downregulation kinetics, cluster 3 contained genes transiently upregulated, and cluster 4 contained genes upregulated, with some genes showing early expression, transient downregulation, and then upregulation (Fig. 4d). The heatmap at the bottom left shows the mean expression of genes in each transcriptional cluster, and the bottom right panel shows the divergence in Jaccard similarity, emphasizing a transition along the erythroid differentiation trajectory.

Here, we presented a step-by-step protocol to apply CellRouter to analyze a mouse single-cell RNA-Seq dataset profiling hematopoietic stem cell differentiation to multiple lineages. CellRouter is a multifaceted single-cell analysis platform that explores the subpopulation structure of single-cell omics data to reconstruct complex single-cell trajectories in population snapshots or nonsteady-state biological systems, such as in tissues containing a continuum of cell types and states, during differentiation or in response to stimuli. CellRouter uses a network representation of cell-cell relationships learned from a low-dimensional embedding. This network and the underlying subpopulation structure are used as a map of cell fate transitions. Utilizing concepts from flow networks to connect cells in different locations of this map (subpopulations), CellRouter allows the reconstruction of complex single-cell trajectories and the study of expression dynamics in samples with diverse cell types, branching points, and cell-state transitions such as during differentiation, cell reprogramming, or oncogenic transformation. Moreover, integration with gene regulatory networks enhances trajectory analysis by allowing identification of putative regulators of cell-state transitions.

3 Notes

1. No Clear Trajectory After Dimensionality Reduction

   Dimensionality reduction depends on the starting gene set. Identification of the most informative genes is an important step in single-cell trajectory analysis. In our experience, although single-cell trajectories may be clearly observed in two- or three-dimensional coordinates, this might not always be the case. Using all genes for dimensionality reduction may make it difficult to visualize potential bifurcations or rarer cell-state transitions. One simple strategy to identify genes for dimensionality reduction is using principal component analysis (PCA). After PCA, genes with high absolute loadings are potentially more discriminative compared to all genes. To include variation from different components, genes with high absolute loadings across a number of principal components can be identified.

   Another strategy is by identifying subpopulation-specific gene signatures or the most variable genes and using them for dimensionality reduction. Among the R packages that can be used for this task, we usually use scLVM [23] or Seurat [24] to identify the most variable genes. Moreover, other dimensionality reduction techniques can also be explored. In this chapter, we used t-SNE, but other alternatives are also possible. For example, in some datasets, a clearer trajectory structure is observed in the PCA space [25]. Diffusion components have also been proven very powerful for trajectory identification [12, 26].

2. No Clear Starting Subpopulation(s)

   In addition, a clear starting subpopulation, which will be at the beginning of the reconstructed trajectories, could be hidden within the diversity of cell types or states present in the data. Prior knowledge of marker genes and pairwise differential expression analysis are useful at this stage. CellRouter provides built-in functions to identify subpopulation-specific gene signatures and visualize the expression distribution of selected genes as well as visualization of gene expression patterns on dimensionality reduction maps. Moreover, in stem cell datasets, we also suggest using StemID [27] to identify the stem cell or multipotent population within a mixture of cell types present in the sample. Differential expression analysis is an active area of research in single-cell genomics, and sophisticated methods have been developed for this task [28]. Therefore, additional differential expression tests might be required other than the one available through CellRouter.

3. Cell Cycle Effects

   Single-cell data suffers from both technical and biological variations. Depending on the biological system and diversity of cell types or states, cell cycle can be a strong source of confounding effects—cells clustering together due to the cell cycle phase instead of the underlying biology. Cell cycle is an intrinsic component of the process in differentiating systems and should be accounted in a different way in this case. Cell cycle effects can strongly influence dimensionality reduction, subpopulation identification, and construction of the kNN graph used by CellRouter. One commonly used strategy is to identify a cell cycle signature and then identify other genes that correlate with this signature and remove these genes from the gene list used for dimensionality reduction [29]. Another strategy, implemented in scLVM [23] and Seurat [24], is to correct the gene expression matrix by modeling the cell cycle as a latent variable model, as in scLVM, or one of many other models available in Seurat to "regress out" the cell cycle. Indeed, Seurat provides helpful functions for quality control and also to take into account technical variation, including differences in library complexity, content of mitochondrial genes, and cell cycle effects for differentiating and non-differentiating biological systems.

References

1. Kumar P, Tan Y, Cahan P (2017) Understanding development and stem cells using single cell-based analyses of gene expression. Development 144(1):17–32. http://dev.biologists.org/lookup/doi/10.1242/dev.133058
2. Wagner A, Regev A, Yosef N (2016) Revealing the vectors of cellular identity with single-cell genomics. Nat Biotechnol 34(11):1145–1160. http://www.nature.com/nbt/journal/v34/n11/full/nbt.3711.html?cookies=accepted%5Cn, http://www.nature.com/nbt/journal/v34/n11/pdf/nbt.3711.pdf
3. Tanay A, Regev A (2017) Scaling single-cell genomics from phenomenology to mechanism. Nature 541(7637):331–338. http://www.nature.com/doifinder/10.1038/nature21350
4. Stubbington MJT, Rozenblatt-rosen O, Regev A (2017) Single-cell transcriptomics to explore

the immune system in health and disease. Science 63(Mi):58–63
5. Bendall SC, Davis KL, Amir EAD, Tadmor MD, Simonds EF, Chen TJ et al (2014) Single-cell trajectory detection uncovers progression and regulatory coordination in human b cell development. Cell 157 (3):714–725. https://doi.org/10.1016/j.cell.2014.04.005
6. Trapnell C, Cacchiarelli D, Grimsby J, Pokharel P, Li S, Morse M et al (2014) The dynamics and regulators of cell fate decisions are revealed by pseudotemporal ordering of single cells. Nat Biotechnol 32(4):381–386. http://www.ncbi.nlm.nih.gov/pubmed/24658644
7. Shin J, Berg DA, Zhu Y, Shin JY, Song J, Bonaguidi MA et al (2015) Single-cell RNA--seq with waterfall reveals molecular cascades underlying adult neurogenesis. Cell Stem Cell 17(3):360–372. https://doi.org/10.1016/j.stem.2015.07.013
8. Lummertz da Rocha E, Rowe RG, Lundin V, Malleshaiah M, Jha DK, Rambo CR et al (2018) Reconstruction of complex single-cell trajectories using CellRouter. Nat Commun 9 (1):892
9. Qiu X, Mao Q, Tang Y, Wang L, Chawla R, Pliner HA et al (2017) Reversed graph embedding resolves complex single-cell trajectories. Nat Methods 14(10):979–982. http://www.nature.com/doifinder/10.1038/nmeth.4402
10. Setty M, Tadmor MD, Reich-Zeliger S, Angel O, Salame TM, Kathail P et al (2016) Wishbone identifies bifurcating developmental trajectories from single-cell data. Nat Biotechnol 34(April):1–14. http://www.nature.com/doifinder/10.1038/nbt.3569
11. Chen J, Schlitzer A, Chakarov S, Ginhoux F, Poidinger M (2016) Mpath maps multi-branching single-cell trajectories revealing progenitor cell progression during development. Nat Commun 7(May):11988. http://www.nature.com/doifinder/10.1038/ncomms11988
12. Haghverdi L, Büttner M, Wolf FA, Buettner F, Theis FJ (2016) Diffusion pseudotime robustly reconstructs lineage branching. Nat Methods 13(10):845–848. http://www.nature.com/doifinder/10.1038/nmeth.3971. https://doi.org/10.1101/041384%5Cn
13. Marco E, Karp RL, Guo G, Robson P, Hart AH, Trippa L et al (2014) Bifurcation analysis of single-cell gene expression data reveals epigenetic landscape. Proc Natl Acad Sci U S A 111(52):5643–5650
14. Weinreb C, Wolock S, Tusi BK, Socolovsky M, Klein AM (2017). Fundamental limits on dynamic inference from single cell snapshots. bioRxiv. 170118. http://www.biorxiv.org/content/early/2017/08/23/170118
15. Sun N, Yu X, Li F, Liu D, Suo S, Chen W et al (2017) Inference of differentiation time for single cell transcriptomes using cell population reference data. Nat Commun 8(1):1–12. https://doi.org/10.1038/s41467-017-01860-2
16. Nowakowski TJ, Bhaduri A, Pollen AA, Alvarado B, Mostajo-radji MA, Di LE et al (2017) Spatiotemporal gene expression trajectories reveal developmental hierarchies of the human cortex. Science 358 (December):1318–1323
17. Velten L, Haas SF, Raffel S, Blaszkiewicz S, Islam S, Hennig BP et al (2017) Human haematopoietic stem cell lineage commitment is a continuous process. Nat Cell Biol 19 (4):271–281
18. See P, Dutertre CA, Chen J, Günther P, McGovern N, Irac SE et al (2017) Mapping the human DC lineage through the integration of high-dimensional techniques. Science 356 (6342):eaag3009
19. Olsson A, Venkatasubramanian M, Chaudhri VK, Aronow BJ, Salomonis N, Singh H et al (2016) Single-cell analysis of mixed-lineage states leading to a binary cell fate choice. Nature 537(7622):698–702. http://www.nature.com/doifinder/10.1038/nature19348
20. Van Der Maaten LJP, Hinton GE (2008) Visualizing high-dimensional data using t-sne. J Mach Learn Res 9:2579–2605. http://www.ncbi.nlm.nih.gov/entrez/query.fcgi?db=pubmed&cmd=Retrieve&dopt=AbstractPlus&list_uids=7911431479148734548related:VOiAgwMNy20J
21. Levine JH, Simonds EF, Bendall SC, Davis KL, Amir EAD, Tadmor MD et al (2015) Data-driven phenotypic dissection of AML reveals progenitor-like cells that correlate with prognosis. Cell 162(1):184–197. https://doi.org/10.1016/j.cell.2015.05.047
22. Shekhar K, Lapan SW, Whitney IE, Tran NM, Macosko EZ, Kowalczyk M et al (2016) Comprehensive classification of retinal bipolar neurons by single-cell transcriptomics. Cell 166 (5):1308–1323.e30. https://doi.org/10.1016/j.cell.2016.07.054
23. Buettner F, Natarajan KN, Casale FP, Proserpio V, Scialdone A, Theis FJ et al (2015) Computational analysis of cell-to-cell heterogeneity in single-cell RNA-sequencing

data reveals hidden subpopulations of cells. Nat Biotechnol 33(2):155–160

24. Satija R, Farrell JA, Gennert D, Schier AF, Regev A (2015) Spatial reconstruction of single-cell gene expression data. Nat Biotechnol 33(5):495–502. https://doi.org/10.1038/nbt.3192
25. Tirosh I, Izar B, Prakadan SM, Wadsworth MH, Treacy D, Trombetta JJ et al (2016) Dissecting the multicellular ecosystem of metastatic melanoma by single-cell RNA-seq. Science 352(6282):189–196. http://science.sciencemag.org.gate2.inist.fr/content/352/6282/189.abstract
26. Nestorowa S, Hamey FK, Sala BP, Diamanti E, Shepherd M, Laurenti E et al (2016) A single-cell resolution map of mouse hematopoietic stem and progenitor cell differentiation. Blood 128(8):20–32
27. Grün D, Muraro MJ, Boisset JC, Wiebrands K, Lyubimova A, Dharmadhikari G et al (2016) De novo prediction of stem cell identity using single-cell transcriptome data. Cell Stem Cell 19(2):266–277
28. Soneson C, Robinson MD (2018) Bias, robustness and scalability in single-cell differential expression analysis. Nat Methods. http://www.nature.com/doifinder/10.1038/nmeth.4612
29. Tusi BK, Wolock SL, Weinreb C, Hwang Y, Hidalgo D, Zilionis R, et al. (2018) Population snapshots predict early haematopoietic and erythroid hierarchies. Nature. http://www.nature.com/doifinder/10.1038/nature25741

Chapter 10

Gene Regulatory Networks from Single Cell Data for Exploring Cell Fate Decisions

Thalia E. Chan, Michael P. H. Stumpf, and Ann C. Babtie

Abstract

Single cell experimental techniques now allow us to quantify gene expression in up to thousands of individual cells. These data reveal the changes in transcriptional state that occur as cells progress through development and adopt specialized cell fates. In this chapter we describe in detail how to use our network inference algorithm (PIDC)—and the associated software package NetworkInference.jl—to infer functional interactions between genes from the observed gene expression patterns. We exploit the large sample sizes and inherent variability of single cell data to detect statistical dependencies between genes that indicate putative (co-)regulatory relationships, using multivariate information measures that can capture complex statistical relationships. We provide guidelines on how best to combine this analysis with other complementary methods designed to explore single cell data, and how to interpret the resulting gene regulatory network models to gain insight into the processes regulating cell differentiation.

Key words Information theory, Gene regulation, Network inference, Cell development, Single cell

1 Introduction

Cell development requires precise spatial and temporal control of gene expression. Networks of transcriptional activators and repressors ensure cells differentiate to form the correct cell types at the right times and locations to allow the development of complex multicellular organisms. Experimental techniques now allow us to assay cell transcriptional states at the single cell level for up to thousands of cells (Fig. 1a, b) [15, 44, 47]. These data provide unprecedented insight into the changes that cells undergo during cell fate decision events and differentiation, and they have motivated the development of many new computational and statistical tools. Technical noise and biological heterogeneity pose challenges for data analysis, and a number of specialized tools have been developed that enable the use of single cell transcriptomic data to learn more about cell development (or other biological processes) [2, 3, 21, 33, 44, 56, 60]. Many of these methods allow us to

Patrick Cahan (ed.), *Computational Stem Cell Biology: Methods and Protocols*, Methods in Molecular Biology, vol. 1975, https://doi.org/10.1007/978-1-4939-9224-9_10,

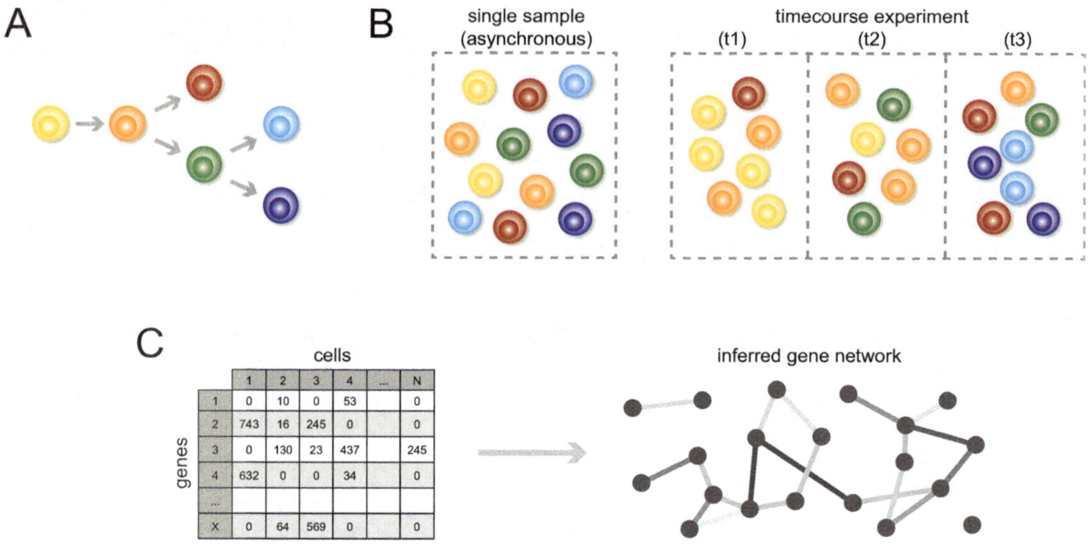

Fig. 1 Gene regulatory network inference from single cell transcriptomic data. (**a**) Cell development is accompanied by precise spatial and temporal changes in gene expression. We can use single cell data to study these changes and explore which changes in gene expression are associated with different stages of development or different cell fates. (**b**) These data may be collected from a single sample comprising cells at many different developmental stages (i.e., a population of cells developing asynchronously), or as part of a timecourse experiment where cells are sampled at a series of times after triggering differentiation. In the latter scenario, the data will still be cross-sectional (since cells must be lysed to quantify the mRNA) and experimental timepoints will not correlate perfectly with developmental stages due to biological heterogeneity and cells progressing at different rates. (**c**) Network inference algorithms aim to infer putative regulatory relationships between genes by detecting statistical dependencies in the observed gene expression values. These are represented as a network where each node is a gene, and edges indicated inferred dependencies between pairs of genes

characterize the changes in gene expression patterns that accompany cell differentiation, and identify (groups of) genes that play important roles within specific developmental routes. Increasingly, there is interest in using such data to derive statistical and mathematical models of the regulatory mechanisms controlling cell developmental processes [2, 8, 34, 37, 52, 60].

Gene regulatory network (GRN) inference algorithms aim to identify functional biological relationships between genes from the observed gene expression patterns, and represent these data in the form of a graph (where nodes represent genes, and edges indicate interactions between genes), Fig. 1c. Many inference methods have been developed in the context of bulk transcriptomic data [28, 36, 38, 53], but single cell data offer new features that we can exploit in order to learn about these regulatory relationships. In particular, single cell datasets are much larger (with sample sizes of hundreds or thousands of cells), and provide inherent variability in expression (due to biological heterogeneity within a cell population). Of course, there are also additional challenges associated with analyzing such data, e.g., confounding sources of heterogeneity and high levels of technical noise, that we need to be aware of [3, 15, 44]. Approaches to determine the structure of GRNs from single cell data include methods to infer both mechanistic and statistical models (reviewed in [2]): the former represent interactions using either Boolean logic or ordinary differential equations (ODEs), whereas the latter detect statistical dependencies between genes (generally using correlation or information theory).

Here, we describe a method to infer GRNs from single cell transcriptomic data that detects statistical dependencies between genes using multivariate information measures [8]. Such dependencies are interpreted as putative functional relationships—i.e., genes that regulate one another, or are subject to the same regulation (co-regulated). We assume that these regulatory relationships give rise to detectable statistical dependencies between those genes' expression states. Correlation is often used to identify statistical dependencies, but this is limited to detecting linear or monotonic relationships, whereas information theoretical measures are able to detect more complex non-linear dependencies [30, 53].

Several network inference algorithms make use of the pairwise information measure mutual information (MI), a symmetrical measure that quantifies the extent to which the knowledge of one variable reduces uncertainty about the other; or the extent to which the expression of one gene constrains (and is constrained by) the expression of another gene [53]. MI and other information measures are calculated from the probability distributions of the variables involved (discussed further in Subheading 3.3.1) and, in GRN inference, are made more accurate by the larger sample sizes of single cell data. This increased accuracy means that we can now feasibly estimate multivariate information measures between groups of three or more variables, allowing us to better distinguish between direct and indirect relationships, a longstanding problem in network inference. One such measure is partial information decomposition (PID), a multivariate information measure that quantifies different types of information between a group of variables [48, 59]. Most interestingly, PID can distinguish between information that is unique to a pair of variables within the group, and information shared redundantly by multiple variables.

PIDC (PID and context) is an algorithm that infers a network between a set of genes by calculating the proportional unique contribution (PUC) between each gene pair [8]. PUC is a heuristic measure, based on PID, that captures the unique information between a pair of genes within the context of all possible gene triplets involving the pair, and can be interpreted as evidence for a direct functional relationship. GRN inference using MI can be improved by transforming the MI score between a pair of genes according to where it lies on the distribution of all MI for each gene in the pair [12, 57], referred to as network context. PIDC improves on MI-based inference algorithms for detecting direct functional relationships within large gene expression datasets by using both PUC and network context [8].

NetworkInference.jl is an open-source software package that infers gene regulatory networks using PIDC. The package is written in the Julia programming language [4], chosen for its speed and straightforward syntax, so that a user can infer networks for hundreds to thousands of genes within minutes, with no need for a high level of computational expertise. Given a single cell data file, the package scores the edges between all possible gene pairs and writes them to file as a fully connected, weighted, undirected network. As with other information theoretic network inference packages, the user may choose whether and how to normalize the data prior to network inference, and how to interpret the results: often a threshold score will be set in order to classify gene pairs as connected by an edge or unconnected. In Subheading 3 we discuss ways to process data and set thresholds, and provide a more detailed guide to using our software.

2 Materials

- NetworkInference.jl is hosted on Github. Source code and documentation can be found at: https://github.com/Tchanders/NetworkInference.jl.
- Further tutorials, including worked examples and sample data and network files, can be found at: https://github.com/Tchanders/network_inference_tutorials/.
- Julia can be downloaded from: https://julialang.org/downloads/.
- Documentation for Julia, including a guide to getting started, is available from: https://docs.julialang.org/en/stable/.
- There are various ways to interact with Julia, the simplest of which is via the terminal. For more information, see: https://docs.julialang.org/en/stable/manual/interacting-with-julia/.
- For further network analysis using Julia, we recommend LightGraphs.jl. Documentation can be read at: http://juliagraphs.github.io/LightGraphs.jl/latest/.

3 Methods

In this chapter, we outline a method designed to infer putative functional relationships between genes from single cell transcriptomic data. Our approach starts with the measured expression levels of a set of genes in each cell, and produces a ranked list of putative pairwise gene–gene relationships that can be represented as an undirected, weighted network. We discuss how to prepare and select suitable expression datasets in Subheadings 3.1 and 3.2, before discussing the theory and practical implementation of our algorithm in Subheadings 3.3 and 3.4. We finish with a brief discussion on analyzing the inferred networks to identify which genes may play important roles in regulating cell differentiation processes (Subheadings 3.5 and 3.6).

3.1 Data Preparation and Normalization

Before starting any downstream analyses such as network inference or clustering, we first need to obtain a matrix comprising normalized gene expression levels for each cell in a given population. Several experimental approaches are available to obtain transcriptomic data at the single cell level, with dataset sizes dependent on the chosen method (*see* **Note 1**). In all cases, single cells must be isolated and lysed, before generating cDNA libraries from the cellular mRNA by reverse transcription and amplification, and quantifying expression using quantitative PCR (qPCR) or RNA sequencing (RNAseq). The technical aspects of single cell experiments (including quality control steps) and the methods used to quantify expression levels in each cell have been extensively reviewed elsewhere, so we refer the reader to these articles (e.g., [3, 15, 44]).

Regardless of the experimental protocol used to obtain relative expression levels, we need to ensure that the data are normalized appropriately. Only a fraction of the transcripts in a given cell will be successfully captured during the reverse transcription step and contribute to the detected expression levels. Cell-to-cell differences in the efficiencies of reverse transcription and quantification mean that these data must be normalized to remove cell-specific biases before we can meaningfully compare expression levels between individual cells. Broadly speaking, we can either apply strategies similar to those used for bulk transcriptomic studies or use exogenous spike-ins. In the first case, we calculate scaling factors for each cell based on the assumption that some sets of genes will be expressed at a constant level in all cells (often either so-called housekeeping genes or by assuming that most genes are not differentially expressed); in some cases total mRNA content is instead used to calculate scaling factors. However, the validity of these assumptions is often questioned in the context of single cell data due to the expected biological heterogeneity in cell populations. As a result, spike-in RNA is frequently used for normalization in single cell experiments. This involves adding mixtures of RNA molecules of known sequences and quantities to each cell lysate; these are processed in parallel to the endogenous RNA and act as internal controls. Again, this normalization method requires some assumptions—namely, that equal amounts of spike-ins are added to each lysate, and that they are reverse transcribed and quantified with the same efficiency as the endogenous RNA. Normalization methods and assumptions have been discussed in detail in a number of recent papers and reviews (e.g., [3, 15, 26, 44, 51]); the most appropriate choice of method will depend on the exact biological experiments performed.

3.2 Data Selection

The newest experimental technologies can provide data for thousands of genes and cells. However, we recommend selecting a subset of these data for network inference to make it practical to infer networks within a reasonable timeframe, but also to aid interpretation of the resulting networks. Our algorithm is implemented in the Julia package NetworkInference.jl; the Julia language was chosen for its speed, making it feasible to infer networks for hundreds to thousands of genes and cells within minutes (Subheading 3.4.7). We recommend limiting the number of genes either by selecting those most likely to be informative about the biological process of interest or based on statistical properties (*see* **Note 2**). We also suggest using data from at least a few hundred cells to infer a given network, as network reconstruction performance improves with higher numbers of cells. However, with larger datasets, it may be useful to infer networks using carefully selected subpopulations of cells in order to derive additional biological insight (e.g., the temporal dependence of inferred interactions between genes). In this section, we discuss how to select suitable groups of genes and cells to analyze. Note, it is of course important (as with any analysis of these data) to ensure that the initial data processing steps include adequate quality control (e.g., to remove poor quality cells with detectable expression of few genes) and consideration of confounding factors (whether technical or biological, e.g., batch effects or cell cycle influences) [3, 6, 15, 44].

3.2.1 Selecting Genes

To select a suitable subset of genes we can use our existing biological knowledge about the process of interest, rely on the observed gene expression levels, or use a combination of these approaches. In the first case, we aim to focus our analysis on those genes that have a known (or suspected) role in the differentiation process we are interested in. Frequently, we may be able to define a suitable list of genes using known functional annotations of genes, and those identified in previous experiments as genes of interest; in the case of single cell qPCR experiments, we only measure expression levels for these selected genes. Of course, a crucial advantage of single cell data is the potential to gain new insights into a given biological system that are not accessible using population averaged data. Relying solely on our existing knowledge to select genes may risk biasing our results towards our current understanding and miss the chance to identify new functional relationships between genes.

Single cell RNAseq experiments provide data on the full transcriptome, but there are several criteria we can use to filter these data and identify subsets of interesting genes. Suitable thresholds for the filtering criteria discussed here will depend on the experimental dataset (e.g., the numbers of cells analyzed, and the observed expression patterns), therefore we cannot recommend precise thresholding values but instead give an overview of the types of criteria you may wish to explore.

First, remove genes that do not show sufficient variability in expression to detect putative relationships. Our algorithm identifies putative functional relationships between genes by detecting statistical dependencies using their observed expression patterns. It is only possible to detect such dependencies when there is variability in the gene expression level within the cell population. In addition, we can only reliably detect dependencies when we observe expression in a reasonable number of cells. Therefore, we recommend a first broad filter of genes by removing those with detectable expression in only a small number of cells. We also remove genes that are only expressed at very low levels in any cell; while they may be of biological interest, it remains difficult to accurately quantify low expression levels using current high throughput single cell approaches.

In addition to these simple filtering metrics, we can use statistical or empirical models to identify genes displaying high variability or differential expression within the analyzed cell populations. In these cases, we make an assumption that genes with potential functional roles in the differentiation process will show variable expression depending on the developmental stage of each cell. Several methods have been proposed to identify genes with higher than expected variability (while accounting for the observed dependencies on mean expression levels) [1, 3, 5]. If we have information about the developmental stage of subsets of cells (e.g., from experimental timepoints, or from clustering or pseudotemporal ordering approaches discussed in Subheading 3.2.2), we can apply statistical tests to identify genes showing differential expression between different cell populations. There are a number of specialized algorithms designed to detect differential expression in single cell data [3, 13, 14, 18, 20, 50], or we can use non-parametric statistical tests (e.g., Kruskal–Wallis or Kolmogorov–Smirnov). Finally, clustering genes with similar expression profiles and selecting exemplar genes to represent each cluster could be another approach to reduce the numbers of variables to suitable levels for network analysis.

3.2.2 Selecting Cells

Our algorithm (like many other network reconstruction approaches) detects probable functional relationships between genes based on observed statistical dependencies. It is therefore important to select populations of cells to analyze where we believe a major source of variability is the biological process of interest (e.g., differentiation).

Sometimes analyzing subsets of larger datasets will provide more biological insight, and ensure interesting relationships are not obscured by other sources of variability. For example, if the dataset comprises cells following several distinct developmental trajectories (e.g., towards alternative cell fates), we could infer networks from the cells following each distinct trajectory in isolation. Each network would then provide information on the potential functional gene interactions that are involved in a given developmental pathway, and allow comparisons between networks involved in alternative cell fates. Equally, it could also be interesting to understand the gene–gene relationships involved at different stages of a given developmental pathway. To do this, we can infer networks from (potentially overlapping) subsets of cells comprising those undergoing earlier or later stages of differentiation. Of course, depending on the experimental design we frequently will not have information about the temporal stage or differentiation route of each cell—instead, we can apply various (mostly unsupervised) learning methods to infer likely developmental stages and pathways (*see* **Note 3**).

Numerous specialized methods have been developed to visualize, cluster, and order cells using single cell transcriptomic data. These methods—and the underlying assumptions—are discussed in detail in many recent reviews (e.g., [2, 3, 44, 56]) but given the variety of options, with little consensus as to the best approach, we recommend considering the alternatives to decide which are most applicable to the specific datasets you are analyzing. Approaches to visualize and cluster these high-dimensional data can help reveal any subpopulation structure in your data, aiding cell selection (e.g., avoiding those that fail to differentiate under the experimental conditions, or those that follow different developmental trajectories). Combining such methods with differential expression tests, or using bi-clustering algorithms (that cluster data according to both gene expression and cellular expression profiles), may also help identify suitable sets of genes to use for network inference. Many "pseudotemporal ordering" algorithms exist that aim to infer the relative developmental stage (and in some cases identify alternative differentiation routes) of cells within a sampled population [16, 25, 39, 42, 49, 58].

3.3 PIDC Algorithm

Once a suitable dataset has been identified and prepared for analysis as discussed in Subheadings 3.1–3.2, we next use the PIDC algorithm to infer a putative GRN. In this section, we outline the relevant background theory about information measures, before explaining how the algorithm works; Subheading 3.4 explains in detail how to use our software package implementing PIDC. For more detail on the development and performance of our method we refer the reader to the original paper introducing PIDC [8].

3.3.1 Information Theory

Information theory provides measures that quantify dependence between variables, while avoiding common assumptions about the nature of the dependence [30, 43, 53]. Such measures are ideal for single cell transcriptomic data because datasets are noisy and relationships between genes are not necessarily linear or monotonic, meaning that they can be overlooked by measures such as Pearson correlation. Dependence between variables is quantified by considering the information content (defined below) of their probability distributions. The probability distribution for a gene summarizes the probabilities that it will be expressed at various levels in a cell (as the probability distribution for the number rolled with a six-sided die is 1/6 for each of the numbers 1–6). Most information measures use discrete probability distributions; these are likely to be different for different genes, and are unknown in advance. This introduces two problems: discretization and probability estimation from observed data.

Discretizing continuous data entails defining a finite number of intervals, or bins, within which different values are considered equivalent. The two aspects of the problem are choosing an appropriate number of intervals and choosing appropriate interval widths. Two broad approaches are fixed-width binning, where every interval is the same size, and adaptive partitioning, where the size of each interval is informed by the density of the observations [30]. In fixed-width binning, the problem becomes choosing how many bins to use; however, there is no consensus as to the ideal number of bins. Adaptive partitioning more closely reflects how the data are distributed; the Bayesian blocks algorithm, for example, finds the optimal number of bins and bin widths for minimizing a cost function [41].

A further problem is estimating the probability distributions for each gene from their observed expression across the cells. The simplest approach, maximum likelihood, is simply to take the observed frequencies as the probabilities, but in some cases it is appropriate to adjust these, for example, based on a prior belief. Our software provides a number of estimators (reviewed in [17]), though it is worth noting that some of these are necessitated by small sample sizes, whereas we find that single cell datasets are large enough to yield good results using the maximum likelihood approach.

At the heart of information theory is information [43]. Consider a gene X, whose expression across the cells falls within discrete intervals with probability distribution $P(X)$. The information associated with each interval x is:

$$I(x) = -\log P(x). \tag{1}$$

Thus an interval with a lower probability has a higher information content; a cell in which the gene has unusually high or low expression is more informative. The base of the logarithm determines the units of information.

The entropy of a variable describes its freedom to adopt different states (i.e., the freedom of a gene to be expressed at different levels), often described as its uncertainty. The entropy of a variable is the mean of the information contained within each state:

$$H(X) = -\sum_{x \in X} p(x)\log p(x). \tag{2}$$

A gene that is expressed at the same level in most cells will have a low uncertainty, and a low entropy, whereas a gene that has an equal probability of being expressed at any level across the cells will have the highest entropy. The joint entropy for two or more variables is similar, but calculated from their joint probability distribution:

$$H(X,Y) = -\sum_{x \in X}\sum_{y \in Y} p(x,y)\log p(x,y). \tag{3}$$

The mutual information (MI) between two variables quantifies the extent to which the knowledge of either variable reduces uncertainty about the other. A reduction in uncertainty is equivalent to a reduction in entropy and can be thought of as one variable constraining the other. One way to express mutual information is as the difference between the joint entropy of the two variables and the sum of their entropies:

$$I(X;Y) = H(X) + H(Y) - H(X,Y) \tag{4}$$

If two genes have independent expression, then their joint entropy should be equal to the sum of their entropies, and their MI zero. If they are perfectly predictive of each other, then each of their entropies, their joint entropy, and their MI should all be equal. The MI between two variables is a widely used measure, and forms the basis of several network inference algorithms (*see* **Note 4**).

Measures for quantifying dependence between systems of three or more variables are more controversial, and several have been defined. Some, such as interaction information, quantify the information between any arbitrarily large group of variables, and are algebraically appealing, but have the ability to be negative, or zero between demonstrably dependent variables, raising questions about their interpretation [48].

Partial information decomposition (PID) solves these problems by allowing for different types of information between systems of three or more variables [48, 59]. For three variables, taking one as the target and the others as sources, PID defines four quantities: the unique information between the first source and the target, i.e., the reduction in uncertainty about the target provided only by knowledge of the first source; the unique information between the second source and the target; the redundancy,

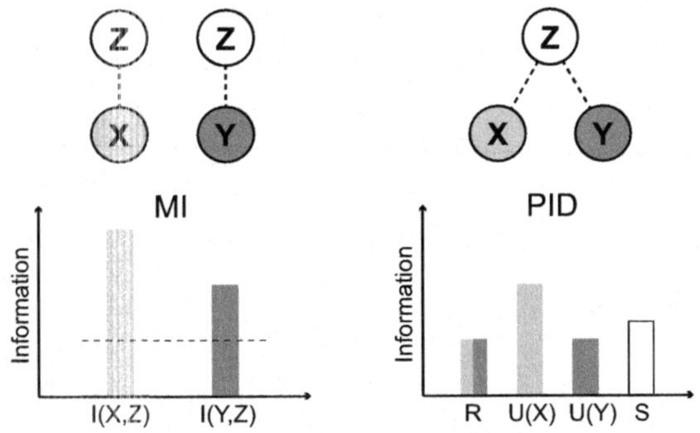

Fig. 2 Mutual information (MI) and partial information decomposition (PID). PID provides a set of information measures that describe the statistical dependencies between groups of three (or more) random variables [48, 59]. In our case these variables, X, Y, Z, are gene expression levels. Using MI, we can characterize the statistical dependencies between any pair of genes, e.g., $I(X, Z)$ quantifies the pairwise relationship between genes X and Z, as illustrated in the left-hand plot. With PID, we can explore statistical dependencies between groups of genes in more detail. If we consider three genes, as illustrated in the right-hand plot, we assign one gene as the target gene (in this case, Z) and can then calculate four measures describing the information provided about gene Z by genes X and Y. Redundancy (R) is the information provided by either X or Y alone, and the unique contributions ($U(X)$ and $U(Y)$) quantify the information that can only be provided by genes X or Y. Note that the MI between a pair of genes is equal to the sum of the redundancy and the corresponding unique contribution (e.g., $I(X, Z) = R + U(X)$ in this figure); the horizontal dashed line in the left-hand plot indicates the portion of $I(X, Z)$ and $I(Y, Z)$ that is accounted for by the redundancy term. Finally, PID also quantifies the information contribution that can only be gained by knowing the state of both X and Y, this is called the synergistic (S) contribution. Each of the three genes can be considered as the target gene in turn, leading to 12 information measures that provide a detailed description of the statistical relationships between these genes. Our PIDC algorithm makes use of the ratio of the unique contributions relative to the MI in order to detect putative functional relationships

which is the reduction in uncertainty provided by knowing either source; and the synergy, the reduction in uncertainty provided only through the knowledge of both sources (Fig. 2). Thus the PID between the source variables X and Z and the target variable Υ is,

$$I(X,Z;\Upsilon) = \text{Redundancy}(\Upsilon;X,Z) + \text{Unique}_Z(\Upsilon;X) + \text{Unique}_X(\Upsilon;Z) + \text{Synergy}(\Upsilon;X,Z), \quad (5)$$

where $\text{Unique}_Z(\Upsilon;X)$ is the unique information between source variable X and target variable Υ when the other source variable is Z. Note that we do not assume that source and target in the information sense correspond to the direction of biological regulation. The terms simply refer to the asymmetry of the PID measures; one gene may be more informative about the other, but it is not clear that the more informative gene would necessarily be the biological regulator. PID is defined for any number of variables; however, the number of terms to calculate increases rapidly, and we find that examining triplets of genes offers a good improvement in network inference, for a reasonable time cost (see also Subheading 3.4.7).

3.3.2 Algorithm

PIDC (PID and context) infers a network of pairwise relationships between a set of genes by examining each pair of genes, X, Y, within the context of every possible gene triple, X, Y, Z, where Z is every gene other than X and Y, in turn. First, for each gene pair X and Y, we calculate the proportional unique contribution (PUC), a heuristic measure based on PID, that utilizes the relationship (Fig. 2):

$$I(X;Y) = \text{Unique}_Z(Y;X) + \text{Redundancy}(Y;X,Z). \tag{6}$$

The MI between genes X and Y is constant across the triples, but, taking X as the first source, and Y as the target, the redundancy and unique information change according to Z, the second source. PUC is a pairwise measure that captures the mean proportion of the MI between X and Y that is accounted for by unique information between X and Y over all triples:

$$u_{X,Y} = \sum_{Z \in S \setminus \{X,Y\}} \frac{\text{Unique}_Z(X;Y)}{I(X;Y)} + \sum_{Z \in S \setminus \{X,Y\}} \frac{\text{Unique}_Z(Y;X)}{I(X;Y)}, \tag{7}$$

where S is the complete set of genes. Note that we take X and Y as the target in turn, since the PID terms are not symmetrical. We interpret a higher PUC as stronger evidence for a direct functional relationship between X and Y.

There are several gene-specific factors that can influence all the PUC scores for a particular gene, such as variability of expression and number of bins into which expression is discretized. A similar phenomenon was noted with MI and was successfully corrected using network context [12, 57], a normalization step whereby the score between a pair of genes is transformed according to where it lies on the distribution of all scores for each gene in the pair. Hence the final score for an edge between a pair of genes is:

$$c = F_X(u_{X,Y}) + F_Y(u_{X,Y}), \tag{8}$$

where $F_X(U)$ is the cumulative distribution function of all the PUC scores involving gene X (Fig. 3).

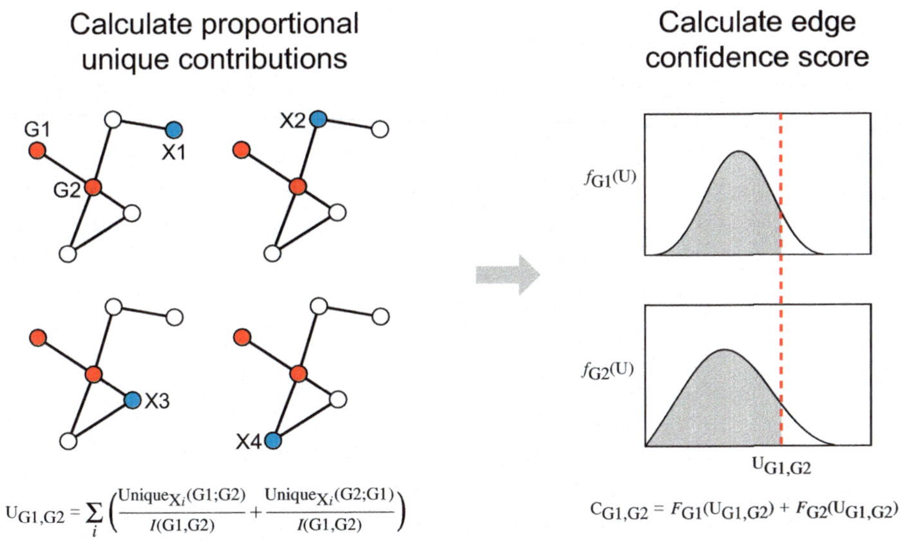

Fig. 3 PIDC network inference algorithm [8]. We first consider every triplet of genes that comprises a given pair of genes (G1, G2) and each other gene (Xi) in our dataset to calculate the proportional unique contribution (PUC, Eq. 7) for the gene pair G1,G2. Next, for each gene, e.g., G1, we construct an empirical probability distribution ($f_{G1}(U)$) from the PUC scores for that gene paired with each other gene in the network. The confidence of an edge between genes G1 and G2 is calculated by adding the cumulative probabilities of the PUC score for that gene pair ($U_{G1,G2}$) given the respective empirical probability distributions for these genes. These confidence scores allow us to rank all possible pairs of genes, and—after choosing a suitable threshold (e.g., top 5% of scores)—we can draw a network illustrating putative functional relationships

3.4 Software Package

NetworkInference.jl is a free open-source software package, implemented in the Julia programming language. In this section we provide instructions for how to install and use our software, see Materials (Subheading 2) for a list of resources. These instructions are correct at the time of writing, but because software versions can change, we advise consulting our up-to-date instructions included with the source code.

3.4.1 Installation

1. Install the latest version of Julia (version 0.6 at the time of writing), following the platform-specific instructions at https://julialang.org/downloads/
2. Start an interactive session by typing into the terminal:

```
julia
```

3. Install NetworkInference.jl and its dependencies by typing into the Julia prompt:

```
Pkg.add(''NetworkInference'')
```

3.4.2 File Formats

The input (data) file should contain a matrix of measurements, such that rows represent genes and columns represent cells. The first line of the file is assumed to contain headers and will be discarded by default. The other lines in the file should correspond to one gene each, and should contain the gene name, followed by the measurements. The delimiter is assumed to be whitespace by default, although this can be overridden (*see* Subheading 3.4.6). Blank lines are ignored.

Each output (network) file represents a fully connected, weighted, undirected network. Each row corresponds to an edge and comprises three tab-separated columns representing (from left to right) the regulating gene, the target gene, and the weight of the edge. The weight is simply the PIDC score, representing the confidence in a regulatory relationship between the two genes, and the rows are arranged in the descending order of weight. Although PIDC infers an undirected network, strictly speaking the file represents a directed network, but with the two edges between the same pair of genes occurring consecutively in each direction, with the same score; this facilitates comparison with directed networks.

3.4.3 Inferring a Network

1. Start an interactive session by typing into the terminal:

```
julia
```

2. To load NetworkInference.jl, type into the Julia prompt:

```
using NetworkInference
```

3. Infer a network using PIDC:

```
network = infer_network(<path to data file>, PIDCNetworkInference())
```

If the dataset is very large, this may take some time—*see* Subheading 3.4.7.

4. Currently `network` is a fully connected, weighted network. For further analyses, you may want to classify the edges, based on their weights, as present in the network and not present. There are two ways to approach this: (1) keep the top proportion, *p*, of edges,

```
adjacency_matrix = get_adjacency_matrix(network, 0.025)
# p = 0.025
```

or (2) choose a minimum weight, *w*, that an edge must have to be considered present in the network (see also Subheading 3.5),

```
adjacency_matrix = get_adjacency_matrix(network, 1.5; absolute = true)
# w = 1.5
```

3.4.4 Further Analysis in Julia

To continue your analysis in Julia, we recommend LightGraphs.jl, which implements the most commonly used network metrics and scales efficiently to large networks. For documentation *see* Subheading 2; here we provide instructions for converting `adjacency_matrix` to a format compatible with LightGraphs.

1. Install LightGraphs.jl, if not yet installed:

```
Pkg.add(''LightGraphs'')
```

2. Load LightGraphs.jl into the current session:

```
using LightGraphs
```

3. Convert `adjacency_matrix` to a graph:

```
graph = LightGraphs.SimpleGraphs.SimpleGraph(adjacency_matrix)
```

3.4.5 Further Analysis in Another Language

If you wish to continue in a different language, write `network` to file (*see* Subheading 3.4.2 for file formats):

```
write_network_file(<path to network file>, network)
```

You can now use your preferred language to load and parse the network file, and perform thresholding and further analyses.

3.4.6 Advanced Options

The PIDC algorithm takes a number of parameters, such as discretization algorithm and probability estimator. We have set defaults for these parameters based on the analysis performed in [8], but it is possible to override them by passing keyword arguments to `infer_network`. A list of keyword arguments and their options follows, along with some guidance for overriding the defaults and an example of how to use keyword arguments.

The following keyword arguments can be passed to `infer_network`:

- `delim` Column delimiter in the data file.
  - `false` (default) Delimiter is whitespace.
  - Any single-character delimiter, between single quotes, e.g., `delim= ,`
- `discretizer` Algorithm for discretizing the data. Discretization is performed separately for each gene, but using the same algorithm.
  - `"bayesian_blocks"` (default) Adaptive discretization method that finds the optimal partitioning and number of bins. Bayesian blocks may fail for a gene with no variation, or a gene that is expressed in very few cells, in which case the software falls back to the uniform width method. We do advise, however, that genes with no variation or expression

in very few cells be filtered out prior to network inference, since they are uninformative.

  - "uniform_width" Partitions the data into equal-width bins. Much quicker than Bayesian blocks, so may be a preferable choice for datasets with tens of thousands of cells.

- estimator Method for estimating the probability distribution. As with discretization, estimation is performed separately for each gene, using the same estimator.
  - "maximum_likelihood" (default) The probability distribution is equivalent to the observed frequencies. This is recommended over the Bayesian estimators, due to large single cell dataset sizes, a lack of prior knowledge about single cell data, and the computational efficiency of this method.
  - "shrinkage" A Bayesian estimator that shrinks the observed frequencies towards a prior (uniform by default).
  - "dirichlet" Another Bayesian estimator, also with a uniform prior by default.
- number_of_bins The number of bins for discretizing the data. This is only used when the discretizer is uniform width, otherwise it is ignored.
  - 10 (default)
  - Any positive integer. Note that with a larger number of bins, PIDC will take longer to run, since more calculations must be performed.
- out_file_path Path to the output network file.
  - "" (default) No file will be written
  - Any string, between double quotation marks, e.g., out_file_path = "network.txt"

In the following example, a network is inferred from a comma-separated data file named "data.csv," and written to the file "network.txt" using uniform width discretization with five bins and the shrinkage estimator (keyword order is unimportant):

```
infer_network(''data.csv'', PIDCNetworkInference(); discretizer = ''uniform_width'',
number_of_bins = 5, estimator = ''shrinkage'', delim = ',')
```

Note that the first two arguments are not keyword arguments. These two arguments specify the path to the data file and the algorithm, in that order, and are not optional. We require that the algorithm be specified because the package offers three algorithms in addition to PIDC: MINetworkInference( ) (rank the edges according to their MI), CLRNetworkInference( ) (as with

MI, but with the network context step), and `PUCNetworkInference( )` (rank the edges by PUC score, without the network context step). The relative accuracy and speed of these algorithms is discussed in [8] and Subheading 3.4.7.

3.4.7 Timings

Here we discuss time taken to infer a network, factors affecting this, and some ways to speed up network inference. Plots indicating how inference time scales for specific settings and dataset sizes can be found in Fig. 7 of [8].

PIDC scales cubically with the number of genes in the dataset because it considers gene triplets, whereas MI-based algorithms tend to scale quadratically. For datasets with up to several hundreds of genes, we offset this by using Julia, which is faster than languages such as R, Python, and MATLAB, particularly when running highly customized code [4]. As a result, for a simulated dataset with 1000 genes and 700 cells, it took a similar time to infer a PIDC network using NetworkInference.jl and a MI-based network using the R package, MINET [32] (a few hundred seconds).

Several factors affect the time taken to infer a PIDC network, in addition to the number of genes in the dataset. Although discretization is only performed once for each gene (so time to discretize scales linearly with the number of genes), the time taken to discretize is also influenced by the number of cells; this is particularly noticeable with Bayesian blocks discretization. Furthermore, the number of bins heavily affects the time taken to calculate all the PIDC scores, and cannot be controlled when using Bayesian blocks. Hence for a very large dataset with several thousands of genes and/or tens of thousands of cells, it is possible to speed up network inference by using uniform width discretization with a small number of bins (e.g., fewer than 10); however, it is worth considering how this could affect the accuracy of the inferred network, and whether it would be more appropriate to filter the dataset to fewer genes.

A further way to improve the speed of network inference is to run NetworkInference.jl on multiple cores, in parallel. Another advantage of using Julia is that parallelization is simple, for both the programmer and the user. The number of cores available will depend on your system, but to infer a network using four cores, simply perform the steps outlined in Subheading 3.4.3, replacing the command in step (1) with:

```
julia -p 4
```

Note that using multiple cores incurs an overhead due to data sharing between the cores, so using four cores will not result in a network being inferred in a quarter of the time. An example of the relationship between number of cores and speed of network inference can be found in [55].

Table 1
Time taken to infer a network from a qPCR dataset with 69 genes and 508 cells, using the four algorithms available in NetworkInference.jl, and assortativity of the resulting networks, keeping the top 2.5% of edges

Algorithm	Time (s)	Assortativity
PIDC	1.82	0.37
PUC	1.74	0.25
CLR	1.62	0.29
MI	1.59	0.16

Assortativity measures the extent to which similarly labeled nodes preferentially connect to each other, with 0 indicating no preference, and 1 indicating that all edges connect two nodes with the same label

While it is important to understand how software scales, particularly with growing RNAseq datasets, we emphasize that PIDC is extremely quick for smaller datasets, such as those typically produced by single cell qPCR experiments. Table 1 shows the time taken to infer networks using the four different algorithms implemented in NetworkInference.jl, for mouse embryonic stem cells undergoing neural differentiation (69 genes and 508 cells) [46]. Functional labels for the genes were published alongside the data, and we estimate the accuracy of the inferred networks using assortativity, a measure of the extent to which nodes (genes) with the same label preferentially connect to each other [35]. In agreement with [8], PIDC produced the most accurate network, while the increased time cost is negligible for a dataset of this size.

3.5 Network Construction and Thresholding

The output from the PIDC inference algorithm is a ranked list of all gene pairs, ordered by the calculated confidence level for each potential pairwise interaction. Note that the score is undirected, so each gene pair is listed twice (i.e., gene A–gene B and gene B—gene A) with identical scores. This information can be represented as a fully connected graph with a node for each gene and edges weighted according to their confidence scores. However, this is visually difficult to interpret—particularly with large numbers of genes—so we recommend restricting the number of edges to those with the highest scores. The simplest way of doing so is to select the top ranked edges, e.g., by setting a threshold to select only the top 2.5 or 5% of edges (*see* Subheading 3.4.3), and assume that these are the most likely to be functionally relevant interactions. This makes it much easier to visualize the most important interactions and detect patterns, e.g., identifying groups of highly connected genes (*see* Subheading 3.6).

Of course, we could potentially adopt more sophisticated ways to classify edges as putative functional relationships, but this is a current area of active research and beyond the scope of this article. Briefly, we may wish to make use of prior information about the network (perhaps by inflating the edge scores of known or suspected interactions), or using an empirical Bayes approach that aims to control the false discovery rate using empirical estimates for a suitable null hypothesis distribution [9, 11].

3.6 Network Interpretation and Analysis

An inferred network represents putative functional relationships between genes, which may be regulatory or co-regulatory in nature (*see* **Note 6**). These inferences are based on identifying statistical dependencies between gene expression states across a population of cells. It is important to be aware that there must be variability in gene expression levels to detect such dependencies, and we should aim to ensure that this variability is associated with the biological process of interest by selecting suitable data (*see* Subheading 3.2). We should view these networks as the collection of interactions that are active (and driving changes in expression) during a developmental process, rather than the complete set of interactions between genes that could occur under a range of different scenarios. In addition, some regulatory relationships may not be detectable due to experimental limitations, e.g., unreliable quantification of low expression levels (*see* **Note 6**). Common to many methods designed to analyze gene expression data, we are of course making the assumption that the observed changes in mRNA levels reflect changes in protein abundance (as the functional interactions are at the protein level). This is not necessarily the case, and thus we will miss some functional relationships (e.g., regulation by post-translational modifications) that are not observable in mRNA quantities.

For datasets comprising sufficiently large numbers of cells, we recommend inferring several networks using carefully selected subsets of cells (*see* Subheading 3.2.2 and **Note 5**). This can help us to derive additional biological insight into a given developmental process—such as identifying interactions that are specific to a certain temporal stage or cell fate—and potentially help identify interactions obscured in more complex datasets. In the example shown in Fig. 4, we illustrate how to use overlapping subsets of cells to determine temporal specificities, with some interactions more important at earlier or later stages of stem cell differentiation towards neural progenitor states.

As well as visually inspecting any inferred networks, we can use the huge array of statistical methods developed for graph analysis to study their structure in more detail. *Graph edit distances*, *community detection*, and using global (e.g., *graph spectra*, *network communities*, and *assortativity measures*) and local/node-specific network statistics (e.g., *degree*, *betweenness centrality*) can serve as useful starting points to investigate the likely functional relevance of inferred interactions and networks. Discussing these in detail is

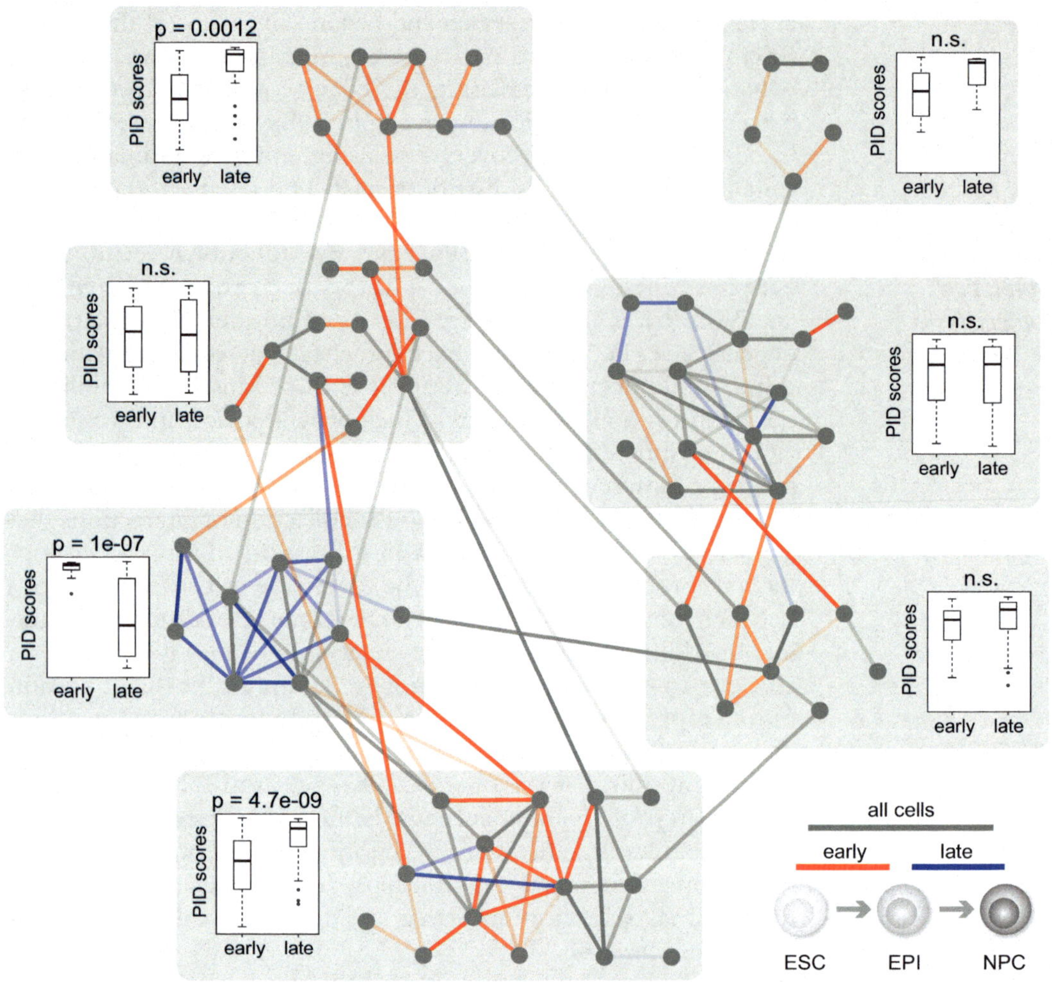

Fig. 4 Gene regulatory network inferred using PIDC algorithm for neuronal differentiation. Expression of selected genes was quantified by Stumpf et al. [46] using single cell qPCR for cells differentiating from a pluripotent embryonic stem cell (ESC) state, via an epiblast (EPI) state, towards a committed neuronal progenitor cell (NPC) state. Network nodes indicate each gene, while edges indicate inferred statistical dependencies between genes (for gene details, see the original paper describing this work [46]). The network shown was inferred using data from all cells, with edge opacity indicating the strength of the dependency. Networks were also inferred using overlapping subsets of the data (comprising cells undergoing the earlier or later stages of differentiation)—edges that were also detected in just one of these temporal specific networks are colored red or blue to indicate early or late specificity, respectively. We analyzed the structure of the network using an unsupervised community detection algorithm [10, 22] that detects groups of highly connected genes—the detected clusters are indicated using grey shaded boxes. Finally, to see whether particular clusters of genes play a more important role during early or later stages of development, we compared PIDC scores for all intra-cluster edges for the early/late networks. These PIDC scores are indicated by the box-plots adjacent to each gene cluster; for three clusters we detected significant differences in PIDC scores between the early and late networks indicating temporal specificity of these gene cluster interactions (statistical significance calculated using a Wilcoxon rank-sum test, p-values are indicated above the box-plots with "n.s." indicating "not significant," see Stumpf et al. [46] for further details)

beyond the scope of this chapter, but we refer the reader to articles such as [40, 45] for an overview of some of the techniques available. In our example in Fig. 4, we used a community detection algorithm [10, 22] to identify which groups of genes were highly connected—in some cases, these groups also showed temporal specificity, allowing us to determine groups of genes and regulatory interactions which we propose to play important roles at distinct developmental stages (for more detail, see [46]). In cases where we infer multiple networks, we could use measures designed to quantify similarity (or discrepancies) between inferred networks, e.g., allowing us to explore the similarities in regulatory interactions associated with different cell fates.

Networks inferred using the approach described in this chapter of course only represent our hypotheses about gene interactions. Like any mathematical model, these structures should be subject to continual refinement and improvement as new information becomes available. We can of course also aim to update our hypotheses by incorporating information from other types of data (e.g., transcription factor binding). We can use the set of putative interactions—bearing in mind the limitations of which interactions are inferrable (*see* **Note 6**)—to suggest candidate (groups of) genes that play important roles during specific developmental transitions. Network inference thus offers a powerful method to explore gene functional relationships that is complementary to other specialized methods for analyzing single cell data, and can help guide the design of further experiments to explore the roles of specific genes and putative interactions in more detail.

4 Notes

1. Single cell qPCR uses gene-specific primers to quantify expression levels for 10s–100s of selected genes for a few hundred cells. RNAseq aims to provide data for the complete transcriptome (i.e., 10,000s of genes) with throughput in terms of numbers of cells dependent on the exact protocol: plate-based and microfluidic approaches tend to generate data for hundreds of cells, while the newest droplet-based technologies assay expression levels for up to thousands of individual cells [3, 19, 24, 27]. The latter experiments use "unique molecular identifiers"—short sequences used to barcode individual transcripts during the reverse transcription step that allow quantification and help reduce noise.
2. The algorithm has complexity $O(n^3)$ with regard to the number of genes (since we examine each possible triplet of genes to calculate our pairwise gene interaction scores). Due to this, and our ability to draw meaningful results from the inferred

networks, we suggest limiting the data to a few hundred genes at most. The algorithm is much more scalable with respect to the number of cells, with complexity dependent on the exact discretization algorithm used (Subheading 3.4.7). We expect the performance of the algorithm (in terms of its ability to capture true functional relationships) to improve with more observations since estimates of multivariate information measures become more accurate under this scenario. In our experience, we have found it sufficient to analyze data from a few hundred cells in order to derive informative network structures. We would not recommend using much smaller datasets (e.g., fewer than 100 cells), although this will also depend on other data features such as the level of technical noise. We discuss in **Note 5** how to assess whether the inferred network structures are robust by using approaches such as subsampling.

3. We may wish to use these methods even when we do have relevant experimental data if we do not believe it is reliable. For example, cells are often sampled at several timepoints during an experiment after triggering differentiation, but the experimental sampling times may not correlate well with developmental stage due to biological heterogeneity (e.g., different rates of developmental progression, or cells failing to respond as expected).

4. Several algorithms have been developed to infer putative regulatory relationships from single cell transcriptomic data (as reviewed in [2]). Many rely on simple correlation-based measures of statistical dependence, but these are limited to identifying linear (or monotonic) pairwise relationships between genes—potentially too simplistic an approach for complex biological interactions. Other methods aim to develop mechanistic models of GRNs (e.g., based on Boolean logic), which have the advantage of providing richer detail about the nature of the interactions, but tend to require stronger mechanistic assumptions and be limited to models comprising small numbers of genes.

   Numerous information theoretic-based algorithms for network inference have been applied in the context of population-based gene expression data [53]. Information-based measures are appealing to use for identifying putative biological relationships as they characterize statistical dependencies between variables (genes) without making assumptions about the nature of these dependencies. However, most existing algorithms rely on MI to detect pairwise relationships. Relevance networks use pairwise MI values directly to rank the statistical significance of each potential network edge (i.e., each gene–gene pair), with a single threshold to determine which edges are included in the final network [7]. ARACNE is a popular algorithm that aims to build on this approach by detecting (and removing) potential indirect interactions between genes using the data processing inequality (DPI) [29]. The DPI states that the MI between two variables (X,Y) that are only connected indirectly via a third variable (Z) will be lower than the MI between either X or Y and the linking variable Z. Other algorithms consider the influence of more extensive parts of the network. MRNET uses a greedy algorithm to identify a minimally redundant but maximum explanatory set of variables that may interact with each gene [31]. The context likelihood relatedness (CLR) algorithm takes into account the broader network context by comparing the MI between a given pair of genes to the MI values of each of those genes with all other network genes (thus taking into account the variability of each gene, and calling edges based on gene-specific threshold values) [12, 57]. A few algorithms make use of multivariate information measures (in particular, conditional mutual information, interaction information, or related concepts) for network inference [23, 53, 54, 57, 61]; however, several of these measures do not have clear interpretations [48]. The recently proposed PID aims to provide a multivariate information measure that overcomes some of these difficulties and provides a detailed description of the statistical dependencies between three or more variables [59]; these are the measures applied in our PIDC algorithm. Notably, calculating these multivariate information measures relies on estimates of high-dimensional probability distributions (for three or more variables); the larger datasets afforded by single cell transcriptomic technologies are thus critical to the success of this approach. We provide a detailed performance comparison in this context of PIDC with other popular information theoretic-based algorithms in the original paper describing our method [8].

5. As mentioned in Subheading 3.2, we recommend using a few hundred cells to infer a network, with performance increasing with larger datasets. However, using subsets of the data can provide additional biological insight (*see* Subheading 3.2.2). We cannot give a precise threshold for the required number of cells as the network reconstruction performance will depend on too many other factors unique to each experiment (e.g., noise, variability of gene expression, complexity of dependencies, etc.). However, a subsampling approach can provide a means to assess the stability of the inferred networks—i.e., repeat the inference process with randomly sampled subsets of cells and compare the resulting networks. In our experience, the inferred networks will remain reasonably consistent until too few cells are included in the subsample (at which point the network structures vary greatly between different subsamples);

this heuristic approach can provide an indication of a suitable minimum threshold for the number of cells used to calculate a network. We often use overlapping subsets of cells to allow us to determine the temporal dependency or cell fate dependency of network structures (e.g., Fig. 4) while ensuring that each subset used for network inference is sufficiently large.

6. Using statistical dependencies to identify putative functional relationships is a very effective way to explore gene expression data, but there are certain limitations and assumptions to such approaches. We can only detect a pairwise relationship when there is variation in the expression level of both genes (thus allowing us to observe a statistical dependency). This means that we can only expect to detect interactions that are actively regulating expression levels under the experimental conditions we have chosen—varying those conditions may (and likely will) lead to variation in the inferred network structure.

   There will be some interactions that—despite being active under our chosen experimental conditions—will not lead to observable statistical dependencies. For example, any regulation that leads to rapidly changing or fluctuating levels of gene expression is unlikely to be detectable from single cell "snapshot" style data; more sustained changes in gene expression are needed to observe the associations between genes across the cell population. Additionally, current technologies do not allow accurate quantification of low levels of expression, so interactions between genes that are only weakly expressed will not be detectable (i.e., if expression levels are comparable in magnitude to levels of technical noise). Finally, interactions that are active but not causing variation in expression under the experimental conditions may not be detectable (e.g., if one gene is maintained at a high expression level across the whole cell population, and inhibits expression of another).

   Equally, if there are other biological processes (such as variation in cell cycle) that are driving transcriptional changes during our experiment—in addition to the developmental process of interest—we may detect additional interactions not directly associated with cell differentiation. There are some proposed approaches to remove cell cycle induced variation (e.g., [6]), but the network context step of the PIDC algorithm should also help with this issue as it aims to detect the most important interactions for each gene in turn (rather than simply detecting the strongest interactions across the whole network).

   The inferred interactions may be regulatory (e.g., a transcription factor—target gene relationship) or co-regulatory (where the expression of two genes is under control of the same regulators resulting in coordinated expression patterns). Additional information from other types of experimental data (e.g., transcription factor binding sites) may help to resolve some of these cases, but despite being an indirect relationship, such co-regulatory relationships will still be informative about the groups of genes involved in controlling cell differentiation and influencing cell fates.

Acknowledgements

This work was supported by a Biotechnology and Biological Sciences Research Council (BBSRC) DTP Studentship to T.E.C., and a BBSRC Future Leaders Fellowship (grant reference BB/N011597/1) to A.C.B. We thank Joe Greener, Gal Horesh, and Ananth Pallaseni for sharing code with us, Suhail Islam for computing support, and Ben MacArthur, Patrick Stumpf, and members of the theoretical systems biology group for useful discussions.

References

1. Andrews TS, Hemberg M (2018) M3Drop: dropout-based feature selection for scRNASeq. Bioinformatics bty1044. https://doi.org/10.1093/bioinformatics/bty1044
2. Babtie AC, Chan TE, Stumpf MPH (2017) Learning regulatory models for cell development from single cell transcriptomic data. Curr Opin Syst Biol 5:72–81
3. Bacher R, Kendziorski C (2016) Design and computational analysis of single-cell RNA-sequencing experiments. Genome Biol 17:63
4. Bezanson J, Edelman A, Karpinski S, Shah VB (2014) Julia: a fresh approach to numerical computing. arXiv 1411.1607
5. Brennecke P, Anders S, Kim JK, Kołodziejczyk AA, Zhang X, Proserpio V, Baying B, Benes V, Teichmann SA, Marioni JC, Heisler MG (2013) Accounting for technical noise in single-cell RNA-seq experiments. Nat Methods 10(11):1093–1095
6. Buettner F, Natarajan KN, Casale FP, Proserpio V, Scialdone A, Theis FJ, Teichmann SA, Marioni JC, Stegle O (2015) Computational analysis of cell-to-cell heterogeneity in single-cell RNA-sequencing data reveals hidden subpopulations of cells. Nat Biotechnol 33(2):155–160
7. Butte AJ, Tamayo P, Slonim D, Golub TR, Kohane IS (2000) Discovering functional relationships between RNA expression and chemotherapeutic susceptibility using relevance networks. Proc Natl Acad Sci 97 (22):12182–12186. https://doi.org/10.1073/pnas.220392197
8. Chan TE, Stumpf MPH, Babtie AC (2017) Gene regulatory network inference from single-cell data using multivariate information measures. Cell Syst 5(3):251.e3–267.e3
9. Chan TE, Pallaseni A, Babtie AC, McEwen K, Stumpf MPH (2018) Empirical Bayes meets information theoretical network reconstruction from single cell data. bioRxiv. https://doi.org/10.1101/264853
10. Delvenne JC, Yaliraki SN, Barahona M (2010) Stability of graph communities across time scales. Proc Natl Acad Sci 107 (29):12755–12760. https://doi.org/10.1073/pnas.0903215107
11. Efron B (2010) Large-scale inference. Empirical Bayes methods for estimation, testing, and prediction. Cambridge University Press, Cambridge
12. Faith JJ, Hayete B, Thaden JT, Mogno I, Wierzbowski J, Cottarel G, Kasif S, Collins JJ, Gardner TS (2007) Large-scale mapping and validation of Escherichia coli transcriptional regulation from a compendium of expression profiles. PLoS Biol 5(1):e8–e13
13. Fan J, Salathia N, Liu R, Kaeser GE, Yung YC, Herman JL, Kaper F, Fan JB, Zhang K, Chun J, Kharchenko PV (2016) Characterizing transcriptional heterogeneity through pathway and gene set overdispersion analysis. Nat Methods 13(3):241–244

14. Finak G, McDavid A, Yajima M, Deng J, Gersuk V, Shalek AK, Slichter CK, Miller HW, McElrath MJ, Prlic M, Linsley PS, Gottardo R (2015) MAST: a flexible statistical framework for assessing transcriptional changes and characterizing heterogeneity in single-cell RNA sequencing data. Genome Biol 16:278
15. Grün D, van Oudenaarden A (2015) Design and analysis of single-cell sequencing experiments. Cell 163(4):799–810
16. Haghverdi L, Büttner M, Wolf FA, Buettner F, Theis FJ (2016) Diffusion pseudotime robustly reconstructs lineage branching. Nat Methods 13(10):845–848. https://doi.org/10.1038/nmeth.3971
17. Hausser J, Strimmer K (2009) Entropy inference and the James-Stein estimator, with application to nonlinear gene association networks. J Mach Learn Res 10:1469–1484
18. Kharchenko PV, Silberstein L, Scadden DT (2014) Bayesian approach to single-cell differential expression analysis. Nat Methods 11 (7):740–742
19. Klein AM, Mazutis L, Akartuna I, Tallapragada N, Veres A, Li V, Peshkin L, Weitz DA, Kirschner MW (2015) Droplet barcoding for single-cell transcriptomics applied to embryonic stem cells. Cell 161 (5):1187–1201
20. Korthauer KD, Chu LF, Newton MA, Li Y, Thomson J, Stewart R, Kendziorski C (2016) A statistical approach for identifying differential distributions in single-cell RNA-seq experiments. Genome Biol 17:222. https://doi.org/10.1186/s13059-016-1077-y
21. Kumar P, Tan Y, Cahan P (2017) Understanding development and stem cells using single cell-based analyses of gene expression. Development 144(1):17–32
22. Lambiotte R, Delvenne JC, Barahona M (2014) Random walks, Markov processes and the multiscale modular organization of complex networks. IEEE Trans Netw Sci Eng 1 (2):76–90. https://doi.org/10.1109/tnse.2015.2391998
23. Liang KC, Wang X (2008) Gene regulatory network reconstruction using conditional mutual information. EURASIP J Bioinform Syst Biol 2008(1):253894. https://doi.org/10.1155/2008/253894
24. Liu S, Trapnell C (2016) Single-cell transcriptome sequencing: recent advances and remaining challenges. F1000Research 5:182. https://doi.org/10.12688/f1000research.7223.1
25. Lönnberg T, Svensson V, James KR (2017) Single-cell RNA-seq and computational analysis using temporal mixture modelling resolves Th1/Tfh fate bifurcation in malaria. Science 2 (9):eaal2192
26. Lun ATL, Calero-Nieto FJ, Haim-Vilmovsky L, Göttgens B, Marioni JC (2017) Assessing the reliability of spike-in normalization for analyses of single-cell RNA sequencing data. Genome Res 27:1795-1806. https://doi.org/10.1101/gr.222877.117
27. Macosko EZ, Basu A, Satija R, Nemesh J, Shekhar K, Goldman M, Tirosh I, Bialas AR, Kamitaki N, Martersteck EM, Trombetta JJ, Weitz DA, Sanes JR, Shalek AK, Regev A, McCarroll SA (2015) Highly parallel genome-wide expression profiling of individual cells using nanoliter droplets. Cell 161(5):1202–1214
28. Marbach D, Costello JC, Küffner R, Vega NM, Prill RJ, Camacho DM, Allison KR, DREAM5 Consortium, Kellis M, Collins JJ, Stolovitzky G (2012) Wisdom of crowds for robust gene network inference. Nat Methods 9 (8):796–804
29. Margolin AA, Nemenman I, Basso K, Wiggins C, Stolovitzky G, Favera R, Califano A (2006) ARACNE: an algorithm for the reconstruction of gene regulatory networks in a mammalian cellular context. BMC Bioinf 7 (Suppl 1):S7–S15
30. McMahon SS, Sim A, Johnson R, Liepe J, Stumpf MPH (2014) Information theory and signal transduction systems: from molecular information processing to network inference. Semin Cell Dev Biol 35:98–108. https://doi.org/10.1016/j.semcdb.2014.06.011
31. Meyer PE, Kontos K, Lafitte F, Bontempi G (2007) Information-theoretic inference of large transcriptional regulatory networks. EURASIP J Bioinform Syst Biol 2007 (1):79879
32. Meyer PE, Lafitte F, Bontempi G (2008) minet: a R/Bioconductor package for inferring large transcriptional networks using mutual information. BMC Bioinf 9(1):461–510
33. Moignard V, Göttgens B (2016) Dissecting stem cell differentiation using single cell expression profiling. Curr Opin Cell Biol 43:78–86
34. Moignard V, Woodhouse S, Haghverdi L, Lilly AJ, Tanaka Y, Wilkinson AC, Buettner F, Macaulay IC, Jawaid W, Diamanti E, Nishikawa SI, Piterman N, Kouskoff V, Theis FJ, Fisher J, Göttgens B (2015) Decoding the

regulatory network of early blood development from single-cell gene expression measurements. Nat Biotechnol 33(3):269–276. https://doi.org/10.1038/nbt.3154

35. Newman MEJ (2002) Mixing patterns in networks. arXiv.org (2 Pt 2), 026,126
36. Oates CJ, Mukherjee S (2012) Network inference and biological dynamics. Ann Appl Stat 6 (3):1209–1235. https://doi.org/10.1214/11-AOAS532
37. Ocone A, Haghverdi L, Mueller NS, Theis FJ (2015) Reconstructing gene regulatory dynamics from high-dimensional single-cell snapshot data. Bioinformatics 31(12):i89–i96. https://doi.org/10.1093/bioinformatics/btv257
38. Penfold CA, Wild DL (2011) How to infer gene networks from expression profiles, revisited. Interface Focus 1(6):857–870
39. Rizvi AH, Camara PG, Kandror EK, Roberts TJ, Schieren I, Maniatis T, Rabadan R (2017) Single-cell topological RNA-seq analysis reveals insights into cellular differentiation and development. Nat Biotechnol 35 (6):551–560
40. Salter Townshend M, White A, Gollini I, Murphy TB (2012) Review of statistical network analysis: models, algorithms, and software. Stat Anal Data Min ASA Data Sci J 5 (4):243–264
41. Scargle JD, Norris JP, Jackson B, Chiang J (2013) Studies in astronomical time series analysis. VI. Bayesian block representations. Astrophys J 764:167. https://doi.org/10.1088/0004-637X/764/2/167
42. Setty M, Tadmor MD, Reich-Zeliger S, Angel O, Salame TM, Kathail P, Choi K, Bendall S, Friedman N, Pe'er D (2016) Wishbone identifies bifurcating developmental trajectories from single-cell data. Nat Biotechnol 34(6):637–714
43. Shannon CE (1948) A mathematical theory of communication. Bell Syst Tech J 27 (3):379–423. http://dx.doi.org/10.1002/j.1538-7305.1948.tb01338.x
44. Stegle O, Teichmann SA, Marioni JC (2015) Computational and analytical challenges in single-cell transcriptomics. Nat Rev Genet 16 (3):133–145
45. Stumpf MPH, Stumpf MPH, Kelly W, Thorne TW, Wiuf C, Wiuf C (2007) Evolution at the system level: the natural history of protein interaction networks. Trends Ecol Evol 22 (7):366–373
46. Stumpf PS, Smith RCG, Lenz M, Schuppert A, Müller FJ, Babtie A, Chan TE, Stumpf MPH, Please CP, Howison SD, Arai F, MacArthur BD (2017) Stem cell differentiation as a non-Markov stochastic process. Cell Syst 5 (3):268.e7–282.e7
47. Tanay A, Regev A (2017) Scaling single-cell genomics from phenomenology to mechanism. Nat Biotechnol 541(7637):331–338
48. Timme N, Alford W, Flecker B, Beggs JM (2014) Synergy, redundancy, and multivariate information measures: an experimentalist's perspective. J Comput Neurosci 36(2):119–140. https://doi.org/10.1007/s10827-013-0458-4
49. Trapnell C, Cacchiarelli D, Grimsby J, Pokharel P, Li S, Morse M, Lennon NJ, Livak KJ, Mikkelsen TS, Rinn JL (2014) The dynamics and regulators of cell fate decisions are revealed by pseudotemporal ordering of single cells. Nat Biotechnol 32(4):381–U251. https://doi.org/10.1038/nbt.2859
50. Vallejos C (2016) Beyond comparisons of means: understanding changes in gene expression at the single-cell level. Genome Biol 17:70. https://doi.org/10.1186/s13059-016-0930-3
51. Vallejos CA, Risso D, Scialdone A, Dudoit S, Marioni JC (2017) Normalizing single-cell RNA sequencing data: challenges and opportunities. Nat Methods 14(6):565–571
52. van Dijk D, Sharma R, Nainys J, Yim K, Kathail P, Carr AJ, Burdziak C, Moon KR, Chaffer CL, Pattabiraman D, Bierie B, Mazutis L, Wolf G, Krishnaswamy S, Pe'er D (2018) Recovering gene interactions from single-cell data using data diffusion. Cell 174 (3):716-729
53. Villaverde A, Ross J, Banga J (2013) Reverse engineering cellular networks with information theoretic methods. Cells 2(2):306–329
54. Villaverde AF, Ross J, Morán F, Banga JR (2014) MIDER: network inference with mutual information distance and entropy reduction. PLoS ONE 9(5):e96,732
55. Villaverde AF, Becker K, Banga JR (2017) PREMER: a tool to infer biological networks. IEEE/ACM Trans Comput Biol Bioinform 15 (4):1193–1202. https://doi.org/10.1109/TCBB.2017.2758786
56. Wagner A, Regev A, Yosef N (2016) Revealing the vectors of cellular identity with single-cell genomics. Nat Biotechnol 34(11):1145–1160
57. Watkinson J, Liang KC, Wang X, Zheng T, Anastassiou D (2009) Inference of regulatory gene interactions from expression data using three-way mutual information. Ann N Y Acad

Sci 1158(1):302–313. https://doi.org/10.1111/j.1749-6632.2008.03757.x

58. Welch JD, Hartemink AJ, Prins JF (2016) SLICER: inferring branched, nonlinear cellular trajectories from single cell RNA-seq data. Genome Biol 17(1):106

59. Williams PL, Beer RD (2010) Nonnegative decomposition of multivariate information. arXiv.org, arXiv:1004.2515

60. Woodhouse S, Moignard V, Göttgens B, Fisher J (2016) Processing, visualising and reconstructing network models from single-cell data. Immunol Cell Biol 94(3):256–265

61. Zhao J, Zhou Y, Zhang X, Chen L (2016) Part mutual information for quantifying direct associations in networks. Proc Natl Acad Sci USA 113(18):5130–5135

Chapter 11

Reconstructing Gene Regulatory Networks That Control Hematopoietic Commitment

Fiona K. Hamey and Berthold Göttgens

Abstract

Hematopoietic stem cells (HSCs) reside at the apex of the hematopoietic hierarchy, possessing the ability to self-renew and differentiate toward all mature blood lineages. Along with more specialized progenitor cells, HSCs have an essential role in maintaining a healthy blood system. Incorrect regulation of cell fate decisions in stem/progenitor cells can lead to an imbalance of mature blood cell populations—a situation seen in diseases such as leukemia. Transcription factors, acting as part of complex regulatory networks, are known to play an important role in regulating hematopoietic cell fate decisions. Yet, discovering the interactions present in these networks remains a big challenge. Here, we discuss a computational method that uses single-cell gene expression data to reconstruct Boolean gene regulatory network models and show how this technique can be applied to enhance our understanding of transcriptional regulation in hematopoiesis.

Key words Gene regulatory network, Single cell, Pseudotime, Transcriptional regulation, Hematopoiesis, Hematopoietic stem/progenitor cell, Boolean network

1 Introduction

Due to their high turnover, hematopoietic cells require constant replacement in order to sustain the blood system throughout adult life. Extensive research over the past 60 years has revealed that hematopoietic stem and progenitor cells (HSPCs) at different stages of commitment can be found in the mammalian bone marrow, with cells from specific populations specified toward one or more of the more than ten different lineages of mature blood cells. This led to the popular view of a hematopoietic hierarchy, where hematopoietic stem cells (HSCs) can differentiate into increasingly specialized progenitor cell populations (Fig. 1a). Due to the accessibility of material and the existence of cell surface marker-based sorting strategies, these hematopoietic populations can be readily isolated and studied and so are well-characterized in comparison to

Patrick Cahan (ed.), *Computational Stem Cell Biology: Methods and Protocols*, Methods in Molecular Biology, vol. 1975, https://doi.org/10.1007/978-1-4939-9224-9_11,

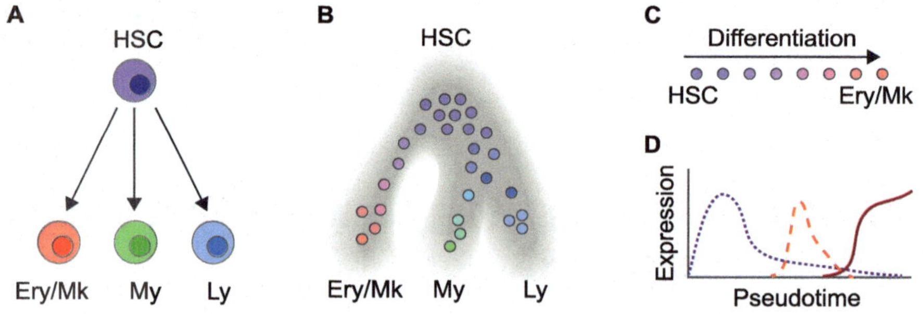

Fig. 1 Single-cell snapshot gene expression data can be used to reconstruct the transcriptional landscape of hematopoiesis. (**a**) Hematopoietic stem cells (HSCs) reside at the apex of the hematopoietic hierarchy. These cells can differentiate toward all of the different blood lineages. Ery, erythroid; Mk, megakaryocytic; My, myeloid; Ly, lymphoid. (**b**) By sampling single cells from the bone marrow and profiling their gene expression, it is possible to capture cells at different stages of differentiation and build up a picture of the underlying transcriptional landscape. (**c**) Cells can be computationally ordered based on similarities in their transcriptional profile. This ordering, often described as "pseudotime," aims to recapitulate molecular changes during differentiation. (**d**) Pseudotime orderings can be used to investigate the sequence of expression changes during differentiation for different markers

stem and progenitor cells in many other adult systems. The hematopoietic system, therefore, presents attractive opportunities for studying stem cell differentiation.

1.1 Transcription Factors as Key Regulators of Cell Fate Decisions

The hematopoietic system must maintain an appropriate balance of all mature cell types, as dysregulation of hematopoietic cell fate decisions is linked to potentially fatal blood disorders such as acute myeloid leukemia. It is therefore vital to understand how blood stem cells regulate their decision to differentiate toward alternative fates. Across many biological systems, it is well established that expression of genes encoding transcription factors plays an important role in determining cell fate decisions [1]. In particular, transcriptional regulation plays a central role in determining cell fate decisions and executing lineage differentiation in the blood [2].

1.2 Discovering Gene Regulatory Networks

In their role of driving specific lineage choices, transcription factors act as components of complex regulatory relations known as gene regulatory networks. While experimental approaches have revealed numerous gene interactions with crucial roles in hematopoiesis, relying purely on such time-consuming and expensive investigation to discover regulatory relationships is not feasible. Instead, the increased availability of high-quality gene expression data sets has led many researchers to use such data as a starting point for computationally discovering regulatory relationships across a variety of biological systems [3].

1.3 Modeling Gene Regulatory Networks

As well as providing a means for discovery, gene regulatory network models enable networks to be simulated in an attempt to better understand the role of gene regulation in a system. One approach used for modeling transcriptional regulatory networks is Boolean abstraction, which has proved popular in studies of the blood system. Here, gene expression levels are converted to ON/OFF expression, and regulation between genes is described as logical rules. Literature curation has been used to build Boolean regulatory network models for common myeloid progenitors [4], for lymphoid progenitors [5], and in blood stem cells [6]. Boolean models present a powerful tool for network modeling as they easily scale to large numbers of components, the effect of network perturbations can be easily simulated in silico, and network simulation relates to experimental approaches that readily translate into laboratory-based experiments such as gene knockouts.

1.4 The Importance of Looking at the Single-Cell Level

The majority of existing models of blood transcriptional regulatory networks are based on either literature curation or experimental approaches using bulk expression data. However, investigation at the single-cell level is essential, as considering only bulk expression data averages over often heterogeneous populations, where heterogeneity in gene expression levels is linked to cell fate choices [7]. The power of single-cell data in uncovering regulatory relationships is increasingly being recognized, for example, correlation analysis of single-cell qRT-PCR data has been successfully used to identify novel regulatory relationships between transcription factors [8, 9]. In systems other than adult hematopoiesis, single-cell expression data have been used to infer Boolean network models in embryonic stem cells [10] and embryonic blood development [11].

In this chapter, we discuss one approach for reconstructing Boolean gene regulatory network models of a hematopoietic differentiation process using snapshot single-cell gene expression data [12].

2 Materials

2.1 Single-Cell Gene Expression Data

The input for this network inference method is a set of single-cell gene expression values for cells at different stages of a differentiation process of interest. As the method involves calculation of correlations between genes, it is beneficial to have a high dynamic range for the majority of genes measured. Additionally, when modeling the Boolean network, genes must ultimately be discretized into "ON" or "OFF" states. This requires data with high sensitivity for detecting whether a gene is expressed in a given cell. High-quality single-cell quantitative real-time PCR (qRT-PCR) data fit both of these requirements. In our case study, we used data originally published in [7] and supplemented with additional populations in [12]. These data sampled cells from hematopoietic stem and progenitor cells using 12 different sorting strategies to capture cells at different stages of commitment toward mature blood lineages (Fig. 1b).

2.2 Ordering Cells Through Differentiation

Pseudotime describes the concept of ordering single-cell molecular profiles based on similarities in the expression states of individual cells. The motivation behind this is the idea that cells close together in terms of differentiation will have smaller changes in gene expression than those further apart (Fig. 1c). This then allows properties such as changes in gene expression levels to be visualized across differentiation (Fig. 1d). Monocle [13] and Wanderlust [14] were the first algorithms designed to order single-cell data into trajectories. Since then, improved versions of these algorithms, as well as entirely new approaches, have been published, able to cope with complex branching trajectories [15–17]. When analysis for our case study was carried out, these later algorithms were not available; hence, we applied the Wanderlust algorithm [14] to our data.

2.3 SAT Solver

The network inference method we describe here involves a step of searching a space of possible Boolean functions to find those with the highest score measuring agreement with the data. With even a relatively small number of genes, the space of Boolean functions becomes incredibly large, so in order to make this computationally tractable, we encode the problem as a Boolean satisfiability problem to find the rules on a per gene basis. We implemented this using the Python Z3 solver (https://github.com/Z3Prover/z3/), which provides an efficient search for the highest scoring functions.

3 Methods

It has long been recognized that gene expression data provides a powerful way to computationally discover transcriptional regulatory relationships, for example, by considering correlations between genes. More recently, large single-cell gene expression data sets have proved an effective starting point for building transcriptional networks. Here, we describe a method for constructing a Boolean network model from single-cell data.

3.1 Data Preprocessing

Single-cell gene expression measurements from cells at different stages of some differentiation process provide the input for the network inference algorithm (Fig. 2a). Single-cell qRT-PCR gene expression measurements are well suited to this analysis. Data can either be acquired pre-normalized from published studies, or Ct values can be transformed to ΔCt values by normalization against housekeeping genes on a per cell basis. Any housekeeping genes, as well as genes that fail quality control due to technical issues, should then be excluded from downstream analysis.

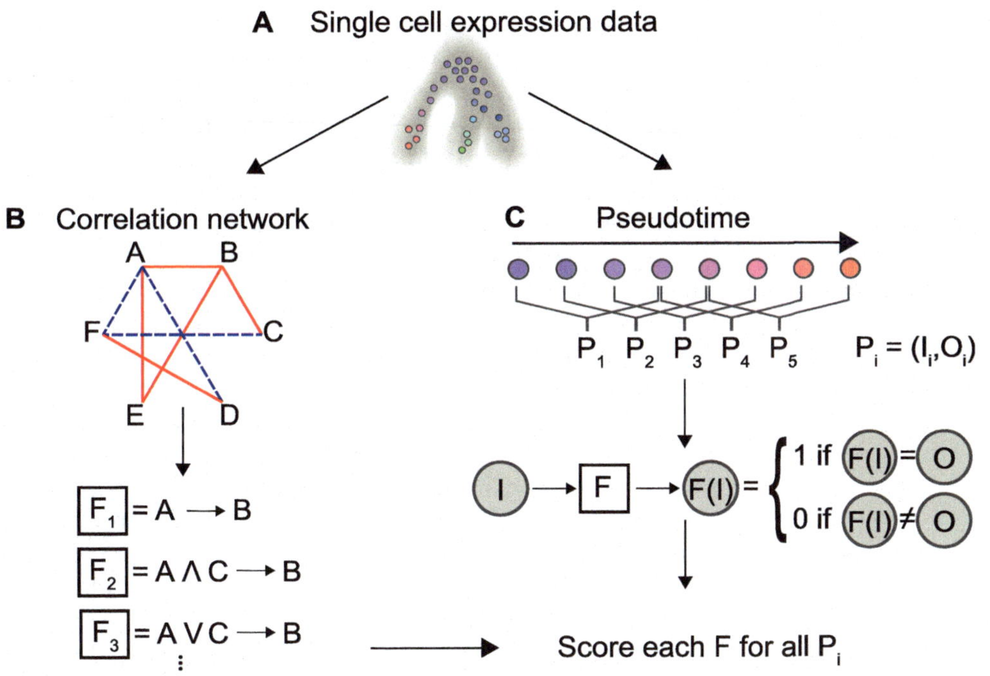

Fig. 2 Gene regulatory network models can be inferred from single-cell gene expression data. (**a**) The network inference method is based on single-cell gene expression profiles of cells at different stages of differentiation. (**b**) A set of possible Boolean logic functions describing the regulatory rules of each gene are found by first identifying strong positive and negative correlations between pairs of genes. This then generates a list of possible functions for each gene. Here, A–F represent a set of genes. (**c**) Cells are ordered in pseudotime, and pairs $P_i = (I_i, O_i)$ constructed by taking cells a fixed distance apart in the pseudotime ordering. These cells act as input and output to a Boolean function *F*. The functions generated by the correlation network are scored against all of the pseudotime pairs

3.2 Identifying Possible Gene Regulation

The first step in our network inference method is to identify potential regulatory relationships between pairs of transcription factor-encoding genes. This is done by calculating pairwise partial correlation coefficients across the whole data set (*see* **Note 1**). Correlation coefficients are filtered to retain pairs with significant interaction, for example, using a threshold of p-value <0.01. Links between the gene pairs are then ranked by the magnitude of the correlation coefficients, and the strongest correlations retained as edges in a gene correlation network. Positive correlation between gene A and gene B is then treated as possible activation of gene A by gene B or of gene B by gene A. Negative correlation is treated as potential repression acting in either direction. Potential activating or repressing relationships are combined with self-activation for each gene, and combinations of these high correlation edges describe a set of possible Boolean functions governing the expression of each gene, with each rule featuring one or more regulators (Fig. 2b). To reduce the search time of our algorithm, we restrict the search to functions of the form $F = F_1 \wedge \neg F_2$ with each F_i a Boolean function made from AND and OR gates with at most two inputs per gate. F_1 represents the activating part of the function, consisting of at most four activating transcription factors for a gene, and F_2 the repressing part of the function, formed from at most two repressing transcription factors.

3.3 Identification of Differentiation Trajectories

While high correlation between a pair of genes can be an indication of a regulatory relationship, additional information is required to determine if the regulation is direct, establish the direction of regulation between genes, and understand if this part of a regulatory event that requires the involvement of multiple transcription factors. This can be done using expression data from different time points or perturbed experimental conditions, yet in many cases, these data are not available and would be costly and time-consuming to generate. Instead, we exploit the concept of pseudotime [13, 14], by using this computational ordering of cells as the basis for a scoring mechanism that is applied to the set of possible Boolean functions from the correlation network (Fig. 2c). For this we need to order cells from the most immature population in the data through to the mature cells along one or more trajectories. A separate ordering is required for each "end point" corresponding to a different mature population. The first goal is to identify the cells representing the "start" and "end" cells for each trajectory. Here, we enlist the help of diffusion maps, a visualization tool which has been successfully applied to single-cell gene expression data to capture structure related to differentiation within the data [18]. If cells have been isolated from different time points, or are known to be different types (e.g., were captured using different surface marker-based sorting strategies in the blood), then this can be used to guide the selection of start and end cells for each trajectory. Alternatively, several pseudotime inference algorithms now provide methods for inferring the tip positions of trajectories automatically (*see* **Note 2**). In the case of data sets where cells can branch to multiple cells fates, it is also necessary to identify which cells lie on the differentiation trajectory toward a specific end cell. In our case study below, we followed a previously described method [19] to assign cells to branches. Again, some pseudotime algorithms now provide an inbuilt method for this.

3.4 Selecting Activating or Repressing Edges in the Network Model

So far the single-cell gene expression data has provided both a set of possible Boolean functions describing the regulation of each gene and an ordering of cells from the start to the end of a differentiation trajectory. We next describe how to use the gene changes along the pseudotime ordering to score the functions on a per gene basis, to identify functions that best fit the data. Gene expression in each cell along the trajectory is first discretized into "ON" or "OFF" expression states, by setting any detected values of gene expression to 1 and any undetected values to 0. Each pair of cells positioned k steps apart in the pseudotime ordering is then treated as an input-output pair $P_i = (I_i, O_i)$ for a Boolean function, where $[I_i]_g$ indicates the binary expression of gene g in input cell I_i. Each function F for a gene g is given a score $S(F) = \sum_i s_i(F)$ where $s_i = \begin{cases} 1, & \text{if } [F(I_i)]_g = [O_i]_g \\ 0, & \text{otherwise} \end{cases}$ for the pseudotime pairs $P_i = (I_i, O_i)$. This calculates the number of times the value of the gene g as predicted by F applied to I_i equals the value of g in the corresponding output cell O_i. The top scoring functions for each gene can then be considered the "best" functions for that gene. To identify the top scoring functions, we encoded the problem as a Boolean satisfiability problem and provide Python code that can be run to identify rules for each gene (https://github.com/fionahamey/Pseudotime-network-inference). *See* **Note 3** for a discussion of simplifying the set of rules returned by the above method.

3.5 Limitations

As with all computational inference methods, the output from this approach can only be considered as a tool for hypothesis generation, either for identifying potential novel regulations or utilizing the resulting Boolean network to simulate the outcome of scenarios such as overexpression and/or knockdown of specific genes in the network. Consequently, some form of experimental validation is nearly always necessary when applying such methods.

One major limitation of an approach based on qRT-PCR data is that the network model will always be restricted to the genes that have been chosen for profiling in the experiment and passed quality control measures. As a result, especially when using previously published data sets perhaps not originally designed for a network inference study, important regulators may in fact be missing from the network (*see* **Note 4** for a discussion of selecting genes). Nevertheless, our approach provides a valuable way for identifying ways in which known regulators act as part of a network to control cell fate decisions.

3.6 Case Study on Hematopoietic Stem Cell Differentiation

To demonstrate our network inference method, we decided to investigate transcriptional regulation of cell fate decisions in murine hematopoietic stem and progenitor cells. We used previously published single-cell qRT-PCR data from 12 different hematopoietic cell-sorting strategies that profiled long-term HSCs, finite self-renewal HSCs, multipotent progenitors, pre-megakaryocyte-erythroid progenitors, megakaryocyte-erythroid progenitors (MEPs), granulocyte-monocyte progenitors, lymphoid-primed multipotent progenitors (LMPPs), and common myeloid progenitors [7, 12]. These data measured the expression of 42 genes including 32 transcription factors in each cell. For network inference we chose to focus on two trajectories, one from HSCs to MEPs and the other from HSCs to LMPPs. Start and end cells for trajectories were selected based on the diffusion map analysis of these data (calculated using the *destiny* R package [20]). We followed a previously described method [19] to select cells on two branches from HSCs to MEP and LMPP cells. This was done by constructing a $k = 30$ nearest-neighbor graph on coordinates of the single-cell profiles in the first four diffusion components. Each branch was then identified by taking the 100 nearest neighbors of all cells lying on the shortest path between the start and end cells. The Wanderlust algorithm [14] was used to order cells along this path, assigning a pseudotime value to each individual cell. A partial correlation network was calculated on the transcription factor-encoding genes with edges between the 100 strongest correlating pairs (*see* **Note 5** for a discussion of heuristics to speed up the analysis). This correlation network and the pseudotime ordering were used as inputs to our network inference algorithm, which was applied separately to the two trajectories. A simplified network can be viewed in Fig. 3, which indicates the activation or repression between transcription factors identified in our model.

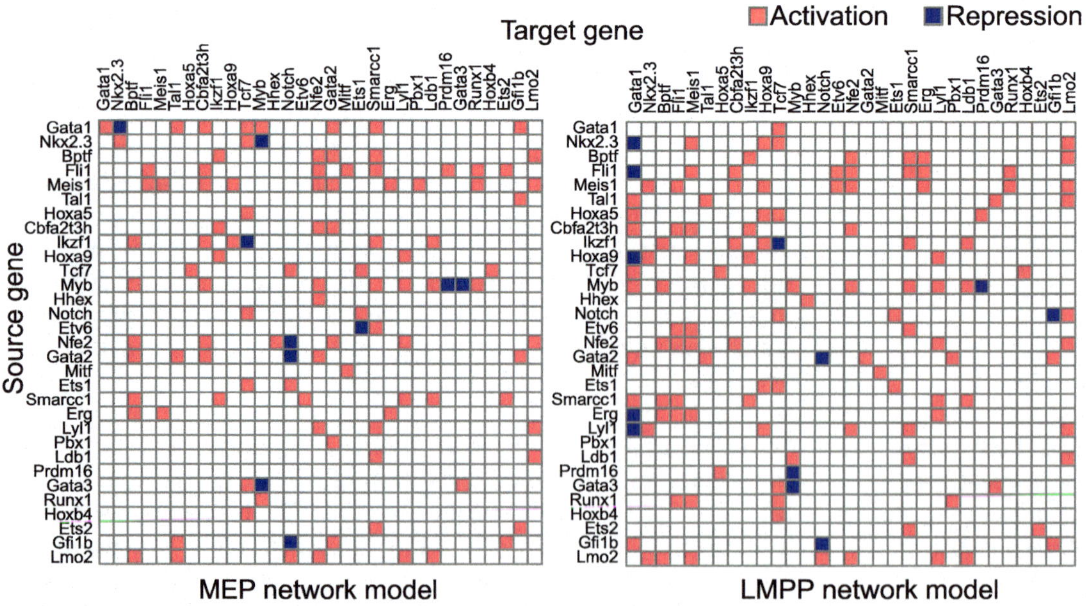

Fig. 3 Alternative gene regulatory network models for differentiation toward megakaryocyte/erythroid progenitors (MEPs) and lymphoid-primed multipotent progenitors (LMPPs) can be identified using single-cell snapshot data from mouse bone marrow. Heatmaps indicate interactions present in simplified network models, showing activation or repression from source gene (rows) to target gene (columns)

4 Notes

Here, we discuss some points for applying the algorithm for network inference to a single-cell data set.

1. *Input for Python Code of Network Inference Algorithm*

   Code for our network inference algorithm can be downloaded from GitHub (https://github.com/fionahamey/Pseudotime-network-inference) and requires three input files: the matrix of binary gene expression, the pseudotime ordering of cells on a differentiation trajectory, and a list of the possible activators and repressors of each gene. Instructions of the format of these files and the parameters for the algorithm are provided on the GitHub page.

2. *Pseudotime Method*

   Although we used the Wanderlust algorithm [14] for ordering cells in pseudotime, many more pseudotime inference algorithms have recently been published and would also be suitable for this analysis. In particular, users may find algorithms such as diffusion pseudotime [15], which can automatically detect trajectory tips within a data set, useful in cases where the start or end of trajectories is not as well understood as in hematopoiesis. The Python code provided for network inference takes an ordered list of cells along a trajectory as input and, hence, is compatible with a pseudotime ordering from any algorithm.

3. *Simplifying the Resulting Rule Set*

   In many cases, particularly when a gene had a high number of regulators, the algorithm returns several rules with equal scores. We first remove any rule with self-activation that does not score higher than the same rule disregarding self-activation. Groups of rules are then simplified to reduce the overall number of rules, discarding those contained within other rules in the set to give the smallest and simplest set of rules. For example, $\{A \rightarrow C, B \rightarrow C, A \vee B \rightarrow C\}$ would be simplified to $A \vee B \rightarrow C$, whereas $\{A \rightarrow C, A \wedge B \rightarrow C\}$ would be reduced to only $A \rightarrow C$.

4. *Choice of Gene Sets for Pseudotime Ordering and Network Inference*

   In the above case study in mouse bone marrow, 42 genes were measured by single-cell qRT-PCR, 32 of which were transcription factors. For transcriptional regulatory network model inference, we were only interested in the expression of transcription factor-encoding genes and so limited the analysis to the set of transcription factors. However, for diffusion map visualization and pseudotime ordering, the continuous expression levels of all genes except housekeeping genes were used, as this gene set demonstrated improved separation of different cell types.

5. *Speeding Up the Analysis*

   If a gene has many activators or repressors, it can vastly increase the search time to identify the high scoring functions. To reduce this time, some parameters can be altered when running the algorithm. Firstly, the maximum numbers of permitted activators or repressors can be reduced in order to search for simpler rules. Secondly, the threshold and threshold step size can be changed. The algorithm works by first searching for any rules that have agreement with a user-defined threshold, for example, 90%, of the pseudotime input-output pairs. If no rules are found, the percentage is lowered, and the algorithm iterates. If there are a very large number of rules above the starting threshold, the algorithm will search for a long time to find all of these, but we are only interested in the highest scoring rule of this set. Therefore, it would reduce the search time to start with a much higher threshold. The threshold lowering step size can also be altered in an attempt to limit the number of rules above the newly lowered threshold.

Acknowledgments

Work in the author's laboratory is supported by grants from Wellcome, Bloodwise, Cancer Research UK, and NIH-NIDDK and core support grants by Wellcome to the Cambridge Institute for Medical Research and Wellcome-MRC Cambridge Stem Cell Institute. F.K.H. is a recipient of a Medical Research Council PhD Studentship.

References

1. Davidson EH, Peter IS (2015) Genomic control process, 2nd edn. Academic Press, Oxford
2. Göttgens B (2015) Regulatory network control of blood stem cells. Blood 125:2614–2620
3. Marbach D, Costello JC, Küffner R et al (2012) Wisdom of crowds for robust gene network inference. Nat Methods 9:796
4. Krumsiek J, Marr C, Schroeder T, Theis FJ (2011) Hierarchical differentiation of myeloid progenitors is encoded in the transcription factor network. PLoS One 6:e22649
5. Collombet S, van Oevelen C, Sardina Ortega JL et al (2017) Logical modeling of lymphoid and myeloid cell specification and transdifferentiation. Proc Natl Acad Sci 114:5792–5799
6. Bonzanni N, Garg A, Feenstra KA et al (2013) Hard-wired heterogeneity in blood stem cells revealed using a dynamic regulatory network model. Bioinformatics 29:i80–i88
7. Wilson NK, Kent DG, Buettner F et al (2015) Combined single-cell functional and gene expression analysis resolves heterogeneity within stem cell populations. Cell Stem Cell 16:712–724
8. Moignard V, Macaulay IC, Swiers G et al (2013) Characterization of transcriptional networks in blood stem and progenitor cells using

high-throughput single-cell gene expression analysis. Nat Cell Biol 15:363–372

9. Pina C, Teles J, Fugazza C et al (2015) Single-cell network analysis identifies DDIT3 as a nodal lineage regulator in hematopoiesis. Cell Rep 11:1503–1510
10. Xu H, Ang Y-S, Sevilla A et al (2014) Construction and validation of a regulatory network for pluripotency and self-renewal of mouse embryonic stem cells. PLoS Comput Biol 10:e1003777
11. Moignard V, Woodhouse S, Haghverdi L et al (2015) Decoding the regulatory network of early blood development from single-cell gene expression measurements. Nat Biotechnol 33:269–276
12. Hamey FK, Nestorowa S, Kinston SJ et al (2017) Reconstructing blood stem cell regulatory network models from single-cell molecular profiles. Proc Natl Acad Sci 114:5822–5829
13. Trapnell C, Cacchiarelli D, Grimsby J et al (2014) The dynamics and regulators of cell fate decisions are revealed by pseudotemporal ordering of single cells. Nat Biotechnol 32:381–386
14. Bendall SC, Davis KL, Amir E-AD et al (2014) Single-cell trajectory detection uncovers progression and regulatory coordination in human B cell development. Cell 157:714–725
15. Haghverdi L, Büttner M, Wolf FA et al (2016) Diffusion pseudotime robustly reconstructs lineage branching. Nat Methods 13:845–848
16. Setty M, Tadmor MD, Reich-Zeliger S et al (2016) Wishbone identifies bifurcating developmental trajectories from single-cell data. Nat Biotechnol 34:637–645
17. Qiu X, Mao Q, Tang Y et al (2017) Reversed graph embedding resolves complex single-cell trajectories. Nat Methods 14:979–982
18. Haghverdi L, Buettner F, Theis FJ (2015) Diffusion maps for high-dimensional single-cell analysis of differentiation data. Bioinformatics 31:1–10
19. Ocone A, Haghverdi L, Mueller NS, Theis FJ (2015) Reconstructing gene regulatory dynamics from high-dimensional single-cell snapshot data. Bioinformatics 31:i89–i96
20. Angerer P, Haghverdi L, Büttner M et al (2016) destiny: diffusion maps for large-scale single-cell data in R. Bioinformatics 32:1241–1243

Chapter 12

Investigating Cell Fate Decisions with ICGS Analysis of Single Cells

Nathan Salomonis

Abstract

Advances in single-cell transcriptome profiling have contributed to new insights into the cellular states and underlying regulatory networks that govern lineage commitment. Such cell states include multipotent progenitors that can manifest as mixed-lineage patterns of gene expression at a single-cell level. Multipotent and other self-renewing progenitors are often difficult to isolate and characterized by subtle transcriptional differences that are challenging to define. This chapter examines the application of newly developed analytical tools to define heterogeneity in diverse stem cell and multipotent progenitor populations from single-cell RNA-Seq data. In addition to the methodology and output of these approaches, we explore their application to diverse single-cell technologies (e.g., Fluidigm, Drop-Seq, 10× Genomics Chromium) and their usability by computational and non-computational biologists. We focus specifically on the use of one tool, called Iterative Clustering and Guide-gene Selection (ICGS), which has been shown to uncover novel committed, transitional, and metastable progenitor cell states. As a component of the AltAnalyze toolkit, ICGS provides advanced methods to evaluate cellular heterogeneity in combination with regulatory prediction, pathway, and alternative splicing analyses. We walk through the individual steps required to perform these analyses in hematopoietic and embryonic kidney progenitor datasets in both graphical user and command-line interfaces. By the end of this chapter, users should be able to analyze their own single-cell RNA-Seq data and obtain deeper insights into the regulatory biology of the discovered cell states.

Key words Unsupervised, Single cell, RNA-Seq, Stem cell, Multipotent

1 Introduction

Single-cell RNA sequencing (scRNA-Seq) has begun to revolutionize the study of mammalian cell development by enabling unsupervised global molecular analyses of thousands of cells. The insights gained from this approach include the discovery of hidden cell states within accepted homogenous cell populations, novel intermediate progenitors with multipotent potentials, and uncharacterized developmental trajectories [1–7]. Single-cell profiling technologies continue to evolve at an astonishing pace, with the analysis of thousands to hundreds of thousands of single cells becoming commonplace [8, 9]. When coupled with new emerging

Patrick Cahan (ed.), *Computational Stem Cell Biology: Methods and Protocols*, Methods in Molecular Biology, vol. 1975, https://doi.org/10.1007/978-1-4939-9224-9_12,

technologies, such as precision genome editing, scRNA-Seq has the potential to uncover new discrete cellular and molecular mechanisms at an unprecedented scale [10, 11].

1.1 Approaches to Characterize Single-Cell Stem Cell Progenitor Populations

With the explosion of new applications for scRNA-Seq comes the important need for reusable and robust analytical solutions. Diverse computational approaches have been described to address these challenges including pseudotemporal ordering for developmental cell profiling [2, 12, 13], rare cell state identification [14–16], cell cycle effects exclusion [17, 18], batch/depth effects removal [19], and entropy-based analyses for identifying the earliest progenitor states [20, 21]. Although many of the described scRNA-Seq approaches can resolve distinct cell populations, they are significantly different in their ability to resolve subtle heterogeneity, identify transitions between cell populations, and find novel rare populations [3, 14, 22]. Transitional states include fine transitions between cellular developmental time-points and stable and metastable multipotent progenitor states (Fig. 1a) [3]. As recently demonstrated, single-cell analysis has the potential to uncover multipotent progenitors that may be defined by as few as a single unique gene but that coincidently express lineage defining genes from multiple other states [3, 23].

1.2 ICGS: Accurate Detection of Rare, Transitional, and Multipotent Cell States

Computationally, the identification of transitional and mixed-lineage cell states from single-cell genomics data remains a significant challenge. While many approaches are explicitly designed to capture discrete cell states and defining molecules (e.g., Seurat [19], Monocle [2], and RaceID [14]), such approaches may not be well-suited to detect transitional and mixed-lineage states.

A new computational approach called Iterative Clustering and Guide-gene Selection (ICGS) was developed to reproducibly and automatically identify rare, common, and novel transitional cell states, efficiently remove cell-cycle effects, predict likely cell types, and identify key transcriptional regulators [3]. A unique strength of this approach is that it is able to simultaneously identify continuums of transitioning and discrete cell states in a fully unsupervised manner (Fig. 1b, c) [3, 24–29]. The approach accomplishes this task by identifying predominant gene expression signatures over multiple rounds of iteration, representative guide-genes (i.e., transcription factors) delineating cell states, optionally excluding cell cycle-associated patterns, and ordering of the cells and genes using a sophisticated hybrid clustering approach, called HOPACH [30]. An example of such observations is shown with the analysis of single cells comprising distinct phases of embryo division (Fig. 1c). In addition to identifying and ordering the known underlying cell states, transitions in gene expression are evident, and potential transcriptional regulators (guide-genes) are highlighted. ICGS incorporates a database of known empirically determined cell

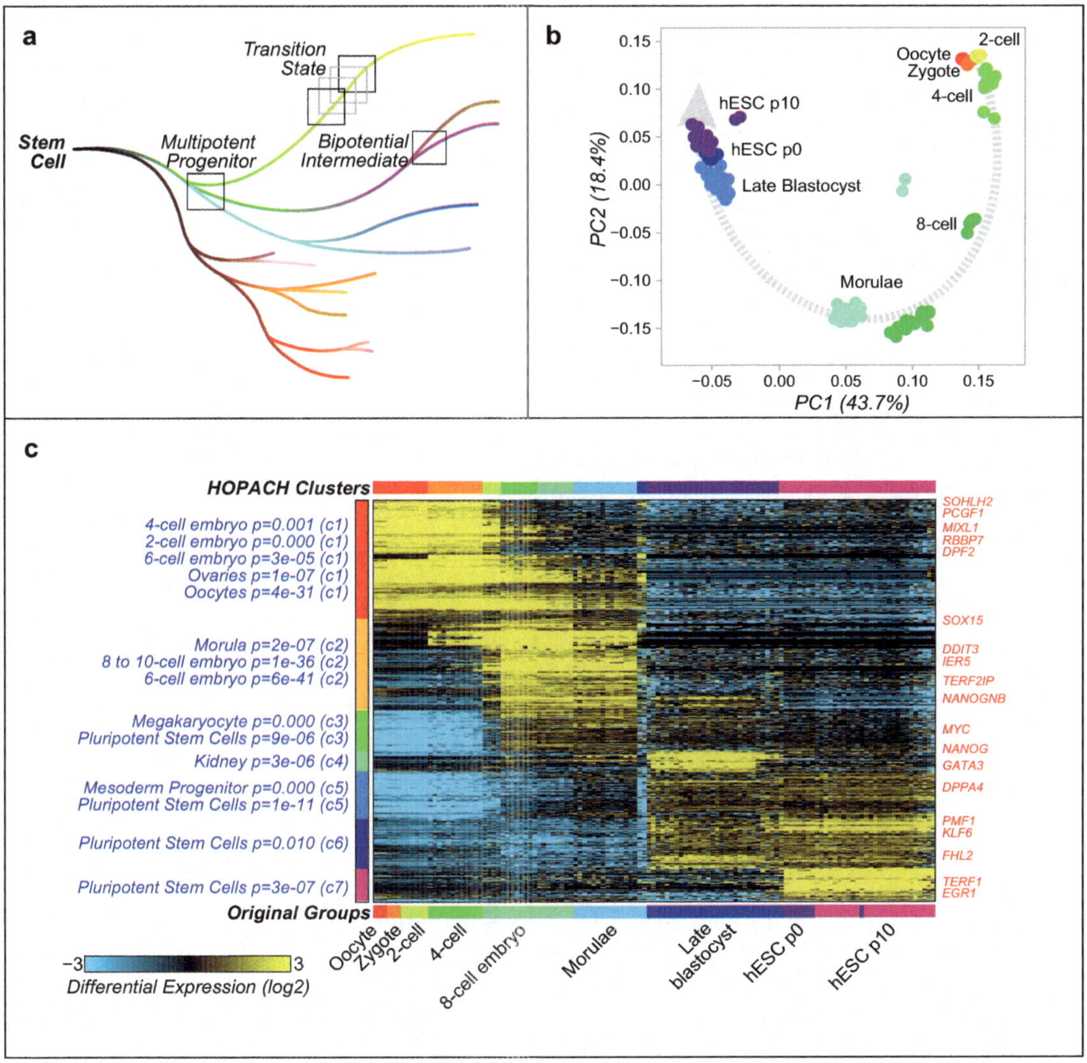

Fig. 1 Uncovering development transitions from single-cell RNA-Seq data. (**a**) Cartoon illustrating hypothetical types of cell-state transitions that can be identified from single-cell analyses. (**b**) PCA plot and (**c**) unsupervised ICGS analysis of scRNA-Seq data of human early stages of embryogenesis [37]. The PCA arrow indicates developmental time. Blue text indicates cell type and tissue gene-sets enriched among ICGS gene clusters. ICGS guide-gene transcription factors are shown red to the right of ICGS heatmap

markers, derived through extensive reanalysis of existing microarray datasets, which can provide evidence of known cell types (i.e., early embryo versus pluripotent stem cells) [3]. In its development, ICGS was benchmarked against multiple scRNA-Seq analysis tools for multiple diverse scRNA-Seq datasets and applied to address diverse research challenges [3, 25]. The final output of ICGS displays all dynamically regulated and correlated genes, along with all cells in a continuum of transitional and distinct cell states (Fig. 1c). By focusing on correlated genes, rather than on

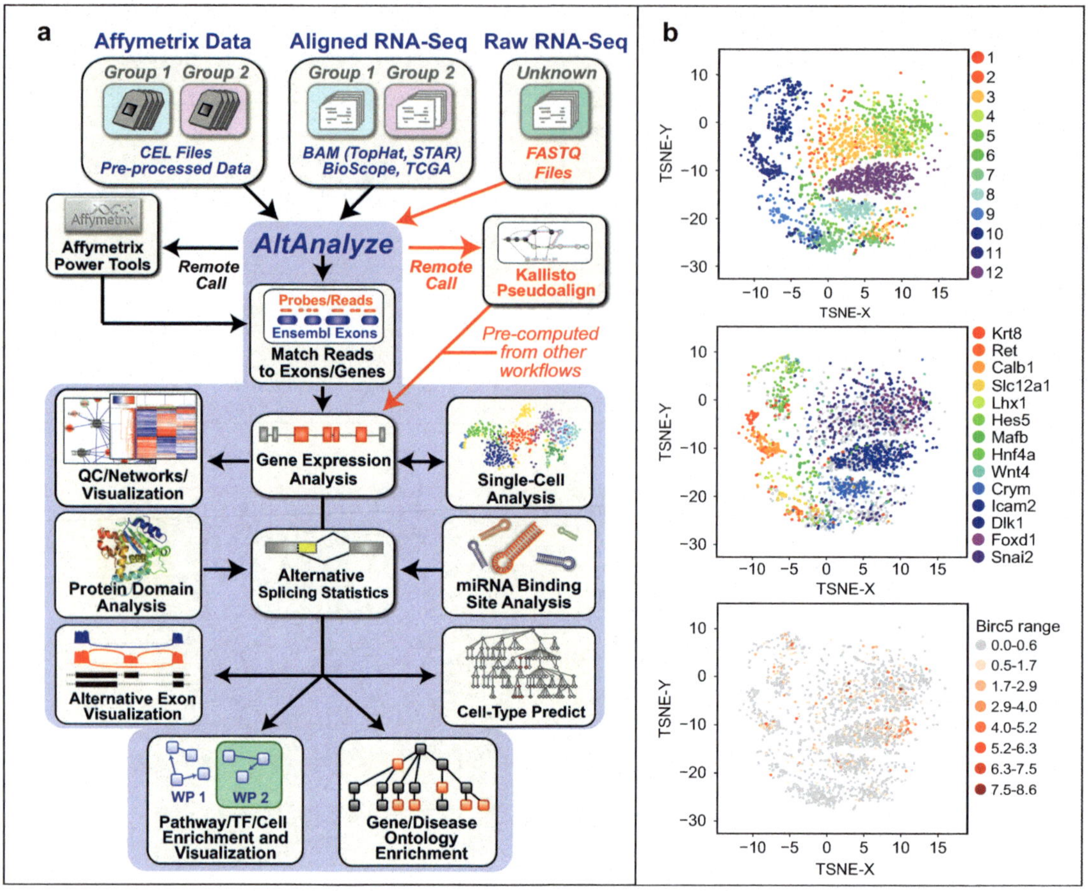

Fig. 2 Using AltAnalyze to perform multidimensional analyses of scRNA-Seq Data. (**a**) Graphical outline of the conventional and single-cell RNA-Seq workflows available from AltAnalyze. These steps include gene, splicing, pathway, network, quality control, and lineage prediction analyses from diverse input file formats. (**b**) Example of interactive data visualization in AltAnalyze, using the **Dimensionality Reduction** interface. T-SNE visualization of embryonic kidney Chromium scRNA-Seq data is shown following ICGS analysis (top panel, colored by ICGS clusters) and queried for the expression of known lineage markers (middle panel) or a single gene (cell-cycle marker, lower panel)

defining cell clusters to then obtain genes, ICGS is able to readily identify transitional states. From the derived cell states, genes with the greatest specificity to those states are also returned and visualized using the AltAnalyze embedded method MarkerFinder. To enable the use of this toolkit by the non-bioinformatics research community, ICGS is integrated with newly developed methods for ultrafast RNA-Seq pseudo-alignment (Kallisto) that are seamlessly run in series with ICGS (Fig. 2a) [31]. When run on a standard desktop or laptop computer, ICGS is compatible with both small and large scRNA-Seq datasets (>100,000 cells), including Fluidigm, Drop-Seq, 10× Genomics Chromium, and Index-Sorting platforms. When applied to larger datasets (>10,000 cells), ICGS applies intelligent down-sampling of the dataset to perform the unsupervised analysis, while remaining fast and scalable. Additional single-cell analysis options, downstream or upstream of ICGS, are currently supported in AltAnalyze (PCA loading gene analysis, interactive t-SNE/UMAP visualization). ICGS can process most scRNA-Seq datasets in under 1 h on a conventional laptop computer.

In conjunction with initial unsupervised single-cell analyses, AltAnalyze implements powerful validated workflows to evaluate differential expression and alternative splicing analyses for over 60 different species (Fig. 2a). These workflows can be run in a fully automatic manner on the command line or through the graphical user interfaces. To gain additional insights into single-cell populations, these tools can be used to interactively query these datasets by visualizing the expression of individual or groups of genes on PCA, t-SNE or UMAP projections (Fig. 2b), refine gene expression clustering results, perform targeted gene-set enrichment analyses, or examine regulatory and protein interactions among population-specific genes. In this chapter, we will download multiple types of scRNA-Seq datasets, including raw sequencing data and process these in ICGS and AltAnalyze to discover predominant progenitor cell populations.

2 Methods

2.1 Downloading AltAnalyze

AltAnalyze provides a number of diverse and flexible interfaces for single-cell RNA-Seq analysis. These include an advanced graphical user interface (GUI) and command-line interfaces. A single-code base provides these interfaces with the option of running either GUI or command line from the same software. This software can be run locally or remotely, such as in the case of a computational cluster or cloud-based environment. Below we walk through the use of these different interfaces with four different types of data inputs: (1) preprocessed gene expression text file, (2) 10× Genomics Chromium sparse-matrix outputs, (3) raw sequence FASTQ files, and (4) already aligned BAM files. Importantly, combined gene and splicing single-cell analyses are supported for FASTQ and BAM files in an automated manner. Please check the AltAnalyze website (http://www.altanalyze.org) to determine if additional input file types are supported, as the AltAnalyze software is continually updated.

1. Standalone Binary Application

   Pre-compiled versions of AltAnalyze are provided for the Mac OS X and Windows operating systems. These executable versions are available from the AltAnalyze website (http://

www.altanalyze.org). We always recommend using with the latest version of the provided software (version 2.1.2 at the time of writing). The compiled versions are compatible with most 64-bit standard computer configurations. After extracting AltAnalyze to a location of interest, open the program folder and find the AltAnalyze executable file (green icon). Double click this icon to start the program.

2. Python Package Installer

   AltAnalyze can be installed using the standard Python installer available on machines that already have Python. AltAnalyze requires Python version 2.7 and is not currently compatible with Python 3. If you have a Mac or Unix-based machine, Python is already installed. For Windows, if you are new to Python and command-line software, *see* https://docs.python.org/2/using/windows.html and follow the instructions to add Python to your system path. To install AltAnalyze and associated dependencies (e.g., Numpy, Matplotlib, Scipy), enter the terminal command:

```
$ sudo pip install AltAnalyze
```

   To open the AltAnalyze GUI, simply type "AltAnalyze" or "altanalyze" in any terminal window or use the command-line options described below. *Note*: for all of the command-line examples below, replace "python AltAnalyze.py" with "AltAnalyze."

3. Python Source Code

   In addition to the above options, the latest AltAnalyze source code can be directly obtained from GitHub at https://github.com/nsalomonis/altanalyze. With this option, you will need to install all necessary Python dependencies. Once installed, AltAnalyze can be run by changing directories to the AltAnalyze Python source code folder. Unmet Python dependencies are printed to the terminal immediately when AltAnalyze is run. To open the AltAnalyze GUI, simply type "python AltAnalyze.py" in any terminal window or use the command-line options described below.

2.2 Installation

1. Installing R

   Although R is not required to run ICGS, it is highly recommended. If AltAnalyze has been installed on Windows using the *standalone binary application*, an included version of R is already provided. On other operating systems, make sure R is installed for the local user after downloading the latest version of R from https://www.r-project.org. ICGS uses the

HOPACH clustering algorithm by default, which improves cluster definitions and ordering relative to the Python methods provided. For Windows users, installing AltAnalyze from the Python installer or source code may additionally need to add R to their system path. To *see* if R is installed, open a terminal window and type R then return.

2. Installing the AltAnalyze Database

   AltAnalyze requires the user download a species gene database for all analyses. These databases are available through software directly and require an Internet connection. Multiple downloads may be required when running for the first time.

3. Graphical User Interface

   The user should be immediately prompted with the option to download a species database after selecting *Begin Analysis*. If problems occur, *see* **Note 1**. Different genome versions will be compatible with different AltAnalyze database versions (Ensembl centric—http://ensembl.org). Additional details on database installation can be found at http://altanalyze.readthedocs.io/en/latest/RunningAltAnalyze.

4. Command-Line Interface

   Performing operations in AltAnalyze on the command line is one of the most efficient ways to use the software. Often only a single command is required. For example, to download the Mouse mm10 compatible database for Ensembl version 72 using the Python package, type:

```
$ AltAnalyze --update Official --species Mm --platform
RNASeq --version EnsMart72
```

   For the GitHub source code, simply substitute "AltAnalyze" with "python AltAnalyze.py" after changing directories to the AltAnalyze program folder:

```
$ python AltAnalyze.py --update Official --species Mm --
platform RNASeq --version EnsMart72
```

   To run the distributed Mac binary application as a command-line application rather than as a GUI, substitute "AltAnalyze" with "AltAnalyze.app/Contents/MacOS/AltAnalyze":

```
$ AltAnalyze.app/Contents/MacOS/AltAnalyze --update
Official --species Mm --platform RNASeq --version EnsMart72
```

For distributed Windows binaries, substitute "AltAnalyze" with "AltAnalyze.exe":

```
$ AltAnalyze.exe --update Official --species Mm --platform
RNASeq --version EnsMart72
```

Note that additional databases (e.g., RNA-Seq specific database) will be installed automatically the first time certain analyses are performed (e.g., ICGS).

2.3 Analysis of scRNA-Seq

1. Performing an Unsupervised Single-Cell RNA-Seq Analysis

   As previously noted, scRNA-Seq analyses can be performed with ICGS or using alternative workflows in AltAnalyze (*see* **Notes 1–4**). Below, we walk through the main workflows, focused on understanding stem and progenitor cell heterogeneity.

2. ICGS Analysis of a Preprocessed Expression File

   Often, already normalized gene expression values will be provided by the authors of existing studies or from other analysis pipelines. When processing such files, ICGS is compatible with a *tab-delimited text file* where the first column of data is gene identifiers and the first row consists of cell/barcode identifiers (**Dataset #1**). Additional columns or rows can result in a failure to properly analyze the data. The below example walks through a standard analysis of scRNA-Seq data. It is important to note that *AltAnalyze automatically determines whether the gene expression data is log or non-log* and converts non-log data to log2 during the analysis. More details and video tutorials of such an analysis can be found at http://altanalyze.org/ICGS.html.

3. Example Dataset 1

   Murine 382 bone marrow progenitors with ~3 million reads/cell [3]. The expression values are RSEM-derived transcript per million (TPM) estimates with technical outliers excluded.

   Note: this datasets is also provided with the software in the DemoData folder.
   http://altanalyze.org/Data/Mm-BoneMarrow-GSE70245.zip

2.3.1 Command-Line Example

The following command will produce all ICGS standard results using the programmatic defaults. The analysis should take ~10–30 min, depending on the machine. The below options include the exclusion of cell cycle effects (--excludeCellCycle conservative) and analysis of all genes (--restrictBy None) rather than protein coding (--restrictBy protein_coding).

```
$ python AltAnalyze.py --runICGS yes --platform RNASeq --
species Mm --excludeCellCycle conservative --expdir Mm-
BoneMarrow-GSE70245/BoneMarrow-scRNASeq.txt --output Mm-
BoneMarrow-GSE70245/
```

The same command is shown below with detailed flags for each of the major algorithm options:

```
$ python AltAnalyze.py --runICGS yes --platform RNASeq --
species Mm --column_method hopach --column_metric cosine --
row_method hopach --ExpressionCutoff 1 --FoldDiff 4 --
SamplesDiffering 4 --restrictBy None --excludeCellCycle
conservative --removeOutliers no --expdir Mm-BoneMarrow-
GSE70245/BoneMarrow-scRNASeq.txt --output Mm-BoneMarrow-
GSE70245/ --rho 0.4 --expname BoneMarrow-scRNASeq
```

See Subheading 2.5 for a description of the outputs produced from this command. Alternative programmatic options are provided in **Note 2**.

2.3.2 Graphical User-Interface Example

The GUI in AltAnalyze provides an intuitive way for computational and non-computational biologists to process their single-cell data. Importantly, the command line and GUI should produce identical results when applied to the same input. However, the GUI provides some additional options for post-processing of the ICGS results, including the automatic production of gene network analysis results, pairwise group differential expression, and pathway analyses. Below are the steps required to analyze preprocessed gene expression data.

1. Analysis Steps:
   - (a) After starting the GUI, select "Begin Analysis." In the main menu, **Select vendor/datatype** as **Other ID** and then select the primary identifier type as **Symbol** from the **Select Platform** option. Other identifier types can be selected from the menu as needed.
   - (b) On the next screen, select the **Process Expression file** option to load our preprocessed TPM file (**Dataset #1**). Select the example input file "BoneMarrow-scRNASeq. txt" in the unzipped dataset archive using the option **Select a gene expression file**. You can set the output directory as this same folder or another directory of interest.
   - (c) In the following window, called **Expression Analysis Parameters**, you can leave all of the defaults or change some to speed up the analysis (Fig. 3b). By selecting "no" for the options **Perform expression clustering and visual QC** and **Perform cell profiling with LineagePro-filer** and **decide later** for **Analyze ontologies and path-ways with GO-Elite**, you will significantly reduce the overall runtime.
   - (d) If the pathway analysis option is selected, a window called **Pathway Analysis Menu** will appear next. In this menu,

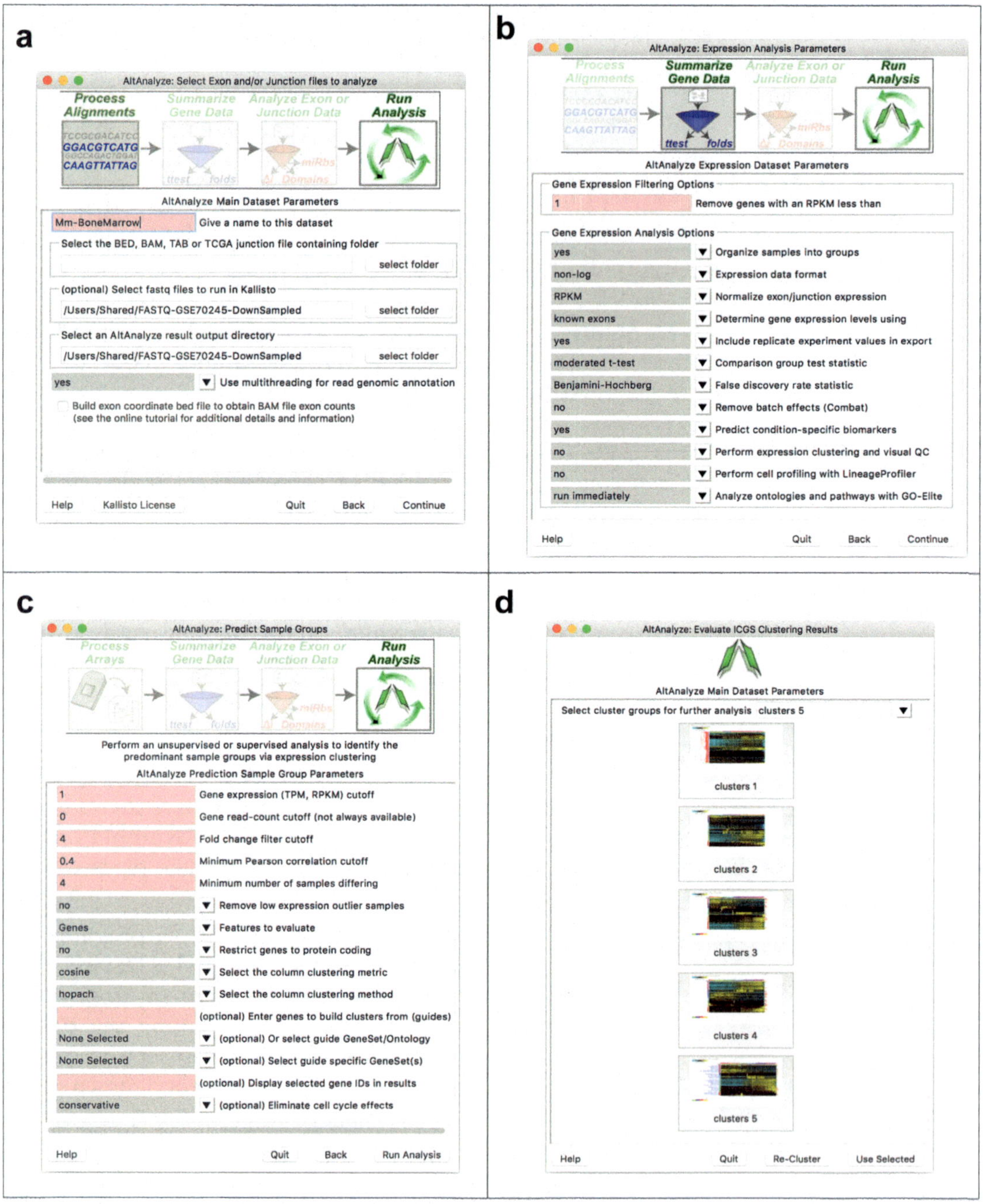

Fig. 3 Performing ICGS using the AltAnalyze graphical user interface. Steps for the analysis of scRNA-Seq data from FASTQ files are shown. (**a**) Interface for loading primary sequencing results (FASTQ, BAM, Cell Ranger). (**b**) Expression analysis parameters window. (**c**) ICGS parameters window. (**d**) Final ICGS clustering results interface (murine bone marrow progenitor scRNA-Seq)

you can indicate the statistical stringency of downstream pairwise group comparison analyses. For example, setting **rawp** to **adjp** will require an FDR adjustment for all tests. If you wish to have the software automatically output all biological pathways as graphical maps, you can select the option to **Visualize all over-represented WikiPathways**. Select **Pathways** from **Only analyze the following resource(s)**.

(e) In the **Assign Files to a Group Annotation** window, a priori assigned groups can be designated for individual cells (not required). Because we included an AltAnalyze groups file alongside the expression file, these annotations will already be pre-populated. Select the option **Run de novo cluster prediction (ICGS) to discover groups**, at the top of this menu.

(f) In the ICGS parameters menu (Fig. 3c), select **Conservative** from **Eliminate cell cycle effects**. If cell cycle effects remain in the final results, you can rerun with the **Strict** option, then select **Run Analysis** (runtime 5–20 min).

(g) Once ICGS has finished running, a series of heatmap thumbnails will appear that can be selected to review the different clustering results produced by ICGS (Fig. 3d).

- At this point, the user can quit the software and explore the results of the ICGS results in more detail (**ICGS** folder in the results directory or **AltAnalyzeViewer** application), rerun the ICGS analyses with different parameters (**Re-Cluster** button), continue with all pairwise group comparison analyses (**Use Selected** button), or perform targeted analyses in AltAnalyze (*see* Subheading 2.5).
- For the **Use Selected** option, the **Select cluster groups for further analysis** option must be set to the ICGS output of interest before proceeding (e.g., cluster 5 aka Guide-3 results). The cell cluster groups (e.g., hopach clusters) indicated in the heatmap images will be used as the defined cell groups. Supervised group comparisons will be run between adjacent clusters after selecting. Users can optionally modify the resultant groups and comparison files before proceeding (*see* **Note 4**). Once completed, the final results (e.g., differential expression, QC, MarkerFinder, network analysis, pathway enrichment, visualized WikiPathways, alternative splicing results) can be navigated from the results folder or from the accompanying **AltAnalyzeViewer** application.

2.4 ICGS Analysis of 10× Chromium Sparse Matrix, FASTQ, and BAM Files

scRNA-Seq datasets generated in house or by core facilities will often not include already-processed gene expression files. Rather, these data will consist of raw sequence files, such FASTQ, genome-aligned BAM format files, or gene count matrices. Fortunately, AltAnalyze has diverse and sophisticated methods to process these and several other diverse input file formats that use the same analyses interfaces and steps outlined in the previous section. In the following section, we describe the steps required for the analyses of these data.

1. Example Dataset 2

   Day 14.5 mouse embryonic kidney collected using the 10× Genomics Chromium platform [24]. This data was generated using the Cell Ranger workflow (10× Genomics, Pleasanton, CA). http://altanalyze.org/Data/Mm-e14.5_Kidney-GSE104396.zip
2. Example Dataset 3

   Selected murine bone marrow progenitor FASTQ files that have been down-sampled for this tutorial [3]. http://altanalyze.org/Data/FASTQ-GSE70245-DownSampled.zip
3. Example Dataset 4

   Selected murine bone marrow progenitor BAM files (genome version mm10 aligned) derived from the Dataset #3 files [3]. http://altanalyze.org/Data/BAM-GSE70245-DownSampled.zip

2.4.1 Source Data Command-Line Examples

The following commands will produce all ICGS results with default options. Analysis times will vary based on the inputs and machine (10 min–1 h).

ICGS of a 10× Chromium Sparse Matrix file

```
$ python AltAnalyze.py --runICGS yes --platform RNASeq --
species Mm --excludeCellCycle conservative --
ChromiumSparseMatrix demo_data /input/matrix.mtx --output
demo_data/input/ --expname "e14.5-Kidney"
```

ICGS of RNA-Seq FASTQ files

```
$ python AltAnalyze.py --runICGS yes --platform "RNASeq" --
species Mm --excludeCellCycle conservative --output
"demo_data/FASTQ-GSE70245-DownSampled" --expname "Mm_HSCs"
--fastq_dir "demo_data/FASTQ-GSE70245-DownSampled/" --
runKallisto yes
```

ICGS of RNA-Seq BAM files

```
$ python AltAnalyze.py --runICGS yes --platform RNASeq --
species Mm --excludeCellCycle conservative --BAM_dir
"demo_data/BAM-GSE70245-DownSampled" --output
"demo_data/BAM-GSE70245-DownSampled" --expname "Mm_HSCs"
```

Pairwise ICGS Population Comparisons

```
$ python AltAnalyze.py --species Mm --platform RNASeq --
expdir "demo_data/ExpressionInput/exp.Mm_HSCs.txt" --
groupdir "demo_data/ExpressionInput/groups.Mm_HSCs.txt" --
compdir "demo_data/ExpressionInput/comps.Mm_HSCs.txt" --
output "demo_data/"
```

2.4.2 Source Data Graphical User-Interface Examples

1. Analysis Steps
   (a) For each of the RNASeq platforms mentioned, open AltAnalyze and **Select vendor/datatype** as **RNASeq**. With 10× Genomics Chromium data, additionally **Select Platform** as 10× Genomics aligned. When proceeding to the next screen, you will be prompted to allow AltAnalyze to download the full RNASeq database, which is required.
   (b) In the next menu, select the first option **Process RNA-seq reads** or **Process Chromium Matrix**. In next menu, you will be asked to (1) name this dataset, (2) select the folder/file of interest, and (3) select an output directory (Fig. 3a). Name the dataset as desired. For BAM files, select the first folder selection option as indicated. For FASTQ files, select the second folder selection option as indicated. In each case, the user will select the folder containing their FASTQ or BAM files and either the same or a different folder as the output. For Chromium data, select the matrix.mtx file output by the Cell Ranger software. This must be in the same folder with two other files: genes.tsv and barcodes.tsv. These files are provided together in the **filtered_gene_bc_matrices** folder from Cell Ranger.
   (c) The remaining steps are virtually identical to those in the prior section (Subheading 2.3.2, **steps c–g**). For the ICGS parameters, please note that the defaults will be changed for 10× Genomics Chromium data. These are Droplet-based sequencing recommended options for platforms with decreased cell-library sequencing depth. The analysis of FASTQ and BAM files will also include alternative splicing results saved to the folder AltResults/AlternativeOutput, following **Select cluster groups for further analysis**. More

information on these splicing analyses and interfaces for visualization can be found from the AltAnalyze documentation.

2.5 Identifying and Interrogating Distinct Progenitor Populations

Once the full ICGS workflow is run, dozens of useful analysis results will be generated that can be used to understand cell fate decisions from progenitor populations. In this section, we walk through the major analysis outputs and graphing functions relevant to stem cell and progenitor biology. The full results from each of the above analyses can be found at: http://www.altanalyze.org/Data/examples.html

2.5.1 Gene Counts and Normalized Expression Values

When analyzing FASTQ, BAM, and sparse-matrix files, gene and isoform (Kallisto) counts will be exported to the output directory along with summary alignment statistics (Kallisto). These results are saved to the folder ExpressionInput with the prefix "counts." Gene expression estimates as TPM (Kallisto), RPKM (BAM), and normalized counts (counts per 10,000 UMIs) are saved in the same folder with the prefix "exp." For BAM files, two expression files are written, with the steady-state file reporting gene-level expression values and the other with all exon and junction-specific values. ICGS will save new expression values reordered based on the ICGS clusters with corresponding group names based on the user-selected (GUI) or Guide-3 results (command line). ICGS also saves a UMAP cell coordinate file that allows for dynamic visualization of any gene in the ICGS "exp." file in a UMAP plot (Subheading 2.5.4). A filtered expression file with the prefix "filteredExp" is saved from the first step of ICGS with only expressed and variable genes can be used for downstream analyses (*see* Subheading 2.6).

2.5.2 Defining Stem and Progenitor Markers from ICGS Cell States

When the full-ICGS analysis pipeline is run, the AltAnalyze MarkerFinder algorithm will assign each expressed gene in the dataset to an ICGS assigned cell state and display the top 60 genes for each state in the MarkerFinder heatmap. MarkerFinder results can be found for two outputs of ICGS: 1) those in the ICGS-NMF folder (ICGS version 2.0 primary output) and 2) DataPlots/MarkerFinder (ICGS version 1.0 primary output). The comprehensive gene results are saved to the file **MarkerGenes.txt** file in the ICGS-NMF folder and in the **AllGenes_correlations-ReplicateBased.txt**, in the folder ExpressionOutput/MarkerFinder. Each gene can only be assigned to a single cell state with a specificity rank based on a Pearson correlation-coefficient value (rho) and p-value. In the case of our bone marrow progenitor experiment (**Dataset #1**), the top 50 ranked MarkerFinder genes of cluster 1 (HSCP-1) are statistically enriched among known markers of leukocyte specification (e.g., *Satb1*, *Flt3*, *Notch2*, *Mn1*, *Dntt*, *Cd72*), which are distinct from cluster 2 (HSCP-2), which is defied by markers for hematopoietic stem cells (e.g., *Gata2*, *Ifitm1*, *Hlf*, *Malat1*, *Mpl*). Because the MarkerFinder heatmap displays maximally distinct genes for each cluster, subtle heterogeneity between distinct ICGS states will be significantly highlighted by this approach and can lead to the identification of novel stem cell specification factors. This approach is also applicable to multi-lineage progenitors, such as those defined in cluster 5 of **Dataset #1**. Although, this population of multi-lineage progenitors are defined by the coincident expression of alternative cell fate specifiers, such as *Gata2* (HSCP), *Gata1* (Eryth), *Itg2ab* (Meg), *Cebpe* (Gran), *Irf8* (Mono), and *Mpo* (Myeloid), MarkerFinder highlights a number of unique genes in this population [3]. To create MarkerFinder heatmaps with differing number of genes, use the command:

```
$ python AltAnalyze.py --update markers --platform "RNASeq"
--species Mm --geneRPKM 1 --genesToReport 200 --expdir Mm-
BoneMarrow-GSE70245/ExpressionInput/exp.BoneMarrow-
scRNASeq-ICGS.txt
```

2.6 Dynamic Pathway Analysis and Exploration of ICGS Clusters

Although ICGS associates each gene cluster with possible cell type or tissue-specific genes, discrete or novel cellular subtypes will not likely be comprehensively described by these gene-sets. To understand the biology of such populations at a deeper level, the user can apply a myriad of gene-set and pathway enrichment analyses to these ICGS results. Each graphical output produced by AltAnalyze has an associated data file generated that allows users to quickly interrogate these results. In the case of ICGS heatmaps, users can select the associated text files, re-cluster these, or perform pathway analyses, among other options. From the AltAnalyze main menu, select the species of interest and then the **Additional Analyses** option on the following screen. Select **Hierarchical Clustering**, and then select the ICGS text file from the **ICGS** or **ICGS-NMF** folder (output directory). Change the default clustering options for **Cluster rows** and **Cluster columns** to "no" to maintain the existing cell, gene, and cluster order. Finally, under **(optional) Perform GeneSet cluster enrichment**, select a gene-set category of interest (i.e., Gene Ontology, WikiPathways, MergedTFTargets, MouseImmuneCells) [32]. Running this analysis will create an interactive heatmap with enriched gene-set categories to the left of the heatmap as shown in Fig. 1c. A description of these different gene-set categories can be found at http://www.genmapp.org/go_elite/. Once run, a new window on the machine will open with the enrichment analysis results (**GO-Elite_results/ pruned-results_z-score_elite.txt**). Additionally, selecting any of the blue enriched terms will display associated genes and a network of genes to terms for that gene cluster. Selecting any gene will open the default web browser with links to additional information. To interactively explore all of the genes in this heatmap, select the option **Open**

heatmap in TreeView in the heatmap. Diverse functions are provided through this interactive viewer to customize these visualizations (http://altanalyze.readthedocs.io/en/latest/Heatmaps/). To perform pathway enrichment analyses on an ICGS heatmap via the **command line**, use:

```
$ python AltAnalyze.py --image hierarchical --row_method
None --column_method None --color_gradient
yellow_black_blue --species Mm --platform RNASeq --display
True --clusterGOElite WikiPathways --normalization median -
input "e14.5-Kidney/ICGS/Clustering-exp.e14.5-Kidney-Guide3
Pcbp4-hierarchical_euclidean_correlation.txt"
```

2.7 Dimensionality Reduction Analysis and Gene Visualization

When running the full downstream ICGS workflow, UMAP and PCA results will be automatically exported to the DataPlots directory. These results can be interactively explored in multiple ways. First, both UMAP plots and PCA plots can be overlaid with gene expression profiles from one or more genes. As shown in Fig. 2b, this can include viewing associated biological groups, when a "groups." prefix file is present in the same folder as a selected expression file. To obtain these views, return to the **Additional Analyses** menu and select **Dimensionality Reduction**. PCA (SVD), UMAP or t-SNE visualization can be chosen using the **Algorithm to use** option. Genes to visualize can be selected by pasting genes from Excel or typing those genes (space separated, case sensitive) into the **Enter a gene to color the PCA** field. For t-SNE or UMAP, selecting the ICGS exp. prefix file in the ExpressionInput folder will allow the user to quickly load existing t-SNE/ UMAP plot coordinates and visualize one or more genes. This workflow can be used to evaluate the specificity of MarkerFinder or other candidate genes to specific cell states. For more information *see* http://altanalyze.readthedocs.io/en/latest/PCA/. To generate a gene colored PCA, UMAP or t-SNE plot on the command line, use:

```
$ python AltAnalyze.py --image "t-SNE" --input
"demo_data/ExpressionInput/exp.e14.5-Kidney-ICGS.txt" --
zscore no --labels no --plotType 2D --genes "Calb1 Ret
Gdnf" --species Mm --platform RNASeq --separateGenePlots
True --display True
```

2.8 Protein and Regulatory Network Analysis

The full ICGS workflow will output a series of gene networks saved to the output folder **GO-Elite/regulated/networks** (Fig. 4). In the case of our bone marrow progenitors (Dataset #1), these networks include highlighted putative transcriptional regulators based on ChIP-Seq and ChIP-chip and genetic perturbation studies (PAZAR and Amadeus databases) and putative protein-protein interactions [32]. These networks are constructed by the

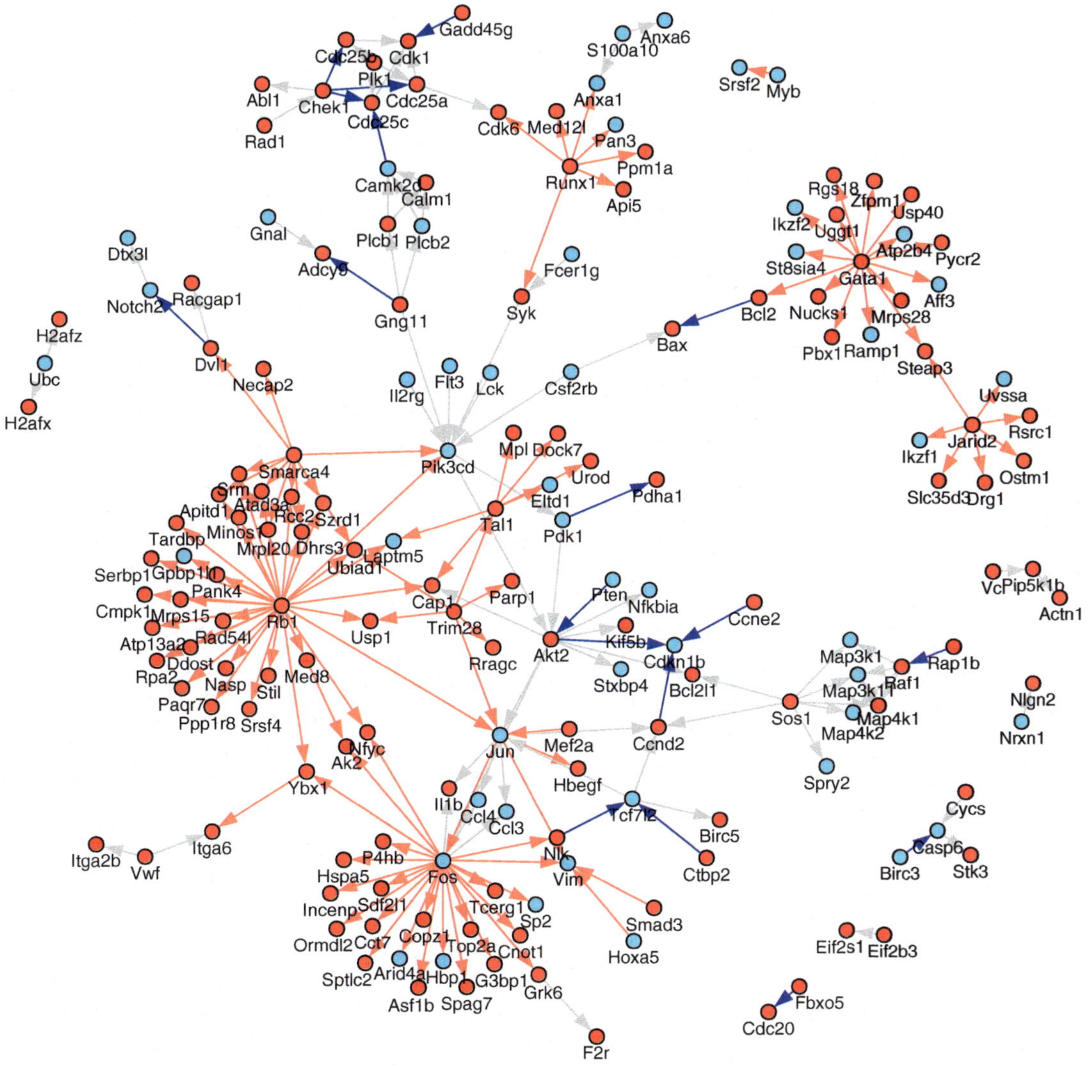

Fig. 4 Inferring regulatory and signaling networks from scRNA-Seq comparison analysis. Automated output from AltAnalyze's NetPerspective algorithm, comparing ICGS cluster 3 (Meg) versus cluster 2 (HSCP-2) differentially expressed genes (murine bone marrow progenitors). Genes in red indicate upregulation in megakaryocyte progenitors, while blue indicates downregulation relative to HSCP-2 (hematopoietic stem cell progenitors). Red arrows indicate putative transcriptional regulatory interactions based on ChIP-Seq or ChIP-chip data, while blue arrows indicate inhibitory interactions and arrows/lines indicate protein interactions

NetPerspective module of AltAnalyze (Additional Analyses > Network Visualization) using the differentially expressed genes from adjacent ICGS cell states [33]. For example, in megakaryocyte versus hematopoietic stem cell progenitors (HSCP-2), we observe several transcription factors and predicted targets previously associated with megakaryocyte specification (e.g., *Gata1*, *Runx1*, *Jarid2*, *Tal1*, *Mpl*). These networks further integrate with signaling networks associated with cell cycle induction, a process required for these cells to become polyploid [34]. As such, these networks can highlight novel downstream targets of well-established key cellular regulators. To generate such networks from the command line, run:

```
$ python AltAnalyze.py --image network --species Mm --
degrees direct --interactionDirs main --input demo_data/GO-
Elite/regulated/GE.3_vs_2-fold2.0_rawp0.05.txt --output
demo_data/GO-Elite/regulated
```

2.9 Secondary Methods to Assess Hidden Cellular Heterogeneity in AltAnalyze

In the hematopoietic and other cell systems, emerging evidence suggests that very subtle transcriptomic differences in a small handful of genes can define functional heterogeneity in certain stem and progenitor populations [35, 36]. As such, unsupervised single-cell analyses, including ICGS, may be underpowered to identify coherent co-regulated gene-sets that define distinct transcription stem cell states. In this section, we outline alternative methods to uncover hidden cellular heterogeneity using (1) a priori defined population comparisons, (2) supervised gene analyses, or (3) PCA-loading gene analyses in AltAnalyze.

2.9.1 Direct Cell Population Comparison

To perform supervised comparisons between independently obtained scRNA-Seq datasets, AltAnalyze has two built in methods to compare scRNA-Seq datasets: 1) cellHarmony and 2) direct comparison of ICGS clusters. cellHarmony is a cell-matching algorithm designed to identify a cell's most similar analogue in a distinct scRNA-Seq dataset and find differentially expressed genes in each cell population. More on this method and how to run can be found online (http://altanalyze.org/cellHarmony/). To compare already ICGS analyzed scRNA-Seq datasets we can also use the standard differential comparison workflow in AltAnalyze. First, the user must combine datasets of interest. If analyzing BAM or FASTQ files, these files can simply be moved to a single directory. For existing expression files (*see* Subheading 2.5.1), the user will need to merge these files prior to analysis. To merge similarly analyzed dataset without presumed batch effects, use the **Additional Analyses** menu option **Merge Files** and select the different tab-delimited expression files of interest to intersect. By default, only matching genes from the selected files will be intersected and exported to a new combined file in which each cell IDs indicate the name of the file from which it originated. To perform this analysis on the command line, run:

```
$ python AltAnalyze.py --accessoryAnalysis MergeFiles --
input "data/exp.WT.txt" --input "data/exp.WT.txt " --output
"data/"
```

From the header of this file, create a group file indicating which cells come from which experiment (*see* http://altanalyze.readthedocs.io/en/latest/ManualGroupsCompsCreation). To compare these cells from these separate datasets, repeat **steps a–e** from Subheading 2.3.2, without performing ICGS. The primary differential expression analysis results can be found in the ExpressionOutput directory in the DATASET file, along with MarkerFinder results for the assigned groups (*see* Subheading 2.5.2). *See* Subheading 2.4.1 for associated pairwise group comparison command-line arguments.

2.9.2 Supervised Guide-Gene Correlation Clustering

In many cases, valuable prior knowledge may already exist to guide the discovery of hidden transcription heterogeneity. In such cases, customized options exist in ICGS and in the AltAnalyze hierarchical clustering menu to exploit prior knowledge. In the ICGS window (Fig. 3c, Subheading 2.3.2, **step e**), both custom genes and stored gene-sets (e.g., Gene Ontology) can be used to guide the ICGS results. For example, reported key specification regulators can be pasted into the ICGS entry box **(optional) Enter genes to build clusters from (guides)**. Simply paste a list of genes from Excel or as a space separated list into this entry field and apply the existing defaults, prior to running ICGS. This option will still run ICGS but initiates the analysis with only genes correlated to the user-supplied input list. This custom guide-gene correlated set replaces the default all-pairwise gene correlation step of ICGS derives. A similar analysis can be run with AltAnalyze stored gene-sets (e.g., Gene Ontology categories) that can be selected through this same interface.

ICGS generates a filtered expression file with only variable and expressed genes which is saved with the prefix **filteredExp** in the folder ExpressionInput. This file can be used for additional supervised expression analyses in the **Hierarchical Clustering** menu under **Additional Analyses**. From this interface, users can cluster their **filteredExp** or other expression file for only selected input gene(s) by pasting those genes from Excel or manually typing them into the (optional) **Type a gene to get top correlated** field. When specifying a Pearson correlation cutoff (e.g., 0.4) into **(optional) Get all correlated genes with rho >**, other genes with correlated or anticorrelated expression to the input genes will also be included in the resulting heatmap. Rather than the user supplying their own genes, gene-sets can be obtained from the **(optional) Select GeneSet/Ontology to filter** and **(optional) Select a specific GeneSet** options. The **Additional correlation options** menu has more options available to control which genes are reported (e.g., positive correlations only) to further customize these results. All generated heatmaps are saved to the DataPlots folder in which the input file is present. These analyses can also be easily run on the command line as well using:

```
$ python AltAnalyze.py --image hierarchical --platform
RNASeq --row_method hopach --column_method hopach --
row_metric correlation --column_metric cosine --
color_gradient yellow_black_blue --species Mm --platform
RNASeq --normalization median --contrast 4 --genes "Mecom
Irf8 Klf1 Gfi1 Satb1 Mpl amplify positive" --rho 0.4 --
display True --input
"demo_data/ExpressionInput/filteredExp.WT.txt"
```

2.10 Unsupervised PCA-Loading Gene Analysis

With datasets characterized by subtle transcriptional heterogeneity, it can also be beneficial to apply alternative unsupervised variance analysis approaches. In Subheading 2.9.2, we described the creation of a **filteredExp** file from ICGS as well as methods for supervised gene-clustering analyses. These two principles can also be applied in combination with the results from a principal component analysis (PCA) to identify transcriptomic heterogeneity in your scRNA-Seq data. To perform such an analysis, return to the **Additional Analyses** and **Dimensionality Reduction** menu and select the **filteredExp** file as input. By simply typing a custom name into the field **(Optional) Store top 200 component genes as**, those genes associated with the most common cell-cell variation in the filteredExp file will be stored for hierarchical clustering analysis. These genes comprise the top 100 correlated and 100 anticorrelated for the top 4 principal components. Returning to the **Hierarchical Clustering** menu, select **StoredGeneSets** from **(optional) Select GeneSet/Ontology to filter** and your custom PCA-loading gene-set name from **(optional) Select a specific GeneSet**. When clustered, only the PCA-loading genes will be used. This analysis can also provide secondary confirmation of ICGS discovered cell states. To perform downstream analyses on these results, the user will need to create a "groups" file from the resulting heatmap text file (DataPlots—*see* the row named "column_clusters-flat").

3 Notes

The complexity of supported analyses and software environments can lead to issues in running the workflows or obtaining optimal results. Some of these problems relate to the structure or formatting of the underlying data files, while others relate to inherent bugs or current limitations of the software environment. Other issues include challenges in the identification of subtype cellular heterogeneity. Overcoming these challenges will increase the likelihood of finding biologically meaningful differences to exploit in downstream analyses. In this section, we briefly review solutions to common challenges with the application of ICGS in AltAnalyze.

1. *Problems with the User Interface*

   AltAnalyze's pre-compiled binaries for Mac and Windows are among the most popular downloads. Since each binary is built on a single machine, issues can arise with machines from different countries, certain operating systems, or hardware configurations. If all else fails, the best solution to unexpected errors that arise is to install AltAnalyze from the **Python Package Installer** described in Subheading 2.2. This will ensure the software is built for your specific machine; however, incompatibilities between certain libraries are more likely to occur with this option. If errors occur due to a missing library, attempt to independently install the associated Python library yourself. If the program stalls when running, it is possible a multi-threading processing conflict has occurred. In such cases, force quite the application and review the log file (.log file) and attempt to rerun the analysis. For most analyses, we recommend machines with a minimum of 8GB of RAM and ideally 16GB of RAM. Please feel free to email the AltAnalyze Google Group with any other unresolved issues. If persistent issues occur, try re-installing the software as certain files may have become overwritten.

   Mac OS X Common Issues: The three most common AltAnalyze interface issues for Mac users are (1) OS warns that the software can't be opened because it is from an unidentified developer, (2) AltAnalyze won't open or crashes, and (3) AltAnalyze quits unexpectedly. These issues are specific to the Mac, with issues 1 and 2 specific to the pre-compiled DMG version. Issue #1 occurs because AltAnalyze is not provided through the Apple App Store, but rather by the AltAnalyze development team. To open, simply Ctrl+click on the application icon and select "Open" and then select the "Open" option when prompted. Issue #2 occurs on some machines due to an unknown OS issue but can be easily resolved in most cases by double clicking on the alternative AltAnalyze alias icon found in the same application directory. Issue #3 occurs on certain machines due to specific Python 2 64-bit Mac interface bug that causes the program to crash when the user leaves AltAnalyze for the desktop or another application and returns. The error occurs when the AltAnalyze icon is selected from the task bar on some Macs. To address, use the command ⌘ + tab key combination to switch between applications while AltAnalyze is running.

   Windows Common Issues: AltAnalyze produces a large number of files in a series of nested directories. On Windows, some of these file paths become very long and can cause errors when saving or retrieving certain files. However, storing the initial input files to relatively low versus deep directory path and giving shorter names to the dataset or groups may address this issue.

2. *Running ICGS*

   Because ICGS can be applied to diverse input data files, the default options may not always be optimal. In such situations, modifying the defaults is recommended to aid in the discovery of distinct cell states. Removing outliers is recommended, which can be performed through the software or manually by the user. To automatically **remove low expression outliers**, select this option in the ICGS GUI, or include the command-line argument `--removeOutliers yes`. This option will exclude cells with <500 genes expressed above the indicated expression cutoff. The user can also remove cells based on their own criterion, for example, by filtering on based on cell estimated aligned counts from the Kallisto summary file (ExpressionInput). Simply save a tab-delimited text file with only the cells of interest in the first column and run the below command (output will have the suffix "-filtered.txt"):

```
$ python AltAnalyze.py --accessoryAnalysis FilterFile --
input "data/exp.WT.txt" --filterFile
"data/cells_to_keep.txt"
```

   Other issues that can arise with ICGS include batch effects, and other confounding variables in the underlying data. The differential expression of ribosomal, mitochondrial genome and other noncoding RNAs is often a common variable feature in scRNA-Seq datasets. Similar to cell cycle, these genes can skew the identification of distinct cell states, and hence removal is often recommended. These noncoding and ribosomal genes can be excluded with the option **Restrict gene to protein coding** or the command line flag: `--restrictBy protein_coding (default)`. In terms of potential batch effects between biological replicates or independent single-cell analyses, other external tools are recommended to reduce the occurrence of such effects prior to analysis with ICGS.

3. *Problems with External Input scRNA-Seq Text Files*

   When processing external gene expression results, the user must make sure the input text files are tab-delimited without any unnecessary columns or rows. When processing an externally provided text file, make sure that the gene IDs in the first column only contain a single-gene ID type (e.g., gene symbol, Ensembl gene ID). If counts are present in the text file and not normalized, the user can convert these to normalized gene count values (CPTT) using:

```
$ python import_scripts/DropSeqProcessing.py --i
data/dropseq_counts.txt
```

4. *Exclusion of Cell Doublets or "Contaminating" Cell Populations*

   Many scRNA-Seq platforms readily output cell profiles which are a combination of multiple captured cells. Such doublet cells often increase in frequency with the more cells loaded. If abundantly occurring, ICGS should identify frequent doublets as ICGS clusters which can be differentiated from real cell states based on the lack of uniquely expressed genes in those cell populations by MarkerFinder (*see* Subheading 2.5.2). A contaminating cell population can be subjective. For example, a T-cell capture may include B-cells as contaminants. To exclude, the user can filter the reordered ICGS gene expression file in the ExpressionInput folder (prefix exp. and suffix -ICGS. txt) based on the clusters in the ICGS groups file with the `--accessoryAnalysis FilterFile` command line listed in **Note 2**. In addition to doublet clusters, automated doublet detection and removal tools can be run to identify such outliers. These include the DoubletDecon R package (https://github.com/EDePasquale/DoubletDecon), which works directly with the outputs of ICGS.

References

1. Treutlein B, Brownfield DG, Wu AR, Neff NF, Mantalas GL, Espinoza FH, Desai TJ, Krasnow MA, Quake SR (2014) Reconstructing lineage hierarchies of the distal lung epithelium using single-cell RNA-seq. Nature 509 (7500):371–375. Epub 2014/4/18. https://doi.org/10.1038/nature13173
2. Trapnell C, Cacchiarelli D, Grimsby J, Pokharel P, Li S, Morse M, Lennon NJ, Livak KJ, Mikkelsen TS, Rinn JL (2014) The dynamics and regulators of cell fate decisions are revealed by pseudotemporal ordering of single cells. Nat Biotechnol 32(4):381–386. Epub 2014/3/25. https://doi.org/10.1038/nbt.2859
3. Olsson A, Venkatasubramanian M, Chaudhri VK, Aronow BJ, Salomonis N, Singh H, Grimes HL (2016) Single-cell analysis of mixed-lineage states leading to a binary cell fate choice. Nature 537(7622):698–702. Epub 2016/9/1. https://doi.org/10.1038/nature19348
4. La Manno G, Gyllborg D, Codeluppi S, Nishimura K, Salto C, Zeisel A, Borm LE, Stott SR, Toledo EM, Villaescusa JC, Lonnerberg P, Ryge J, Barker RA, Arenas E, Linnarsson S (2016) Molecular diversity of midbrain development in mouse, human, and stem cells. Cell 167(2):566–80.e19. Epub 2016/10/8. https://doi.org/10.1016/j.cell.2016.09.027
5. Treutlein B, Lee QY, Camp JG, Mall M, Koh W, Shariati SA, Sim S, Neff NF, Skotheim JM, Wernig M, Quake SR (2016) Dissecting direct reprogramming from fibroblast to neuron using single-cell RNA-seq. Nature 534 (7607):391–395. Epub 2016/6/10. https://doi.org/10.1038/nature18323
6. Villani AC, Satija R, Reynolds G, Sarkizova S, Shekhar K, Fletcher J, Griesbeck M, Butler A, Zheng S, Lazo S, Jardine L, Dixon D, Stephenson E, Nilsson E, Grundberg I, McDonald D, Filby A, Li W, De Jager PL, Rozenblatt-Rosen O, Lane AA, Haniffa M, Regev A, Hacohen N (2017) Single-cell RNA-seq reveals new types of human blood dendritic cells, monocytes, and progenitors. Science 356(6335):eaah4573. https://doi.org/10.1126/science.aah4573
7. Shekhar K, Lapan SW, Whitney IE, Tran NM, Macosko EZ, Kowalczyk M, Adiconis X, Levin JZ, Nemesh J, Goldman M, McCarroll SA, Cepko CL, Regev A, Sanes JR (2016) Comprehensive classification of retinal bipolar neurons by single-cell transcriptomics. Cell 166

(5):1308–23.e30. https://doi.org/10.1016/j.cell.2016.07.054

8. Zheng GXY, Terry JM, Belgrader P, Ryvkin P, Bent ZW, Wilson R, Ziraldo SB, Wheeler TD, McDermott GP, Zhu J, Gregory MT, Shuga J, Montesclaros L, Masquelier DA, Nishimura SY, Schnall-Levin M, Wyatt PW, Hindson CM, Bharadwaj R, Wong A, Ness KD, Beppu LW, Deeg J, McFarland C, Loeb KR, Valente WJ, Ericson NG, Stevens EA, Radich JP, Mikkelsen TS, Hindson BJ, Bielas JH (2017) Massively parallel digital transcriptional profiling of single cells. Nat Commun 8:14049. https://doi.org/10.1101/065912
9. Macosko EZ, Basu A, Satija R, Nemesh J, Shekhar K, Goldman M, Tirosh I, Bialas AR, Kamitaki N, Martersteck EM, Trombetta JJ, Weitz DA, Sanes JR, Shalek AK, Regev A, McCarroll SA (2015) Highly parallel genome-wide expression profiling of individual cells using nanoliter droplets. Cell 161 (5):1202–1214. Epub 2015/5/23. https://doi.org/10.1016/j.cell.2015.05.002
10. Adamson B, Norman TM, Jost M, Cho MY, Nunez JK, Chen Y, Villalta JE, Gilbert LA, Horlbeck MA, Hein MY, Pak RA, Gray AN, Gross CA, Dixit A, Parnas O, Regev A, Weissman JS (2016) A multiplexed single-cell CRISPR screening platform enables systematic dissection of the unfolded protein response. Cell 167(7):1867–82.e21. https://doi.org/10.1016/j.cell.2016.11.048
11. Dixit A, Parnas O, Li B, Chen J, Fulco CP, Jerby-Arnon L, Marjanovic ND, Dionne D, Burks T, Raychowdhury R, Adamson B, Norman TM, Lander ES, Weissman JS, Friedman N, Regev A (2016) Perturb-Seq: dissecting molecular circuits with scalable single-cell RNA profiling of pooled genetic screens. Cell 167(7):1853–66.e17. https://doi.org/10.1016/j.cell.2016.11.038
12. Cordero P, Stuart JM (2017) Tracing co-regulatory network dynamics in noisy, single-cell transcriptome trajectories. Pac Symp Biocomput 22:576–587. https://doi.org/10.1142/9789813207813_0053
13. Ji Z, Ji H (2016) TSCAN: Pseudo-time reconstruction and evaluation in single-cell RNA-seq analysis. Nucleic Acids Res 44(13):e117. https://doi.org/10.1093/nar/gkw430
14. Grun D, Lyubimova A, Kester L, Wiebrands K, Basak O, Sasaki N, Clevers H, van Oudenaarden A (2015) Single-cell messenger RNA sequencing reveals rare intestinal cell types. Nature 525(7568):251–255. Epub 2015/8/20. https://doi.org/10.1038/nature14966
15. Kiselev VY, Kirschner K, Schaub MT, Andrews T, Yiu A, Chandra T, Natarajan KN, Reik W, Barahona M, Green AR, Hemberg M (2017) SC3: consensus clustering of single-cell RNA-seq data. Nat Methods 14(5):483–486. https://doi.org/10.1038/nmeth.4236
16. Jiang L, Chen H, Pinello L, Yuan GC (2016) GiniClust: detecting rare cell types from single-cell gene expression data with Gini index. Genome Biol 17(1):144. https://doi.org/10.1186/s13059-016-1010-4
17. Buettner F, Natarajan KN, Casale FP, Proserpio V, Scialdone A, Theis FJ, Teichmann SA, Marioni JC, Stegle O (2015) Computational analysis of cell-to-cell heterogeneity in single-cell RNA-sequencing data reveals hidden subpopulations of cells. Nat Biotechnol 33(2):155–160. Epub 2015/1/20. https://doi.org/10.1038/nbt.3102
18. Barron M, Li J (2016) Identifying and removing the cell-cycle effect from single-cell RNA--Sequencing data. Sci Rep 6:33892. https://doi.org/10.1038/srep33892
19. Satija R, Farrell JA, Gennert D, Schier AF, Regev A (2015) Spatial reconstruction of single-cell gene expression data. Nat Biotechnol 33(5):495–502. Epub 2015/4/14. https://doi.org/10.1038/nbt.3192
20. Grun D, Muraro MJ, Boisset JC, Wiebrands K, Lyubimova A, Dharmadhikari G, van den Born M, van Es J, Jansen E, Clevers H, de Koning EJ, van Oudenaarden A (2016) De novo prediction of stem cell identity using single-cell transcriptome data. Cell Stem Cell 19(2):266–277. Epub 2016/6/28. https://doi.org/10.1016/j.stem.2016.05.010
21. Guo M, Bao EL, Wagner M, Whitsett JA, Xu Y (2017) SLICE: determining cell differentiation and lineage based on single cell entropy. Nucleic Acids Res 45(7):e54. https://doi.org/10.1093/nar/gkw1278
22. Guo M, Wang H, Potter SS, Whitsett JA, Xu Y (2015) SINCERA: a pipeline for single-cell RNA-Seq profiling analysis. PLoS Comput Biol 11(11):e1004575. Epub 2015/11/26. https://doi.org/10.1371/journal.pcbi.1004575
23. Brunskill EW, Park JS, Chung E, Chen F, Magella B, Potter SS (2014) Single cell dissection of early kidney development: multilineage priming. Development 141(15):3093–3101. Epub 2014/7/24. https://doi.org/10.1242/dev.110601
24. Magella B, Adam M, Potter AS, Venkatasubramanian M, Chetal K, Hay SB, Salomonis N, Potter SS (2018) Cross-platform single cell analysis of kidney development shows stromal cells express Gdnf. Dev Biol 434:36–47. https://doi.org/10.1016/j.ydbio.2017.11.006

25. Meyer SE, Qin T, Muench DE, Masuda K, Venkatasubramanian M, Orr E, Suarez L, Gore SD, Delwel R, Paietta E, Tallman MS, Fernandez H, Melnick A, Le Beau MM, Kogan S, Salomonis N, Figueroa ME, Grimes HL (2016) DNMT3A haploinsufficiency transforms FLT3ITD myeloproliferative disease into a rapid, spontaneous, and fully penetrant acute myeloid leukemia. Cancer Discov 6 (5):501–515. Epub 2016/3/27. https://doi.org/10.1158/2159-8290.CD-16-0008
26. Yanez A, Coetzee SG, Olsson A, Muench DE, Berman BP, Hazelett DJ, Salomonis N, Grimes HL, Goodridge HS (2017) Granulocyte-monocyte progenitors and monocyte-dendritic cell progenitors independently produce functionally distinct monocytes. Immunity 47 (5):890–902.e4. https://doi.org/10.1016/j.immuni.2017.10.021
27. Velten L, Haas SF, Raffel S, Blaszkiewicz S, Islam S, Hennig BP, Hirche C, Lutz C, Buss EC, Nowak D, Boch T, Hofmann WK, Ho AD, Huber W, Trumpp A, Essers MA, Steinmetz LM (2017) Human haematopoietic stem cell lineage commitment is a continuous process. Nat Cell Biol 19(4):271–281. https://doi.org/10.1038/ncb3493
28. Pavlicev M, Wagner GP, Chavan AR, Owens K, Maziarz J, Dunn-Fletcher C, Kallapur SG, Muglia L, Jones H (2017) Single-cell transcriptomics of the human placenta: inferring the cell communication network of the maternal-fetal interface. Genome Res 27(3):349–361. Epub 2017/2/9. https://doi.org/10.1101/gr.207597.116
29. Lim CY (2017) Understanding transcriptional regulation through computational analysis of single-cell transcriptomics. University of Cambridge, Cambridge
30. van der Laan MJ, Pollard KS (2003) A new algorithm for hybrid hierarchical clustering with visualization and the bootstrap. J Stat Plan Inference 117(2):275–303
31. Bray NL, Pimentel H, Melsted P, Pachter L (2016) Near-optimal probabilistic RNA-seq quantification. Nat Biotechnol 34 (5):525–527. Epub 2016/4/5. https://doi.org/10.1038/nbt.3519
32. Zambon AC, Gaj S, Ho I, Hanspers K, Vranizan K, Evelo CT, Conklin BR, Pico AR, Salomonis N (2012) GO-Elite: a flexible solution for pathway and ontology over-representation. Bioinformatics 28 (16):2209–2210. Epub 2012/6/30. https://doi.org/10.1093/bioinformatics/bts366
33. McLendon PM, Davis G, Gulick J, Singh SR, Xu N, Salomonis N, Molkentin JD, Robbins J (2017) An unbiased high-throughput screen to identify novel effectors that impact on cardiomyocyte aggregate levels. Circ Res 121 (6):604–616. https://doi.org/10.1161/CIRCRESAHA.117.310945
34. Machlus KR, Italiano JE Jr (2013) The incredible journey: From megakaryocyte development to platelet formation. J Cell Biol 201 (6):785–796. https://doi.org/10.1083/jcb.201304054
35. Cabezas-Wallscheid N, Buettner F, Sommerkamp P, Klimmeck D, Ladel L, Thalheimer FB, Pastor-Flores D, Roma LP, Renders S, Zeisberger P, Przybylla A, Schonberger K, Scognamiglio R, Altamura S, Florian CM, Fawaz M, Vonficht D, Tesio M, Collier P, Pavlinic D, Geiger H, Schroeder T, Benes V, Dick TP, Rieger MA, Stegle O, Trumpp A (2017) Vitamin A-retinoic acid signaling regulates hematopoietic stem cell dormancy. Cell 169(5):807–23.e19. https://doi.org/10.1016/j.cell.2017.04.018
36. Wilson NK, Kent DG, Buettner F, Shehata M, Macaulay IC, Calero-Nieto FJ, Sanchez Castillo M, Oedekoven CA, Diamanti E, Schulte R, Ponting CP, Voet T, Caldas C, Stingl J, Green AR, Theis FJ, Gottgens B (2015) Combined single-cell functional and gene expression analysis resolves heterogeneity within stem cell populations. Cell Stem Cell 16 (6):712–724. https://doi.org/10.1016/j.stem.2015.04.004
37. Yan L, Yang M, Guo H, Yang L, Wu J, Li R, Liu P, Lian Y, Zheng X, Yan J, Huang J, Li M, Wu X, Wen L, Lao K, Li R, Qiao J, Tang F (2013) Single-cell RNA-Seq profiling of human preimplantation embryos and embryonic stem cells. Nat Struct Mol Biol 20 (9):1131–1139. Epub 2013/8/13. https://doi.org/10.1038/nsmb.2660

Chapter 13

Lineage Inference and Stem Cell Identity Prediction Using Single-Cell RNA-Sequencing Data

Sagar and Dominic Grün

Abstract

With the advent of several single-cell RNA-sequencing (scRNA-seq) techniques, it has become possible to gain novel insights into the fundamental long-standing questions in biology with an unprecedented resolution. Among the various applications of scRNA-seq, (1) discovery of novel rare cell types, (2) characterization of heterogeneity among the seemingly homogenous population of cells described by cell surface markers, (3) stem cell identification, and (4) construction of lineage trees recapitulating the process of differentiation remain at the forefront. However, given the inherent complexity of these data arising from the technical challenges involved in such assays, development of robust statistical and computational methodologies is of major interest. Therefore, here we present an in-house state-of-the-art scRNA-seq data analyses workflow for de novo lineage tree inference and stem cell identity prediction applicable to many biological processes under current investigation.

Key words Single-cell RNA sequencing, Single-cell data analysis, Stem cell identification, Lineage inference, Pseudo-temporal ordering, Fate bias, Multipotent, RaceID3, StemID2, FateID

1 Introduction

During embryonic development as well as adult tissue homeostasis and regeneration, cells move from an immature naïve state to a mature functional state through a series of complex transcriptional changes governed by various cell-intrinsic and cell-extrinsic factors. Quantification of genome-wide transcription of individual cells using scRNA-seq provides a unique opportunity to capture and characterize distinct cell states occurring during such biological processes. Moreover, since the process of differentiation is not completely synchronous in complex multicellular organisms, single-cell expression data often comprises of not only immature and mature cell types but also transient cell types with unique transcriptional signature representing the intermediate states of cellular differentiation. Therefore, statistical methods and computational tools de novo or in combination with prior biological

Patrick Cahan (ed.), *Computational Stem Cell Biology: Methods and Protocols*, Methods in Molecular Biology, vol. 1975, https://doi.org/10.1007/978-1-4939-9224-9_13, © Springer Science+Business Media, LLC, part of Springer Nature 2019 knowledge can be used to predict the identity of stem cells and infer the lineage tree by aligning the intermediate cell states along a trajectory, thereby tracking the global gene expression changes occurring during the process of differentiation.

During the last few years, several methods have been developed for lineage inference and stem cell identification from scRNA-seq data [1–7]. Here we describe our in-house step-by-step workflow of relevant aspects of previously published RaceID3, StemID2, and FateID algorithms including the sample R-codes for de novo lineage tree construction and stem cell identity prediction, thereby providing important tools to facilitate scRNA-seq-driven discoveries.

2 Materials

The analysis workflow described in this protocol is implemented in R. Analyses can be performed either on a private workstation or compute servers [8].

1. R (https://www.r-project.org/) and RStudio (optional, https://www.rstudio.com/).
2. RaceID3 and StemID2 algorithms (https://github.com/dgrun/RaceID3_StemID2). The link contains two R files: RaceID3_StemID2_class.R and RaceID3_StemID2_sample.R as well as an example data in .xls format.
3. FateID algorithm available as an R package through CRAN package repository.

3 Methods

This section provides a step-by-step workflow using a previously published dataset where 5-day-old progeny of *Lgr5*$^+$ mouse intestinal cells was sequenced [9]. In order to identify stem cells in the scRNA-seq data, first robust identification of all cell types in the dataset is necessary. This is done by applying the RaceID3 algorithm, which performs k-medoids clustering and a subsequent outlier identification step to recover all cell types, including the rare ones. Afterward, the StemID2 algorithm is used to construct a lineage tree. The StemID2 algorithm is also used to identify the putative stem/progenitor cell cluster(s) in the dataset. Additionally, based on the StemID2-predicted differentiation trajectory, cells in the dataset can be arranged from the progenitor state to the differentiated state, thereby constructing a pseudo-temporal profile of cells recapitulating the process of differentiation [8]. This approach provides an important informative tool to visualize the dynamic transcriptional changes of known and unknown genes occurring during differentiation. Furthermore, we also present another independent approach based on our recently developed FateID algorithm to quantify the level of multipotency based on the entropy level of fate bias. Note that we discuss only the most relevant aspects of our analysis pipeline in this protocol due to space constraints and users are advised to go through the manuals available online to explore the full functionality of our computational tools.

3.1 Package Installation and Data Preprocessing

1. After installing R, install various R packages required for the proper functioning of the RaceID3 and StemID2 algorithms using the following R code:

```
install.packages(c("tsne", "pheatmap", "MASS", "cluster",
"mclust", "flexmix", "lattice", "fpc", "RColorBrewer",
"permute", "amap", "locfit", "vegan", "Rtsne", "scran",
"randomForest","rgl"))
```

2. Additionally, install the following bioconductor packages, scran, DESeq2, and biomaRt, using the following command:

```
source("https://bioconductor.org/biocLite.R")
biocLite("scran")
biocLite("DESeq2")
biocLite("biomaRt")
```

3. Create a directory of your desired name, and copy the following downloaded files using the link https://github.com/dgrun/RaceID3_StemID2 into it:

```
RaceID3_StemID2_class.R,
RaceID3_StemID2_sample.R and
transcript_counts_intestine_5days_YFP.xls
```

4. Load the Rscript `RaceID3_StemID2_class.R` containing all functions required to run the algorithm implemented as an S4 class object `SCseq` using the following command:

```
source("RaceID3_StemID2_class.R")
```

5. Load the example data `transcript_counts_intestine_5days_YFP.xls` as a data frame object in the working environment using the following commands (*see* **Note 1**):

```
expr.data <-
read.csv("transcript_counts_intestine_5days_YFP.xls",
sep="\t", header=TRUE)
rownames(expr.data) <- expr.data$GENEID
```

```
expr.data <-
expr.data[grep("ERCC",rownames(expr.data),invert=TRUE),-
1]
```

6. Create an `SCseq` object called `sc` using the `expr.data` data frame using the following command: `sc <- SCseq(prdata)`.

3.2 Data Normalization and Removal of Unwanted Variability

After having the data in the correct format and an `SCseq` object being created, the next steps are performed by calling the function `filterdata`. It has several important arguments for data normalization and removal of unwanted variability. The steps below provide a brief explanation of the most important arguments.

1. `mintotal` defines the minimum total number of transcripts per cell. Only cells with a total number of transcripts greater than `mintotal` are included in the analysis. It is quite likely to have a few cells displaying very low number of total transcripts due to various technical reasons. Such cells should be discarded from the further downstream analysis.
2. Data normalization can be achieved using two different methods – downsampling and rescaling. If `downsample` is set to `TRUE`, the number of transcripts in each cell will be downsampled to a value of `mintotal`. Rescaling is done by dividing the number of transcript for each gene per cell by the total number of transcripts detected in this cell followed by multiplication with the minimum of the total number of transcripts across all cells present in the analysis after removing cells having less than `mintotal` number of transcripts (*see* **Note 2**).
3. In order to avoid handling zeros present in the expression matrix for further calculations, e.g., log transformation of the expression matrix, a pseudocount of 0.1 is added to the expression data. The data after normalization and pseudocount addition is saved in the `sc@ndata` slot.
4. `minexpr` and `minnumber` arguments define the minimum number of transcript of a particular gene in a cell and minimum number of cells, respectively. Only if a gene is expressed at a value greater than `minexpr` in at least `minnumber` of cells, it is included in the clustering analysis. The expression data containing the remaining genes with their corresponding transcript counts in all the cells included in the analysis is saved in the `sc@fdata` slot.
5. In order to remove further genes, which may artificially influence the clustering (such as analyzing cells from a transgenic mouse, which overexpress a particular gene), the argument `FGenes` can be initialized with gene names whose exclusion from clustering is desired (*see* **Note 3**).

6. Furthermore, instead of excluding a single gene or set of genes using `FGenes`, if removal of all the genes correlating with certain gene(s) is necessary, then argument `CGenes` can be initialized. `CGenes` removes all the genes positively or negatively correlated with gene(s) provided in `CGenes` argument with a threshold defined by the argument `ccor` (*see* **Note 4**).
7. The following R-code can used to initialize all the parameters for data normalization and filtering steps:

```
sc <- filterdata(sc, mintotal=3000, minexpr=5,
minnumber=1, maxexpr=Inf, downsample=FALSE, sfn=FALSE,
hkn=FALSE, dsn=1, rseed=17000, CGenes=NULL,
FGenes=NULL, ccor=.4)
```

8. RaceID3 provides two optional functions to remove unwanted variability: `CCcorrect` and `varRegression`. To use `CCcorrect`, any GO annotation, for example, "cell cycle" and "cell proliferation," can be retrieved using the bioconductor package `biomaRt`. The associated variability to the gene sets is removed using a principal component analysis (PCA) or an independent component analysis (ICA) using argument `mode`. The following commands can be executed to remove the variability associated with "cell cycle" and "cell proliferation" using `biomaRt`:

```
require(biomaRt)
g <- sub("__chr.+","",rownames(sc@fdata))
k <- getBM(attributes = c("external_gene_name",
"go_id", "name_1006"),
filters = "external_gene_name", values = g, mart = mart)
gCC <- name2id( k$external_gene_name[k$name_1006 == "cell
cycle"],rownames(sc@fdata))
gCP <- name2id( k$external_gene_name[k$name_1006 == "cell
proliferation"],rownames(sc@fdata))
vset <- list(gCC,gCP)
x <- CCcorrect(sc@fdata, vset=vset, CGenes=NULL, ccor=.4,
nComp=NULL, pvalue=.05, quant=.01, mode="pca")
```

The function returns a list consisting of three objects. `x$n` returns the number of components that have been removed, `x$pca` contains all the information associated with the PCA analysis, and `x$cor` contains the corrected expression data which has to be used to overwrite the `sc@fdata` slot using the following command:

```
sc@fdata <- x$xcor
```

9. RaceID3 also offers an optional function `varRegression` to remove batch-associated variability through linear regression. The following command can be used to remove batch-associated variability in the given dataset:

```
sc@fdata <- varRegression(sc@fdata,vars)
```

The argument `vars` defines a data frame having batch information as columns and cell names/identification codes as rows. For example, for the given mouse intestinal dataset, the following command can be used to extract the batch information:

```
vars <- data.frame(row.names=names(sc@fdata),
batch=sub("(_|)\crd.+","",names(sc@fdata)))
```

3.3 k-Medoids Clustering, Outlier Identification, and Data Visualization

Once the expression data is normalized and filtered as well as any unwanted variability, if required, is removed, the next step involves a robust clustering of the expression data to identify different cell types. This is achieved using k-medoids clustering. However, standard clustering approaches such as k-means clustering and k-medoids clustering require a previous knowledge of the number of clusters that can be inferred from the dataset, thereby preventing them to identify rare cell types. To circumvent this problem, RaceID3 algorithm includes an extra outlier identification step to identify rare cell types after the initial k-medoids clustering is performed. The outlier identification step is based on the transcript count distribution for each gene in a given cluster. Such distribution is assumed to be negative binomial governed by two parameters: the mean and the dispersion. While the mean is directly calculated as the average expression of the given gene across all the cells in a given cluster, the dispersion parameter is derived by fitting a second-order polynomial regression on the variance-mean relationship for each gene in the whole dataset. Based on this fit, for each gene a probability value is calculated for the observed transcript counts in each cell of a given cluster. If the transcript count of `x` number of genes in a cell have a probability value less than `p`, which is multiple testing corrected by the Benjamini-Hochberg method (across cells), this cell is flagged as an outlier. `x` and `p` can be user-defined keeping certain guidelines in consideration described below. Finally, the outlier cells are merged based on their similarity to each other. After the outlier cells are merged, new cluster centers are defined for the original clusters after removing the outliers and for the newly identified outlier clusters. Subsequently, each cell is assigned to the nearest cluster center. The final clusters are given numerical values in an increasing order. Thereafter, the visualization of cell clusters is required. This goal is achieved via reducing the dimensions of the expression data into two-dimensional space using the t-Distributed Stochastic Neighbor Embedding (tSNE) algorithm, which preserves the local structure of the higher-dimensional data into the two-/three-dimensional space [10]. This section describes the step-by-step workflow of clustering, outlier identification, and data visualization using tSNE in detail.

1. k-medoids clustering is performed using `clustexp` function using the following R-code:

```
sc <- clustexp(sc, clustnr=30, bootnr=50,
metric="pearson", do.gap=FALSE, sat=TRUE,
SE.method="Tibs2001SEmax", SE.factor=.25, B.gap=50,
cln=0, rseed=17000, FUNcluster="kmedoids",
FSelect=TRUE)
```

   The next steps describe the important arguments of `clustexp` function in detail.

2. To perform k-medoids clustering, the expression data saved in `sc@fdata` object is transformed into a distance matrix. The distance matrix is calculated as `1-correlation matrix`, with a cell-to-cell correlation matrix. The method to calculate correlation can be changed using the `metric` function. The default method is `pearson` but can be changed to `spearman`, `logpearson`, `euclidean`, `kendall`, `maximum`, `manhattan`, `canberra`, `binary`, or `minkowski`. The resulting distance matrix is saved in `sc@distances` slot.

3. As mentioned previously, since k-medoids clustering requires the number of clusters to be defined beforehand, the next important step is to determine the number of clusters defining the dataset, which in theory should reflect the number of abundant cell types captured during the scRNA-seq experiment. The algorithm is designed to determine the cluster number using two different approaches: within-cluster dispersion and gap statistic (*see* **Note 5**). By default, the cluster number is determined by the saturation of the average within-cluster dispersion. The cluster number is determined such that adding more clusters only leads to a linear decrease of the within-cluster dispersion. The logarithmic within-cluster dispersion and the change in logarithmic within-cluster dispersion can be visualized as a function of the cluster number using the following two commands, respectively (Fig. 1a, b):

```
plotsaturation(sc,disp=TRUE) and
plotsaturation(sc,disp=FALSE)
```

4. If the saturation of the within-cluster dispersion exhibits further gains at larger cluster numbers (gray dots far above the red

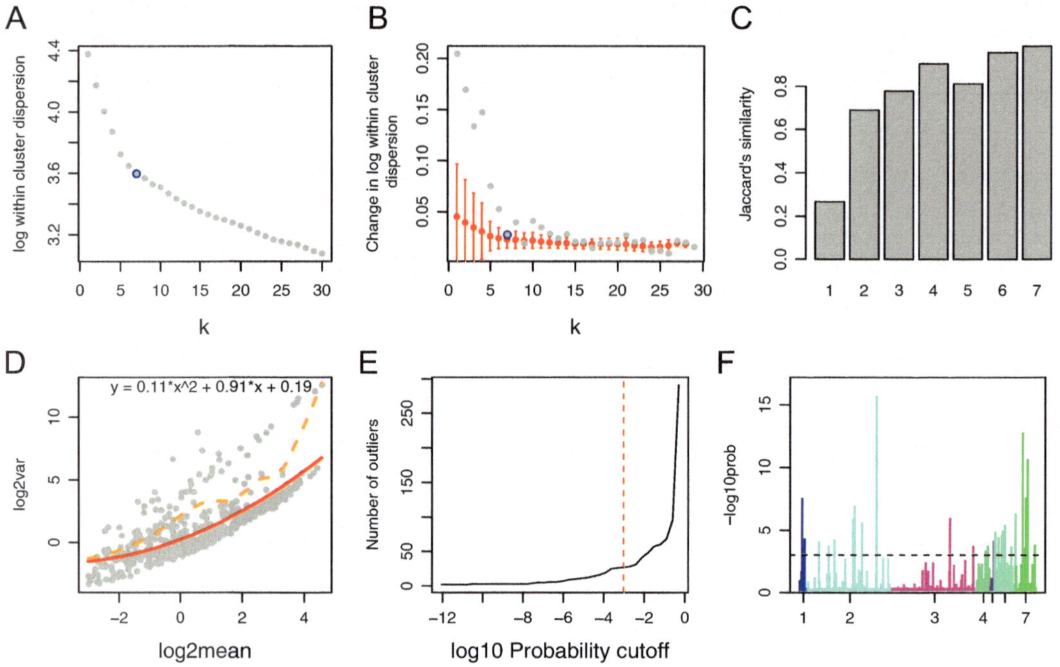

Fig. 1 k-Medoids clustering and outlier identification. (**a**) Log-transformed within-cluster dispersion and (**b**) change in it as a function of the number of clusters. (**c**) Jaccard's similarity bar plot of identified clusters. (**d**) Second-order polynomial regression of the variance on the mean of transcript expression across all cells in logarithmic space. (**e**) Number of outliers as a function of the probability threshold. (**f**) Outlier probabilities of all cells across clusters

error bars in Fig. 1b), the cluster number can be changed manually by setting the argument `cln` to the desired number. A good choice will be the minimal cluster number where a clear saturation is apparent.

5. Further, to assess the robustness of the inferred clusters, there are two methods implemented in the algorithm: Jaccard's similarity and silhouette plot. In this protocol chapter, only Jaccard's similarity score is explained. To plot a bar chart depicting Jaccard's similarity score for each cluster, use the following command (Fig. 1c): `plotjaccard(sc)`.
6. Jaccard's similarity score is proportional to the stability of the clusters. Stable clusters should have a score greater than 0.6. Clusters with lower values can be resolved by the outlier identification procedure. If too many clusters (more than two) have low values, a smaller number of clusters should be considered by manually setting the argument `cln` to the desired number.
7. After k-medoids clustering is performed, the unique feature of the RaceID3 algorithm enables the identification of outliers in the given scRNA-seq dataset. As mentioned before, the outlier identification step consists of three steps—calibration of a

background model, calculation of probability value for each gene in every cell of a given cluster based on which a cell is identified as an outlier, and the final clustering of the outlier cells based on their similarity. These steps are executed using the following R-code:

```
sc <- findoutliers(sc, outminc=5,outlg=2,probthr=1e-3,
thr=2**-(1:40), outdistquant=.95)
```

`outminc` defines the minimum transcript count for the identification of outlier genes, `outlg` is the minimum number of genes required to classify a cell as an outlier, `probthr` is outlier probability threshold only below which a cell is considered an outlier for a given gene, and `outdistquant` defines the quantile of the distance distribution for all pairs of cells within the original clusters after outlier removal. Outlier cells having distances lower than the `outdistquant` value are grouped together in the outlier clusters.

8. After executing `findouliers` with desired parameters, the quality of the background model fit can be visualized by plotting the second-order polynomial regression of the variance on the mean in logarithmic space using the following command (Fig. 1d): `plotbackground(sc)`. Furthermore, the number of outliers as a function of probability values can be plotted using (Fig. 1e) `plotsensitivity(sc)`. To ensure maximum sensitivity of the method, it is recommended to choose a `probthr` value, which covers the tail of this distribution as outliers. To visualize the outlier probabilities of all cells in the dataset as a bar plot, the following command can be used (Fig. 1f): `plotoutlierprobs(sc)`.
9. After the optimization of clustering including the identification of rare cells/clusters using RaceID3, the high-dimensional expression data need to be reduced to two dimensions for cluster visualization. To achieve this, RaceID3 implements tSNE algorithm using the following command:

```
sc <- comptsne(sc,rseed=15555,sammonmap=FALSE,initial
_cmd=TRUE)
```

10. After the tSNE map is computed, it can be visualized by the following two commands: `plottsne(sc,final=FALSE)` and `plottsne(sc,final=TRUE)`. If `final` is set to `FALSE`, the original k-medoids clusters are shown, whereas if it is set to `TRUE`, the outlier clusters/cells are also depicted in the tSNE map representation (Fig. 2a, b).

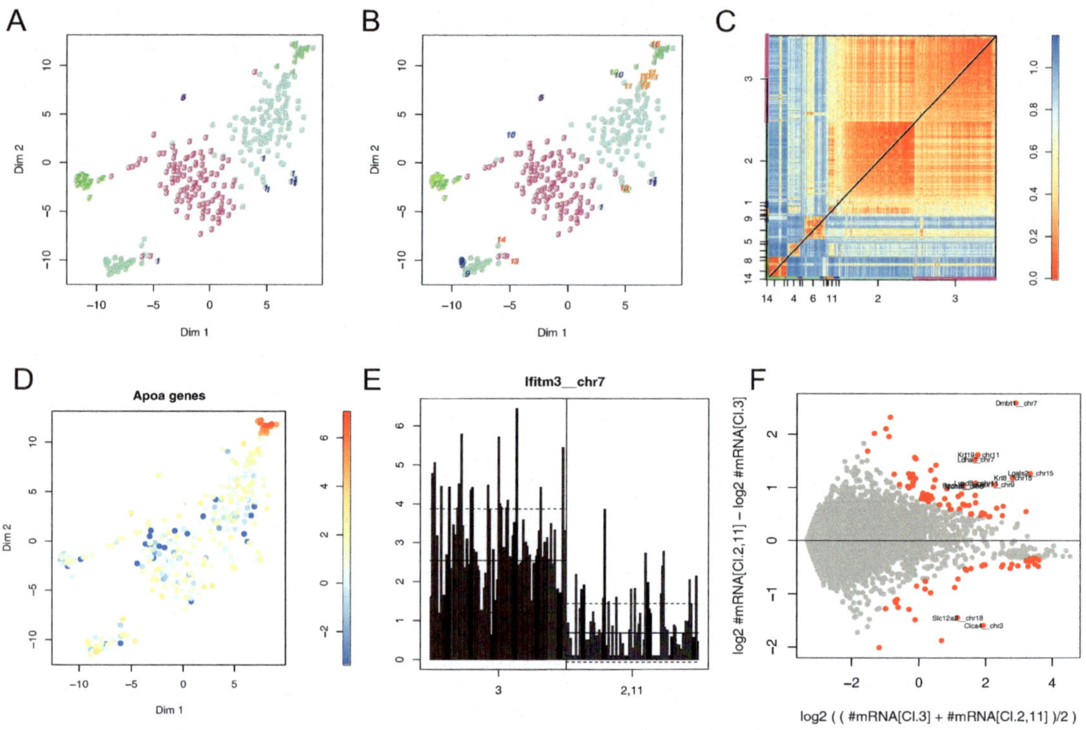

Fig. 2 Visualization and analysis of final clusters. tSNE map representation (**a**) before and (**b**) after outlier identification. (**c**) Heatmap representation of final clusters based on the cell-to-cell distance matrix. (**d**) tSNE map showing the log-transformed expression of *Apoa* genes in the dataset. (**e**) Expression of *Ifitm3* gene in cluster 3 versus clusters 2 and 11 calculated by `diffgenes` function. (**f**) MA plot showing differentially expressed genes between cluster 3 and 2 and 11 calculated by the function `diffexpnb`

11. The clusters can also be plotted in a heatmap representation of the cell-to-cell distances using the following command (Fig. 2c):

```
x <- clustheatmap(sc,final=TRUE,hmethod="single")
```

If `final` is set to `FALSE`, only k-medoids clusters prior to outlier identification are plotted (not shown).

12. In order to visualize the number of transcripts of a gene or set of genes on the tSNE map representation, the following command can be used:

```
plotexptsne(sc,g, logsc=FALSE)
```

`g` represents a character vector containing gene name similar to the row names of the expression matrix. If `logsc` argument is set to `TRUE`, the number of transcripts is log-transformed. The expression pattern of, e.g., all *Apoa* genes can be plotted using the following command (Fig. 2d):

```
g <- c("Apoa1__chr9", "Apoa1bp__chr3", "Apoa2__chr1",
"Apoa4__chr9", "Apoa5__chr9")
plotexptsne(sc,g,n="Apoa genes",logsc=TRUE)
```

3.4 Gene Expression Analysis of the Clusters

In order to identify genes characterizing a cluster, there are three functions provided in the RaceID3 algorithm. They are named `clustdiffgenes`, `diffgenes`, and `diffexpnb`. Below we provide an explanation of each one of them and the corresponding arguments they require for their proper execution.

1. The function `clustdiffgenes` compares each cluster with all the other clusters present in the dataset to find the differentially expressed genes and can be executed using the following code: `cdiff <- clustdiffgenes(sc,pvalue=.01)`. It returns a list containing a data frame for each cluster, consisting of the following five columns: `mean.ncl` (the mean expression of all cells not in the cluster), `mean.cl` (the mean expression of all cells in the cluster), `fc` (fold-change: ratio of `mean.cl` and `mean.ncl`), `pv` (p-value for differential expression between all cells in a cluster and all remaining cells), and `padj` (the Benjamini-Hochberg corrected false discovery rate). Only genes with `pv` lower than the argument `pvalue` are included in the output.
2. The tables containing information about the differentially expressed genes can be exported as `.xls` files using the following command:

```
for ( n in names(cdiff) ) write.table( data.frame(GENEI-
D=rownames(cdiff[[n]]), cdiff[[n]]), paste( paste("cell_-
clust_diff_genes",sub("\cr.","\cr_",n), sep="_"), ".xls",
sep=""), row.names=FALSE, col.names=TRUE, sep="\t", quo-
te=FALSE)
```

3. In order to specifically examine expression differences between two clusters or two sets of clusters, the function `diffgenes` can be used using the following command:

```
d <- diffgenes(sc,cl1=3,cl2=c(2,11),mincount=1)
```

The arguments `cl1` and `cl2` are numerical vectors defining the cluster numbers to be compared. For instance, in the above code, we compare cluster 3 against clusters 2 and 11 in the example dataset. Only genes with more than `mincount` transcript in at least a single cell of `cl1` or `cl2` are evaluated. The function computes the z-score for expression differences between the two groups using the average of the standard deviation in `cl1` and `cl2`. If `cl1` or `cl2` contains only a single cell, the standard deviation is estimated by the square root of the mean. The function returns a list of five elements: (1) z, a vector of z-scores in decreasing order with genes upregulated in cl1 appearing at the top of the list; (2) cl1, a data frame with expression values for cells in cl1; (3) cl1n, a vector of cluster numbers in cl1; (4) cl2, a data frame with expression values for cells in cl2; and (5) cl2n, a vector of cluster numbers in cl2. The function `plotdiffgenes` can be used to plot expression in the two groups for a given gene (here first gene with the highest z-score, Fig. 2e) using the following command:

```
plotdiffgenes(d,gene=names(d$z)[1])
```

The last function `diffexpnb` provides a tool to perform differential expression analysis between two groups. This function implements an approach similar to DESeq [11] by which negative binomial distributions are fitted to transcript counts in two populations, representing, for example, two different cell types. These two groups can also be two different clusters identified by RaceID3. The code to execute the function is as follows:

```
x <- diffexpnb(sc@ndata, A=A, B=B, method="pooled"
,norm=FALSE, DESeq=FALSE, vfit=sc@background$vfit,
locreg=FALSE)
```

In the code above, we utilize the background model generated in the RaceID3 run which is saved in `sc@background $vfit` slot. As input, the normalized expression data frame `sc@ndata` is used. The function has the following other input arguments: (1) A, vector of cell IDs corresponding column names of x, and (2) B, vector of cell IDs corresponding column names of x. For example, to find the differentially expressed genes in cluster 3 versus clusters 2 and 11, A and B argument of the `diffexpnb` function can be initialized with the following command:

```
A <- names(sc@cpart)[sc@cpart == 3]
B <- names(sc@cpart)[sc@cpart %in% c(2,11)]
```

The results can be visualized as an MA plot using the following function (Fig. 2f): `plotdiffgenesnb(x, pthr=.05, lthr=1, mthr=1, show_names=TRUE, padj=TRUE)`.

3.5 Lineage Inference and Stem Cell Identity Prediction

After applying RaceID3 to identify different cell types and characterizing the gene expression patterns, StemID2 can be applied for the inference of differentiation trajectories and the prediction of the stem cell identity. StemID2 starts with embedding the higher-dimensional space of transcript counts for each gene, in which every cell can be represented, into a lower-dimensional space in order to maintain only the number of dimensions necessary to reliably represent all point-to-point distances. For example, only $n - 1$ dimensions are necessary to embed n data points from a high-dimensional space ($>n$ dimensions) with exactly conserved distances for a Euclidean metric. Since RaceID3 uses a correlation-based metric, this task is achieved by embedding this correlation matrix into $k < n - 1$ dimensions, with k being the number of positive eigenvalues of the squared double-centered distance matrix. The distance $d_{i,j}$ between cells i and j is defined as $d_{i,j} = 1 - \rho_{i,j}$, where $\rho_{i,j}$ is the Pearson's correlation coefficient of these cells calculated based on the transcriptome. The major improvement of StemID2 over the first version is a fast mode (*see* **Note 6**), which allows running the entire algorithm on large datasets in substantially reduced computational time at a cost of a minor loss in sensitivity.

To derive the differentiation trajectories, the medoid m_i of cluster i is connected to the medoids m_j of all other clusters j ($j = 1, \ldots, i - 1, i + 1, \ldots, N$) in the embedded space. Subsequently, for each cell k in cluster i, the vector $z_{i,k} = y_{i,k} - m_i$ connecting its position $y_{i,k}$ to m_i is projected onto each link $l_{i,j} = m_j - m_i$ between cluster i and j ($j = 1, \ldots, i - 1, i + 1, \ldots, N$), i.e., the component of this vector (anti-) parallel to each connection is calculated. Projections $p_{k,i,j}$ correspond to the cosine of the angle $\alpha_{k,i,j}$ between $z_{i,k}$ and $l_{i,j}$ times the length of $l_{i,j}$ and are computed based on the dot product of the two vectors:

$$p_{k,i,j} = |l_{i,j}| \cdot \cos \alpha_{k,i,j} = \frac{z_{i,k} \cdot l_{i,j}}{|z_{i,k}|}$$

If the vector component is antiparallel to a projection, it will be negative. The respective cell is then assigned to the connection with the longest projection using the coordinate computed from the projection. This procedure is repeated for every cell in each cluster. To determine connections with significantly more assigned cells than expected by chance, the computation is repeated after randomizing the cell positions in the embedded space. Randomization is performed by sampling new cell positions from a uniform interval with boundaries given by the real data for each embedded dimension. Cluster centers are kept constant to maintain the topology of the configuration. Outgoing and incoming links are distinguished for the p-value calculation, i.e., for each cluster, it is computed how many of its cells are assigned to each link to another cluster. The distribution of expected cells on each outgoing link is sampled by repeating the randomization procedure many times (*see* below). A p-value for each link is derived as the quantile of this distribution corresponding to the actual number of cells on the link. In general, a cluster can have an enriched outgoing link, which is at the same time a depleted incoming link. We consider a link significantly enriched if this is true for either the outgoing or the incoming link.

To compute a p-value, the sampling is repeated sufficiently often. For instance, if a p-value threshold of $p < 0.01$ is chosen to assign significance to a link, the randomization is repeated 2000 times to calculate the 1%-quantile with sufficient confidence. For lower p-values the number of randomizations needs to be increased. The ensemble of significant connections can be interpreted as a predicted lineage tree comprising all commonly used differentiation trajectories of a system.

To predict the stem cell identity from the lineage tree, StemID2 assumes stem cells to be multipotent and tries to identify the cell population with highest degree of multipotency [12]. For this purpose, the algorithm computes a score called StemID2 score, s_k, for each cluster using the following formula:

$$s_k = l_k \cdot \Delta E_k$$

where l_k defines the number of significant links of cluster k and ΔE_k is the median delta-entropy. The first parameter, l_k, takes into account the number of links each cluster has to the other clusters. The higher the number of links, the more likely it is that such a cell type can fluctuate toward a diversity of cell fate biases, whereas a cell type with fewer links is much more canalized. The second parameter, ΔE_k, deals with the uniformity of transcriptome as calculated by Shannon's entropy [13]. Here we hypothesize that stem cells possess higher uniformity in their transcriptome, whereas mature cells express relatively smaller number of genes albeit at a higher level which are important for their cell type-specific functions. For example, pancreatic beta cells express high levels of insulin.

The entropy E_j of cell j is computed as:

$$E_j = \sum\nolimits_{i=1}^{N} p_{i,j} \log_N p_{i,j}$$

where $p_{i,j} = n_{i,j}/N$ *and* $n_{i,j}$ equals the number of transcripts of gene i in cell j. N equals the total number of transcripts in each cell. Next, the median delta-entropy ΔE_k is computed for each cluster k, using the following formula:

$$\Delta E_k = \text{median}_{j \in k}(E_j) - \min_l(\text{median}_{j \in l}(E_j))$$

Below we provide a step-by-step protocol with the R-code to construct a lineage tree and predict the cluster with higher multipotency, thereby reflecting its stemness.

1. In order to use StemID2, it is required to run RaceID3 before. Like RaceID3, StemID2 is also implemented as an S4 class object named `Ltree` and requires the `SCseq` object, `sc`, to be created previously (*see* Subheading 3.1). The following command can be used for initialization: `ltr <- Ltree(sc)`.
2. The entropy of each cell is calculated using the function `compentropy`, and the following command can be executed for this calculation: `ltr <- compentropy(ltr)`.
3. The embedding of the correlation matrix into a lower-dimensional space and the calculation of the projections of each cell onto all inter-cluster links are performed by the function `projcells` using the following command:

```
ltr <- projcells(ltr,cthr=5,nmode=FALSE)
```

   `cthr` defines the minimum number of cells that a cluster should comprise of in order to be included in the lineage tree inference. If the argument `nmode` is set to `TRUE`, the assignment to inter-cluster links for each cell is not done based on the longest projection but based on identifying the cluster (other than the cluster the cell belongs to) that contains the nearest neighbor of the cell, i.e., the cell with the most similar transcriptome. The coordinate on the assigned link is still derived based on the projection. Default is `FALSE`.
4. In order to analyze the ratio between the cell-to-cell distances in the embedded space and the input space, a histogram can be plotted using the following command (Fig. 3a):

```
plotdistanceratio(ltr)
```

5. The randomization is performed using the function `projback`. The following command can be executed for this purpose:

```
ltr <- projback(ltr,pdishuf=2000,nmode=FALSE,fast=FALSE,
rseed=17000)
```

   The argument `pdishuf` defines the number of randomizations to be performed. Importantly, if `fast` is set to `TRUE`, StemID2 operates in the fast mode (*see* **Note 6**). The fast mode is recommended for large datasets (>3000 cells).
6. To compute the lineage tree, the function `lineagetree` can be executed as follows:

```
ltr <- lineagetree(ltr,pthr=0.01,nmode=FALSE,fast=FALSE)
```

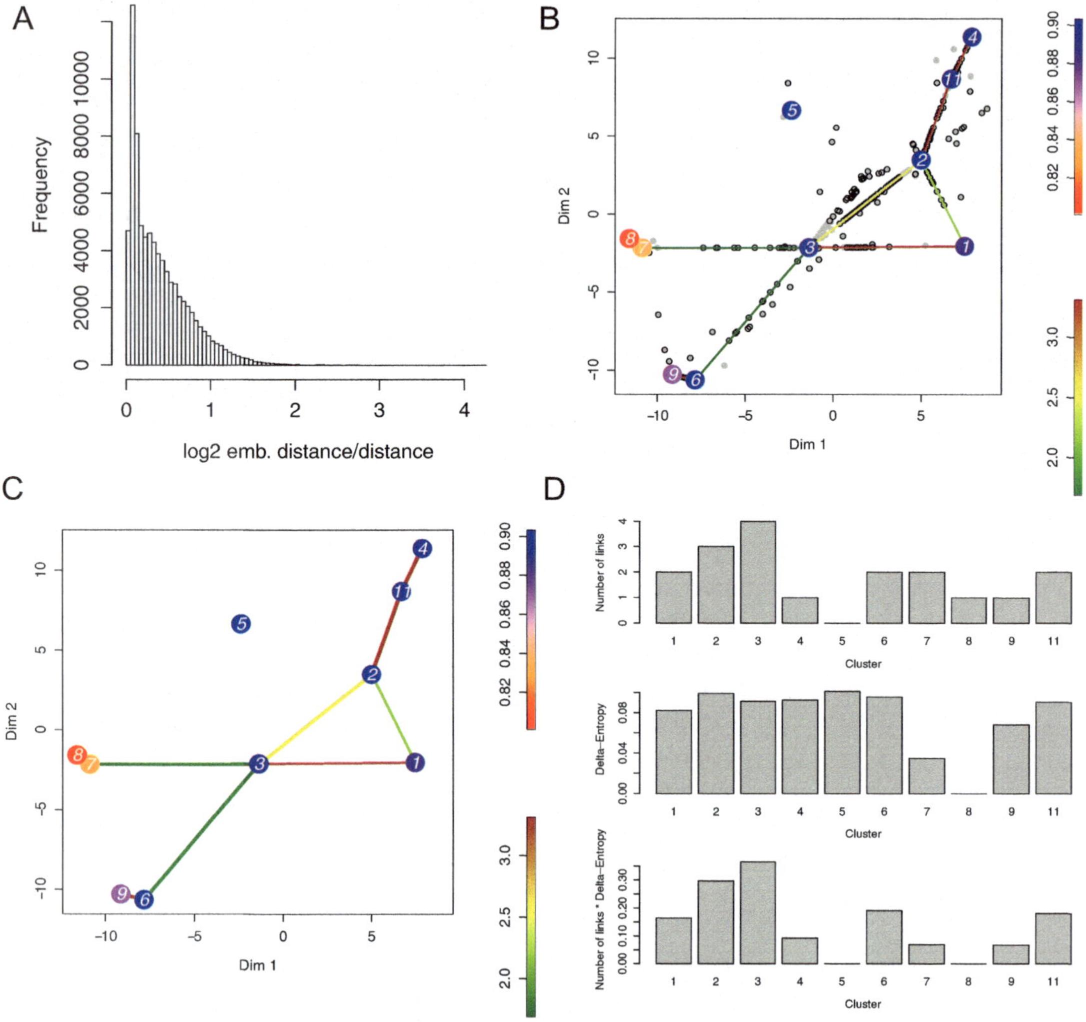

Fig. 3 Lineage tree inference and stem cell identity prediction. (**a**) Histogram depicting the $\log_2$ratio of the cell-to-cell distances in the lower-dimensional embedded space and the higher-dimensional input space. Values close to zero indicate a perfect conversion of the distances in the embedded space. Inferred lineage tree (**b**) with cells and (**c**) without cells. Cells are circled in back to indicate a statistically significant projection coordinate compared to the randomized background model. The color of the links indicates the $-\log_{10}p$-value of the link, and the color of the vertices indicates the delta-entropy. The width of the links indicates the link score. (**d**) Bar plots showing the number of links, the delta-entropy, and the StemID2 score for all the clusters. Cluster 3 shows the highest StemID2 score

where pthr corresponds to the p-value threshold, which is used to determine whether the magnitude of an observed cell coordinate on a link is significantly larger than observed for the randomized background distribution. This criterion is not used to infer significance of a link but shown in a graphical representation of the tree.

7. Afterward, the function `comppvalue` has to be executed in order to identify connections at a given level of significance. The following R-code can be executed for this purpose:

```
ltr <- comppvalue(ltr,pethr=0.01,nmode=FALSE,fast=FALSE)
```

   `pethr` corresponds to a *p*-value threshold to determine if a given link is populated by number of cells significantly larger than expected for the randomized background distribution.

8. Finally the inferred lineage tree can be visualized with the function `plottree` using the following command (Fig. 3b, c):

```
plottree(ltr,showCells=TRUE,nmode=FALSE,scthr=.3)
```

   `showCells` is a logical argument; if set to `TRUE`, single cells will also be shown in the plotted lineage tree. Otherwise, only the significant branches of the tree will be shown, and `scthr` can be defined as a link score, which can be given any value between 0 and 1. The link score corresponds to `1 -- fraction of link not covered by a cell`. Link score close to zero corresponds to a situation where most cells on a link reside close to the connected cluster center, while a score close to one arises in a situation where the link is covered uniformly with cells. At higher values of `scthr`, only links are shown that represent higher confidence predictions for actual differentiation trajectories.

9. Finally, the StemID2 score is computed by executing the function `compscore` using the following R-code: `x <- compscore (ltr,nn=1,scthr=.3)`.

10. The StemID2 score can be plotted by the function `plotscore` using the following command: `x <- plotscore(ltr,nn=1, scthr=.3)`. The function plots three histograms: the number of links, the delta-entropy, and the StemID2 score (Fig. 3d).

3.6 Pseudo-Temporal Ordering

StemID2 can also be used for the visualization and analysis of pseudo-temporal gene expression changes. Below we explain the steps necessary to create such pseudo-temporal profiles using example dataset provided with the StemID2 algorithm.

1. In the first step, the StemID2 object `ltr` and the information about the cells along a StemID2-predicted differentiation trajectory are provided using the following R-code:

```
n <- cellsfromtree(ltr,c(3,2,11,4))
```

   Here, as an example, the differentiation trajectory predicted by StemID2 on the example data is used. Cluster

3, which has the highest StemID2 score, is used as a starting point, while cluster 4 is considered an endpoint.

2. Afterward, normalized expression data stored in the slot `sc@ndata` is given as an input to the function `filterset` in order to remove lowly expressed genes. The function `filterset` can be executed using the following command:

```
fs <- filterset(ltr@sc@ndata,n=n$f,minexpr=2,minnumber=1)
```

The function removes those genes, which are expressed at a value less than `minexpr` in at least `minnumber` of cells.

3. Now a self-organizing map (SOM) for grouping similar expression profiles into modules is created using the function `getsom`. The function requires the filtered expression data generated in the previous step and few other arguments such as number of nodes of the SOM (`nb`), window-size (`k`) for calculating the running mean of the expression values in the ordered cells, and also an option to generate such profiles using local regression (`locreg`). The following command is used to create a SOM: `s1d <- getsom(fs,nb=1000,k=5, locreg=TRUE,alpha=.5)`.

4. In order to group the nodes of the SOM into larger modules and to produce additional z-score transformed and binarized pseudo-temporal expression profiles, function `procsom` is executed using the following code:

```
ps <- procsom(s1d,corthr=.85,minsom=3)
```

`corthr` defines the minimum correlation threshold required for merging the neighboring two nodes, and `minsom` defines the minimum number of transcripts that a node should have to be included in the map.

5. Now the plots can be visualized using the `plotheatmap` function. First a color scheme corresponding to the particular cluster number is defined using the following two commands: `fcol <- ltr@sc@fcol` and `y <- ltr@sc@cpart[n$f]`. The `plotheatmap` function provides visualization of following four different types of pseudo-temporal plots ordered by SOM—(1) average z-score of all modules, (2) z-score profile of each gene, (3) normalized expression profile of each gene, and (4) binarized expression profile of each gene. The following commands can be executed to visualize the ordered plots (Fig. 4a–d):

```
plotheatmap(ps$nodes.z, xpart=y, xcol=fcol, ypart=unique
(ps$nodes), xgrid=FALSE, ygrid=TRUE, xlab=FALSE)
```

```
plotheatmap(ps$all.z, xpart=y, xcol=fcol, ypart=ps$nodes,
xgrid=FALSE, ygrid=TRUE, xlab=FALSE)
plotheatmap(ps$all.e, xpart=y, xcol=fcol, ypart=ps$nodes,
xgrid=FALSE, ygrid=TRUE, xlab=FALSE)
plotheatmap(ps$all.b, xpart=y, xcol=fcol, ypart=ps$nodes,
xgrid=FALSE,ygrid=TRUE, xlab=FALSE)
```

6. In order to visualize the dynamics of gene expression changes of a particular node or an individual gene, the following command can be used:

```
g <- names(ps$nodes)[ps$nodes == 1]
```

where g extracts the genes present in node 1. Now it can be plotted using the following command (Fig. 4e):

```
plotexpression(fs, y, g, n$f, k=25, col=fcol, name="Node
1", cluster=FALSE, locreg=TRUE, alpha=.5, types=NULL)
```

To visualize the expression profile changes of a particular gene, e.g., Clac4 here, the following command can be used (Fig. 4f):

```
plotexpression(fs, y, "Clca4__chr3", n$f, k=25, col=fcol,
cluster=FALSE, locreg=TRUE, alpha=.5, types=NULL)
```

3.7 Prediction of Stem Cell Identity Using Fate Bias Probabilities

As clear from the above section, StemID2 implementation relies on the clustering analysis performed by RaceID3. Although RaceID3 has an outlier identification step to refine the clustering, it is plausible that clustering analysis may not reveal the transcriptional differences, which are subtle and gradual in nature. For example, homogenous progenitor populations can be transcriptionally primed toward different lineages, but these differences could be so subtle that clustering analysis is not sensitive enough to detect such early priming and fate biases in the multipotent cells. As an attempt to overcome this challenge, we have recently developed an algorithm called FateID to quantify fate bias manifested in terms of subtle lineage-specific transcriptional characteristics within a multipotent progenitor population [8].

FateID utilizes prior knowledge to identify committed stages of all lineages and tracks differentiation trajectories backward in time, in order to predict the likelihood of multipotent progenitors to give rise to each lineage. FateID utilizes random forests, a supervised learning method known for its robustness to overfitting, to learn the identity of cells in a test set given the training data. FateID starts with target clusters of committed cell populations, identified by clustering or marker gene analysis. Cells in the

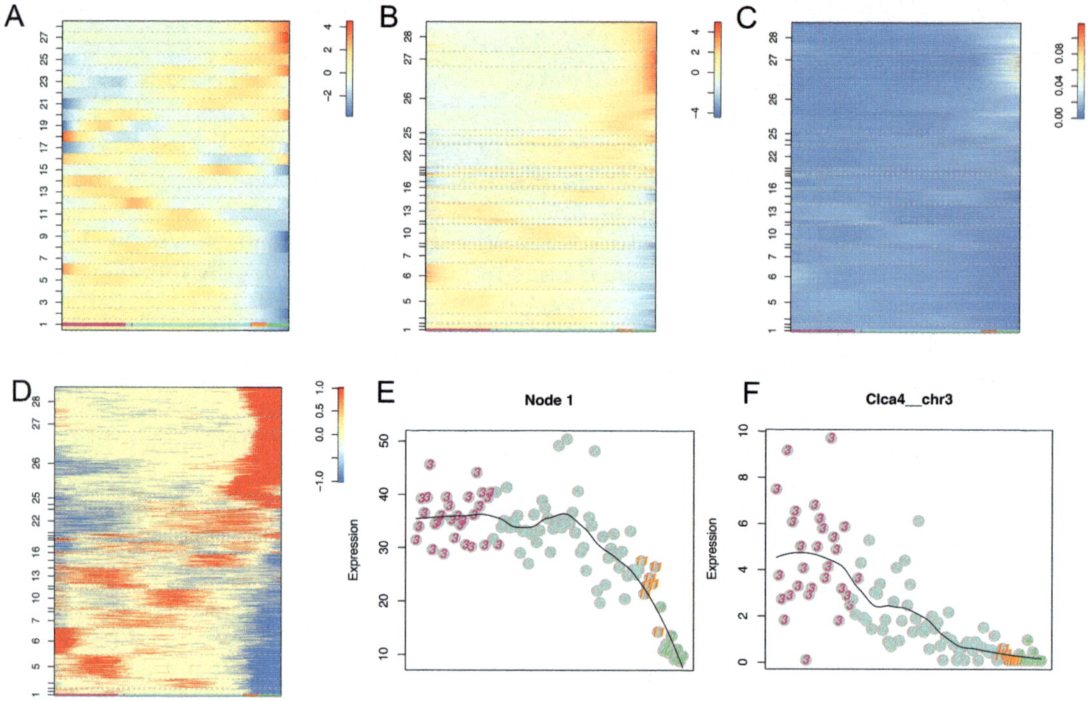

Fig. 4 Pseudo-temporal ordering. Heatmaps showing (**a**) average z-score of all modules, (**b**) z-scores of all genes, (**c**) expression profile of each gene, and (**d**) binarized expression profile of each gene ordered by SOM. (**e**) Pseudo-temporal expression profiles of all the genes in module 1 and (**f**) *Clac4*, a stem cell marker gene

local neighborhood of each of these target clusters are selected to build a test set. Next, a defined number of cells from the target clusters, most similar to the test set, are used as training set to derive the probability of each cell in the test set to belong to any of the target clusters based on random forests votes. Cells with a significant bias toward a particular lineage become part of the respective target cluster and are utilized for classification in the next iteration. FateID proceeds until the fate biases of all cells have been inferred.

In principle, the fate bias probabilities calculated for each cell using FateID can be used to predict the level of multipotency in a given population of cells. For example, if a cell has equal probabilities of differentiation toward all target clusters, it is more likely to be more multipotent compared to a cell which has a high fate bias toward one particular lineage. Furthermore, based on fate biases for the different target clusters, an entropy value for each cell is calculated. An entropy value of the fate biases will indicate the level of multipotency with larger fate bias entropies corresponding to more multipotent cell states.

Below we describe a step-by-step approach to calculate the fate bias and entropy based on these values to predict the level of multipotency of the cells present in the scRNA-seq dataset.

1. The FateID algorithm demands three important inputs in order to calculate fate bias, (1) the expression dataset `x`, (2) a vector `y` containing the information about the partitioning of cells into various clusters, and (3) information about the target clusters `tar`, which represent the mature cell types in the dataset (*see* **Note 7**). The following commands can be executed to initialize these three objects:

```
y <- sc@cpart[sc@cpart %in% c(4,11,2,3,7,8,6,9)]
x <- sc@fdata[,colnames(sc@fdata) %in% names(y)]
tar <- c(8,9,4)
```

   Please note that we have used the RaceID3 cluster partitioning as an input to the FateID algorithm. Moreover, one can also choose the final number of clusters to be included in the analysis. Here we have included clusters 4, 11, 2, 3, 7, 8, 6, and 9 in the further analysis and set clusters 8, 9, and 4 as target clusters representing the most mature cell types.

2. The next step of the algorithm utilizes `fateBias` function to calculate the fate biases. The following command can be executed to achieve this goal:

```
fb <- fateBias(x, y, tar, z=NULL, minnr=5, minnrh=10,
nbfactor=5, use.dist=FALSE, seed=12345, nbtree=NULL)
```

3. In order to visualize the high-dimensional data into two- and three-dimensional space (`k=2` for 2D and `k=3` for 3D), the algorithm utilizes various dimensional reduction algorithms (argument `m`) such as tSNE (`tsne`), classical multidimensional scaling (`cmd`), locally linear embedding (`lle`), and diffusion maps (`dm`). Dimensional reduction can be achieved using the following code:

```
dr <- compdr(x, z=NULL, m=c("tsne","cmd","dm","lle"), k=2,
lle.n=30, dm.sigma="local", dm.distance="euclidean", tsne.
perplexity=30, seed=12345)
```

4. The low-dimensional data can be now visualized, for example, using the t-SNE coordinates through the following command (Fig. 5a): `plotFateMap(y,dr,k=2,m="tsne")`.

5. The following command can be used to visualize the fate bias in the desired dimensional reduction method map for target cluster 9 (Fig. 5b): `plotFateMap(y,dr,k=2,m="tsne", fb=fb,g="t9")`.

   The additional argument `g` provides the name of the target cluster concatenated with the letter `t`. To plot the fate bias for

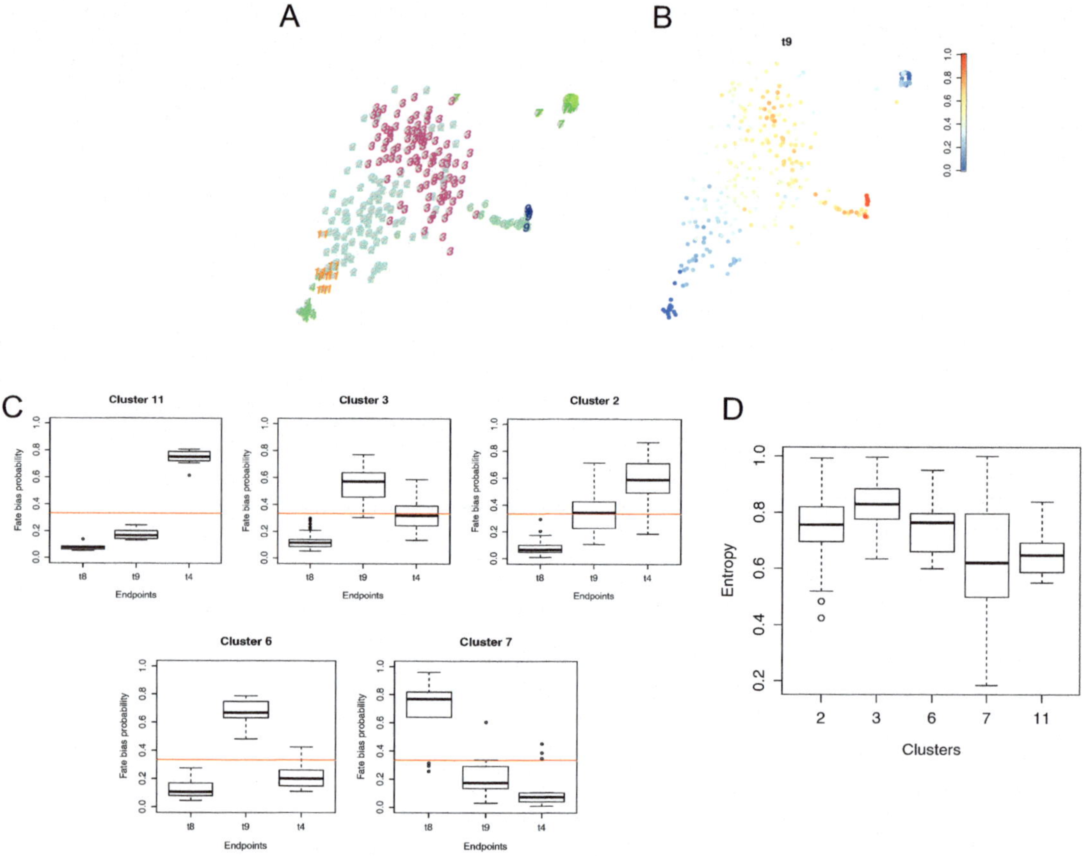

Fig. 5 Identifying multipotency levels of clusters using FateID. (**a**) tSNE representation of the clusters included in the FateID analysis. (**b**) tSNE representation of fate bias of cells toward cluster 9. Color-coding shows the fate probabilities on a scale of 0–1. (**c**) Boxplots depicting the fate probabilities of the progenitor clusters toward different mature states. The horizontal red line depicts an equal fate bias toward the three mature states (1/3 = 0.33). (**d**) Boxplot showing the entropy values calculated based on the fate bias. Cluster 3 shows the highest entropy level

clusters 8 and 4, `g` is initialized as "`t8`" and "`t4`," respectively (not shown).

6. Next, the fate bias boxplots of each cluster included in the analysis toward the target clusters can be visualized using the following command (Fig. 5c):

```
data.fate.bais <- as.data.frame(cbind(y,fb$probs))
rc <- ceiling(sqrt(length(unique(data.fate.bais$y))))
par(mfrow =c(rc,rc))
for(i in c(1:length(unique(data.fate.bais$y))))
 {
boxplot(subset(data.fate.bais, data.fate.bais$y %in%
unique(data.fate.bais$y)[i])[,-1], main =
```

```
paste("Cluster",unique(data.fate.bais$y)[i], sep = " " ),
xlab="Endpoints", ylab="Fate bias probability",
ylim=c(0,1))
abline(h = 1/length(tar), col="red")
}
```

7. In order to calculate the entropy based on fate bias probabilities, the following command can be executed: `E <- plotFateMap(y,dr,k=2,m="tsne",g="E",fb=fb)`. Finally, to visualize the entropy boxplots for each cluster included in the analysis, use the following command (Fig. 5d):

```
data.fate.bais.ent <- as.data.frame(cbind(y,E))
boxplot(data.fate.bais.ent$E~data.fate.bais.ent$y,
xlab="Clusters", ylab="Entropy")
```

4 Notes

1. The first column of the `.xls` example file contains unique gene ids and has to be assigned as row names of the R data frame object. Also, the gene ids of the non-endogenous spike-in RNAs starts with "ERCC," and these transcripts have to be removed.
2. We recommend using downsampling only if batch effects due to differences in complexity between distinct libraries in the dataset are observed. Normalization by rescaling is in general more sensitive.
3. We routinely initialize `FGenes` with *Hspa1a*, *Hsp90aa1*, *Jun*, *Fos*, *Zfp36*, *Junb*, *Fosb*, *Socs3*, *Atf3*, *Jund*, *Hspe1*, *Hspa1b*, *Hsp90ab1*, *Hspb1*, *Egr1*, *Hspa8*, *Cebpd*, and *Cebpb*, since these genes were recently shown to be dissociation-induced [14].
4. Sometimes user may want to remove cell cycle-associated variability, and in this case, `CGenes` can be initialized with genes such as *Pcna* and *Mki67*.
5. If the argument `do.gap` is set to `TRUE`, the algorithm determines the number of clusters as the first local maximum of gap-statistic. The graph showing the value of gap-statistic as a function of the number of clusters can be plotted using the following command: `plotgap(sc)`. Since gap-statistic calculation requires longer computational time, we recommend users to opt within-cluster dispersion approach to choose a desired cluster number.

6. In this mode, the p-value is not computed by performing a large number of randomizations. Here, an expression table is first generated, which contains for each cluster a large number of cells (corresponding to the size of the entire dataset). Subsequently, the positions of these cells are randomized within the boundaries dictated by the actual data and projected on the links between the cluster of origin an all other clusters and assigned to the link with the longest projection. Next, a probability to fall on each link is derived from the occupancy of this link, and the number of cells on that link in the actual data is evaluated using a binomial test with this probability parameter as input. This way, a p-value for all links is derived in a much more efficient way than the multiple shuffling-based approach. However, and in contrast to the multiple shuffling-based approaches, the underlying model does not imply the correlation structure between the occupancy of links and is thus less precise than the original approach.
7. If a partitioning into cell types and states from a prior clustering analysis is not available, FateID can derive a partitioning based on marker gene information which defines the most mature cell types using the function `FMarker`.

References

1. Bendall SC, Davis KL, Amir el AD, Tadmor MD, Simonds EF, Chen TJ, Shenfeld DK, Nolan GP, Pe'er D (2014) Single-cell trajectory detection uncovers progression and regulatory coordination in human B cell development. Cell 157(3):714–725
2. Trapnell C, Cacchiarelli D, Grimsby J, Pokharel P, Li S, Morse M, Lennon NJ, Livak KJ, Mikkelsen TS, Rinn JL (2014) The dynamics and regulators of cell fate decisions are revealed by pseudotemporal ordering of single cells. Nat Biotechnol 32(4):381–386
3. Marco E, Karp RL, Guo G, Robson P, Hart AH, Trippa L, Yuan GC (2014) Bifurcation analysis of single-cell gene expression data reveals epigenetic landscape. Proc Natl Acad Sci U S A 111(52):E5643–E5650
4. Shin J, Berg DA, Zhu Y, Shin JY, Song J, Bonaguidi MA, Enikolopov G, Nauen DW, Christian KM, Ming GL, Song H (2015) Single-cell RNA-Seq with waterfall reveals molecular cascades underlying adult neurogenesis. Cell Stem Cell 17(3):360–372
5. Qiu X, Mao Q, Tang Y, Wang L, Chawla R, Pliner HA, Trapnell C (2017) Reversed graph embedding resolves complex single-cell trajectories. Nat Methods 14(10):979–982
6. Guo M, Bao EL, Wagner M, Whitsett JA, Xu Y (2017) SLICE: determining cell differentiation and lineage based on single cell entropy. Nucleic Acids Res 45(7):e54
7. Grun D, Muraro MJ, Boisset JC, Wiebrands K, Lyubimova A, Dharmadhikari G, van den Born M, van Es J, Jansen E, Clevers H, de Koning EJP, van Oudenaarden A (2016) De novo prediction of stem cell identity using single-cell transcriptome data. Cell Stem Cell 19(2):266–277
8. Herman JS, Sagar, Grun D (2018) FateID infers cell fate bias in multipotent progenitors from single-cell RNA-seq data. Nat Methods 5:379–386. https://doi.org/10.1101/218115
9. Grun D, Lyubimova A, Kester L, Wiebrands K, Basak O, Sasaki N, Clevers H, van Oudenaarden A (2015) Single-cell messenger RNA sequencing reveals rare intestinal cell types. Nature 525(7568):251–255

10. van der Maaten L, Hinton G (2008) Visualizing data using t-SNE. J Mach Learn Res 9:2570–2605
11. Anders S, Huber W (2010) Differential expression analysis for sequence count data. Genome Biol 11:R106
12. Ridden SJ, Chang HH, Zygalakis KC, MacArthur BD (2015) Entropy, ergodicity, and stem cell multipotency. Phys Rev Lett 115:208103
13. Shannon CE (1948) A mathematical theory of communication. Bell Syst Tech J 27:379–423, 623–656.
14. van den Brink SC, Sage F, Vértesy A, Spanjaard B, Peterson-Maduro J, Baron CS, Robin C, van Oudenaarden A (2017) Single-cell sequencing reveals dissociation-induced gene expression in tissue subpopulations. Nat Methods 14:935–936

Part III

Metabolism

Chapter 14

Dynamic Network Modeling of Stem Cell Metabolism

Fangzhou Shen, Camden Cheek, and Sriram Chandrasekaran

Abstract

Stem cell metabolism is intrinsically tied to stem cell pluripotency and function. Yet, understanding metabolic rewiring in stem cells has been challenging due to the complex and highly interconnected nature of the metabolic network. Genome-scale metabolic network models are increasingly used to holistically model the metabolic behavior of various cells and tissues using transcriptomics data. However, these powerful approaches that model steady-state behavior have limited utility for studying dynamic stem cell state transitions. To address this complexity, we recently developed the dynamic flux activity (DFA) approach; DFA is a genome-scale modeling approach that uses time-course metabolic data to predict metabolic flux rewiring. This protocol outlines the steps for modeling steady-state and dynamic metabolic behavior using transcriptomics and time-course metabolomics data, respectively. Using data from naive and primed pluripotent stem cells, we demonstrate how we can use genome-scale modeling and DFA to comprehensively characterize the metabolic differences between these states.

Key words Dynamic network modeling, Genome-scale metabolic models (GEMs), Constraint-based modeling, Flux balance analysis (FBA), Stem cells, Metabolism, Metabolomics

1 Introduction

The human genome contains more than 13,000 metabolic reactions [1]. The enormous size and interconnectedness of the metabolic network make the interpretation of metabolic data highly challenging. This complexity has so far limited system-level understanding of stem cell metabolism.

Genome-scale metabolic models (GEMs) have been recently developed that enable unbiased interpretation of high-throughput data [2]. GEMs are mathematical representations of biochemical reactions that can be used for predicting biochemical properties of a system from omics data [3]. GEMs are manually curated and represent the interconnection of metabolites, proteins, and reactions in a biological system via reaction stoichiometries, compartments, and gene-protein-reaction associations [4, 5].

High-throughput transcriptomics and proteomics data can be integrated with GEMs to identify metabolic reactions that are active

Patrick Cahan (ed.), *Computational Stem Cell Biology: Methods and Protocols*, Methods in Molecular Biology, vol. 1975, https://doi.org/10.1007/978-1-4939-9224-9_14,

in a system. Such transcriptomics-constrained models have proved useful for predicting metabolic behavior of human normal and cancer cells at steady state [2, 6–8]. These models overcome many limitations of existing informatics approaches that rely on pre-defined list of pathways or gene sets. Due to the highly interconnected nature of the metabolic network, the underlying assumption behind pathway enrichment analysis—that each pathway is independent of all other pathways—does not hold for metabolism [9]. Adjacent pathways on the network can influence each other's activity, and altered flux activity can arise due to perturbations upstream or downstream of a given pathway. Further, the expression levels of individual enzymes in a pathway do not change coherently as a single unit and can have different levels of activity. Genome-scale network models are able to integrate and reconcile differences at both the individual reaction level and the network level in order to interpret cellular behavior.

This protocol will describe the steps for integrating transcriptomics and proteomics data with GEMs. We will then describe the more challenging task of modeling metabolomics data using representative data from stem cells. Analysis of gene expression or metabolomics data can provide a snapshot of cellular metabolism, but it is limited because gene or metabolite changes are not often consistent with the flux changes in metabolic reactions. Further, existing approaches to integrate omics data rely on the steady-state assumption and hence are unable to capture dynamic processes such as stem cell differentiation.

To address these challenges, we recently developed a dynamic genome-scale modeling approach called the dynamic flux activity (DFA) that utilizes time-course metabolomics data to predict metabolic flux [10]. The accumulation or depletion of a metabolite over time is used by DFA as evidence for overall increased or decreased flux activity of the set of reactions involving the metabolite. By integrating several metabolite-level changes with a genome-scale metabolic model, this approach can identify reactions with differential activity to describe the dynamic patterns in stem cell metabolism.

2 Materials

These approaches are implemented in MATLAB (http://www.mathworks.com). The Gurobi mathematical programming solver is utilized to solve the optimization problems (http://www.gurobi.com). The datasets for this tutorial are available at www.sriramlab.org/software.

2.1 COBRA Toolbox

The COnstraint-Based Reconstruction and Analysis (COBRA) toolbox is a software package intended to process and analyze metabolic network models [11, 12]. This package can be downloaded from http://opencobra.github.io and has both a MATLAB and Python version [13].

2.2 Metabolic Model

The RECON1 model is a comprehensive human metabolic model and has been widely applied for interpreting metabolism of pathological and normal states [14]. The RECON1 model can be downloaded from the BiGG database [15]. The original model needs to be modified to include an objective function and specific nutrient uptake rates to accurately reflect stem cell metabolism (*see* **Note 1**). The updated model is available for download in the supplementary materials.

2.3 Transcriptomics for Stem Cell

Transcriptomics data was obtained from the GEO (Gene Expression Omnibus) database. Here we use transcriptomics profiling of induced pluripotent stem (iPS) cells derived from mouse embryonic fibroblasts (MEF) with Lin28a and Lin28b double knockout [16]. GEO accession of this dataset is GSE67568.

2.4 Time-Course Metabolomics for Stem Cell

Time-course metabolomics data of naïve and primed murine pluripotent stem cells were used to create dynamic genome-scale metabolic models of each cell state [10]. The metabolomics data were collected at four different time points.

3 Methods

3.1 Metabolic Network and Flux Balance Analysis

Flux balance analysis (FBA) is a widely used technique for analyzing GEMs by calculating the reaction rates, or fluxes, through a biochemical network. The technique relies on two basic assumptions: mass balance and cellular optimization. The quasi-steady-state assumption (QSSA) posits that because metabolism is fast and approximately unchanging, there will be no buildup of metabolites at steady state [17]. In other words, the fluxes into or out of a metabolite sum to zero. The second assumption, cellular optimization, states that cells maximize their fitness, which is usually represented biochemically as a biomass equation. Based on these assumptions, FBA can predict metabolic phenotypes that optimize flux state by maximizing biomass or other cellular objectives. These constraints enable the calculation of robust predictions of metabolic flux without requiring kinetic parameters [18, 19].

For a metabolic network with M metabolites and N reactions, the FBA formulation is shown below:

$$\text{Maximize} \quad Z = c \cdot v \tag{1}$$

$$\text{Subjectto} \quad \sum_{j=1}^{N} S_{ij} v_j = 0, \quad i = 1, \ldots, M \tag{2}$$

$$lb_j \leq v_j \leq ub_j, \quad j = 1, \ldots, N \tag{3}$$

where S is the stoichiometric matrix of the metabolic network and S_{ij} represents the stoichiometric coefficient of metabolite i for reaction j. The flux vector v is a vector of all flux values, where v_j represents the flux value corresponding to the jth reaction. lb_j and ub_j are the lower bound and upper bound for the flux of the jth reaction. The coefficient vector c is a series of weights that, when multiplied by the reaction fluxes, defines the optimization function Z.

The product $(S \cdot v)$ is a vector that represents the accumulation or consumption rate of each metabolite. It is set to zero in FBA due to the quasi-steady-state assumption. Constraints can be imposed on each reaction as the lower and upper bounds of v, which are related to the turnover number of the corresponding enzyme and its abundance [2].

The equation $S \cdot v = 0$ combined with the lower and upper bound constraints for the reactions defines a solution space for the vector v. In order to identify a biologically meaningful flux state from the vast solution space, an optimization problem is solved that maximizes an objective function Z. The most commonly used function is the rate of biomass formation, which generally provides a good approximation of cellular behavior [20, 21]. Solving this optimization problem will provide the flux distribution of all reactions in network that maximizes the objective [2]. Many constraints such as substrate uptake rates or secretion rates can also be imposed, further restricting the possible flux state of the network [22]. For example, since glucose is a commonly used carbon source in culture, the upper bound for D-glucose intake can be set to a non-zero value, while uptake of other metabolites not in the culture media is set to zero. All other reactions in the model are left unconstrained (i.e., default bounds from −1000 to 1000) if no information is available on their activity.

Given the large solution space, multiple flux solutions can achieve the same optimal biomass objective. Among these solutions, the solution with the lowest sum of all fluxes is frequently chosen—this is based on the principle of parsimonious enzyme usage, which assumes that cells maximize their objective with the least amount of enzyme (*see* **Note 2**).

3.2 Integrating Metabolic Network with Transcriptomics or Proteomics

Transcriptomics or proteomics data can be used to further constrain the solution space of the network. The constrain_flux_regulation algorithm takes as input a list of genes or reactions that are upregulated, a list of genes or reactions that are downregulated, and a genome-scale metabolic model. The output is a metabolic flux state that is consistent with the transcriptomics or proteomics constraints.

Similar to FBA, this algorithm maximizes an objective function. In this case, in addition to maximizing the biomass function, this algorithm also maximizes the number of active reactions with proteomic or transcriptomic evidence that they are upregulated and minimizes the flux through reactions that are downregulated. The mathematical formulation used to maximize the number of reactions with upregulated evidence is shown below and is similar to the iMAT approach [23]:

$$\text{Maximize} \quad \sum_{i=1}^{N} (t_i + r_i), \quad i = 1, \ldots, N \tag{4}$$

$$\text{Subject to} \quad v_i - \left(epsilon_i + M\right) t_i \geq -M \tag{5}$$

$$t_i = 0 \text{ or } 1 \tag{6}$$

$$v_i + \left(epsilon_i + M\right) r_i \leq M \tag{7}$$

$$r_i = 0 \text{ or } 1 \tag{8}$$

where *epsilon* is the extent of flux through upregulated reactions, which can be a fixed positive value for all upregulated reactions or set independently for each reaction (*see* **Note 3**). v_i is the flux value of ith reaction. t_i and r_i are the indicators of whether the reaction is active with a flux value at least $epsilon_i$ in positive or negative direction. If $t_i = 1$, the reaction is active in positive direction. If $r_i = 1$, the reaction is active in negative direction. M (default = 10,000) is a constant that is much larger than *epsilon* (default = 1E-3). The objective of this formulation is to maximize the number of active reactions, corresponding to the upregulated genes.

This algorithm does not turn off downregulated reactions but rather minimizes the flux through them. The mathematical formulation is similar to the PROM algorithm [24]. The strength of downregulation can be a fixed value or set independently for each reaction (*see* **Note 3**).

The mathematical formulation used for minimizing flux through the downregulated reactions is shown below:

$$\text{Minimize} \quad \sum_{i=1}^{K} \kappa(\alpha_i + \beta_i), \quad i = 1, \ldots, K \tag{9}$$

$$\text{Subjectto} \quad v_i \geq lb_i' - \alpha_i \tag{10}$$

$$v_i \leq ub_i' + \beta_i \tag{11}$$

$$\alpha_i \geq 0, \quad \beta_i \geq 0 \tag{12}$$

where lb'_i and ub'_i are constraints based on transcriptional regulation (default value set to zero), α_i and β_i are positive constants that represent deviation from those constraints, and *kappa* (κ) represents the penalty for such deviations. The higher the value of κ, the stronger the constraint on the system based on transcriptomics. lb_i and ub_i by default are set to zero, and hence this function minimizes flux through downregulated reactions.

Syntax

```
>> [flux, growth, objective] = constrain_flux_regulation(mod-
el, onreactions, offreactions, kappa, rho, epsilon, mode)
```

Input Data

onreactions, offreactions—contains the set of genes or reactions that are upregulated (*onreactions*) or downregulated (*offreactions*). If gene symbols are provided, then the ***mode*** variable must be set to 0; otherwise, it should be set to 1 (default, for reactions).

Upregulated and downregulated genes can be identified from transcriptomics or proteomics data through several strategies such as using fold change or *t*-test. Once a list of differentially expressed genes has been identified, they can be overlaid onto the metabolic model. Please make sure that the gene identifiers match the corresponding identifiers for the genes in the metabolic model (*see* **Note 4**).

If given a list of genes rather than reactions, the algorithm matches the genes to the corresponding metabolic reactions using the gene-protein-reaction relationships in the model. This is specifically achieved through the *deleteModelGenes* function in COBRA toolbox. Direct input of up- and downregulated reactions is also accepted.

Optional Inputs

rho, kappa—gives the relative weight for on- and off-reactions, respectively. Default value = 1, implying equal weight for on- and off-reactions and growth objective. This value should be reduced for large networks or optimizations involving large number of reactions (rho = 1E-3 recommended for 100+ reactions).

These two variables can also be set to a different value for each reaction. In this case, the number of elements of kappa and rho should match the number of reactions that are down- or upregulated (*see* **Note 3**).

epsilon—gives the minimum flux for on-reactions. Default value = 1E-3, implying a flux of either greater than +1E-3 or less than −1E-3 for reactions predicted to be upregulated. This value can also be set to a different value for each upregulated reaction. In this case, the number of elements in epsilon should match the number of reactions that are upregulated. For example, this value can be set proportional to the extent of upregulation of a gene's expression (*see* **Note 3**).

mode—describes the input list: genes (mode = 0) or reactions (default; mode = 1).

Output

flux—gives the metabolic flux state of all the reactions. An optimization problem is solved that maximizes the consistency with the input data. In addition, the growth rate is also maximized. Further, the flux state that has the minimum sum of all fluxes is provided as output (similar to Parsimonious FBA or P-FBA).

growth—is the metabolic objective of the model if provided (usually the biomass).

objective—is the optimization solver objective (should be non-zero).

Example: Overlaying Transcriptomics Data from Pluripotent Stem Cells

This example data has genes up- and downregulated in murine naive and primed pluripotent stem cells. The example datasets for this tutorial is available at www.sriramlab.org/software.

```
>> load("naive_genelist.mat") % this file has the differentially expressed genes in naïve cells (lin28 double knockout) relative to wild type cells from transcriptomics data
>> load("model_MGSA.mat") % this file has the human metabolic model
>> genelistup = intersect(genelist_up, metabolicmodel.genes); %get the genes in the metabolic model
>> genelistdown = intersect(genelist_down, metabolicmodel.genes);
Run the code with the transcriptomics data from naive cells. (Settings: onreactions—genelistup, offreactions—genelistdown, kappa = 1, rho = 1, epsilon = 1E-3 and mode = 0)
>> [fluxstate_naive,grate_naive,solverobj_naive]=constrain_flux_regulation(model_MGSA, genelistup, genelistdown, 1, 1, 1E-3, 0);
```

The resulting growth rate (grate_naive) should be 1.13, and the fluxstate_naive variable should contain the metabolic flux through all reactions in the model.

Note that all individual constraints can be violated to maximize the overall fit to the omics data. For example, some “on-reactions” can be predicted to be off if this maximizes the overall consistency with the remaining up- and downregulated reactions.

3.3 Interpreting Time-Course Metabolomics Using GEMs

Time-course metabolomics can be integrated with GEMs to analyze metabolic activity in dynamic systems. Here we describe the steps for using the dynamic flux activity (DFA) algorithm, a constraint-based modeling method that identifies the metabolic state that best fits the time-course metabolomics data. The algorithm takes in as input time-course metabolomics data and a genome-scale metabolic model. The algorithm outputs a metabolic model with time-course metabolic constraints overlaid on it.

DFA has a similar formulation to FBA, but the right-hand side vector *epsilon* (in Eq. 14) is set to a value proportional to the measured rate of change of each metabolite from the metabolomics data. In addition, due to the noise in metabolomics data, the constraints on rate of change of metabolites are made flexible in order to have a feasible solution from the model. Hence, we create two pseudo-reactions (α, β) that have positive flux, representing deviation from the measured experimental data. DFA minimizes the sum of α and β, which results in a flux distribution that is most consistent with the metabolomics data. The formulation is shown as below:

$$\text{Minimize} \quad \sum_{j=1}^{N} \left(\alpha_j + \beta_j\right), \quad j = 1, \ldots, N \tag{13}$$

$$\text{Subject to} \quad \sum_{i=1}^{M} S_{ij} v_j + \alpha_j - \beta_j = epsilon_i, \quad i = 1, \ldots, M \tag{14}$$

$$\alpha_j \geq 0, \quad \beta_j \geq 0 \tag{15}$$

where S_{ij} and v_j are the same representations as in FBA (*see* Subheading 3.1). $epsilon_i$ is a constant proportional to the rate of change of metabolite levels over time for metabolites measured in the metabolomics assay (for calculation of *epsilon*, *see* Subheading 3.3.2). For the remaining metabolites in the model not measured using metabolomics, the $epsilon_i$ is set to zero.

To maximize the objective function and minimize the sum of α and β simultaneously, the objective function coefficients for metabolomics constraints are set as *-kappa*, where kappa is a parameter with positive value. *kappa* represents the strength of the metabolomics constraints relative to the biomass objective. In addition, we minimize the sum of absolute flux values through other reactions (*see* **Note 2**). The objective function coefficients for these constraints are set as *-kappa2*. *kappa* and *kappa2* are optional inputs and will change the constraints in model (*see* **Note 5**).

3.3.1 Match Metabolites from Metabolomics to Model

Since some metabolites have more than one synonym, the first step is to match the metabolites in metabolomics data to the metabolic model.

1. MetaboAnalyst (http://www.metaboanalyst.ca/) is a web repository of informatics tools for metabolomics analysis [25]. The *Compound ID Conversion* tool, which can be found in *Other Utilities* page, maps metabolite IDs between databases like HMDB [26], KEGG [27], PubChem [28], ChEBI [29],

and MetLin [30]. If an automatic search for a metabolite returns no results, click the *View* link to perform an approximate search, and then manually select the correct match if found.

2. Using a programming script like Python, annotations of metabolites can be extracted from the original SBML [31] file for the metabolic model. In Python, *lxml* package is recommended for processing XML files. Annotations of metabolites in RECON1 are primarily from HMDB and ChEBI.
3. Both MetaboAnalyst and SBML extraction are used for matching metabolites in the metabolomics data to the metabolic model. Metabolites can be easily matched with Excel or Python scripts after collecting the annotations. In this case, annotations from HMDB and ChEBI were used to match the metabolites. A manual matching of metabolite names may be necessary if there are unmatched metabolites after this automated approach (*see* **Note 6**).
4. Finally, we have to determine the location of metabolites in the metabolic model. Since subcellular compartment-specific information is lost during metabolomics measurement, it is assumed that the measured metabolites represent the sum total of those in the cytoplasm, nucleus, and mitochondrial compartments in the model. For intracellular metabolites, we search for positions of matched metabolites in cytosol, which is marked as "[c]" in RECON1 model, as well as in mitochondria marked as "[m]" and nucleus marked as "[n]." For extracellular metabolites, we search for positions of matched metabolites in extracellular space which is marked as "[e]."

3.3.2 Calculate and Normalize Metabolite Rate of Change

In Subheading 3.3, we introduced the right-hand side vector *epsilon*, which is the normalized rate of change of metabolite concentration in the DFA formulation. By using *polyfit* function with degree of 1 in MATLAB, time-course metabolomics is fit to a linear model for each matched metabolite. Values in *epsilon* are set as the proportion of slope and intercept of these linear models:

$$epsilon_i = \frac{slope_i}{intercept_i} \tag{16}$$

where *intercept* and the *slope* are the outputs of *polyfit* function. *slope* represents the rate of change of each metabolite over time, and the *intercept* represents the metabolite levels at time = 0. The vector of *epsilon* values is then normalized by dividing by the maximum of the absolute values. The resulting values of *epsilon* are between −1 and 1, but relative magnitude and direction of change of metabolites are maintained.

Table 1
Example of DFA input file format

Metabolite	Position in model			Time points/ h			
				0	0	0.5	0.5
1,3-Diphosphoglycerate	0	0	0	0.018	0.017	0.833	0.834
2-Deoxyglucose-6-phosphate	0	0	0	1.444	1.436	1.300	1.352
2-Hydroxyglutarate	195	196	0	1.235	1.359	0.806	0.806
3-Phosphoglycerate	171	0	0	0.961	1.195	1.065	1.284
3-Phosphohydroxypyruvate	0	0	0	1.913	3.309	0.467	0.772
3-Phosphoserine	2338	0	0	0.390	0.536	0.856	1.297

$$epsilon'_j = \frac{epsilon_j}{\max(\mathrm{abs}(epsilon))} \tag{17}$$

When the maximum absolute value in *epsilon* is much larger than the average, an advanced normalization approach is recommended (*see* **Note 7**).

Syntax

```
[dynamicmodel, grate_wt, geneko_growthrate, rxnko_growthrate,
slope, solverobj, geneko_growthrate_obj, rxnko_growthra-
te_obj] = flux_activity_coeff(model, timecourse_metabolo-
mics_datafile,sheetname,genedelflag,rxndelflag,kappa,kappa2)
```

Input Data

timecourse_metabolomics_datafile, sheetname—The time-course metabolomics data for each state is provided by the user as an Excel file. The rows correspond to metabolites, and the columns correspond to different time points. In addition, the position in the metabolic model organism must be provided for each of the metabolites. If the data for each condition is present in a different sheet in the same Excel file, the sheet name should also be provided as input for the algorithm. Table 1 shows an example of the input file format.

In Table 1, the metabolite 3-phosphoglycerate is the 171st metabolite in the model. In addition, some metabolites are present in different compartments in the model. In those cases, the most relevant compartment or all the compartments should be used. In this study, metabolites were assumed to arise from the nuclear, cytoplasmic, and mitochondrial compartments. The position for unmatched metabolites should be set to 0; these metabolites are not used in the modeling process.

Optional Inputs

The DFA algorithm can also be used for performing gene or reaction deletion in each condition.

genedeletionflag, reactiondeletionflag—These two parameters take on binary values (0 or 1). 0 indicates that gene or reaction deletion will not be performed in the model, while 1 indicates otherwise.

The DFA algorithm has three optimization goals: maximize the objective function in the metabolic model (*see* Subheading 3.1), maximize the extent of flux distribution match with the metabolomics data (*see* Subheading 3.3), and minimize overall flux of other reactions (*see* **Note 2**). The mathematical formulations are linearly combined by using *-kappa* and *-kappa2* as the weights of these two goals, respectively.

kappa—gives the relative weight for the extent of flux distribution consistency with metabolomics data compared with the weight to maximize biomass function. This value should be reduced when there is no growth in predicted flux distribution or increased if there is very low consistency with the metabolomics (Default *kappa* = 1).

kappa2—gives the relative weight for the extent of minimizing the sum of absolute flux values of other reactions. This value should be increased when the magnitudes of some flux values in the predicted flux distribution are relatively larger than what would be expected biochemically (Default *kappa2* = 1E-3).

Output

dynamicmodel—The output is a metabolic model constrained with time-course metabolomics data.

grate_wt—This has the growth rate of the dynamic model when the objective is set to maximize the growth rate of the cell.

Optional Outputs

geneko_growthrate, rxnko_growthrate—When genedeletionflag or reactiondeletionflag is 1, the growth rates of the gene or reaction deletion analysis are stored in these two variables, respectively.

slope—This has the rate of accumulation or depletion of metabolites from time-course metabolomics data.

solverobj, geneko_growthrate_obj, rxnko_growthrate_obj—In addition to the growth rate, the algorithm also provides the value of the optimal objective function. This can be used to assess the fit to the metabolomics data. Genes or reactions that when deleted lead to a reduction in the objective function suggest that these are necessary for satisfying the metabolic objectives of the cell.

Example: Running Dynamic Metabolic Model for Pluripotent Stem Cells

This example data (available at www.sriramlab.org/software) has time-course metabolomics from murine naive and primed pluripotent stem cells.

Run the code with the metabolomics data from naive cells (Default settings). The file 'naive_prime_input_data_metabolomics.xlsx' contains the time course metabolomics data.

```
>> load("model_MGSA.mat") % this file has the human metabolic model
>> [naivemodel, naive_growthrate] = flux_activity_coeff2(model_MGSA, 'naive_prime_input_data_metabolomics.xlsx', 'naive');
```

The naive_growthrate should be 1.35, and the naivemodel variable has the dynamic model with metabolite constraints.

3.4 Evaluating the Metabolic Flux Predictions by DFA

Isotope tracing can be used to validate the in silico predictions of metabolic flux. In our prior study [10], we found using 13C-glucose and 13C-glutamine tracing that compared to the primed state, the naive state had increased glucose incorporation into the mitochondrial TCA cycle metabolites such as citrate, a-ketoglutarate/glutamate, fumarate, and malate and decreased glucose incorporation into one-carbon metabolism metabolites such as serine, 3-phosphoserine, and methionine, thereby validating the DFA predictions.

Another strategy is to compare the concordance between predictions of the relative sensitivity of the metabolic reactions (i.e., their impact on the cellular objective) between the two conditions and metabolomics data. In our case, metabolites associated with differentially sensitive reactions were also more likely to be differentially labeled between the two states ($p = 0.005$, hypergeometric test), confirming the strong concordance between model predictions and metabolomics datasets.

3.5 Visualizing Metabolic Flux Predictions by DFA

By running DFA analysis in two different conditions, the metabolic flux state can be compared between the two scenarios. The output can then be visualized using online tools like Escher (https://escher.github.io/). Figure 1 illustrates differentially active reactions in amino acid metabolism in the two stem cell fates. Glycine, serine, and glutathione metabolism is more active in primed model compared to naïve model. Figure 2 shows differentially active reactions in nucleotide metabolism in the two models. DFA predicted higher flux through purine synthesis in primed cells compared to naive cells.

4 Notes

1. In the human metabolic model, we blocked the flux through the reactions in the methylglyoxal pathway (MGSA and MGSA2) to allow flux through glycolysis. This correction is

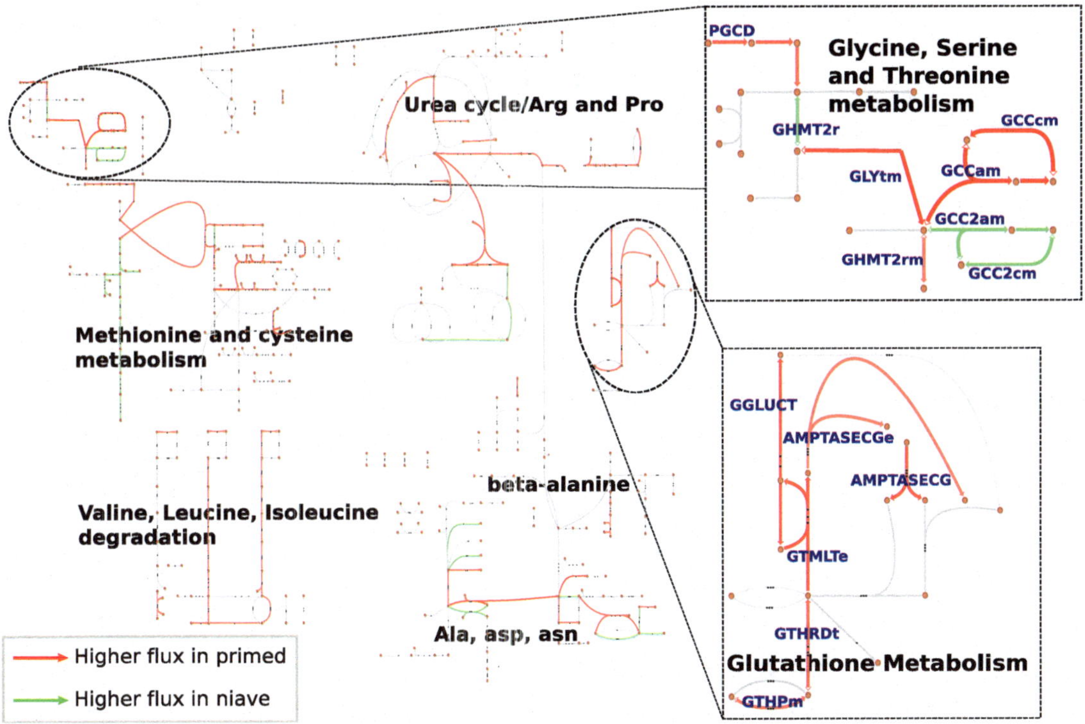

Fig. 1 Comparison of primed and naive model in amino acid metabolism (partial)

necessary to allow flux from glucose to citric acid cycle through glycolysis, which has been observed across several fluxomics studies [32, 33].

2. Due to the large space of flux solutions, it is recommended that an additional constraint is used to minimize the overall flux through the metabolic network. This strategy comes from parsimonious enzyme usage FBA (pFBA), which assumes that cells minimize the total necessary enzyme mass to achieve optimal growth [34].

   Usually we will maximize the biomass growth objective function and minimize the overall flux of other reactions simultaneously. An objective function coefficient for the flux minimization problem is set as -kappa, where kappa is a parameter with positive value. It can be tuned to change the relative weights of overall flux minimization over growth.

3. The constrain_flux_regulation algorithm currently does not currently support providing different weights for each gene. It does, however, support inputting a vector of different weights for each reaction if the algorithm is run with mode = 0. This limitation is due to the difficulty of translating a new weighting scheme from genes to reactions with gene-protein-reaction relationships.

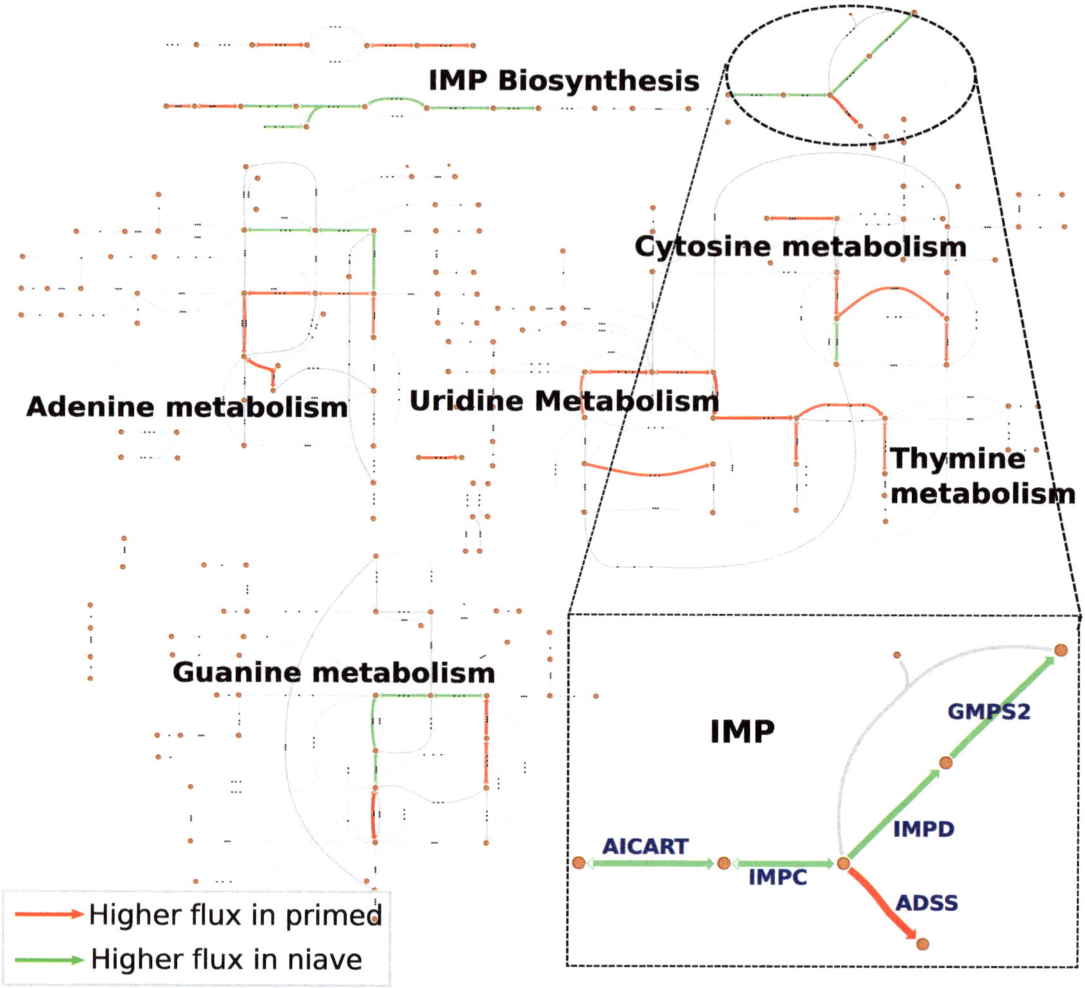

Fig. 2 Comparison of primed and naive model in nucleotide metabolism

4. When gene IDs in transcriptomics or proteomics data are not consistent with gene IDs in the metabolic model, online tools like DAVID Bioinformatics Resources (https://david.ncifcrf.gov) and UniProt (http://www.uniprot.org/uploadlists/) can be used for ID mapping.
5. In DFA algorithm, two parameters, *kappa* and *kappa2*, can be adjusted to change the constraints of model. Overall, larger *kappa* and *kappa2* will lead to lower flux in biomass, and smaller *kappa* and *kappa2* will lead to larger biomass flux.
6. For the metabolites from metabolomics that cannot be matched using MetaboAnalyst, chemical formula of the metabolite can be found in SBML and used as a matching parameter in HMDB. Note that chemical formula in SBML may lack

several H atoms compared to HMDB due to the charge of metabolites in model.

7. When the maximum absolute value of epsilon is very large relative to other epsilon values (>100 fold), the values in epsilon can be divided by a quantile (e.g., 90%) instead of the maximum.

References

1. Brunk E, Sahoo S, Zielinski DC et al (2018) Recon3D enables a three-dimensional view of gene variation in human metabolism. Nat Biotechnol 36:272
2. O'Brien EJ, Monk JM, Palsson BO (2015) Using genome-scale models to predict biological capabilities. Cell 161:971–987. https://doi.org/10.1016/j.cell.2015.05.019
3. Feist AM, Herrgard MJ, Thiele I et al (2009) Reconstruction of biochemical networks in microorganisms. Nat Rev Microbiol 7:129–143. https://doi.org/10.1038/nrmicro1949
4. Bordbar A, Monk JM, King ZA, Palsson BO (2014) Constraint-based models predict metabolic and associated cellular functions. Nat Rev Genet 15:107–120. https://doi.org/10.1038/nrg3643
5. Lewis NE, Nagarajan H, Palsson BO (2012) Constraining the metabolic genotype-phenotype relationship using a phylogeny of in silico methods. Nat Rev Microbiol 10:291–305. https://doi.org/10.1038/nrmicro2737
6. Folger O, Jerby L, Frezza C et al (2011) Predicting selective drug targets in cancer through metabolic networks. Mol Syst Biol 7:501. https://doi.org/10.1038/msb.2011.35
7. Frezza C, Zheng L, Folger O et al (2011) Haem oxygenase is synthetically lethal with the tumour suppressor fumarate hydratase. Nature 477:225–U132. https://doi.org/10.1038/nature10363
8. Uhlen M, Fagerberg L, Hallstrom BM et al (2015) Tissue-based map of the human proteome. Science 80:347. https://doi.org/10.1126/science.1260419
9. Shlomi T, Cabili MN, Herrgard MJ et al (2008) Network-based prediction of human tissue-specific metabolism. Nat Biotechnol 26:1003–1010. https://doi.org/10.1038/nbt.1487
10. Chandrasekaran S, Zhang J, Sun Z et al (2017) Comprehensive mapping of pluripotent stem cell metabolism using dynamic genome-scale network modeling. Cell Rep 21:2965–2977. https://doi.org/10.1016/J.CELREP.2017.07.048
11. Becker SA, Feist AM, Mo ML et al (2007) Quantitative prediction of cellular metabolism with constraint-based models: the COBRA Toolbox. Nat Protoc 2:727–738. https://doi.org/10.1038/nprot.2007.99
12. Schellenberger J, Que R, Fleming RMT et al (2011) Quantitative prediction of cellular metabolism with constraint-based models: the COBRA Toolbox v2.0. Nat Protoc 6:1290–1307. https://doi.org/10.1038/nprot.2011.308
13. Ebrahim A, Lerman JA, Palsson BO, Hyduke DR (2013) COBRApy: COnstraints-Based Reconstruction and Analysis for Python. BMC Syst Biol 7:74. https://doi.org/10.1186/1752-0509-7-74
14. Duarte NC, Becker SA, Jamshidi N et al (2007) Global reconstruction of the human metabolic network based on genomic and bibliomic data. Proc Natl Acad Sci U S A 104:1777–1782. https://doi.org/10.1073/pnas.0610772104
15. King ZA, Lu J, Drager A et al (2016) BiGG models: a platform for integrating, standardizing and sharing genome-scale models. Nucleic Acids Res 44:D515–D522. https://doi.org/10.1093/nar/gkv1049
16. Zhang J, Ratanasirintrawoot S, Chandrasekaran S et al (2016) LIN28 regulates stem cell metabolism and conversion to primed pluripotency. Cell Stem Cell 19:66–80. https://doi.org/10.1016/j.stem.2016.05.009
17. Gunawardena J (2014) Time-scale separation—Michaelis and Menten's old idea, still bearing fruit. FEBS J 281:473–488. https://doi.org/10.1111/febs.12532
18. Kauffman KJ, Prakash P, Edwards JS (2003) Advances in flux balance analysis. Curr Opin Biotechnol 14:491–496. https://doi.org/10.1016/j.copbio.2003.08.001
19. Orth JD, Thiele I, Palsson BO (2010) What is flux balance analysis? Nat Biotechnol 28:245–248. https://doi.org/10.1038/nbt.1614

20. Ibarra RU, Edwards JS, Palsson BO (2002) *Escherichia coli* K-12 undergoes adaptive evolution to achieve in silico predicted optimal growth. Nature 420:186–189. https://doi.org/10.1038/nature01149
21. Feist AM, Palsson BO (2016) What do cells actually want? Genome Biol 17:110. https://doi.org/10.1186/s13059-016-0983-3
22. Reed JL (2012) Shrinking the metabolic solution space using experimental datasets. PLoS Comput Biol 8:e1002662. https://doi.org/10.1371/journal.pcbi.1002662
23. Zur H, Ruppin E, Shlomi T (2010) iMAT: an integrative metabolic analysis tool. Bioinformatics 26:3140–3142. https://doi.org/10.1093/bioinformatics/btq602
24. Chandrasekaran S, Price ND (2010) Probabilistic integrative modeling of genome-scale metabolic and regulatory networks in Escherichia coli and *Mycobacterium tuberculosis*. Proc Natl Acad Sci U S A 107:17845–17850. https://doi.org/10.1073/pnas.1005139107
25. Xia JG, Sinelnikov IV, Han B, Wishart DS (2015) MetaboAnalyst 3.0-making metabolomics more meaningful. Nucleic Acids Res 43:W251–W257. https://doi.org/10.1093/nar/gkv380
26. Wishart DS, Feunang YD, Marcu A et al (2018) HMDB 4.0: the human metabolome database for 2018. Nucleic Acids Res 46:D608–D617. https://doi.org/10.1093/nar/gkx1089
27. Kanehisa M, Furumichi M, Tanabe M et al (2017) KEGG: new perspectives on genomes, pathways, diseases and drugs. Nucleic Acids Res 45:D353–D361. https://doi.org/10.1093/nar/gkw1092
28. Kim S, Thiessen PA, Bolton EE et al (2016) PubChem substance and compound databases. Nucleic Acids Res 44:D1202–D1213. https://doi.org/10.1093/nar/gkv951
29. Hastings J, Owen G, Dekker A et al (2016) ChEBI in 2016: improved services and an expanding collection of metabolites. Nucleic Acids Res 44:D1214–D1219. https://doi.org/10.1093/nar/gkv1031
30. Smith CA, O'Maille G, Want EJ et al (2005) METLIN—a metabolite mass spectral database. Ther Drug Monit 27:747–751. https://doi.org/10.1097/01.ftd.0000179845.53213.39
31. Hucka M, Finney A, Sauro HM et al (2003) The systems biology markup language (SBML): a medium for representation and exchange of biochemical network models. Bioinformatics 19:524–531. https://doi.org/10.1093/bioinformatics/btg015
32. Raghevendran V, Gombert AK, Christensen B et al (2004) Phenotypic characterization of glucose repression mutants ofSaccharomyces cerevisiae using experiments with13C-labelled glucose. Yeast 21:769–779. https://doi.org/10.1002/yea.1136
33. Sabra W, Bommareddy RR, Maheshwari G et al (2017) Substrates and oxygen dependent citric acid production by Yarrowia lipolytica: insights through transcriptome and fluxome analyses. Microb Cell Factories 16:78. https://doi.org/10.1186/s12934-017-0690-0
34. Lewis NE, Hixson KK, Conrad TM et al (2010) Omic data from evolved E-coli are consistent with computed optimal growth from genome-scale models. Mol Syst Biol 6:390. https://doi.org/10.1038/msb.2010.47

Chapter 15

Metabolomics in Stem Cell Biology Research

Zhen Sun, Jing Zhao, Hua Yu, Chenyang Zhang, Hu Li, Zhongda Zeng, and Jin Zhang

Abstract

Stem cell research has been greatly facilitated by comprehensive and integrative multi-omics studies. As a unique approach of functional analysis, metabolomics measures many metabolites and activities of metabolic pathways which can directly indicate cellular energetic status, cell proliferation and fitness, and stem cell fate choices such as self-renewal versus differentiation. Here we describe the methods of applying metabolomics, ^{13}C-labeled glucose and glutamine tracing with mouse embryonic stem cells (ES cells), metabolite analysis using mass spectrometry tools, and the following statistical and computational modeling analysis. Integration of these methods into the more common gene expression and epigenetics analysis toolbox will help to generate a more complete picture and in-depth understanding of one's stem cells of interest.

Key words Metabolomics, ^{13}C-tracing analysis, Mass spectroscopy, Metabolite, Metabolic pathway, Statistical analysis

1 Introduction

A current trend in stem cell research is that new discoveries are driven by computational analysis of high-dimensional large data sets collected from stem cell genome, epigenome, transcriptome, and proteome and by utilization of computational tools [1]. These systems biology approaches allow researchers to comprehensively characterize stem cells to understand cellular state and cell fate choices and how stem cells respond to certain genetic manipulation or environmental insults. However, information from epigenome, transcriptome, and proteome is indirect to indicate cellular phenotype, as expression of a gene or a set of genes in a pathway can only imply functional outcome, but does not necessarily mean that the function of that pathway is indeed active or not. Furthermore, acute responses to environment in a time scale of a few minutes

Zhen Sun and Jing Zhao are co-first authors.

Patrick Cahan (ed.), *Computational Stem Cell Biology: Methods and Protocols*, Methods in Molecular Biology, vol. 1975, https://doi.org/10.1007/978-1-4939-9224-9_15,

cannot be captured by gene expression changes. These cases are even more obvious in studies on intracellular metabolic pathways or on how metabolic changes are conferred by environmental insults. In contrast, metabolomics studies can link quantitative features of metabolites to functions of a cell and, therefore, is a direct approach for measuring cellular phenotype. Indeed, accumulating research has shown that virtually all cellular functions are related to and regulated by metabolism, such as cell energetics, cell proliferation, and cell death influenced by anabolic and catabolic metabolism, intracellular signaling, and cell communication mediated by key metabolites, etc. [2–6]. Therefore, there is a strong need in applying metabolomics approaches to enhance our understanding of stem cell biology.

Metabolomics can be generally categorized to untargeted and targeted approaches. Targeted metabolomics are used when the pathway of interest is clear and the study is focused, such as on the TCA cycle [6]. Otherwise, untargeted metabolomics is performed first to unbiasedly screen most significantly meaningful pathways, followed by more focused targeted approach [7, 8]. Depending on the research interest, intracellular pathway can be examined by collecting cell materials, while extracellular metabolome can be collected from stem cell culture media [3], and the latter is sometimes called exometabolomics or metabolic fingerprinting. In addition to profile metabolites in a living cell which only generate a snapshot of the metabolic network, it is also important to use isotope-labeled nutrients such as ^{13}C-glucose to trace metabolic pathway to understand the dynamics of metabolic flux which even more closely reflects metabolic function [6, 9]. In this chapter, we focus on describing the experimental and computational methods of intracellular metabolomics and isotope tracing studies on mouse embryonic stem cells.

2 Materials

ES cell LIF/2i media, PD03259010 (STEMCELL Technologies), CHIR99021 (STEMCELL Technologies), DMEM/F12 (11320-033, Gibco), neurobasal medium (21103-049, Gibco), N2 and B27 supplements (Life Technologies), L-glutamine (Gibco), and LIF (PEPRO TECH).

Isotope tracing media, D-glucose (U-13C6, 99%, Cambridge Isotope), glucose-free LIF/2i media, L-glutamine (U-13C6, 99%, Cambridge Isotope), glutamine-free media, mouse embryonic fibroblast (MEF) media, high-glucose DMEM (11960-069, Gibco), FBS (10099-141, Gibco), nonessential amino acids (GNM71450, GENOM), metabolite extraction solution, 80% methanol solution, HPLC-grade methanol, and water.

Penicillin/streptomycin (Gibco), 0.1% gelatin (Solarbio), 0.25% trypsin/EDTA (wt/vol).

5500 QTRAP triple quadrupole tandem mass spectrometry (AB/SCIEX).

Prominence HPLC System (Shimadzu).

MultiQuant 2.1 software (AB/SCIEX).

Human Metabolome Database: http://www.hmdb.ca/.

3 Methods

3.1 Mouse ES Cell Preparation for Metabolomics Studies

To prepare ES cell LIF/2i media, 1 μM PD03259010 and 3 μM CHIR99021 were supplemented into a 1:1 mix of DMEM/F12 and neurobasal medium containing N2 and B27 supplements (1:100), 2 mM L-glutamine, and LIF (1000 U/mL). To prepare isotope tracing media, D-glucose (U-13C6, 99%) is added to glucose-free LIF/2i media, and the final concentration is 21.25 mM, or L-glutamine (U-13C6, 99%) is added to glutamine-free media at the final concentration 3.25 mM.

Mouse ES cells grow on gelatinated MEF plates in LIF/2i media to maintain its self-renewal and pluripotency at the best condition [10]. To make the plate, we add 1 mL 0.1% gelatin into one well of 6-well plates to cover the entire culture surface. Incubate the plates for 30 min at RT. Then place a vial of frozen MEF in 37 °C water bath for about 1 min until the cells completely thaw. Then transfer it to 9 mL MEF cultured media (DMEM containing 10% FBS) in 15 mL centrifuge tube. Centrifuge for 2 min at 1100 rpm, and aspirate supernatant without touching the cell pellets. Gently resuspend and dispense the MEF into gelatin-coated plates at a density of 5×10^4 cells per well.

The next day, remove the medium of mES cell plate that is ready to be split, and rinse the cells with PBS. Apply 200 μL trypsin/EDTA solution per well, and incubate the plate in a 37 °C incubator for 5 min until cells completely round up. Then, add 1 mL MEF cultured media into the above well and transfer it to a centrifuge tube, and centrifuge at 1100 rpm for 2 min. After centrifugation, aspirate supernatant, and resuspend mES cells in LIF/2i medium. Then seed cells in the MEF feeder-coated plate as described above at a ratio of 1:10–1:20.

3.2 ^{13}C Tracing in Cultured Mouse ES Cells

For ^{13}C-tracing experiments (Fig. 1), ^{13}C-labeled nutrients such as glucose, pyruvate, or amino acid are dissolved in the corresponding media to culture mES cells for a certain period of time. We typically prepare mES cells on 6-well plates coated with MEF feeder cells at the density of 10^5 cells per well, cultured in LIF/2i medium. To start labeling, discard the previous medium, and change to the fresh

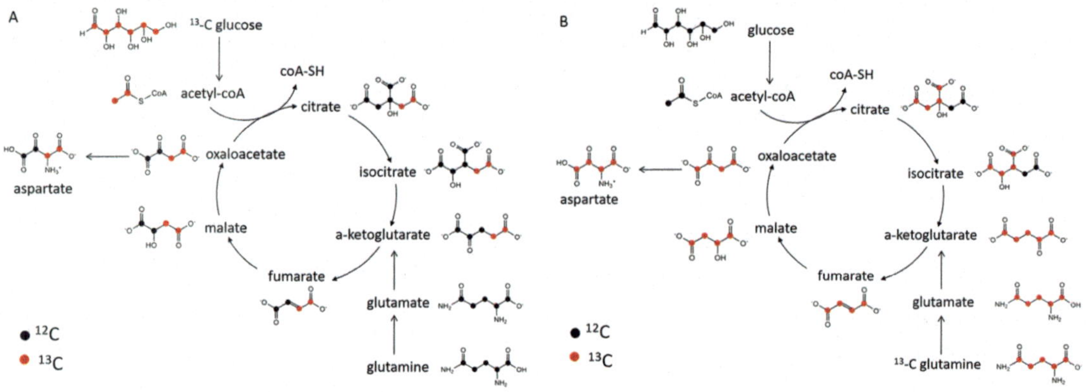

Fig. 1 Schematic picture showing the fate of ^{13}C in ^{13}C-labeled glucose and glutamine tracing analysis for the TCA cycle metabolic pathway. (**a**) $[U^{13}C_6]$-D-glucose has all its ^{12}C (black) replaced with ^{13}C (red). Through the glycolysis pathway, two ^{13}Cs are in acetyl-CoA, and after acetyl-CoA enters into the TCA cycle, the two ^{13}Cs are passed along the TCA cycle intermediate metabolites. (**b**) ^{13}Cs in $[U^{13}C_5]$-L-glutamine enter the TCA cycle after conversion to a-ketoglutarate. Four ^{13}Cs are passed along the TCA cycle intermediate metabolites

LIF/2i medium supplemented with D-glucose (U-^{13}C6, 99%, the final concentration is 21.25 mM) or L-glutamine (U-^{13}C6, 99%, the final concentration is 3.25 mM). Cells are then cultured for a certain period of time before metabolites are extracted. As different metabolic pathway has different kinetics, it is highly recommended to take multiple time points to find the time that ^{13}C labeling reaches to steady state [6] (*see* **Note 1**).

3.3 Metabolite Extraction from Mouse ES Cells

Certain metabolites are not stable, and in order to consistently extract metabolites with high quality, it is highly recommended to use standard operating protocols (SOP) when handling different types of cells or tissues [11]. In the case of mouse ES cells, we prepare extraction solution by adding 8 mL of HPLC-grade methanol to 2 mL of LC/MS-grade water to make an 80% methanol solution. The solution can be stored in −80 °C freezer for further use.

At the time of collection, mES cells are trypsinized for 5 min at 37 °C followed by 1 mL MEF cultured media to terminate digestion. To avoid MEF contamination, we deplete MEF cells from mouse ES cells (a process that we refer to as "deMEF") for 45–60 min at 37 °C, by letting the cells sit still in a new culture plate. MEFs will adhere to the bottom of the plate before the ES cells. For isotope tracing analysis, the procedures of deMEF can be performed in the presence of the same ^{13}C-containing media.

After deMEF, we collect the culture supernatant containing mES cells into conical tubes, centrifuge for 2 min at 1100 rpm, and then resuspend in the proper PBS and counted by hemocytometer. Usually a minimum of 10^6 cells are needed for downstream mass spectrometry experiments. Centrifuge proper number of cells for 2 min at 1100 rpm, and aspirate supernatant completely. Cell pellets are snap-frozen with 500 μL 80% (vol/vol) methanol (cooled to −80 °C) on dry ice, followed by incubation at −80 °C from 15 min to overnight. Lastly, to separate dissolved metabolites from cell debris, centrifuge the tubes at 14,000 × *g* for 15 min at 4 °C to pellet the cell debris. The supernatant is evaporated with a speed vacuum at 30 °C for about 4 h (depend on the volume of extraction solution). Dried samples can be stored at −80 °C for mass spectrometry analysis.

It is likely that deMEF procedures may change certain metabolites. Therefore, an alternative approach for sample collection and metabolite extraction is that splitting mouse ES cells onto a gelatinated plate without MEF for one passage before collecting (cells tend to grow worse without MEF for multiple passages). At the time of collection, aspirate off media and directly transfer the 6-well culture plate onto dry ice, quickly add 1 mL cold 80% methanol, and incubate in a − 80 freezer for at least 15 min, before scraping off and aspirating cells into a collection tube on dry ice, followed by the same procedures for metabolite extraction as above. These procedures can best preserve the quality of metabolites.

3.4 Data Collection with Mass Spectrometry

After metabolite extraction, selection of an optimal analytical platform is critical for measuring maximal number of small molecules with highest efficiency and quality, because there are significant differences among chemical properties and complementary capability for the separation and detection methods of each type of instrument. So far, chromatographic technique coupled with high-resolution mass spectrometry (MS) has been a definite pillar in untargeted metabolomics study after longtime arguments of the trends and prospects for future applications between MS and nuclear magnetic resonance (NMR) [12–14], mainly because of the extremely high sensitivity and rapid update of MS technology in recent years. In targeted metabolomics study, triple quadrupole tandem mass spectrometry has highest priority owing to its accurate quantification of metabolite targets [15]. A comprehensive comparison of UPLC-MS-based techniques is provided in Table 1.

Though the total number of metabolites existing in a certain biological sample is still unknown, usually hundreds of types of metabolites from tens of thousands of MS features can be detected in a single high-resolution MS-based experimental run. This makes peak detection and extraction of rich metabolite features not easy to attain, especially for matching peaks and retention time alignment across samples [16–18]. A series of chemometrics methods and programming platforms were developed in the last decades, including XCMS, mzMine, mzMatch, apLCMS, Maven, MetaboAnalyst, and many other softwares from instrumental manufacturers such as SIEVE (www.thermo.com), MassHunter (www.agilent.com), and

Table 1
Advantages and disadvantages for UPLC-MS-based techniques in metabolomics studies

No.	Indices	HPLC-MS-based instrumental types				
		TOF[a]	Quadrupole	Ion trap	Orbitrap	FT-ICR[b]
1	Resolution	Good	Not bad	Not bad	Good	Excellent
2	Dynamic range	Good	Excellent	Not bad	Not bad	Not bad
3	Sensitivity	Excellent	Excellent	Excellent	Excellent	Excellent
4	Speed	Excellent	Good	Excellent	Good	Not bad
5	Expense ($, k)	150–300	100–150	100–150	500–800	1000–1200
6	Maintenance	Average	Average	Average	Average	Expensive

[a] *TOF* time of flight
[b] *FT-ICR* Fourier-transform ion cyclotron resonance

MarkerLynx (www.waters.com) [16, 19–22]. The purpose of this step is to get a peak table including the whole information of MS features of all analyzed samples. Using XCMS as an example, it can be applied for simultaneous peak detection to identify hundreds of endogenous metabolites, peak matching to correct variation of different samples, and nonlinear retention time alignment to match time shift across samples [16, 19, 20]. After this, a table with MS features as rows and samples as columns will be obtained as input for statistical data processing. The chemical complexity of each sample and richness of metabolite features obtained from high-resolution MS makes the peak table unreliable in different methods mentioned above, and this is still a challenge to translate detected peaks to metabolite features in metabolomics study.

3.5 Data Analysis of Metabolomics Studies

Data collection is followed by statistical analysis methods, some of which are unique to metabolomics data. Generally, the procedure can be summarized as shown in Fig. 2. Data preprocessing is the first and vital step to enhance data quality and usability, including normalization, scaling for samples and for features, respectively, filling of missing data, data transformation, and others.

It should be noted that performance of the complicated methods for modeling in the next steps may be completely different with different data preprocessing strategies. For instance, seven different methods for data pre-treatment were comprehensively compared with different outcomes, including range scaling, Pareto scaling, autoscaling, vast scaling, centering, log transformation, and power transformation, and the first two methods were recommended for metabolomics data analysis [23]. After this, exploratory analysis can be applied to investigate similarities and differences among samples. Using the widely used principal component analysis (PCA) as an

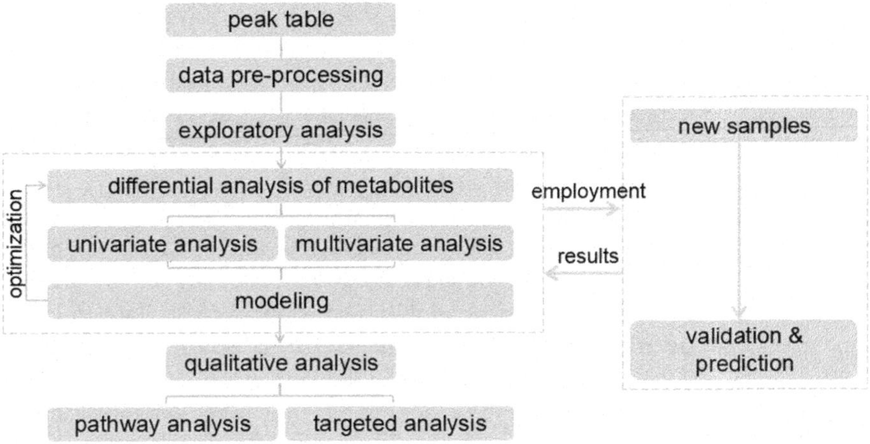

Fig. 2 Flowchart of data analysis in metabolomics study. Raw peak table data generated from mass spectrometry analysis are applied to unbiased exploratory analysis, followed by identification of metabolite ID and qualitative analysis. Then more specific pathway analysis or targeted analysis can be used for the metabolites/pathways of one's interest

example, the raw data in high-dimensional space can be reduced to limited dimensions with no significant loss of data information [24, 25]. This is achieved through reconstructing the original variables to principal components (PCs) along the directions with maximum variation of variance. The scores and loadings of PCA analysis can be utilized to introduce the characteristics of analyzed data, that is, the peak table with both information of observations and features of metabolites.

In additional, evaluation of data quality can be partially obtained by using the results from PCA analysis, along with many other methods such as correlation analysis. Generally, distribution of quality control (QC) samples in the scores plot of PCA results and variation of internal standards (IS) are applied to assess the reasonableness and potential experimental deviation [26, 27]. Also, many researchers contributed to calibrate non-biological effects, including batch effects, sampling sequence, and others. These methods include centering, scaling, quantile, ratio-based, linear normalization, linear regression, IS calibration, and batch normalizer [28]. Very recently, a method to calibrate gross and systematic errors was developed by Xu et al. with good performance [29].

Exploratory statistical analysis includes differential analysis of metabolites, biomarker discovery, as well as univariate analysis and multivariate quantitative modeling. Univariate analysis includes many methods for hypothesis test, which is to find the features with statistical difference among sample classes, such as paired and unpaired *t*-test, Wilcoxon rank-sum test, Mann-Whitney test, and ANOVA analysis [30]. Statistical visualization has special importance to find the features with statistical significance, including volcano plot, heatmap, Z-score plot, cloud plot, biplot, and S-plot [31]. Multivariate analysis includes PCA, partial least squares-linear discriminant analysis (PLS-DA), orthogonal PLS-DA (OPLS-DA), support vector machine, random forest and artificial neural network, and others [32–34]. Using PLS-DA as an example, the original data is transformed to a series of latent variable for linear discriminant analysis but with no loss of the structured information. The difference between PLS-DA and PCA is with or without employment of the class variable.

After statistical analysis, to interpret the data, metabolites can be identified by metabolic features' spectral peaks, and pathway enrichment analysis can be achieved with the help of the web-based databases such as Human Metabolome Database (http://www.hmdb.ca/) (*see* **Note 2**). Furthermore, one can also gain insight into metabolomics data by integrating other types of "omics" data, such as gene expression and epigenomic data, which will help to generate a more complete picture and enhance our understanding on biological questions.

4 Notes

1. Typically, in mouse ES cells, ^{13}C from glucose labels glycolysis and pentose phosphate shunt pathway intermediate metabolites to a steady state in a few minutes and labels the TCA cycle within 2 h. If the question is to find the steady-state level of ^{13}C incorporation into a metabolic pathway, it is recommended to take a time point after the incorporation curve reaches to plateau. For instance, when comparing two states of cells (stem cells versus differentiated cells), often time there is significant difference in the steady-state level. In other cases, such as when a metabolic enzyme or regulator is perturbed, the kinetics of ^{13}C incorporation might be altered, whereas the steady-state level is still unchanged; it is recommended to take multiple time points before the curve reaches to the plateau to capture the window with different kinetics.
2. Often times, the statistical and pathway enrichment analysis will generate a list of metabolites or metabolic pathways that are significantly different between two cell state conditions or between your experimental condition and control condition. In order to validate this is truly functionally relevant, it is recommended to design specific isotope tracing analysis based on the metabolic pathways selected from metabolomics analysis. For instance, if the mitochondrial TCA cycle metabolites are enriched in the altered metabolites, it is recommended to perform the ^{13}C-glucose or ^{13}C-glutamine analysis as shown in

Fig. 1. If the altered metabolites are in the serine/threonine amino acid metabolism pathway, one can use ^{13}C-labeled serine to validate this pathway.

References

1. Bian Q, Cahan P (2016) Computational tools for stem cell biology. Trends Biotechnol 34 (12):993–1009
2. Zhang J et al (2011) UCP2 regulates energy metabolism and differentiation potential of human pluripotent stem cells. EMBO J 30:4860–4873
3. Zhang J et al (2016) LIN28 regulates stem cell metabolism and conversion to primed pluripotency cell stem cell article LIN28 regulates stem cell metabolism and conversion to primed pluripotency. Cell Stem Cell 19:66–80
4. Folmes CD et al (2011) Somatic oxidative bioenergetics transitions into pluripotency-dependent glycolysis to facilitate nuclear reprogramming. Cell Metab 14:264–271
5. Nagaraj R et al (2017) Nuclear localization of mitochondrial TCA cycle enzymes as a critical step in mammalian zygotic genome activation. Cell 168:210–223.e11
6. Carey BW et al (2015) Intracellular α-ketoglutarate maintains the pluripotency of embryonic stem cells. Nature 518:413–416
7. Panopoulos AD et al (2012) The metabolome of induced pluripotent stem cells reveals metabolic changes occurring in somatic cell reprogramming. Cell Res 22(1):168–177
8. Sperber H et al (2015) The metabolome regulates the epigenetic landscape during naive-to-primed human embryonic stem cell transition. Nat Cell Biol 17:1523–1535
9. Chandrasekaran S et al (2017) Comprehensive mapping of pluripotent stem cell metabolism using dynamic genome-scale network modeling. Cell Rep 21(10):2965–2977
10. Weinberger L et al (2016) Dynamic stem cell states: naive to primed pluripotency in rodents and humans. Nat Rev Mol Cell Biol 17:155–169
11. Yuan M et al (2012) A positive/negative ion-switching, targeted mass spectrometry-based metabolomics platform for bodily fluids, cells, and fresh and fixed tissue. Nat Protoc 7:872–881
12. Theodoridis GA et al (2012) Liquid chromatography-mass spectrometry based global metabolite profiling: a review. Anal Chim Acta 711:7–16
13. Zhang T, Watson DG (2015) A short review of applications of liquid chromatography mass spectrometry based metabolomics techniques to the analysis of human urine. Analyst 140 (9):2907–2915
14. Roux A et al (2011) Applications of liquid chromatography coupled to mass spectrometry-based metabolomics in clinical chemistry and toxicology: a review. Clin Biochem 44(1):119–135
15. Siskos AP et al (2017) Interlaboratory Reproducibility of a Targeted Metabolomics Platform for Analysis of Human Serum and Plasma. Anal Chem 89(1):656–665
16. Myers OD et al (2017) Detailed investigation and comparison of the XCMS and MZmine 2 chromatogram construction and chromatographic peak detection methods for preprocessing mass spectrometry metabolomics data. Anal Chem 89(17):8689–8695
17. Tian T-F et al (2016) Web server for peak detection, baseline correction, and alignment in two-dimensional gas chromatography mass spectrometry-based metabolomics data. Anal Chem 88(21):10395–10403
18. Zhang JQ et al (2009) Review of peak detection algorithms in liquid-chromatography-mass spectrometry. Curr Genomics 10 (6):388–401
19. Smith CA et al (2006) XCMS: processing mass spectrometry data for metabolite profiling using nonlinear peak alignment, matching, and identification. Anal Chem 78(3):779–787
20. Mahieu NG, Genenbacher JL, Patti GJ (2016) A roadmap for the XCMS family of software solutions in metabolomics. Curr Opin Chem Biol 30:87–93
21. Hu M et al (2016) Optimization of LC Orbitrap HRMS acquisition and MZmine 2 data processing for nontarget screening of environmental samples using design of experiments. Anal Bioanal Chem 408(28):7905–7915
22. Katajamaa M, Miettinen J, Oresic M (2006) MZmine: toolbox for processing and

visualization of mass spectrometry based molecular profile data. Bioinformatics 22 (5):634–636
23. van den Berg RA et al (2006) Centering, scaling, and transformations: improving the biological information content of metabolomics data. BMC Genomics 7:142
24. Rui Alves M (2012) Evaluation of the predictive power of biplot axes to automate the construction and layout of biplots based on the accuracy of direct readings from common outputs of multivariate analyses: 1. Application to principal component analysis. J Chemom 26 (5):180–190
25. Jiang Q, Yan X (2013) Weighted kernel principal component analysis based on probability density estimation and moving window and its application in nonlinear chemical process monitoring. Chemom Intell Lab Syst 127:121–131
26. Gika HG et al (2012) A QC approach to the determination of day-to-day reproducibility and robustness of LC-MS methods for global metabolite profiling in metabonomics/metabolomics. Bioanalysis 4(18):2239–2247
27. Godzien J et al (2014) Controlling the quality of metabolomics data: new strategies to get the best out of the QC sample. Metabolomics 11 (3):518–528
28. Wang S-Y, Kuo C-H, Tseng YJ (2013) Batch normalizer: a fast total abundance regression calibration method to simultaneously adjust batch and injection order effects in liquid chromatography/time-of-flight mass spectrometry-based metabolomics data and comparison with current calibration methods. Anal Chem 85(2):1037–1046
29. Zhao Y et al (2016) A novel strategy for large-scale metabolomics study by calibrating gross and systematic errors in gas chromatography-mass spectrometry. Anal Chem 88 (4):2234–2242
30. Vinaixa M et al (2012) A guideline to univariate statistical analysis for LC/MS-based untargeted metabolomics-derived data. Meta 2 (4):775–795
31. Patti GJ et al (2013) A view from above: cloud plots to visualize global metabolomic data. Anal Chem 85(2):798–804
32. Wiklund S et al (2008) Visualization of GC/TOF-MS-based metabolomics data for identification of biochemically interesting compounds using OPLS class models. Anal Chem 80(1):115–122
33. Shen X et al (2016) Normalization and integration of large-scale metabolomics data using support vector regression. Metabolomics 12 (5):89
34. Li H, Liang Y, Xu Q (2009) Support vector machines and its applications in chemistry. Chemom Intell Lab Syst 95(2):188–198

Part IV

Assessing and Improving Cell Fate Engineering

Chapter 16

Molecular Interaction Networks to Select Factors for Cell Conversion

John F. Ouyang, Uma S. Kamaraj, Jose M. Polo, Julian Gough, and Owen J. L. Rackham

Abstract

The process of identifying sets of transcription factors that can induce a cell conversion can be time-consuming and expensive. To help alleviate this, a number of computational tools have been developed which integrate gene expression data with molecular interaction networks in order to predict these factors. One such approach is Mogrify, an algorithm which ranks transcriptions factors based on their regulatory influence in different cell types and tissues. These ranks are then used to identify a nonredundant set of transcription factors to promote cell conversion between any two cell types/tissues. Here we summarize the important concepts and data sources that were used in the implementation of this approach. Furthermore, we describe how the associated web resource (www.mogrify.net) can be used to tailor predictions to specific experimental scenarios, for instance, limiting the set of possible transcription factors and including domain knowledge. Finally, we describe important considerations for the effective selection of reprogramming factors. We envision that such data-driven approaches will become commonplace in the field, rapidly accelerating the progress in stem cell biology.

Key words Cell fate, Cell reprogramming, Gene expression, Regulatory networks, Transcriptional regulation, Transdifferentiation, Transcription factor

1 Introduction

The ability to transdifferentiate between cell types by overexpression of transcription factors is a phenomenon that is widely studied and is beginning to make an impact in the clinic. Despite the intense interest in this area, many groups are still using trial and error-based approaches in order to discover sets of transcription factors that can induce a cell conversion. This process is both time-consuming and expensive and subsequently has led to the development of a number of computational techniques to help facilitate the discovery. In general these approaches utilize gene expression data [1] and in some cases also combine gene regulatory information [2, 3] in order to predict the required factors for any given cell

Patrick Cahan (ed.), *Computational Stem Cell Biology: Methods and Protocols*, Methods in Molecular Biology, vol. 1975, https://doi.org/10.1007/978-1-4939-9224-9_16,

conversion. The motivation for these approaches is to make the identification of sets of transcription factors easier and in doing so facilitate experimental progress. One such approach is Mogrify, a technique which utilizes gene expression data and gene regulatory interactions to predict sets of transcription factors for cell conversion. These predictions can be tailored to specific experimental requirements, for instance, limiting the set of possible transcription factors, selectively removing or adding specific factors based on experimental evidence, or specifying the maximum number of transcription factors. The process of how these predictions are made involves several key steps: (1) calculating differential gene expression, (2) scoring each transcription factor based on its predicted regulatory influence, and (3) selecting transcription factors such that their regulatory control is not heavily redundant. In the following sections, we will discuss some important features of cell conversions, the Mogrify algorithm and the associated web resource. As with all computational and experimental protocols, understanding each of these aspects and the context in which they have been developed will provide the maximum utility.

1.1 Differentiation, Reprogramming, and Transdifferentiation: A Short History

Much of our understanding of cell fate changes originally came from studying the natural process of differentiation; early embryonic cells progressively confine their lineage toward specific cell types in response to external and internal cues. The earliest studies in this area focused on identifying populations of cells that maintain this ability to differentiate, the first of which in 1961. In this experiment, bone marrow cells were extracted and implanted into spleen samples taken from irradiated mice. These cells then formed "pluripotent colony-forming units" which could be made to differentiate into different cell types depending on their environment [4]. Two decades later, the isolation of an embryonic stem cell (ESC) line from mouse embryos which was capable of differentiating into any other cell type was another major milestone [5]. The derivation of ESCs allowed scientists to begin to understand and experiment with the biology of differentiation. With the knowledge that it is possible to isolate cells that are pluripotent came the question of whether the genetic material required for this behavior is lost from somatic cells or is simply silenced during development. An important discovery that helped to address this came with the development of somatic cell nuclear transfer (SCNT). This experimental procedure involves taking the nucleus of a somatic cell and inserting it into an enucleated egg cell, following which the somatic cell is reprogrammed to pluripotency. SCNT was first used to clone amphibians [6] and later (and more famously) Dolly the sheep [7], but what these experiments demonstrated was that the ability to differentiate was not lost in somatic cells but must be silenced epigenetically. Consequently, the possibility that with the right stimulus it would be possible to reactivate this ability was opened up. Several years later, this was proven to be the case when Yamanaka and colleagues discovered a set of four transcription factors which, if overexpressed in fibroblasts, resulted in induced pluripotent stem cells or iPS cells [8]. What is particularly striking about this reprogramming is that very few transcription factors are required in order to induce a huge change in the cell state. As well as conversions to pluripotency, a number of direct conversions between two somatic cell types, known as transdifferentiations, have been demonstrated (*see* Table 1 for a summary). The first such discovery was made by Harold Weintraub who discovered that the overexpression of a single transcription factor (MyoD) in fibroblasts was enough to convert them into a myoblast state [9]. Each of these transdifferentiations also only require a relatively low number of transcription factors. However, the identification of the correct set is a huge combinatorial challenge given the large number of possible transcription factors. Thus far the discovery of these factors has come about from a process of trial and error. Research groups focused on a single cell type and overexpress many different sets of transcription factors until an effective combination is found. This process is both expensive and time-consuming. In the field of computational stem cell biology, the guiding principle is that the identification of these factors should be done in a data-driven way, thereby greatly reducing the need for trial and error.

1.2 Gene Expression Profiles

Despite possessing nearly the same DNA, many different types of cells exist in the human body, carrying out distinct biological functions. Each cell type possesses a unique gene expression profile, expressing a specific set of key genes which are necessary to perform the cell's function (Fig. 1).

Numerous gene expression profiling techniques have been developed to measure the activity of thousands or tens of thousands of genes simultaneously. For example, microarrays measure the genome wide gene expression via fluorescence proportional to the concentration of different mRNAs in a population of cells. More modern methods such as RNA sequencing (RNA-seq) and capped analysis of gene expression (CAGE) measure gene expression by counting the number of short mRNA fragments associated with individual genes (*see* Subheading 2.1 for more details).

1.3 Differential Gene Expression

To identify factors for cell reprogramming, we need to first establish differences in gene expression arising from various cell types. A gene is said to be differentially expressed between condition A and B if there is significant difference in the amount of the mRNA produced under the different conditions. Tools for statistically testing whether a gene is differentially expressed vary depending on the experimental technique. This is because the observed gene expression from various experimental techniques is best

Table 1
List of transdifferentiations of human cell types

Source cell type	Target cell type	Set of TF	Reference
Fibroblasts	Adipocytes	PRDM16, CEBPB	Kajimura et al. 2009 [10]
Fibroblasts	Astrocytes	NFIA, NFIB, SOX9	Caiazzo et al. 2015 [11]
Fibroblasts	Cardiac muscle cells	GATA4, MEF2C, TBX5, miR-133	Muraoka et al. 2014 [12]
Fibroblasts	Cardiac progenitors	ETS2, MESP1	Islas et al. 2012 [13]
Fibroblasts, cardiac fibroblasts	Cardiomyocytes	GATA4, MEF2C, TBX5, MESP1, MYOCD	Wada et al. 2013 [14]
Fibroblasts	Cardiomyocytes	GATA4, HAND2, TBX5, MYOCD, miR-1, miR-133	Nam et al. 2013 [15]
Fibroblasts	Cardiomyocytes	GATA4, MEF2C, TBX5, ESRRG, MESP1, MYOCD, ZFPM2	Fu et al. 2013 [16]
Placenta cells	Chondrocytes	BCL6, T, c-MYC, MITF, BAF60C	Ishii et al. 2012 [17]
Fibroblasts	Dopaminergic neurons	MASH1, NGN2, SOX2, NURR1, PITX3	Liu et al. 2012 [18]
Fibroblasts	Dopaminergic neurons	ASCL1, BRN2, MYT1L, LMX1A, FOXA2	Pfisterer et al. 2011 [19]
Fibroblasts	Endothelial cells	ETV2, FLI1, GATA2, KLF4	Wong and Cooke 2016 [20]
Peripheral blood monocytes	Erythroid lineage cells	PU-1-specific siRNA	Nouri et al. 2016 [21]
Neural progenitor cells (iPSC-derived)	Excitatory neurons	NGN2	Ho et al. 2016 [22]
Embryonic stem cells	Extraembryonic endoderm cells	GATA6	Wamaitha et al. 2015 [23]
Fibroblasts	Glutamatergic neurons	ASCL1, BRN2, MYT1L, NEUROD1	Pang et al. 2011 [24]
Dermal microvascular endothelial, umbilical vein endothelial cells	Hematopoietic multipotent progenitor cells	FOSB, GFI1, RUNX1, SPI1	Sandler et al. 2014 [25]
Fibroblasts	Hematopoietic progenitor cells	OCT4	Szabo et al. 2010 [26]

(continued)

Table 1 (continued)

Source cell type	Target cell type	Set of TF	Reference
Fibroblasts	Hepatocytes	HNF1A, HNF4A, HNF6, CEBPA, ATF5, PROX1, c-MYC	Du et al. 2014 [27]
Umbilical cord mesenchymal stromal cells	Islet cells	PDX1	Wang et al. 2014 [28]
Fibroblasts	Keratinocytes	SOX9, FOS, REL, CDH1, MAFB, FOXQ1	Rackham et al. 2016 [3]
B cells	Macrophages	CEBPA	Rapino et al. 2013 [29]
Fibroblasts	Medium spiny neurons	BCL11B, DLX1, DLX2, MYT1L, miR-9/9*, miR124	Victor et al. 2014 [30]
Fibroblasts	Melanocytes	MITF, SOX10, PAX3	Yang et al. 2014 [31]
Keratinocytes	Microvascular endothelial cells	SOX17, SMAD1, IRF1, TAL1, TCF7L1	Rackham et al. 2016 [3]
Mesenchymal stem cells	Motor neurons	OLIG2, HB9	Park et al. 2012 [32]
Fibroblasts	Myoblasts	MYOD1	Lattanzi et al. 1998 [33]
Proximal tubule (HK2) cell line	Nephron progenitor cells	SIX1, SIX2, OSR1, EYA1, HOXA11, SNAI2	Hendry et al. 2013 [34]
Fibroblasts	Neural crest cells	SOX10	Kim et al. 2014 [35]
Fibroblasts	Neural precursor cells	SOX2, FOXG1, BRN2	Lujan et al. 2012 [36]
Fibroblasts	Neural progenitor cells	OCT4	Mitchell et al. 2014 [37]
Peripheral blood cells	Neural progenitor cells	OCT4	Lee et al. 2015 [38]
Cord blood CD34+ cells	Neural stem cells	OCT4	Liao et al. 2015 [39]
Fibroblasts	Neuronal cells	ASCL1, SOX2, NGN2	Zhao et al. 2015 [40]
Astrocytes	Neurons	NEUROD1	Guo et al. 2014 [41]
Fibroblasts	Neurons	ASCL1, BRN2, MYT1L	Vierbuchen et al. 2010 [42]

(continued)

Table 1
(continued)

Source cell type	Target cell type	Set of TF	Reference
Fibroblasts	Neurons	MYT1L, BRN2, miR-124	Ambasudhan et al. 2011 [43]
Fibroblasts	Neurons	miR-124, BRN2, MYT1L, SH2B1	Hsu et al. 2014 [44]
Fibroblasts	Neurons	miRNA-302/367, miRNA-9/9*, miRNA-124	Zhou et al. 2015 [45]
Pericytes	Neurons	SOX2, MASH1	Karow et al. 2012 [46]
Fibroblasts	Osteoblasts	RUNX2, SP7, OCT4, MYCL	Yamamoto et al. 2015 [47]
Pancreatic alpha cells	Pancreatic beta cells	HNF4A	Sangan et al. 2015 [48]
Liver cells	Pancreatic cells	PDX1, PAX4, MAFA	Berneman-Zeitouni et al. 2014 [49]
Keratinocytes	Pancreatic islets insulin+ cells	PDX1	Mauda-Havakuk et al. 2011 [50]
Pluripotent stem cells	Pancreatic islets insulin+ cells	MAFA	Chiou et al. 2011 [51]
Fibroblasts	Pluripotent stem cells	OCT4, SOX2, KLF4, c-MYC	Takahashi et al. 2007 [52]
Fibroblasts	Pluripotent stem cells	OCT4, SOX2, NANOG, LIN28	Yu et al. 2007 [53]
Fibroblasts	Pluripotent stem cells	OCT4, SOX2, miR-302-367	Deng et al. 2015 [54]
Neural stem cells	Pluripotent stem cells	OCT4	Kim et al. 2009 [55, 56]
Fibroblasts	Retinal epithelial cells	RAX, CRX, PAX6, MITF, OTX2, NRL, KLF4, C-MYC	Zhang et al. 2014 [57]
Fibroblasts	Retinal pigmented epithelium cells	PAX6, LHX2, OTX2, SOX9, MITF, SIX3, ZNF92, GLIS3, FOXD1	D'Alessio et al. 2015 [1]
Pluripotent stem cells	Retinal pigmented epithelium cells	OTX1, RX1, SIX3	Li et al. 2016 [58]
Fibroblasts	Spinal motor neurons	NEUROD1, ASCL1, BRN2, MYT1L, LHX3, HB9, ISL1, NGN2	Son et al. 2011 [59]

(continued)

Table 1
(continued)

Source cell type	Target cell type	Set of TF	Reference
Fibroblasts	Thymic epithelial cells	FOXN1	Bredenkamp et al. 2014 [60]
Fibroblasts	Trophoblast progenitors	CDX2, EOMES, ELF5	Chen et al. 2013 [61]
Amniotic cells	Vascular endothelial cells	ETV2, FLI1, ERG1	Ginsberg et al. 2012 [62]

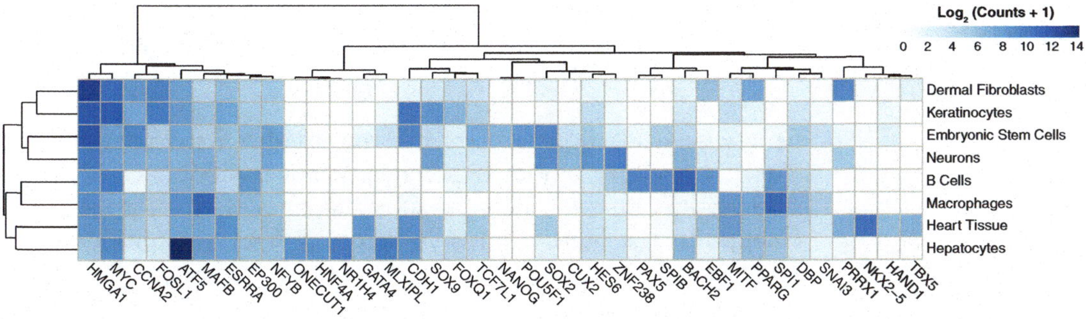

Fig. 1 Different cell types possess different gene expression profiles, illustrated here with the log-transformed normalized counts of selected transcription factors (TFs) across eight different cell types and tissues. The selected TFs are the top five TFs predicted by Mogrify for conversion to the cell type or tissue of interest

approximated by different statistical distributions. As a result, the underlying assumptions of the statistical model have to be compatible with this distribution.

The Mogrify algorithm uses count-based gene expression profiles obtained from the CAGE technique (*see* Subheading 2.1 for more details). Since the gene expression profile is in the form of integer counts, a statistical tool that describes this using a discrete probability distribution is crucial to ensure the accuracy of downstream analysis. As such, we elected to use the DESeq tool [63], modelling the gene expression using a negative binomial distribution. The negative binomial distribution is ideal for count data, which spans over an unbounded positive range, since its sample variance exceeds the sample mean. The DESeq analysis pipeline comprises three main steps (Fig. 2). First, the library size of each sample is estimated which is then used to normalize the counts. Second, DESeq shrinks the observed variability between replicates, given by the dispersion, since this is often exaggerated in small sample sizes ($n \approx 3$). Sharing information across genes, the relationship between the mean counts and dispersion is fitted, and the

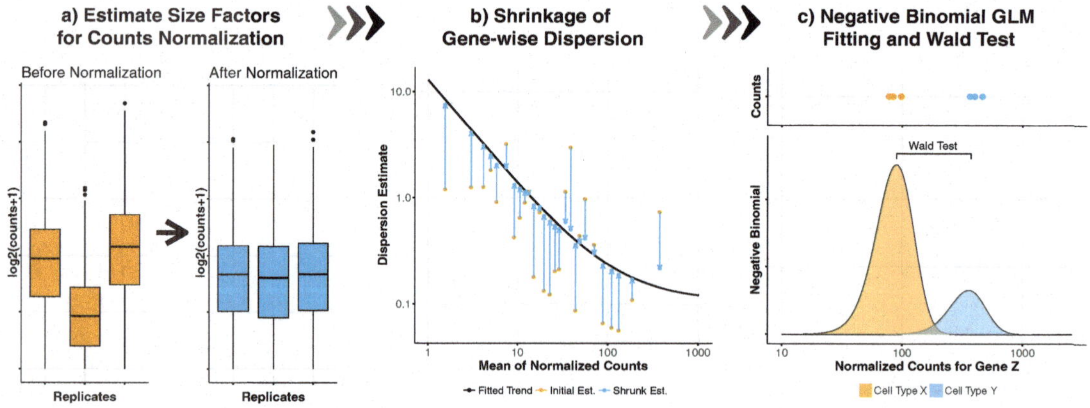

Fig. 2 The DESeq pipeline for differential gene expression analysis comprises three main steps. (**a**) The raw counts are first normalized for library size, followed by (**b**) dispersion shrinkage which reduces the observed variability between replicates. (**c**) Finally, a negative binomial GLM is fitted, and the Wald statistical test is performed to determine if genes are differentially expressed

gene-wise dispersions are shrunk toward the fitted relationship. Third, a negative binomial generalized linear model (GLM) with a logarithm link to the normalized counts is fitted to explain changes in the gene expression across different conditions. The Wald statistical test is performed on the fitted GLM coefficients to determine if the log-transformed fold change (LFC) in the gene expression is statistically significant. Genes are then deemed to be differentially expressed based on cutoffs on the false discovery rate (FDR) from the statistical test and the magnitude of the LFC. *See* **Note 1** for a discussion of the impact of replicates on predictive power.

1.4 Transcription Factor-Mediated Transcriptional Regulation

Differences in gene expression are primarily mediated by transcription factors. These are proteins that control the initiation of transcription of DNA into pre-mRNA by binding onto the DNA in the regulatory elements of a gene (i.e., a promoter or enhancer). In human cells, there are general transcription factors (GTFs) that recruit the RNA polymerase to the promoter region of a gene in order to initiate transcription. These GTFs, along with the RNA polymerase, form the basic transcriptional machinery and are needed to transcribe any gene. However, the majority of transcription factors are not GTFs. Instead, they bind to DNA in a sequence-specific manner, allowing them to target only a specific subset of genes. These specific DNA sequences are known as transcription factor binding sites (TFBS). These TFBS are often represented by position-weighted matrices (PWM), which encode the preference for the four different nucleotides at each position in the binding site (Fig. 3). TFBS are often identified from ChIP-seq data, which pinpoints the regions of the genome that a transcription factor interacts with. From these ChIP-seq regions, the PWMs

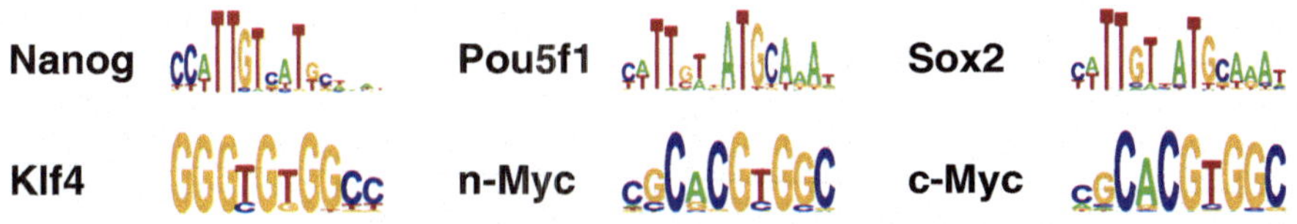

Fig. 3 Position-weighted matrices for pluripotency transcription factors, which give the nucleotide binding preference at each position in the binding site, taken from ref. 66

can be constructed for a given transcription factor. However, due to the quality of currently available antibodies, it may not be possible to perform ChIP-seq experiments on all known transcription factors. Furthermore, some transcription factors may bind to certain DNA architectures (i.e., secondary structures) or exhibit partial TFBS motif recognition, making the construction of PWM impossible [64]. Consequently, PWMs are only known for 476 transcription factors, which is less than a third of all known transcription factors [65].

Once a transcription factor binds to the promoter of a gene, it can either help recruit the transcriptional machinery, i.e., the GTFs and RNA polymerase increasing gene expression (a transcriptional activator), or block their attachment (a transcriptional repressor). It should be noted that a transcription factor can be either an activator or repressor for different genes in the same cell type.

1.5 Gene Regulatory Networks

The control of gene expression and protein abundance is a highly complex process, with various levels of interactions and controls. The most common way to represent these interactions is using networks. Briefly, a network (or graph) consists of nodes and edges. The nodes represent the possible interaction partners, which can be genes or proteins. The edges describe the interactions between partners and can be either directed or undirected. A directed edge implies that there is some directionality involved in the interaction, for instance, NANOG regulates SOX2 expression. An undirected edge has no such constraint and describes a mutual interaction such as protein-protein interaction. The interpretation of a network changes greatly depending on the data that it encodes; however, there are certain common features. For instance, network edges can either be binary, i.e., present or absent, or they can carry some information about the strength or probability of the edge. Where available, this edge level information is particularly important to effectively use gene regulatory networks since it allows for the filtering of edges with low confidence or strength.

Once constructed it is also possible to use the network to identify characteristics of individual nodes based on their involvement in the network. For instance, the number of incoming or outgoing connections that a given node has (i.e., its indegree and outdegree) is a useful proxy for the importance of the node in a

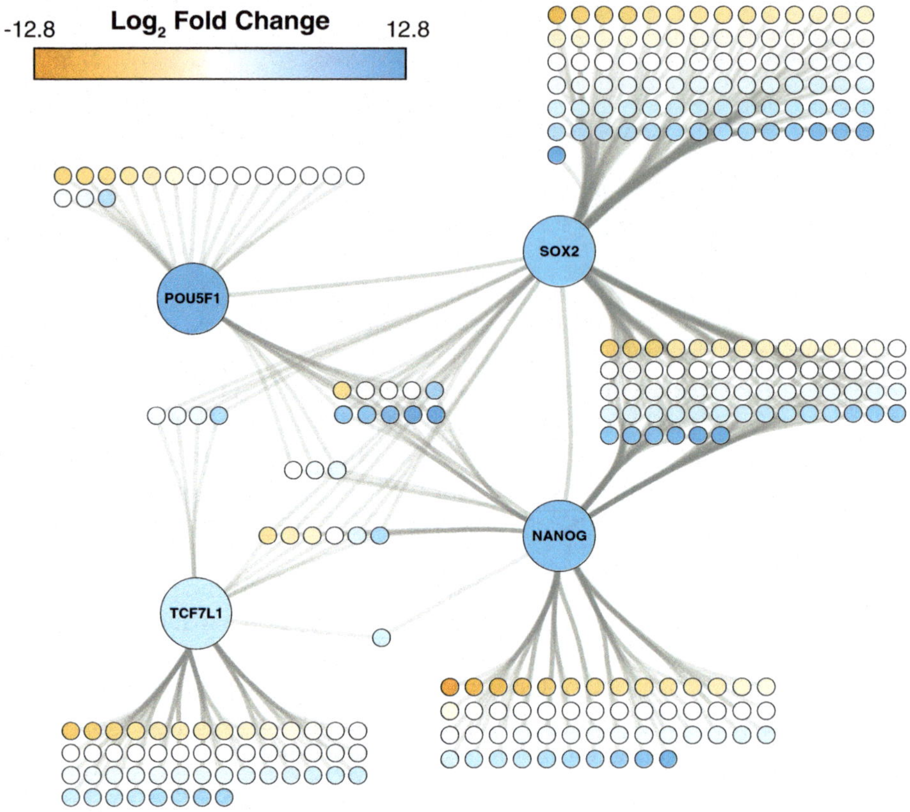

Fig. 4 An example of an undirected gene regulatory network obtained from the STRING database [67] showing the interaction partners of NANOG, SOX2, POU5F1, and TCF7L1 which are the top four most important transcription factors predicted by Mogrify in H9 embryonic stem cells. All the nodes are colored based on the log2 fold change of the gene expression level between embryonic stem cells and dermal fibroblasts

network. The higher the outdegree of a node, the more regulatory connections it has, and as a result, it has a higher potential in providing influence to other nodes. These high outdegree nodes are sometimes referred to as hubs of the network, and in a gene regulatory network, they are often transcription factors. For example, in embryonic stem cells, the transcription factors NANOG, SOX2, and POU5F1 are often referred to as hubs since they are involved in a great number of interactions important in pluripotency (Fig. 4).

2 Materials

Being a data-driven approach, the Mogrify algorithm draws information from extensive databases that are publicly available. The following section outlines how the gene expression and gene regulatory information are integrated in order to produce the Mogrify predictions.

2.1 FANTOM5 Gene Expression Data

Since all cell types are primarily a product of their gene expression, an important aspect in identifying reprogramming factors is a high-quality gene expression resource. For the Mogrify approach, we use the human gene expression atlas generated by the FANTOM5 consortium [68]. This extensive gene expression dataset contains 700 libraries from 173 human cell types and 134 tissues analyzed by the CAGE technique. The CAGE technique gives a snapshot of transcription focused on the transcription start sites (TSS) by capturing and sequencing the 5′ end of mRNAs. These reads are subsequently mapped to the promoters of genes in order to estimate the gene expression level (Fig. 5). This is in contrast to conventional RNA-seq which produces RNA fragments that map to the entire length of the transcript. We chose the FANTOM5 gene expression atlas as it contains a large number of primary cells, i.e., pure cell populations obtained directly from human donors. These samples are biologically similar to the resident cells in the body, which is crucial in the context of cell reprogramming. Furthermore, all the samples were prepared at the same facility, reducing the possibility of batch effects which can confound any downstream analysis.

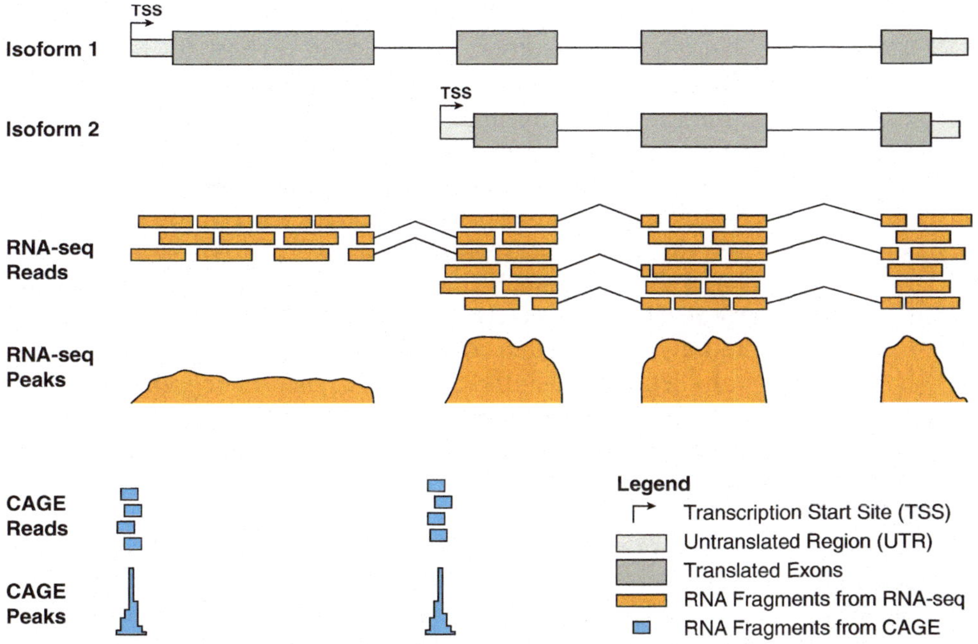

Fig. 5 Comparison of RNA-seq and CAGE in measuring the gene expression of a gene with two isoforms. In RNA-seq, short RNA fragments of roughly 100 nucleotides are obtained which can map to any segment of the transcript for both isoforms. In contrast, CAGE only captures very short RNA fragments of approximately 27 nucleotides corresponding to the transcription start sites of each transcript

2.2 Curated List of Transcription Factors

Transcription factors make a natural choice for driving cell reprogramming as they can regulate transcription of many genes and strongly influence the cell state. Thus, it is important for computational approach to first identify the set of putative transcription factors. However, the definition of a transcription factor is not commonly agreed upon. They can be defined based on either the ability to regulate transcription, the presence of DNA-binding domains, or the localization of the protein in the nucleus. In the Mogrify algorithm, we curated a list of transcription factors making use of resources such as gene ontology (GO) [69, 70], TF-class [71], and TFCat [72]. This approach was taken to maximize the number of possible transcription factors taken into consideration. However, in the future a universal definition of transcription factors will greatly benefit the reprogramming field.

2.3 STRING Protein-Protein Association Network

To construct the transcriptional regulatory network for each transcription factor, an extensive database of the interactions between the genes is necessary. Proteins are the major product of gene expression, and the interactions between expressed proteins can alter the physiology of the cell and in turn have a knock-on effect on the transcriptional regulation. Thus, we used the STRING database [67], consisting of direct (physical) and indirect (functional) associations, in order to provide interactions beyond the transcriptional level.

Each interaction in the STRING database is given seven evidence scores based on a variety of information, including (1) actual experiments, (2) curated biological pathways, (3) automated text mining of literature, and (4) gene coexpression data. The remaining three genome-based evidence scores are less relevant to human cells, involving the proximity of genes in the genome neighborhood, association of pairs of proteins with fusion gene orthologs, and the phylogenetic distribution of protein orthologs. From these seven sources, a final combined score is computed. In the Mogrify approach, we selected interactions with a combined score >500 and an experimental score >0, corresponding to a medium-high confidence in the interaction. Note that the regulatory edges in the STRING network are undirected as the interactions are inferred from functional association of their corresponding protein products.

2.4 Motif Activity Response Analysis-Based Regulatory Network

In addition to the STRING database, we used the Motif Activity Response Analysis (MARA), a computational method to generate a regulatory network describing the direct interaction of transcription factors with DNA [73]. MARA identifies the key regulators, in this case transcription factors, which account for changes in gene expression across different samples. For each gene, MARA uses a linear model to fit the gene expression against the number of occurrences of known TFBS motifs in the promoter region of the gene. The fitted coefficients represent the motif activity, and a corresponding z-score is computed which indicates the overall importance of the motif. Notably, it is not the model's aim to accurately fit the gene expression but to determine the influence of each motif in explaining the gene expression. Since different transcription factors can share the same binding motif, MARA also correlates the gene expression of each of the transcription factors with the motif activities in order to identify the most likely regulator.

MARA was performed on the gene expression data from the FANTOM5 project described above. The large number of samples ensures that the z-scores obtained from MARA are robust. A z-score cutoff of >3 was chosen to only select the significant regulatory interactions. We note that MARA requires the binding motif of the transcription factor, and these are only known for less than a third of known transcription factors (Subheading 1.4), limiting the scope of the network. Furthermore, the regulatory edges in the MARA-based network are directed, unlike STRING, and this can be taken into account in the Mogrify algorithm.

3 Methods

The Mogrify algorithm comprises three main steps in order to identify transcription factors that can induce cell fate conversion. For each cell type and tissue, we first perform a tree-based differential expression from which a "gene score" is calculated for each gene. The regulatory influence of each transcription factor is then computed using this gene score and information from the STRING and MARA interaction networks. The transcription factors are then ranked based on both their gene score and network influence. Using the ranked list, we choose a list of transcription factors that maximizes the regulatory coverage of cell-specific genes.

3.1 Tree-Based Differential Expression

In order to understand the transcriptional regulatory network in each cell type, we need to establish the importance of each gene in controlling cell identity. We perform differential gene expression analysis to find genes that are specifically expressed in each particular cell type. To this end, we use DESeq due to the suitability of its statistical model in handling count data, discussed earlier in Subheading 1.3. Furthermore, DESeq runs quickly, allowing for the analysis of a large number of cell types in reasonable time. DESeq also allows for datasets without any replicates, which is common for the tissue samples in the FANTOM5 gene expression atlas.

Differential expression analysis identifies the differences in gene expression between two conditions, namely, the foreground and background. The foreground corresponds to the samples of the cell type of interest, while the background includes a diverse set of other

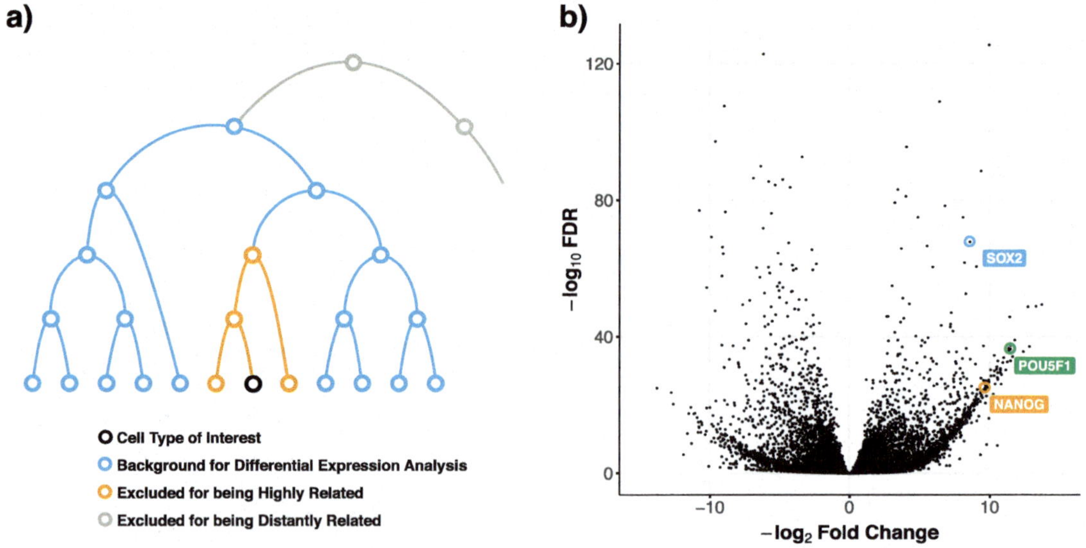

Fig. 6 (**a**) To select an appropriate background for differential gene expression, we use the FANTOM5 cell type ontology and excluded the highly related and distantly related cell types from the background. (**b**) Example result from the tree-based differential expression for H9 embryonic stem cells. The important pluripotency transcription factors, which are colored and labelled, possess a large LFC and a small FDR

cell types. By comparing to a diverse background, as opposed to a pairwise test, the differentially expressed genes will be those which are specific to the cell type of interest. In this respect, the choice of background is crucial since the gene specificity information obtained here underlies the accuracy of the later calculations. Inclusion of sample types that are highly similar to the cell type of interest makes it difficult to distinguish differentially expressed genes. For example, if we are interested in foreskin fibroblast, the inclusion of dermal fibroblast or other fibroblast cell types in the background would greatly reduce the number of differentially expressed genes. In contrast, adding cell types that are distantly related or of a different type introduces bias into the gene expression, which decreases the statistical power of the test. For instance, comparing primary cell types with tissues will not provide a clear indication of the differentially expressed genes. To ensure a moderate similarity between the samples in the foreground and the background, we used a tree-based background selection based on the FANTOM5 cell type ontology (Fig. 6a) [68]. A breaking point is chosen near the top of the cell ontology tree to exclude distantly related cell types and highly similar samples in the same clade as the cell type of interest are also removed.

Using the tree-based method to select the background, DESeq is performed on the FANTOM5 gene expression atlas, from which we obtain a log-fold change (LFC) in gene expression and a FDR for each gene in each cell type/tissue. We then define a nonnegative gene score $G_s(x)$ that indicates the importance of the gene in the cell type of interest:

$$G_s(x) = |L_s(x)| \times \left(-\log_{10} P_s(x)\right)$$

where $L_s(x)$ is the LFC in the gene expression of gene x in cell type/tissue s and $P_s(x)$ is the FDR for gene x in cell type/tissue s. The gene score reflects two characteristics for highly important genes (Fig. 6b): these genes have a high relative level of gene expression (given by a large LFC) and are specifically expressed in the cell type of interest (i.e., a small FDR). In the gene score formula, we chose to discard information regarding the directionality of LFC. This is because the role of a transcription factor can vary from gene to gene, increasing the expression of one while decreasing the other. Furthermore, the distribution of the number of cell types in the cell ontology tree is uneven, resulting in a nonuniform background in the differential gene expression analysis. Thus, the gene scores are not comparable across different cell types. Instead, the transcription factors will be compared based on the ranking of their gene scores and network regulatory scores.

We would like to highlight that previous approaches exist which aim to find genes that are highly expressed and specific in a particular cell type of interest. In particular, D'Alessio et al. employed an entropy-based approach using the Jensen-Shannon divergence [1].

3.2 Transcriptional Regulatory Influence of Transcription Factors

Ideally, we want to select transcription factors that not only have a high gene score but also regulate other genes that are important in determining cell identity. Thus, we calculate how well the cell-specific genes are regulated by each transcription factor by considering its local regulatory neighborhood using the STRING and MARA networks. The two interaction networks complement each other in that the STRING network provides the interactions between the genes' protein products while the MARA network describes directly the TF-DNA binding interactions.

For each network n, a score $N_s^n(x)$ is calculated for each transcription factor x, expressed as a weighted sum of gene scores:

$$N_s^n(x) = \sum_{v \in V^n(x)} G_s(v) \times \frac{1}{l^n(v)} \times \frac{1}{o^n(v)}$$

where gene v belongs to the set of nodes $V^n(x)$ that make up the local subnetwork of the transcription factor. The subnetwork extends up to three levels (or number of steps) from the transcription factor of interest. The level that gene v is away from x in network n is defined as $l^n(v)$ and the outdegree of the parent of gene v in network n as $o^n(v)$. Since the gene score indicates the importance of a gene in controlling cell identity, the network score

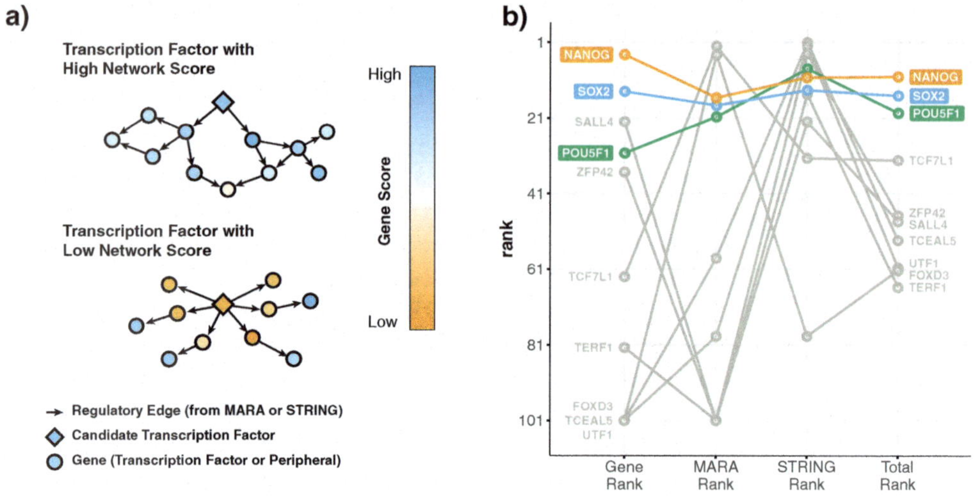

Fig. 7 (**a**) Transcription factors with a high regulatory influence or network score regulate genes with high gene scores, namely, the genes that are important to cell identity. (**b**) Distribution of the different ranks for the top 10 transcription factors in embryonic stem cells. An important transcription factor does not only possess a high gene score but also regulate other important genes via the MARA and STRING networks, as demonstrated by the pluripotency factors

represents the overall importance of the genes regulated by a given transcription factor adjusted for the network topology (Fig. 7a).

The topology of the network is accounted for in two aspects; first, we expect a gene to be less tightly regulated the further it is from the transcription factor of interest. Thus, we introduce a distance-based weighting where the gene influence decreases inversely with the number of steps away from the transcription factor. This also explains why we only need to consider the local neighborhood, which is defined as being up to three steps in the Mogrify algorithm. Second, we want to select transcription factors that regulate only the genes that are important in governing cell identity and not general transcription factors. To incorporate this regulatory specificity, we included a connectivity-based weighting which depends on the reciprocal of the outdegree of a gene's parent. Thus, the children (or interaction partners) of highly ubiquitous transcription factors are penalized, which in turn prevents the inflation of the network scores of these non-specific factors. It should be emphasized that the outdegree of the gene itself is not used since it is insufficient to prevent such factors from receiving an artificially high network score. It is likely that a gene which is two or three steps away from the transcription factor of interest can be regulated having multiple parents in the intermediate layers. For the Mogrify algorithm, we pick only the parent with the smallest outdegree since this represents a "path of least resistance." Overall, the formula of the network score reflects the diminishing influence with distance as well as regulatory specificity.

By comparing the values of $G_s(x)$, a gene rank is assigned, indicating the relative importance of the transcription factor in governing cell identity. Similarly, a MARA rank and STRING rank are obtained by comparing the respective network scores (Fig. 7b). Since we do not observe significant differences in the scores beyond the top 100 transcription factors, these factors all receive the same rank (rank = 101). A total rank is then calculated by summing all ranks taking into account of their gene expression and network regulatory influence. This total rank will be used to determine the order in which transcription factors are selected for cell reprogramming.

Transcriptional regulatory network information has also been incorporated into other programs that predict factors for cell reprogramming. For instance, the CellNet method constructs gene regulatory networks from a large number of microarray datasets, and cell-specific subnetworks are extracted in order to predict transcription factors for cell conversion [2].

3.3 Regulatory Coverage of Transcription Factors

Thus far, we have only considered the importance of transcription factors in a cell type of interest, i.e., the target cell type. However, a transcription factor that is important for the target cell type might already be sufficiently expressed in the source cell type. When predicting a conversion, we remove transcription factors in the target cell type ranked list which are already expressed (at >20 TPM) in the source cell type. Subsequently, we have a list of potential transcription factor candidates for cell fate conversion between the source and target cell type, ranked by their importance in controlling the target cell fate.

A unique feature of the Mogrify algorithm is that it avoids selecting transcription factors that are transcriptionally redundant. To this end, we define the combined coverage of a set of transcription factors T on the regulatory network of the target cell type s as:

$$C_s(T) = \frac{\left|\cap_{t \in T} \mathcal{g}_s(t)\right|}{\left|\mathcal{g}_{s,\text{all}}\right|}$$

where $\mathcal{g}_{s,\text{ all}}$ is the set of genes that are differentially expressed in the target cell type s (FDR < 0.05). The set of differentially expressed genes in cell type s that are directly regulated by transcription factor t belonging to the transcription factor set T is defined as $\mathcal{g}_s(t)$, and the vertical bars $|\cdot|$ denote the number of genes in the gene set. Thus, $\cap_{t \in T} \mathcal{g}_s(t)$ is the set of genes that are directly regulated by the transcription factor set T. Biologically, $\mathcal{g}_{s,\text{ all}}$ constitutes the gene regulatory network of the target cell s, and the combined coverage $C_s(T)$ is a proportion from 0 to 1 that indicates how well this network can be controlled by the transcription factor set T. Note that the regulatory edges are obtained by combining both the MARA and STRING networks.

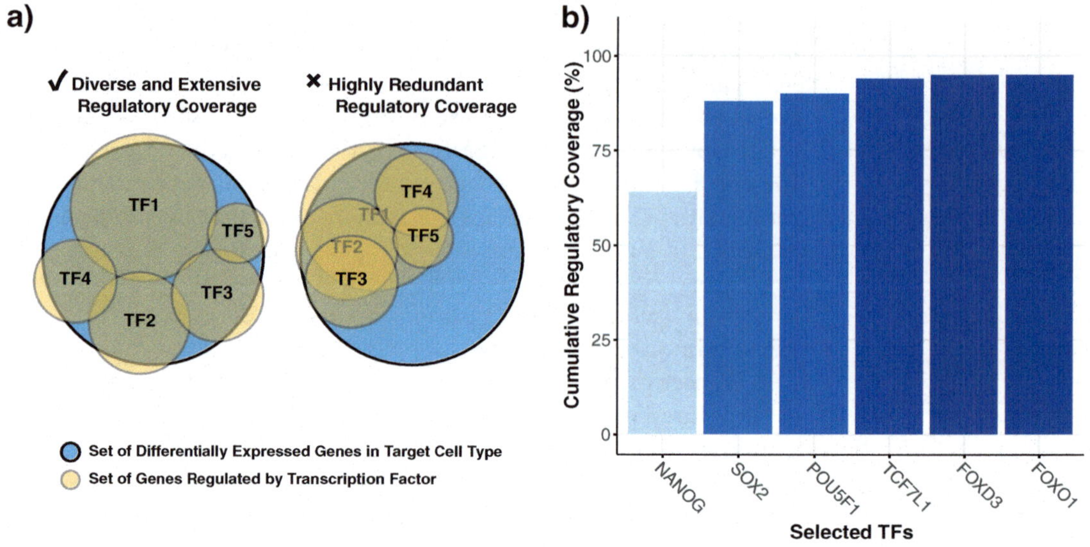

Fig. 8 (**a**) The Mogrify algorithm maximizes the regulatory coverage by removing transcription factors with redundant regulatory coverage. (**b**) Cumulative regulatory coverage of the top six selected nonredundant transcription factors in H9 embryonic stem cells. The coverage reaches 95% with the inclusion of the sixth transcription factor

Throughout the ranked list, a transcription factor is removed if there is a higher-ranking transcription factor that regulates over 98% of the same genes. This ensures that selected factors are diverse in their regulatory spheres of influence (Fig. 8a). The cutoff of 98% is chosen empirically. A lenient cutoff would result in too many selected transcription factors, many of which only marginally improve the combined coverage, while a strict cutoff makes it difficult to select factors. Finally, the combined coverage also provides an indirect measure of the efficacy of selected transcription factor set (Fig. 8b). We hypothesize that the fewer transcription factors required to reach 95% coverage, the more likely the conversion is to take place.

While our algorithm provides a deterministic way to select an ideal transcription factor set, we note that other combinations of factors may also lead to a high combined coverage. This explains why differing cell conversion protocols exist, as shown in Table 1, and the Mogrify algorithm allows users to compare the combined coverage of different transcription factor sets through the mogrify.net resource.

4 Mogrify.net Resource for Cell Reprogramming

We ran the Mogrify algorithm on the FANTOM5 gene expression atlas, and the resulting predictions are consolidated into a publicly

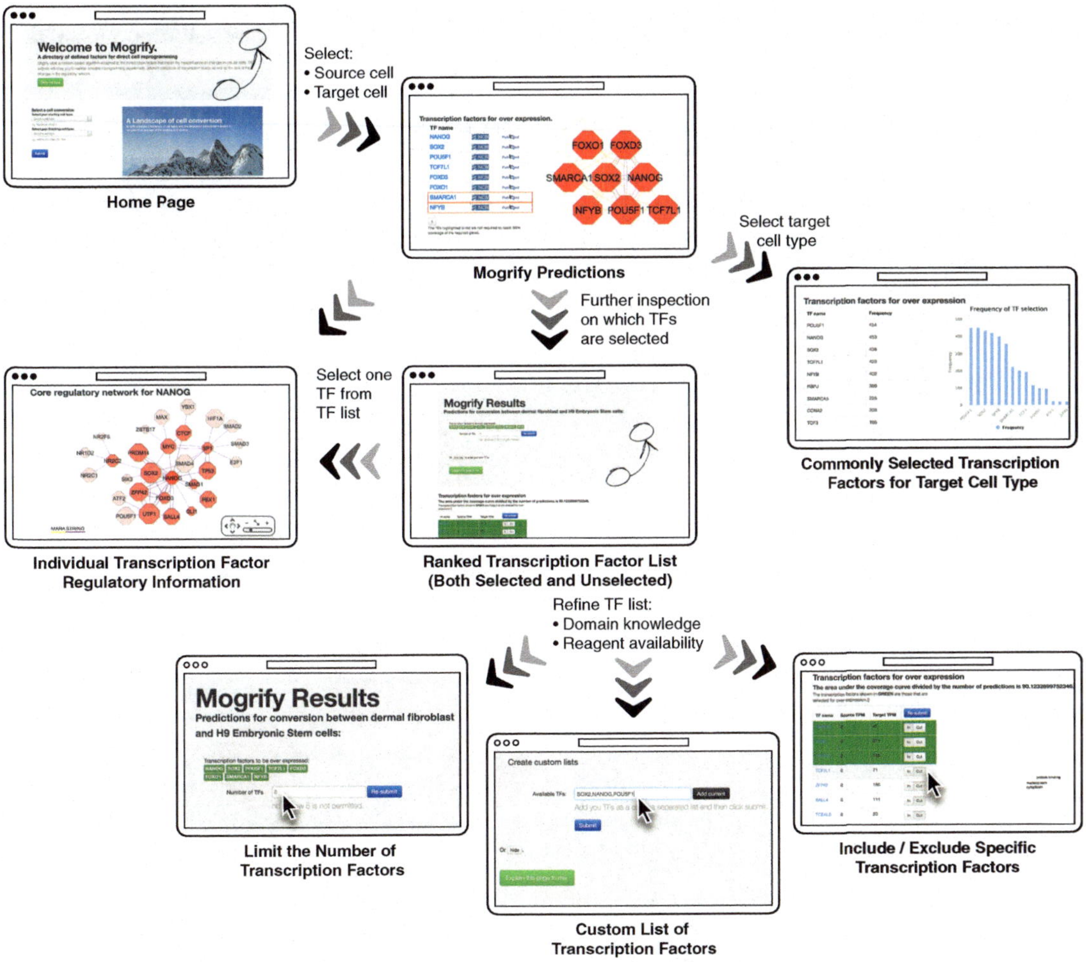

Fig. 9 Sitemap of mogrify.net. From the home page, users can select the source and target cell type which will bring them to the default Mogrify predictions (Subheading 4.1). From there, users can modify or refine the predicted list of transcription factors (Subheading 4.2). Users can also access transcription factor or cell type-specific information (Subheading 4.3)

available online resource: mogrify.net (Fig. 9). Users can access the set of predicted transcription factors to convert between cell types/tissues in the FANTOM5 dataset. The predicted set of factors can also be modified based on domain knowledge and/or reagent accessibility. Furthermore, mogrify.net provides regulatory information for individual transcription factors and the list of factors that are specific to a cell type/tissue of interest.

4.1 Default Mogrify Predictions

On the homepage, users can select the source and target cell type to retrieve the predictions for cell conversion. Furthermore, at the bottom of the homepage, there are links for the most widely studied conversion and the conversions from the Mogrify paper.

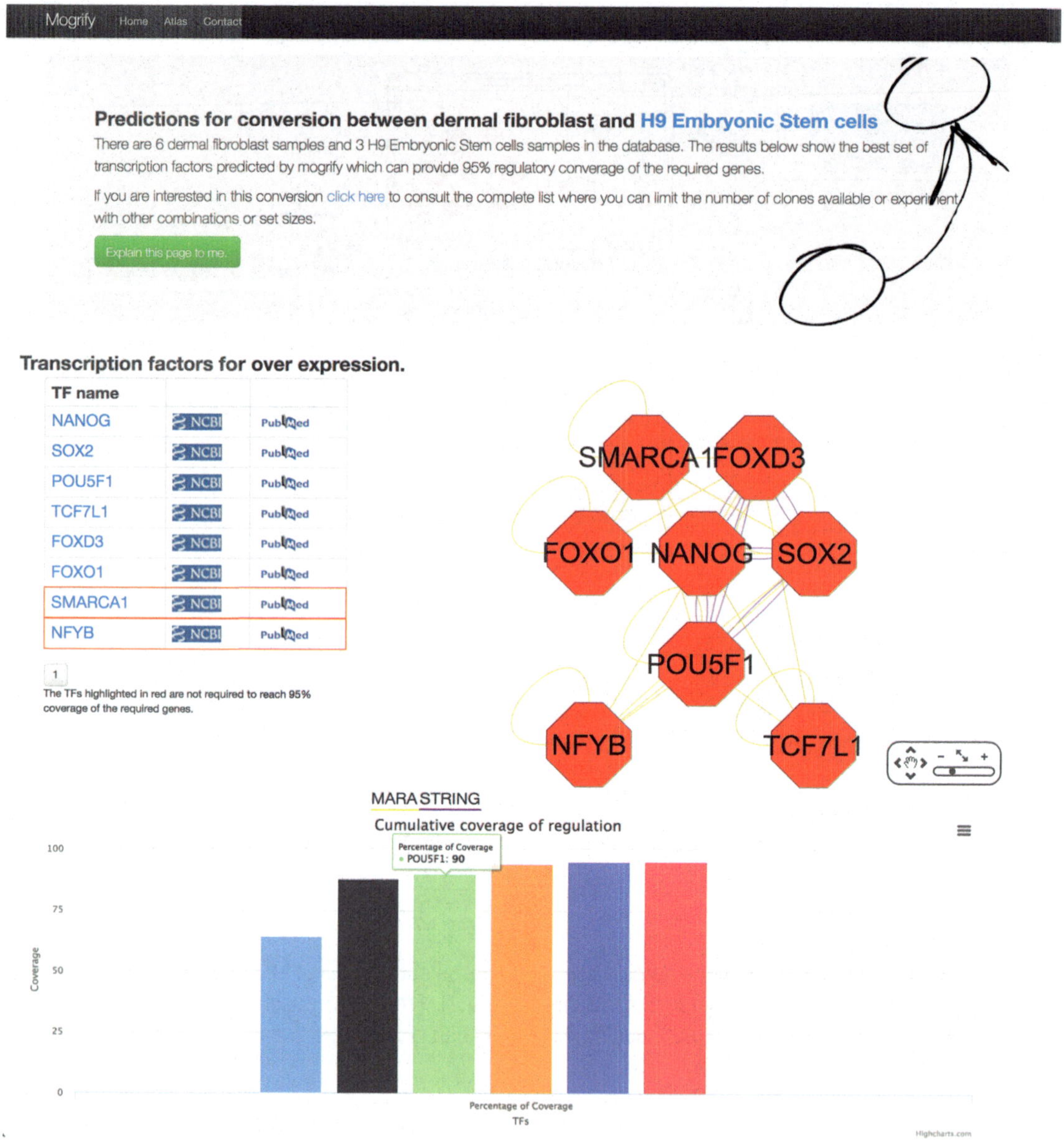

Fig. 10 The default Mogrify predictions for converting dermal fibroblast to embryonic stem cells. The list of predicted transcription factors and their interactions is given, together with the cumulative regulatory coverage of the selected transcription factors

After selecting the source and target cell type, the user will be brought to the default Mogrify predictions (Fig. 10). Here, the user will be presented with (1) the list of predicted transcription factors required to reach 95% coverage, (2) the MARA and STRING interactions between the predicted factors, and (3) their cumulative regulatory coverage. For example, for the conversion between dermal fibroblasts and H9 embryonic stem cells, Mogrify predicts six factors (NANOG, SOX2, POU5F1, TCF7L1, FOXD3, and FOXO1) in order to achieve 95% coverage.

As well as the predictions, the default Mogrify prediction page serves as a starting point to explore other interesting information regarding the cell conversion, which will be covered in the subsequent sections. *See* **Notes 2** and **3** for discussions of selecting starting and target cell types.

4.2 Modifying the Default Mogrify Results

The default list of transcription factors predicted by Mogrify is one of the many possible transcription factor sets which might induce cell reprogramming. Thus, mogrify.net has the option to allow users to refine the default Mogrify prediction (Fig. 11).

One straightforward alteration to the transcription factor set is to change the number of factors. By default, Mogrify gives the top eight predicted factors, to be overexpressed. However, it can be difficult to effectively transfect starting cells with a high number of transcription factors, so users can reduce the number of predicted factors. For example, in the conversion from dermal fibroblast to H9 embryonic stem cells, the factors FOXD3, FOXO1, and SMARCA1 are removed from the predictions when the number is decreased from eight to five (Fig. 11, orange circle). Users may also consider increasing the default number of factors to create a more comprehensive list of candidate factors for initial experimentation.

Another modification is to supply a custom list of transcription factors on which the Mogrify algorithm will be executed. This is useful when users only have reagents for certain factors and the available factors can be given as the custom list. For the same embryonic stem cells example, supplying a limited number of transcription factors yields a different prediction (Fig. 11, blue circle).

Furthermore, users can make adjustments to the Mogrify predictions by including or excluding certain factors. Users can click on the "in" or "out" buttons next to each factor to include or exclude the factor and then recalculate the Mogrify predictions by pressing the "resubmit" button. This allows users to refine the Mogrify predictions based on domain knowledge. For example, TCF7L1 is known to induce differentiation in embryonic stem cells [74]; thus one can exclude it and recompute an alternative set of predictions (Fig. 11, green circle).

Overall, we included several easy-to-operate ways to modify the predicted list of transcription factors based on domain knowledge or reagent availability. This is done in line with Mogrify's goal of simplifying and accelerating the discovery of new cell reprogramming protocols.

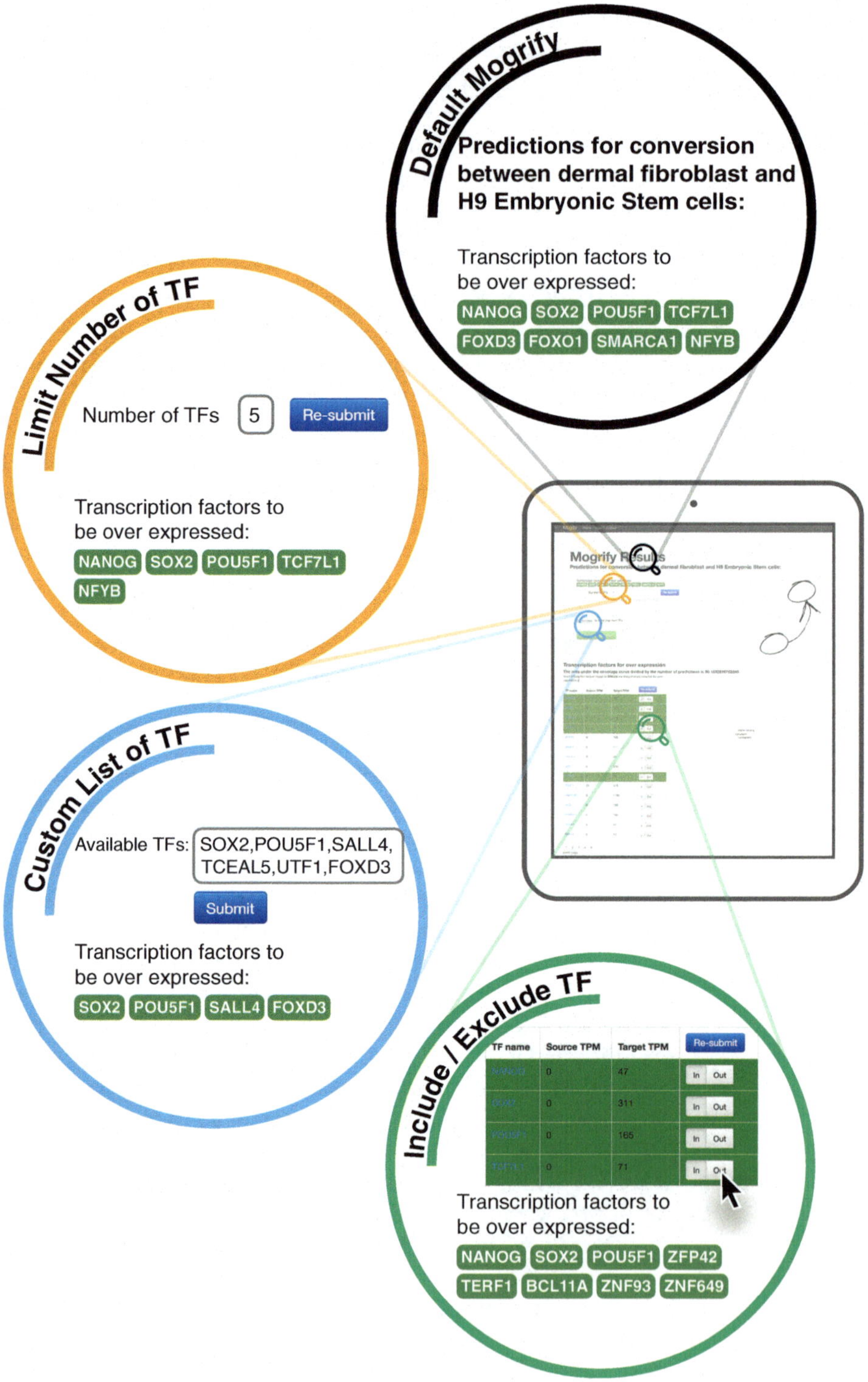

Fig. 11 Users can refine the default Mogrify predictions based on domain knowledge and reagent availability

4.3 Transcription Factor-Specific and Cell Type-Specific Information

From the default Mogrify predictions (Subheading 4.1) or the complete list of transcription factors (Subheading 4.2), users can click on any of the factors to examine their regulatory information. The core regulatory network is presented, which includes other factors that are regulated up to two steps away. The expression levels of the regulatory targets, averaged across replicates, are also tabulated in tags per million (TPM). For example, in embryonic stem cells, we observe that NANOG regulates SALL4 (126 TPM) and ZFP42 (215 TPM), both of which are also known to be factors that are important in embryonic stem cells [75, 76]. Finally, the expression levels of the selected factor across all FANTOM5 samples are plotted. Using the same example, we see that NANOG is highly expressed in the human-induced pluripotent stem cells (hIPS) and H9 embryonic stem cells.

From the default Mogrify prediction, users can also click on the name of the target cell type to access cell type-specific information. Mogrify tabulates the frequency of transcription factors being predicted as one of the top eight factors for each conversion. For example, POU5F1, NANOG, and SOX2 are the top three most frequently predicted factors for embryonic stem cells. This result is not surprising as these factors are well-studied key inducers of pluripotency [53]. *See* **Note 4** for a discussion of how to interpret cases in which the predictions do not reach high cell-type coverage and **Note 5** for a discussion of identifying alternate candidate sets of transcription factors.

5 Notes

From the mogrify.net web resource, users can easily retrieve and refine the predicted list of transcription factors between a large number of cell types and tissues. However, care has to be taken when accessing some predictions. This is largely due to the availability of gene expression data (number of replicates) and also the nature of the samples used to generate the gene expression data. Below are some considerations that should be made depending which cell conversion you are interested in.

1. The accuracy of the Mogrify predictions depends largely on the quality of the input gene expression data. In some instances, there can be very few replicates for a given cell type or tissue ($n = 1$ or $n = 2$). The low number of replicates makes it difficult for the DESeq tool to estimate the gene-wise dispersion and subsequently reduces the reliability of statistical test (Subheadings 1.3 and 3.1). This has an impact on the downstream regulatory influence calculation and ultimately undermines the accuracy of the predictions made by Mogrify. Thus, users should interpret these Mogrify predictions with caution

and could consider using other similar cell types. For example, if a user selects "dermal fibroblast" and "brain-adult" as the starting and finishing cell type, respectively, a warning on the number of replicates is displayed because the "brain-adult" tissue only has two replicates. Users can instead consider using other neuronal cell types with more than three replicates such as "human neuron samples" to obtain more accurate results.

2. The FANTOM5 gene expression atlas is very comprehensive, but there are certain aspects of the samples that should be kept in mind. One such consideration is the inclusion of tissues and the comparison with primary cells. Tissues are composed of a mixture of different cell types, and as a result, the gene expression profile is also heterogeneous. Primary cells are a single cell type, isolated or sorted from a tissue in some way. This can have an adverse effect on their gene expression profile. For instance, one conversion of interest is between cardiac fibroblasts and cardiac myocyte owing to its therapeutic possibilities. In mogrify.net there are two options to generate predictions for this conversion, using adult heart tissue or the primary cardiac myocyte data. The primary cardiac myocytes should provide a purer gene expression profile; however, the cells do not express NKX2.5, a canonical marker for mature cardiac myocyte. This indicates that the profiled cells may have been adversely affected by the sorting or extraction process. Instead, users can consider using the heart tissue or heart valve samples which is composed of cardiac myocyte together with other cell types such as cardiac fibroblasts, endothelial cells, and perivascular cells [77]. However, this may not always be the case, for instance, using the liver tissue to produce predictions for Kupffer cells (which only account for ~15% of liver population) [78] would most likely generate poor predictions.
3. In mogrify.net, some similar cell types are grouped together, for example, there is a grouped cell type named "fibroblast" comprising all fibroblast samples regardless of their source location in the body. It is recommended that users select the specific cell type or tissue that most closely resembles their starting or ending cell population of interest. However, if users are interested in the conversion from fibroblast in general to some other cell type, users can select the grouped cell type "fibroblast" instead. Furthermore, the grouped cell types are useful for understanding the transcriptional regulation that is common to that class of cells. Thus, users should pick the cell type based on the level of abstraction that they are interested in.
4. In order to select the set of transcription factors for reprogramming, Mogrify aims to achieve a combined coverage of 95% (Subheading 3.3). However, it can sometimes be the case that

no set of transcription factors is able to generate a coverage this high. For instance, the set of eight predicted factors for a conversion between "locus coeruleus-adult" and "parotid gland-adult" only has a combined coverage of 50%. This coverage is much lower than the predictions from most other conversions. For instance, the conversion from "dermal fibroblast" to "H9 embryonic stem cells" only requires six transcription factors to reach a coverage of 95%. If the conversion you are interested in also has a coverage problem, it might mean that the quality of the input gene expression data or regulatory network information is not of high enough quality and should be investigated in more detail prior to performing a reprogramming experiment.

5. As mentioned in Subheading 4.2, the default Mogrify predictions is one of many possible transcription factor sets that can induce cell conversion. In mogrify.net, we allow users to refine the list of transcription factors by (1) changing the number of factors, (2) supplying a custom list of factors, and (3) including or excluding certain transcription factors (Subheading 4.2). To get some indication of the efficiency of different transcription factor sets, users can compare their regulatory coverage. However, since many transcription factors have a similar predicted influence and are only included/excluded as a result of very small differences in gene expression or network coverage, there are likely a number of equally effective combinations. This phenomenon is also noticeable in the published factors in Table 1 where for the same conversion small (or sometimes drastic) changes in the transcription factor set can result in the same cell conversion. Nonetheless, Mogrify drastically reduces the number of combinations of transcription factors that need to be tested through the use of gene regulatory information.

References

1. D'Alessio AC, Fan ZP, Wert KJ, Baranov P, Cohen MA, Saini JS, Cohick E, Charniga C, Dadon D, Hannett NM et al (2015) A systematic approach to identify candidate transcription factors that control cell identity. Stem Cell Rep 5:763–775
2. Cahan P, Li H, Morris SA, Lummertz da Rocha E, Daley GQ, Collins JJ (2014) CellNet: network biology applied to stem cell engineering. Cell 158:903–915
3. Rackham OJL, Firas J, Fang H, Oates ME, Holmes ML, Knaupp AS, FANTOM Consortium, Suzuki H, Nefzger CM, Daub CO et al (2016) A predictive computational framework for direct reprogramming between human cell types. Nat Genet 48:331–335
4. Thomas DB (1971) The differentiation of transplanted haematopoietic cells derived from bone marrow, spleen and fetal liver. J Anat 110:297–306
5. Martin GR (1981) Isolation of a pluripotent cell line from early mouse embryos cultured in medium conditioned by teratocarcinoma stem cells. Proc Natl Acad Sci U S A 78:7634–7638
6. Briggs R, King TJ (1952) Transplantation of living nuclei from blastula cells into enucleated frogs' eggs. Proc Natl Acad Sci U S A 38:455–463
7. Campbell KH, McWhir J, Ritchie WA, Wilmut I (1996) Sheep cloned by nuclear transfer from a cultured cell line. Nature 380:64–66

8. Takahashi K, Yamanaka S (2006) Induction of pluripotent stem cells from mouse embryonic and adult fibroblast cultures by defined factors. Cell 126:663–676
9. Lassar AB, Paterson BM, Weintraub H (1986) Transfection of a DNA locus that mediates the conversion of 10T1/2 fibroblasts to myoblasts. Cell 47:649–656
10. Kajimura S, Seale P, Kubota K, Lunsford E, Frangioni JV, Gygi SP, Spiegelman BM (2009) Initiation of myoblast to brown fat switch by a PRDM16–C/EBP-β transcriptional complex. Nature 460:1154
11. Caiazzo M, Giannelli S, Valente P, Lignani G, Carissimo A, Sessa A, Colasante G, Bartolomeo R, Massimino L, Ferroni S et al (2015) Direct conversion of fibroblasts into functional astrocytes by defined transcription factors. Stem Cell Rep 4:25–36
12. Muraoka N, Yamakawa H, Miyamoto K, Sadahiro T, Umei T, Isomi M, Nakashima H, Akiyama M, Wada R, Inagawa K et al (2014) MiR-133 promotes cardiac reprogramming by directly repressing Snail and silencing fibroblast signatures. EMBO J 33:1565–1581
13. Islas JF, Liu Y, Weng K-C, Robertson MJ, Zhang S, Prejusa A, Harger J, Tikhomirova D, Chopra M, Iyer D et al (2012) Transcription factors ETS2 and MESP1 transdifferentiate human dermal fibroblasts into cardiac progenitors. Proc Natl Acad Sci U S A 109:13016–13021
14. Wada R, Muraoka N, Inagawa K, Yamakawa H, Miyamoto K, Sadahiro T, Umei T, Kaneda R, Suzuki T, Kamiya K et al (2013) Induction of human cardiomyocyte-like cells from fibroblasts by defined factors. Proc Natl Acad Sci U S A 110:12667–12672
15. Nam Y-J, Song K, Luo X, Daniel E, Lambeth K, West K, Hill JA, DiMaio JM, Baker LA, Bassel-Duby R et al (2013) Reprogramming of human fibroblasts toward a cardiac fate. Proc Natl Acad Sci U S A 110:5588–5593
16. Fu J-D, Stone NR, Liu L, Spencer CI, Qian L, Hayashi Y, Delgado-Olguin P, Ding S, Bruneau BG, Srivastava D (2013) Direct reprogramming of human fibroblasts toward a cardiomyocyte-like state. Stem Cell Rep 1:235–247
17. Ishii R, Kami D, Toyoda M, Makino H, Gojo S, Ishii T, Umezawa A (2012) Placenta to cartilage: direct conversion of human placenta to chondrocytes with transformation by defined factors. Mol Biol Cell 23:3511–3521
18. Liu X, Li F, Stubblefield EA, Blanchard B, Richards TL, Larson GA, He Y, Huang Q, Tan A-C, Zhang D et al (2012) Direct reprogramming of human fibroblasts into dopaminergic neuron-like cells. Cell Res 22:321–332
19. Pfisterer U, Kirkeby A, Torper O, Wood J, Nelander J, Dufour A, Björklund A, Lindvall O, Jakobsson J, Parmar M (2011) Direct conversion of human fibroblasts to dopaminergic neurons. Proc Natl Acad Sci U S A 108:10343–10348
20. Wong WT, Cooke JP (2016) Therapeutic transdifferentiation of human fibroblasts into endothelial cells using forced expression of lineage-specific transcription factors. J Tissue Eng 7:2041731416628329
21. Nouri M, Deezagi A, Ebrahimi M (2016) Reprogramming of human peripheral blood monocytes to erythroid lineage by blocking of the PU-1 gene expression. Ann Hematol 95:549–556
22. Ho S-M, Hartley BJ, Tcw J, Beaumont M, Stafford K, Slesinger PA, Brennand KJ (2016) Rapid Ngn2-induction of excitatory neurons from hiPSC-derived neural progenitor cells. Methods 101:113–124
23. Wamaitha SE, del Valle I, Cho LTY, Wei Y, Fogarty NME, Blakeley P, Sherwood RI, Ji H, Niakan KK (2015) Gata6 potently initiates reprograming of pluripotent and differentiated cells to extraembryonic endoderm stem cells. Genes Dev 29:1239–1255
24. Pang ZP, Yang N, Vierbuchen T, Ostermeier A, Fuentes DR, Yang TQ, Citri A, Sebastiano V, Marro S, Südhof TC et al (2011) Induction of human neuronal cells by defined transcription factors. Nature 476:220–223
25. Sandler VM, Lis R, Liu Y, Kedem A, James D, Elemento O, Butler JM, Scandura JM, Rafii S (2014) Reprogramming human endothelial cells to haematopoietic cells requires vascular induction. Nature 511:312–318
26. Szabo E, Rampalli S, Risueño RM, Schnerch A, Mitchell R, Fiebig-Comyn A, Levadoux-Martin M, Bhatia M (2010) Direct conversion of human fibroblasts to multilineage blood progenitors. Nature 468:521–526
27. Du Y, Wang J, Jia J, Song N, Xiang C, Xu J, Hou Z, Su X, Liu B, Jiang T et al (2014) Human hepatocytes with drug metabolic function induced from fibroblasts by lineage reprogramming. Cell Stem Cell 14:394–403
28. Wang XL, Hu P, Guo XR, Yan D, Yuan Y, Yan SR, Li DS (2014) Reprogramming human umbilical cord mesenchymal stromal cells to islet-like cells with the use of in vitro-synthesized pancreatic-duodenal homebox 1 messenger RNA. Cytotherapy 16:1519–1527

29. Rapino F, Robles EF, Richter-Larrea JA, Kallin EM, Martinez-Climent JA, Graf T (2013) C/EBPα induces highly efficient macrophage transdifferentiation of B lymphoma and leukemia cell lines and impairs their tumorigenicity. Cell Rep 3:1153–1163

30. Victor MB, Richner M, Hermanstyne TO, Ransdell JL, Sobieski C, Deng P-Y, Klyachko VA, Nerbonne JM, Yoo AS (2014) Generation of human striatal neurons by microRNA-dependent direct conversion of fibroblasts. Neuron 84:311–323

31. Yang R, Zheng Y, Li L, Liu S, Burrows M, Wei Z, Nace A, Herlyn M, Cui R, Guo W et al (2014) Direct conversion of mouse and human fibroblasts to functional melanocytes by defined factors. Nat Commun 5:5807

32. Park H-W, Cho J-S, Park C-K, Jung SJ, Park C-H, Lee S-J, Oh SB, Park Y-S, Chang M-S (2012) Directed induction of functional motor neuron-like cells from genetically engineered human mesenchymal stem cells. PLoS One 7: e35244

33. Lattanzi L, Salvatori G, Coletta M, Sonnino C, Cusella De Angelis MG, Gioglio L, Murry CE, Kelly R, Ferrari G, Molinaro M et al (1998) High efficiency myogenic conversion of human fibroblasts by adenoviral vector-mediated MyoD gene transfer. An alternative strategy for ex vivo gene therapy of primary myopathies. J Clin Invest 101:2119–2128

34. Hendry CE, Vanslambrouck JM, Ineson J, Suhaimi N, Takasato M, Rae F, Little MH (2013) Direct transcriptional reprogramming of adult cells to embryonic nephron progenitors. J Am Soc Nephrol 24:1424–1434

35. Kim YJ, Lim H, Li Z, Oh Y, Kovlyagina I, Choi IY, Dong X, Lee G (2014) Generation of multipotent induced neural crest by direct reprogramming of human postnatal fibroblasts with a single transcription factor. Cell Stem Cell 15:497–506

36. Lujan E, Chanda S, Ahlenius H, Südhof TC, Wernig M (2012) Direct conversion of mouse fibroblasts to self-renewing, tripotent neural precursor cells. Proc Natl Acad Sci U S A 109:2527–2532

37. Mitchell RR, Szabo E, Benoit YD, Case DT, Mechael R, Alamilla J, Lee JH, Fiebig-Comyn-A, Gillespie DC, Bhatia M (2014) Activation of neural cell fate programs toward direct conversion of adult human fibroblasts into tri-potent neural progenitors using OCT-4. Stem Cells Dev 23:1937–1946

38. Lee J-H, Mitchell RR, McNicol JD, Shapovalova Z, Laronde S, Tanasijevic B, Milsom C, Casado F, Fiebig-Comyn A, Collins TJ et al (2015) Single transcription factor conversion of human blood fate to NPCs with CNS and PNS developmental capacity. Cell Rep 11:1367–1376

39. Liao W, Huang N, Yu J, Jares A, Yang J, Zieve G, Avila C, Jiang X, Zhang X-B, Ma Y (2015) Direct conversion of cord blood CD34 + cells into neural stem cells by OCT4. Stem Cells Transl Med 4:755–763

40. Zhao P, Zhu T, Lu X, Zhu J, Li L (2015) Neurogenin 2 enhances the generation of patient-specific induced neuronal cells. Brain Res 1615:51–60

41. Guo Z, Zhang L, Wu Z, Chen Y, Wang F, Chen G (2014) In vivo direct reprogramming of reactive glial cells into functional neurons after brain injury and in an Alzheimer's disease model. Cell Stem Cell 14:188–202

42. Vierbuchen T, Ostermeier A, Pang ZP, Kokubu Y, Südhof TC, Wernig M (2010) Direct conversion of fibroblasts to functional neurons by defined factors. Nature 463:1035–1041

43. Ambasudhan R, Talantova M, Coleman R, Yuan X, Zhu S, Lipton SA, Ding S (2011) Direct reprogramming of adult human fibroblasts to functional neurons under defined conditions. Cell Stem Cell 9:113–118

44. Hsu Y-C, Chen S-L, Wang Y-J, Chen Y-H, Wang D-Y, Chen L, Chen C-H, Chen H-H, Chiu I-M (2014) Signaling adaptor protein SH2B1 enhances neurite outgrowth and accelerates the maturation of human induced neurons. Stem Cells Transl Med 3:713–722

45. Zhou C, Gu H, Fan R, Wang B, Lou J (2015) MicroRNA 302/367 cluster effectively facilitates direct reprogramming from human fibroblasts into functional neurons. Stem Cells Dev 24:2746–2755

46. Karow M, Sánchez R, Schichor C, Masserdotti G, Ortega F, Heinrich C, Gascón S, Khan MA, Lie DC, Dellavalle A et al (2012) Reprogramming of pericyte-derived cells of the adult human brain into induced neuronal cells. Cell Stem Cell 11:471–476

47. Yamamoto K, Kishida T, Sato Y, Nishioka K, Ejima A, Fujiwara H, Kubo T, Yamamoto T, Kanamura N, Mazda O (2015) Direct conversion of human fibroblasts into functional osteoblasts by defined factors. Proc Natl Acad Sci U S A 112:6152–6157

48. Sangan CB, Jover R, Heimberg H, Tosh D (2015) In vitro reprogramming of pancreatic alpha cells towards a beta cell phenotype following ectopic HNF4α expression. Mol Cell Endocrinol 399:50–59

49. Berneman-Zeitouni D, Molakandov K, Elgart M, Mor E, Fornoni A, Domínguez MR, Kerr-Conte J, Ott M, Meivar-Levy I, Ferber S (2014) The temporal and hierarchical control of transcription factors-induced liver to pancreas transdifferentiation. PLoS One 9: e87812
50. Mauda-Havakuk M, Litichever N, Chernichovski E, Nakar O, Winkler E, Mazkereth R, Orenstein A, Bar-Meir E, Ravassard P, Meivar-Levy I et al (2011) Ectopic PDX-1 expression directly reprograms human keratinocytes along pancreatic insulin-producing cells fate. PLoS One 6:e26298
51. Chiou S-H, Chen S-J, Chang Y-L, Chen Y-C, Li H-Y, Chen D-T, Wang H-H, Chang C-M, Chen Y-J, Ku H-H (2011) MafA promotes the reprogramming of placenta-derived multipotent stem cells into pancreatic islets-like and insulin+ cells. J Cell Mol Med 15:612–624
52. Takahashi K, Tanabe K, Ohnuki M, Narita M, Ichisaka T, Tomoda K, Yamanaka S (2007) Induction of pluripotent stem cells from adult human fibroblasts by defined factors. Cell 131:861–872
53. Yu J, Vodyanik MA, Smuga-Otto K, Antosiewicz-Bourget J, Frane JL, Tian S, Nie J, Jonsdottir GA, Ruotti V, Stewart R et al (2007) Induced pluripotent stem cell lines derived from human somatic cells. Science 318:1917–1920
54. Deng W, Cao X, Chen J, Zhang Z, Yu Q, Wang Y, Shao G, Zhou J, Gao X, Yu J et al (2015) MicroRNA replacing oncogenic Klf4 and c-Myc for generating iPS cells via cationized Pleurotus eryngii polysaccharide-based nanotransfection. ACS Appl Mater Interfaces 7:18957–18966
55. Kim JB, Greber B, Araúzo-Bravo MJ, Meyer J, Park KI, Zaehres H, Schöler HR (2009) Direct reprogramming of human neural stem cells by OCT4. Nature 461:649–643
56. Kim JB, Sebastiano V, Wu G, Araúzo-Bravo MJ, Sasse P, Gentile L, Ko K, Ruau D, Ehrich M, van den Boom D et al (2009) Oct4-induced pluripotency in adult neural stem cells. Cell 136:411–419
57. Zhang K, Liu G-H, Yi F, Montserrat N, Hishida T, Esteban CR, Izpisua Belmonte JC (2014) Direct conversion of human fibroblasts into retinal pigment epithelium-like cells by defined factors. Protein Cell 5:48–58
58. Li P, Sun X, Ma Z, Liu Y, Jin Y, Ge R, Hao L, Ma Y, Han S, Sun H et al (2016) Transcriptional reactivation of OTX2, RX1 and SIX3 during reprogramming contributes to the generation of RPE cells from human iPSCs. Int J Biol Sci 12:505–517
59. Son EY, Ichida JK, Wainger BJ, Toma JS, Rafuse VF, Woolf CJ, Eggan K (2011) Conversion of mouse and human fibroblasts into functional spinal motor neurons. Cell Stem Cell 9:205–218
60. Bredenkamp N, Ulyanchenko S, O'Neill KE, Manley NR, Vaidya HJ, Blackburn CC (2014) An organized and functional thymus generated from FOXN1-reprogrammed fibroblasts. Nat Cell Biol 16:902–908
61. Chen Y, Wang K, Gong YG, Khoo SK, Leach R (2013) Roles of CDX2 and EOMES in human induced trophoblast progenitor cells. Biochem Biophys Res Commun 431:197–202
62. Ginsberg M, James D, Ding B-S, Nolan D, Geng F, Butler JM, Schachterle W, Pulijaal VR, Mathew S, Chasen ST et al (2012) Efficient direct reprogramming of mature amniotic cells into endothelial cells by ETS factors and TGFβ suppression. Cell 151:559–575
63. Anders S (2010) Analysing RNA-Seq data with the DESeq package. Mol Biol. Available at: http://owl.fish.washington.edu/halfshell/bu-git-repos/LabDocs/code/DESeq/MANUAL_DESeq.pdf
64. Kulakovskiy IV, Medvedeva YA, Schaefer U, Kasianov AS, Vorontsov IE, Bajic VB, Makeev VJ (2013) HOCOMOCO: a comprehensive collection of human transcription factor binding sites models. Nucleic Acids Res 41: D195–D202
65. Yevshin I, Sharipov R, Valeev T, Kel A, Kolpakov F (2017) GTRD: a database of transcription factor binding sites identified by ChIP-seq experiments. Nucleic Acids Res 45:D61–D67
66. Chen X, Xu H, Yuan P, Fang F, Huss M, Vega VB, Wong E, Orlov YL, Zhang W, Jiang J et al (2008) Integration of external signaling pathways with the core transcriptional network in embryonic stem cells. Cell 133:1106–1117
67. Szklarczyk D, Morris JH, Cook H, Kuhn M, Wyder S, Simonovic M, Santos A, Doncheva NT, Roth A, Bork P et al (2017) The STRING database in 2017: quality-controlled protein--protein association networks, made broadly accessible. Nucleic Acids Res 45:D362–D368
68. FANTOM Consortium and the RIKEN PMI and CLST (DGT), Forrest ARR, Kawaji H, Rehli M, Baillie JK, de Hoon MJL, Haberle V, Lassmann T, Kulakovskiy IV, Lizio M et al (2014) A promoter-level mammalian expression atlas. Nature 507:462–470
69. Ashburner M, Ball CA, Blake JA, Botstein D, Butler H, Cherry JM, Davis AP, Dolinski K, Dwight SS, Eppig JT et al (2000) Gene ontology: tool for the unification of biology. Nat Genet 25:25

70. The Gene Ontology Consortium (2017) Expansion of the gene ontology knowledgebase and resources. Nucleic Acids Res 45: D331–D338
71. Wingender E, Schoeps T, Haubrock M, Dönitz J (2015) TFClass: a classification of human transcription factors and their rodent orthologs. Nucleic Acids Res 43:D97–D102
72. Fulton DL, Sundararajan S, Badis G, Hughes TR, Wasserman WW, Roach JC, Sladek R (2009) TFCat: the curated catalog of mouse and human transcription factors. Genome Biol 10:R29
73. Balwierz PJ, Pachkov M, Arnold P, Gruber AJ, Zavolan M, van Nimwegen E (2014) ISMARA: automated modeling of genomic signals as a democracy of regulatory motifs. Genome Res 24:869–884
74. Cole MF, Johnstone SE, Newman JJ, Kagey MH, Young RA (2008) Tcf3 is an integral component of the core regulatory circuitry of embryonic stem cells. Genes Dev 22:746–755
75. Yuri S, Fujimura S, Nimura K, Takeda N, Toyooka Y, Fujimura Y-I, Aburatani H, Ura K, Koseki H, Niwa H et al (2009) Sall4 is essential for stabilization, but not for pluripotency, of embryonic stem cells by repressing aberrant trophectoderm gene expression. Stem Cells 27:796–805
76. Masui S, Ohtsuka S, Yagi R, Takahashi K, Ko MSH, Niwa H (2008) Rex1/Zfp42 is dispensable for pluripotency in mouse ES cells. BMC Dev Biol 8:45
77. Xin M, Olson EN, Bassel-Duby R (2013) Mending broken hearts: cardiac development as a basis for adult heart regeneration and repair. Nat Rev Mol Cell Biol 14:529–541
78. Vollmar B, Menger MD (2009) The hepatic microcirculation: mechanistic contributions and therapeutic targets in liver injury and repair. Physiol Rev 89:1269–1339

Chapter 17

Computational Analysis of Altering Cell Fate

Hussein M. Abdallah and Domitilla Del Vecchio

Abstract

The notion of reprogramming cell fate is a direct challenge to the traditional view in developmental biology that a cell's phenotypic identity is sealed after undergoing differentiation. Direct experimental evidence, beginning with the somatic cell nuclear transfer experiments of the twentieth century and culminating in the more recent breakthroughs in transdifferentiation and induced pluripotent stem cell (iPSC) reprogramming, have rewritten the rules for what is possible with cell fate transformation. Research is ongoing in the manipulation of cell fate for basic research in disease modeling, drug discovery, and clinical therapeutics. In many of these cell fate reprogramming experiments, there is often little known about the genetic and molecular changes accompanying the reprogramming process. However, gene regulatory networks (GRNs) can in some cases be implicated in the switching of phenotypes, providing a starting point for understanding the dynamic changes that accompany a given cell fate reprogramming process. In this chapter, we present a framework for computationally analyzing cell fate changes by mathematically modeling these GRNs. We provide a user guide with several tutorials of a set of techniques from dynamical systems theory that can be used to probe the intrinsic properties of GRNs as well as study their responses to external perturbations.

Key words iPSC, Transdifferentiation, Cell fate, Dynamical systems, Gene regulatory network

1 Introduction

The probe into the theoretical question of how to reprogram cell fate can be aided by identification and understanding of the gene regulatory network(s) (GRNs) that are implied in the transformation between one phenotype and another. A common conceptualization of this problem is the Waddington view of cellular differentiation [1] in which a ball rolls down a hilly landscape with several different valleys representing the multiple paths that undifferentiated cells may take in acquiring their differentiated identity (Fig. 1). In an energetic landscape sense, differentiation is represented as a spontaneous "downhill" process associated with typical development during an organism's periods of growth and maturation. While it was a long-held consensus that this development was a permanent one-way process, experimental

Patrick Cahan (ed.), *Computational Stem Cell Biology: Methods and Protocols*, Methods in Molecular Biology, vol. 1975, https://doi.org/10.1007/978-1-4939-9224-9_17,

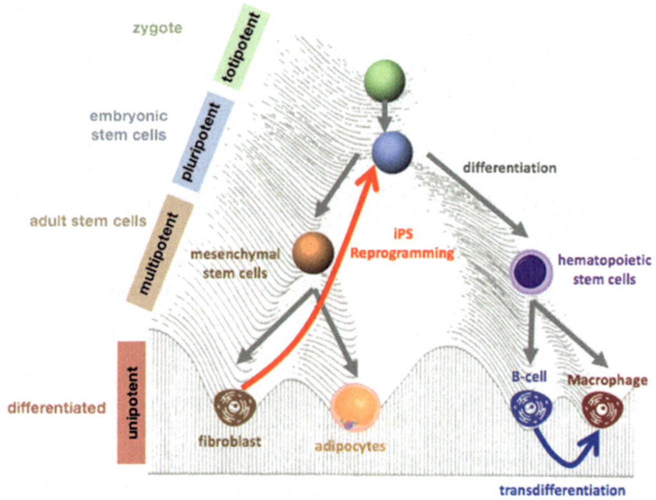

Fig. 1 Waddington landscape of cell differentiation. In this metaphor for differentiation, cell fate specification is akin to a marble rolling down a hill with several different valleys that represent the diverse fates that cells with the same genetic encoding ultimately adopt. Along the way to differentiation, these cells are often lumped under the umbrella term of "stem cells," though they have different potencies depending on their degree of differentiation. Zygotes and embryonic stem cells (ESCs) in the first few divisions after fertilization are totipotent and can give rise to an entire organism including the placenta and umbilical cord. When ESCs become pluripotent, they can continue to give rise to all three germ layers and thus an entire organism. Multipotent cells are adult stem cells that retain some potential for further differentiation, which is typically limited to a particular tissue type (e.g., hematopoietic cells, neural stem cells, mesenchymal stem cells) [2]. iPSC reprogramming as depicted is tantamount to an "uphill" movement from a somatic state (e.g., fibroblast) to the pluripotent state. Transdifferentiation as depicted is tantamount to directly transforming from one lineage to another (e.g., B-cell to macrophage, as shown in Xie et al. [3])

manipulations of cell fate during the second half of the twentieth century (such as somatic cell nuclear transfer [4–7] and transdifferentiation [8, 9]) began chipping away at this notion. In 2006, it was shattered entirely when Yamanaka and colleagues discovered that mouse fibroblasts could be reverted to what was coined an induced pluripotent stem cell (iPSC) state [10], a transition tantamount to an "uphill" climb in the Waddington landscape. Remarkably, this transformation was accomplished using the mere overexpression of a small cocktail of transcription factors (TFs): Oct4, Sox2, Klf4, and c-Myc (also known as "OSKM factors"). It is understood that in successfully reprogrammed cells, the researchers had transformed the phenotype from fibroblast to iPSC by shifting the initial cell state, characterized by low levels of OSKM, to the iPSC state, characterized by higher levels of these factors [11].

While Yamanaka's work and the variations that have followed [12] have been remarkable achievements in their own right, much remains to be answered about the biochemical and biomolecular changes that accompany many reprogramming processes [13–16]. Chief among these questions are those surrounding the low efficiencies, variable quality, and high latency of iPSC formation [17–19] that persist in the laboratory today. Yamanka himself, when describing the first clinical trial with iPSCs in 2016, remarked "what we learned from the first patient, in which we performed autologous iPS cell transplantation, is that the entire process is too expensive. It also took almost a year to make iPS cells from the patient's own cells, to transfer the original cells, and to perform all the rigorous quality control tests. It was too expensive, and it took long" [20]. Even when other cell fate transitions occur with greater efficiency in the laboratory, such as in certain transdifferentiation protocols [3, 21–25], many open questions surrounding the biomolecular changes that take place during the reprogramming remain.

As a result, iPSC reprogramming, and more generally cell fate reprogramming, is an active area of research. From a theoretical standpoint, understanding the complex dynamics associated with the GRNs involved in cell fate determination can be a promising approach to provide insight into current bottlenecks and suggest novel approaches. The task of creating mathematical models of GRN dynamics is today rendered possible by the vast amounts of genomic and molecular data available in the post-genomic era [26]. In particular, for iPSC reprogramming, a key GRN was discovered by Boyer and colleagues in 2005 [27]. This GRN, which is considered to be the core pluripotency network, consists of a "fully connected triad" of mutually activating transcription factors, Oct4, Sox2, and Nanog. Activating this network has been implicated as a signature of achieving pluripotency starting from differentiated cells, and the network itself is known to be part of an extended regulation network involving other TFs, various signaling components, differentiation genes, and chromatin remodelers [27–30]. Other well-studied cell fate reprogramming pathways include transdifferentiation pathways between different phenotypes in the hematopoietic stem cell (HSC) lineage. For instance, in the transdifferentiation between the megakaryocyte lineage and the granulocyte/macrophage lineage, a core GRN involving the transcription factors PU.1 and GATA-1 has been implicated, supporting the fact that the two lineages are characterized by high GATA-1, low PU.1 levels and high PU.1, low GATA-1 levels, respectively [31–33].

When sufficient data about a cell fate reprogramming process exist, mathematical models of the GRN(s) in question may be constructed and thus used as a representation of the reprogramming process itself. The utility of such models lies in the extent to which they can be used to explain or predict experimental outcomes, shed light on biomolecular mechanisms taking place during the reprogramming process, or suggest new informative experiments. There is a plethora of models that have been devised for various cell fate reprogramming contexts. Given their importance at the top of the developmental hierarchy, several models have been constructed for the various TFs involved in establishing mammalian pluripotency, including those composing the core network of Oct4, Nanog, and Sox2 (the "triad") discovered by Boyer and colleagues in 2005. Shortly after Boyer's discovery, a deterministic model of the triad was proposed, which demonstrated bistability [34]. According to that work, when the triad is expressed (the pluripotent state or "on" state), self-renewal genes that establish pluripotency were also shown to be expressed, while differentiation genes were off and vice versa when the triad was not expressed. The model also predicted that if input to the system turned on the switch, the network could persist in the "on" state without sustained input if the binding strength of Nanog to Oct4 and Sox2 was increased or the constitutive transcription level of Nanog was increased. A deterministic model of the triad along with two other TFs, Cdx2 and Gata6, was then proposed [30]. That model showed that antagonisms of Cdx2 and Gata6 with Oct4 and Nanog, respectively, are involved in controlling the murine pluripotent state's differentiation into the trophectoderm and endoderm lineages, respectively. In addition, various reprogramming paths from these differentiated lineages back to the pluripotent state were explored using this model. We have also previously used a mathematical model for the triad to explore questions surrounding the outcome of reprogramming experiments and how to explain their prevalent failures [35, 36]. Olariu and colleagues [37] have studied the impact of demethylation on the iPSC reprogramming process, constructing a deterministic model of Oct4, Nanog, and Tet, which is a protein that demethylates the Oct4 promoter before it can be reactivated. This deterministic model was combined with a stochastic description of methylation and demethylation at the promoters of Oct4 and Nanog to form a multilayered model that underscored the importance of demethylation in the iPSC reprogramming process.

Several mathematical models have also been published for the well-studied network involving the TFs GATA-1 and PU.1, which govern the decision of the multipotent common myeloid precursor (CMP) to differentiate into either the erythroid or myeloid lineages. Huang and colleagues [38] used a differential equation model to mathematically describe the differentiation of a "metastable" CMP state into either a stable erythroid or myeloid lineage by way of stochastic fluctuations or directed deterministic perturbations. At a mathematical level, the bistability underlying this model hinged upon an assumption that the functions describing the interactions between the two TFs were sigmoidal. However, high levels of TF cooperativity, which is typically what gives these functions sigmoidal shape [39], were not consistent with experimental results on how PU.1 and GATA-1 were interacting [40]. Chickarmane and colleagues [30] pointed this out and suggested the need for a potentially undiscovered cofactor mechanism to establish the experimentally observed bistability. In [41], a stochastic model of the GATA-1/PU.1 GRN with the TF GATA-2 was constructed, which showed the existence of three states that correspond to three different blood cell phenotypes that could be reached stochastically from the same starting conditions (as is the case in real-life development). Another example includes the work by Zhou and colleagues [42], which proposes a differential equation model of ten essential genes involved in pancreatic cell differentiation that predicts which artificial perturbations are needed to induce transitions among the various pancreatic cell types.

The purpose of this chapter is to illustrate in detail how practitioners of cell fate reprogramming (including iPSC reprogramming and transdifferentiation) can mathematically model their particular process in question if there is already a known GRN implied in the transformation between the phenotypes of interest. In Subheading 2.1, we discuss how to use such GRNs to model the resulting dynamics of species in the network using ordinary differential equation (ODE) models (Fig. 2, top panel). The validity of such modeling techniques rests on some key assumptions which will be explained in detail as a prerequisite to the approaches applied.

In Subheading 2.2, we introduce the notion of reprogramming in the state-space domain, while in Subheading 2.3 we introduce analysis tools from the fields of dynamical systems [43] that can be used to ***understand the intrinsic properties and behavior of the GRN*** (Fig. 2, middle panel). By moving the reprogramming problem to the state-space domain, it becomes possible to characterize the admissible steady state landscapes of the system, understand how robust these are to parameter changes, and determine which state transitions are possible under the influence of external perturbations. Most importantly, the comparison between the outcome of an experiment and the model prediction using these tools can often reveal biological mechanisms that are necessary to explain the data. Finally, this workflow will culminate in Subheading 2.4, where we provide a discussion of the analysis of how a GRN responds to artificial perturbations, such as those imposed by reprogramming practices (Fig. 2, bottom panel). Questions that can be addressed at this level of analysis include the following: How does the steady state landscape reconfigure after artificial perturbations? What interstate transitions can or cannot be triggered with external

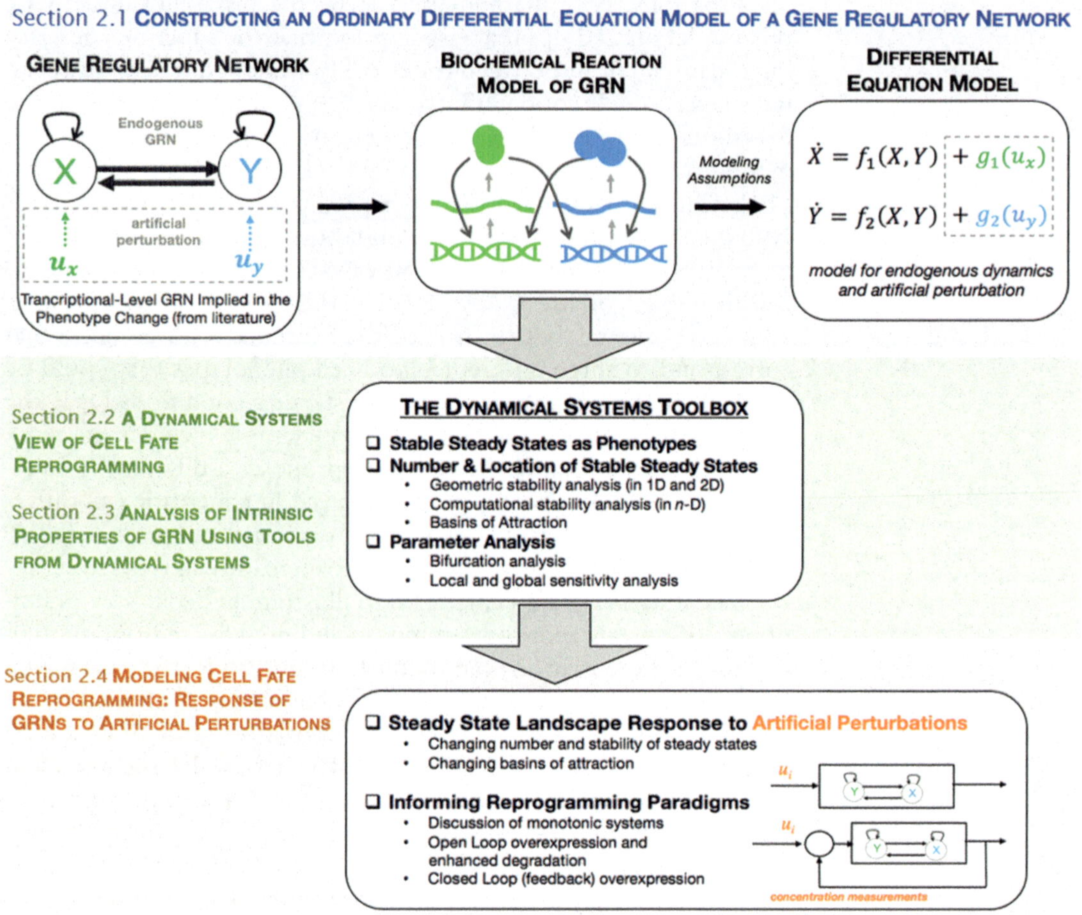

Fig. 2 The dynamical systems approach to cell fate reprogramming. (Top panel) If a known GRN is implied during a phenotype change in a cell fate transformation experiment, a biochemical reaction model of the network can be constructed. Under certain assumptions about this biochemical reaction model that are detailed in Subheading 2.1, ODE models can be constructed that describe the endogenous dynamics of the system and also incorporate the effect of artificial perturbations. (Middle panel) Once an ODE model is constructed, the dynamical systems toolbox can be applied toward the goal of understanding the intrinsic properties of the GRN in question. Notably, these tools seek to understand the number and location of stable steady states implied by the mathematical model, including an assessment of how these depend on the model's parameters. (Bottom panel) The response of the GRN to artificial perturbations is the ultimate goal of formulating the reprogramming problem into these mathematical terms. By understanding how the number, location, and extent of basins of attraction surrounding the stable steady states will change in response to artificial perturbations such as ectopic overexpression, practitioners can understand the advantages and limitations of different reprogramming paradigms, which may be used to inform experiments

perturbations? And, what type of interventions are required to trigger a particular transition?

2 Methods

2.1 Constructing an Ordinary Differential Equation Model of a Gene Regulatory Network

As introduced above, viewing the problem of cell fate reprogramming through the lens of a mathematical model can lend many useful insights into the biology of a given reprogramming practice and potentially inform future experiments. In this section, we show by way of the example in Fig. 2 how to construct an ODE model for a GRN. ODE models provide an appropriate starting point to gauge a deeper understanding of the GRN dynamics, and hence we focus this chapter on them. However, ODE models are suitable only when the molecular counts of species are sufficiently high such that the system can be described in terms of concentrations (*see* [44, 45] for a detailed overview of assumptions behind ODE models of biomolecular systems). In such situations, the ODE models provide an accurate description of the mean behavior of the system dynamics. When one is interested in stochastic properties, such as determining the response of a system to noise, or the system is characterized by low molecular counts, descriptions other than ODEs should be considered. The Chemical Langevin Equation (CLE) is a stochastic differential equation that describes the system dynamics by adding a noise term to the ODE [46]. Just like ODEs, CLEs can be used only when concentration is a meaningful way of describing the system. When this is not the case and molecular counts are very low, the Chemical Master Equation (CME) is the appropriate description [47]. In these cases, the behavior of an ODE model can be dramatically different from the mean behavior of the stochastic model [48].

There are many different ODE models that one could write for a GRN of interest [49]. In our representation, we describe a network involving n TFs, or n "nodes," using n ODEs, with each equation describing the rate of change of the concentration of a single TF. If $\mathbf{X} = (\mathrm{X}_1, \mathrm{X}_2, \ldots, \mathrm{X}_n)$ represents the n nodes in a GRN with concentrations $\boldsymbol{X} = (X_1, X_2, \ldots, X_n)$, respectively, we write an ODE model for their concentrations as follows:

$$\frac{dX_i}{dt} = H_i(\boldsymbol{X}) - \gamma_i X_i,$$

where $H_i(\boldsymbol{X})$ is the production rate of TF X_i that lumps together the effects of all activators and repressors of gene of X_i and γ_i is the decay rate of the TF, which encompasses both active degradation and dilution [44]. We will derive $H_i(\boldsymbol{X})(i = 1, 2)$ for the simple GRN network with nodes X and Y in Fig. 2 using a set of molecular interactions between these TFs themselves, the genes that encode for them, and the mRNA transcripts from which they are translated (we use the nodes X and Y rather than X_1 and X_2 to simplify notation). Though we decide upon these interactions and declare them to be true for illustrative purposes, in general these will have to be determined using a literature search surrounding the particular GRN being studied: different TFs have different binding behaviors to their gene targets, and this is one of the factors that affect the final functional form of $H_i(\boldsymbol{X})$. Moreover, there is also freedom in deciding which molecular species to include in a representation of the interactions based on the degree of detail one is interested in capturing. We begin describing these interactions for what will be our example in Fig. 2 using a biochemical reaction model.

2.1.1 From Literature Data to Biochemical Reaction Model

Formation of Transcription Factor Multimers ("Multimerization")

It is common for TFs in a GRN to act as multimers, or proteins consisting of multiple monomers. In general, a literature search should be conducted to determine the most probable multimerization state of each TF in the network under study. As will be shown, whether a TF is acting as a monomer, dimer, trimer, etc. will be captured by a key parameter during ODE construction. In our running example, for illustrative purposes, we treat TFs X and Y as acting in their homodimerized forms, X_d and Y_d, respectively. The biochemical reactions corresponding to these reversible dimerizations are shown in Table 1a.

Transcriptional Level Activation

In the basic modeling formalism presented here, we assume that regulatory interactions, such as the activations between TFs X and Y in Fig. 2, occur at the transcriptional level. Other regulatory mechanisms are possible, such as regulation of translation rate through microRNAs [50] and regulation of protein activity through covalent modification [51]. Here, for simplicity, we only consider transcriptional regulation since it is typically better characterized experimentally. In particular, we treat the arrow going from node X to node Y (node Y to node X) in the GRN as representing protein product X (Y) activating the promoter of gene Y (X). Likewise, the "autoregulating" arrow going from node X (Y) to X (Y) itself represents the TF X (Y) activating its own promoter.

There are a number of ways two TFs can co-bind to a common promoter [52]:

Independent Binding: the two TFs bind to the promoter on separate sites and do not affect each other's binding/unbinding.

Cooperative Binding: the second TF only binds to the promoter when the first TF is already bound or vice versa.

Competitive Binding: the two TFs compete for the same binding site, which they mutually exclusively occupy. Hence, if one TF is bound to the promoter, the other TF cannot bind.

Table 1
Dimerization, DNA-binding, transcription reactions, translation, decay, and ectopic overexpression reactions

Reaction Type	Biochemical Reaction Model Representation

A. Dimerization

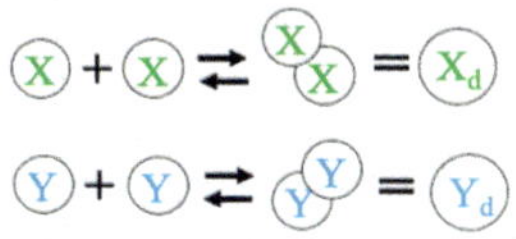

$$X + X \underset{d_x}{\overset{a_x}{\rightleftarrows}} X_d$$

Represents TF X reversibly forming a homo-dimer, X_d. Units of a_X: $(\text{concentration} \cdot \text{time})^{-1}$, d_X: $(\text{time})^{-1}$

$$Y + Y \underset{d_y}{\overset{a_y}{\rightleftarrows}} Y_d$$

Represents Y reversibly forming a homo-dimer, Y_d. Units of a_Y: $(\text{concentration} \cdot \text{time})^{-1}$, d_Y: $(\text{time})^{-1}$

Reaction Type	Biochemical Reaction Model Representation

B. Reversible DNA Binding

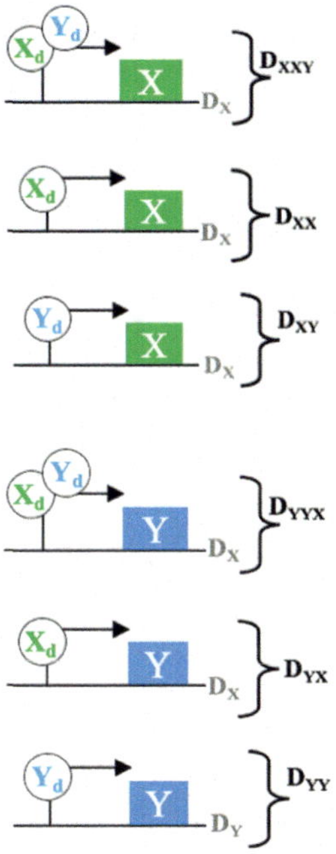

INDEPENDENT ACTIVATION

AUTO-ACTIVATION

$$D_X + X_d \underset{d_1}{\overset{a_1}{\rightleftarrows}} D_{XX}$$

Represents homo-dimer X_d reversibly binding its own free promoter, D_X, to form an activated promoter D_{XX}.

$K_1 = d_1/a_1$

$$D_Y + Y_d \underset{d_4}{\overset{a_4}{\rightleftarrows}} D_{YY}$$

Represents homo-dimer Y_d reversibly binding to the promoter of Y, D_Y, to form activated promoter D_{YY}.

$K_4 = d_4/a_4$

MUTUAL-ACTIVATION

$$D_X + Y_d \underset{d_2}{\overset{a_2}{\rightleftarrows}} D_{XY}$$

Represents homo-dimer Y_d reversibly binding the free promoter of X, D_X, to form an activated promoter D_{XY}.

$K_2 = d_2/a_2$

$$D_Y + X_d \underset{d_3}{\overset{a_3}{\rightleftarrows}} D_{YX}$$

Represents homo-dimer X_d reversibly binding to the promoter of Y, D_Y, to form activated promoter D_{YX}.

$K_3 = d_3/a_1$

CO-ACTIVATION

$$D_{XX} + Y_d \underset{d_5}{\overset{a_5}{\rightleftarrows}} D_{XXY} \qquad D_{XY} + X_d \underset{d_6}{\overset{a_6}{\rightleftarrows}} D_{XXY}$$

(left) Represents homo-dimer Y_d reversibly binding the promoter of X when X_d is already bound (i.e. D_{XX}) to form an independently co-activated promoter D_{XXY}. **(right)** Represents homo-dimer X_d reversibly binding the promoter of X when Y_d is already bound (i.e. D_{XY}) to form an independently co-activated promoter D_{XXY}.

$K_5 = d_5/a_5$, $K_6 = d_6/a_6$

$$D_{YX} + Y_d \underset{d_7}{\overset{a_7}{\rightleftarrows}} D_{YYX} \qquad D_{YY} + X_d \underset{d_8}{\overset{a_8}{\rightleftarrows}} D_{YYX}$$

(left) Represents homo-dimer Y_d reversibly binding the promoter of Y when X_d is already bound (i.e. D_{YX}) to form an independently co-activated promoter D_{YXX}. **(right)** Represents homo-dimer X_d reversibly binding the promoter of Y when Y_d is already bound (i.e. D_{YY}) to form an independently co-activated promoter D_{YYX}.

$K_7 = d_7/a_7$, $K_8 = d_8/a_8$

(continued)

Table 1
(continued)

Reaction Type	Biochemical Reaction Model Representation

C. Transcription

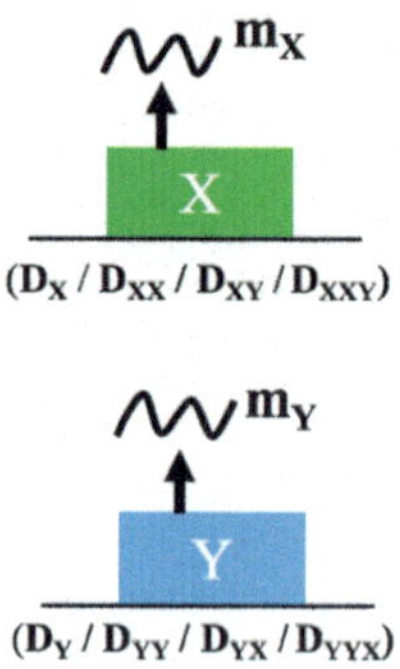

CONSTITUTIVE TRANSCRIPTION	ACTIVATED TRANSCRIPTION	
$D_X \xrightarrow{\alpha_x} m_X + D_X$ $D_Y \xrightarrow{\alpha_y} m_Y + D_Y$	$D_{XX} \xrightarrow{\alpha_x^1} m_X + D_{XX}$ $D_{XY} \xrightarrow{\alpha_x^2} m_X + D_{XY}$ $D_{XXY} \xrightarrow{\alpha_x^3} m_X + D_{XXY}$	$D_{YY} \xrightarrow{\alpha_y^1} m_Y + D_{YY}$ $D_{YX} \xrightarrow{\alpha_y^2} m_Y + D_{YX}$ $D_{YYX} \xrightarrow{\alpha_y^3} m_Y + D_{YYX}$
Transcription of genes X and Y without activation, i.e. "leaky" transcription.	Represents transcription of activated genes of X to form species m_x.	Represents transcription of activated genes of Y to form species m_y.

for activated DNA: $\alpha_x^i, \alpha_y^i > \alpha_{xo}, \alpha_{yo}$,
for repressed DNA: $\alpha_x^i, \alpha_y^i < \alpha_{xo}, \alpha_{yo}$ ($i \in \{1,4\}$).

Units for all rate constants: (unit time)$^{-1}$

Reaction Type	Biochemical Reactions

D. Translation

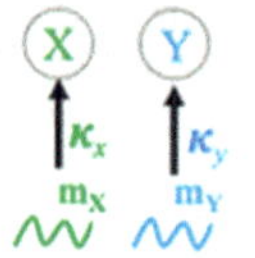

$$m_X \xrightarrow{\kappa_x} m_X + X \qquad m_Y \xrightarrow{\kappa_y} m_Y + Y$$

Represents translation of mRNA transcript of gene X into the protein product X. Represents translation of mRNA transcript of gene Y into the protein product Y.

E. Decay

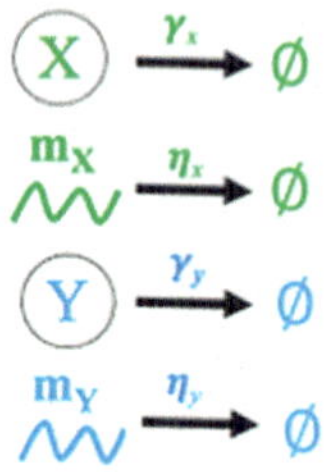

DEGRADATION	DILUTION
$X \xrightarrow{\delta_x} \phi \quad Y \xrightarrow{\delta_y} \phi$ $m_X \xrightarrow{\eta_x} \phi \quad m_Y \xrightarrow{\eta_y} \phi$	$X \xrightarrow{\beta} \phi \quad Y \xrightarrow{\beta} \phi$
Represents the degradation of TFs X and Y and mRNA transcripts m_X and m_Y.	Represents the dilution of TFs X and Y due to cell-division.

In total, the decay rates γ_x and γ_y of TFs X and Y, respectively, are equal to the sum of degradation and dilution rates for each TF: $\gamma_x = \delta_x + \beta, \gamma_y = \delta_y + \beta$.

F. Ectopic Overexpression

$$\phi \xrightarrow{u_x} m_X \qquad \phi \xrightarrow{u_y} m_Y$$

Represents the additional mRNA species m_X and m_Y formed via ectopic overexpression at rates u_x and u_y mRNA species / unit time.

In general, the mode(s) of binding for each TF in a network will have to be determined from the literature. In our model, we treat TFs X and Y as acting independently, which means they can occupy their targets alone or together. These configurations are captured by the specific association and dissociation rate constants in Table 1b, where we have introduced free DNA promoters D_X and D_Y for genes X and Y, respectively. Based on the binding modes shown in Table 1, the promoters of gene X and gene Y can exist as:

1. D_X: this species represents the unbound promoter of gene X.
2. D_{XX}: this species represents D_X bound by the homodimer X_d alone.
3. D_{XY}: this species represents D_X bound by the homodimer Y_d alone.
4. D_{XXY}: this species represents D_X bound by the homodimers X_d and Y_d.
5. D_Y: this species represents the unbound promoter of gene Y.
6. D_{YY}: this species represents D_Y bound by the homodimer Y_d alone.
7. D_{YX}: this species represents D_Y bound by the homodimer X_d alone.
8. D_{YYX}: this species represents D_Y bound by the homodimers X_d and Y_d.

There are a variety of experimental strategies used for the discovery of DNA binding sites for a TF of interest [53]. Some common techniques for genome-wide binding site discovery are ChIP-on-Chip [54] and ChIP-Seq [55]. In ChIP-on-Chip, the TFs of interest are cross-linked to chromatin that is then precipitated using antibodies specific for each TF, hence revealing their binding site to a particular DNA sequence. When done in a high-throughput manner using microarray chips, this is a practical way to reveal binding sites at the genomic scale. In ChIP-Seq, TFs are similarly cross-linked and precipitated, and then the precipitated samples are run through high-throughput sequencing where peak-finding software tools are used to identify the presumed TF binding sites.

Once it is determined that a protein is acting as a TF, elucidating the direction of transcriptional regulation is commonly done by measuring the fold change of transcription in separate experiments in which the TF in question is knocked out or overexpressed. Once again, this can be done in a high-throughput manner using microarray chips or RNA sequencing [56] to determine transcription levels. In addition, experimenters may sometimes find the same TF acting in a context-dependent manner as both an activator and repressor. In such cases, further investigation is needed to

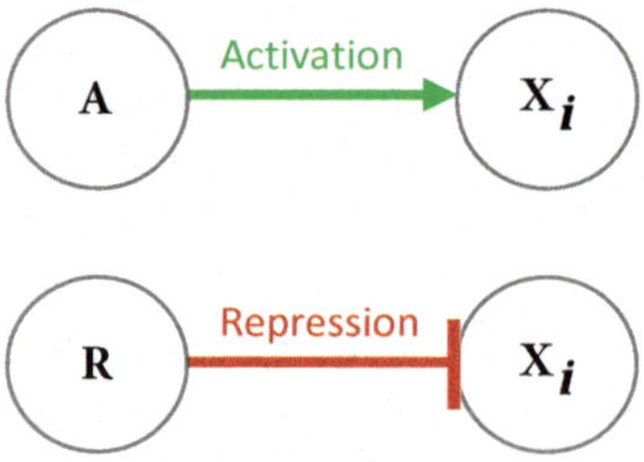

Fig. 3 Representing activation and repression between nodes in a GRN

determine if another mechanism, such as context-dependent cofactors, are influencing the TF's effect on transcription.

At the biochemical reaction level, the representation for activators and repressors (Fig. 3) is similar. A repressor R can bind a free promoter D at its target to form a repressed promoter D_R. The fact that this DNA promoter is repressed rather than activated will be reflected in the fact that the relative transcription rate from the promoter configuration D_R is less than that from the non-repressor-bound promoter, D. Likewise, if A is an activator, the transcription rate from the promoter configuration representing A bound to its target promoter, D_A, will be higher than the transcription rate on the nonactivated target promoter, D. We illustrate these details in the following section for our running example of Fig. 2.

Transcription and Translation

Next, we provide biochemical reaction models for transcription and translation. In prokaryotes and eukaryotes alike, transcription is an elaborate process involving several steps including initiation, elongation, and termination of an mRNA transcript, among others [57]. Likewise, translation consists of several steps including polypeptide formation and elongation and is followed by protein folding before the protein can reach its active form. A full mechanistic model of the transcriptional and translational apparatuses is laid out in Eqs. 10 and 11 of Chapter 2 in [44]. In modeling GRNs for cell fate reprogramming, however, this level of detail is not typically important for capturing the qualitative behavior of a network and its response to external perturbations. As derived in [44], we can treat transcription and translation as single-step reactions with lumped rates that encompass the time it takes for all the steps in the elaborate machinery to complete.

For our particular GRN in Fig. 2, these one-step biochemical reactions are shown in Table 1c, where each DNA promoter state established above produces an mRNA transcript, m_X or m_Y, from the genes of X and Y, respectively. Table 1c distinguishes between constitutive, or "leaky" transcription that represents the basal rate of transcription of a gene without activation or repression, and activated transcription in our example. The degree to which constitutive transcription occurs will have to be determined from the literature for each gene in question; some genes will constitutively be active, while others only transcribe mRNA upon activation. Moreover, the "activating" property of these reactions will be encoded in the fact that the rates of transcription for activated DNA will be greater than the rate of constitutive transcription, as shown in the table.

To summarize Table 1c, the mRNA transcript, m_X, of gene X can be produced in four ways:

- From the constitutive promoter of gene X, D_X, at a rate α_x (unit time)$^{-1}$.
- From the singly activated promoter D_{XX} at a rate α_x^1 of mRNA transcripts/unit time.
- From the singly activated promoter D_{XY} at a rate α_x^2 of mRNA transcripts/unit time.
- From the dually activated promoter D_{XXY} at a rate α_x^3 of mRNA transcripts/unit time.

Because the species D_{XX}, D_{XY}, and D_{XXY} represent activated forms of D_X, the rates of transcription from these species will be greater than from constitutive transcription, i.e., $\alpha_x^1, \alpha_x^2, \alpha_x^3 > \alpha_x$. Since the GRN is symmetric, similar representations hold for the way m_Y is produced.

Note that if these promoters were being repressed rather than activated, the only change would be in the relative rates of transcription between free promoters and repressor-bound promoters, i.e., $\alpha_x^1, \alpha_x^2, \alpha_x^3 < \alpha_x$ and $\alpha_y^1, \alpha_y^2, \alpha_y^3 < \alpha_y$.

Though we began with several DNA promoter states, these ultimately led to the same two mRNA transcripts, m_X and m_Y. At this point under the assumptions we have made, translation can also be represented by a one-step process from these transcripts into their protein products as shown in Table 1d.

Decay

While the reactions from above lead to the production of X and Y, we also model the decay of these proteins as well as their respective mRNA species. In general, these proteins and mRNA species will undergo decay at some rate that is a combination of both active degradation (δ_x and δ_y for proteins X and Y, respectively, and η_x and η_y for mRNA species m_X and m_Y, respectively) and dilution (β), as shown in Table 1e. For proteins X and Y, the total rate of decay is $\gamma_x = \delta_x + \beta$ and $\gamma_y = \delta_y + \beta$, respectively.

In general, the degradation parameters can be determined using the following relations [44]:

$$\gamma_i = \ln(2)/t_{1/2}, \quad \delta_i = \ln(2)/t_{1/2}$$

where $t_{1/2}$ represents the half-lives of each protein or mRNA transcript. The half-life of a species is the time it takes for its concentration to fall to half of its initial concentration [58] and can often be found from the literature. Care should be taken to use the half-life value in a cell environment that most closely resembles the phenotypes in question during the reprogramming process under study, as half-lives can vary as a function of environment [59]. Likewise, the dilution rate β can be calculated using the cell division rate, or doubling time (t_{doubling}), of the cell in which the proteins are contained: $\beta \approx \ln(2)/t_{\text{doubling}}$. The doubling time of prokaryotic cells such as *E. coli* can be on the order of minutes [60], while that of eukaryotic cells can be approximated by the characteristic human cell cycle time of approximately 24 h [61].

Overexpression (Artificial Perturbation)

Thus far we have described the reactions that comprise the endogenous components of our GRN. Ultimately, the reprogramming framework we describe will include artificial perturbations to our endogenous GRN that can be captured in our model as follows. As described in the Introduction, reprogramming is typically done by artificially overexpressing TFs of the GRN through insertion of ectopic DNA [62]. Therefore, we can model this by adding an additional production rate of the TF's mRNA from the ectopic DNA (Table 1f).

As seen above, this is the phase in the modeling process that will require a literature search to determine the multimerization state of the TFs in the GRN, the configurations in which they can bind the DNA they are regulating, as well as their half-lives and dilution rates. The online database bionumbers.org [63] can be a useful source for some of these parameters in many cases.

2.1.2 From Biochemical Reaction Model to ODE

With this biochemical reaction model established, we are ready to construct an ODE model for the GRN. Since ODE models are deterministic, they implicitly assume that the species in our biochemical reactions exist at sufficiently high copy numbers and in well-stirred volumes (Subheading 2.1 in [44]).

The procedure for obtaining the ODE for each species introduced is relatively straightforward, with no expertise in differential equations required: one must simply go through every reaction listed in Table 1 and add appropriate terms to the differential equations of each species involved in that reaction (*see* Subheading 2.1 of [44]). For a given species S,

$$\text{Change in concentration of S} = \frac{d}{dt}(S) = \dot{S}$$
$$= \Sigma \text{ (all biochemical reaction rates involving S)}$$

Doing so for our biochemical reaction network yields the equations in Table 2.

This 14-dimensional ODE model of the GRN contains a large number of parameters that make it impractical to analyze for cell

Table 2
Full 14D model of GRN

$\dot{X}_d = a_x X^2 - d_x X_d - a_1 D_X X_d + d_1 D_{XX} - a_3 D_Y X_d + d_3 D_{YX} - a_6 D_{XY} X_d + d_6 D_{XXY} - a_8 D_{YY} X_d + d_8 D_{YYX}$	(1)
$\dot{Y}_d = a_y Y^2 - d_y Y_d - a_2 D_X Y_d + d_2 D_{XY} - a_4 D_Y Y_d + d_4 D_{YY} - a_5 D_{XX} Y_d + d_5 D_{XXY} - a_7 D_{YX} Y_d + d_7 D_{YYX}$	(2)
$\dot{D}_X = -a_1 X_d D_X + d_1 D_{XX} - a_2 Y_d D_X + d_2 D_{XY}$	(3)
$\dot{D}_Y = -a_4 Y_d D_Y + d_4 D_{YY} - a_3 X_d D_Y + d_3 D_{YX}$	(4)
$\dot{D}_{XX} = a_1 D_X X_d - d_1 D_{XX} - a_5 D_{XX} Y_d + d_5 D_{XXY}$	(5)
$\dot{D}_{XY} = a_2 D_X Y_d - d_2 D_{XY} - a_6 D_{XY} X_d + d_6 D_{XXY}$	(6)
$\dot{D}_{YX} = a_3 D_Y X_d - d_3 D_{YX} - a_7 D_{YX} Y_d + d_7 D_{YYX}$	(7)
$\dot{D}_{YY} = a_4 D_Y Y_d - d_4 D_{YY} - a_8 D_{YY} X_d + d_8 D_{YYX}$	(8)
$\dot{D}_{XXY} = a_5 D_{XX} Y_d - d_5 D_{XXY} + a_6 D_{XY} X_d - d_6 D_{XXY}$	(9)
$\dot{D}_{YYX} = a_7 D_{YX} Y_d - d_7 D_{YYX} + a_8 D_{YY} X_d - d_8 D_{YYX}$	(10)
$\dot{m}_X = \alpha_x D_X + \alpha_x^1 D_{XX} + \alpha_x^2 D_{XY} + \alpha_x^3 D_{XXY} - \eta_x m_X + u_x$	(11)
$\dot{m}_Y = \alpha_y D_Y + \alpha_y^1 D_{YY} + \alpha_y^2 D_{YX} + \alpha_x^3 D_{YYX} - \eta_y m_Y + u_y$	(12)
$\dot{X} = \kappa_x m_X - \gamma_x X$	(13)
$\dot{Y} = \kappa_y m_Y - \gamma_y Y$	(14)

fate reprogramming. However, if we assume that dimerization, DNA binding/unbinding, and mRNA dynamics in Table 1a–c occur sufficiently faster than protein production and decay in Table 1d, e, the temporal derivatives of the respective species can be set to zero, indicating that the species concentration reaches its (quasi) steady state very quickly compared to the rest of the system. This is called the quasi-steady state (QSS) approximation and is widely used in modeling biochemical reaction systems [44]. In our example, this amounts to setting the temporal derivatives in Eqs. 1–12 to zero.

Using the QSS approximation and accounting for the fact that DNA is not destroyed by employing the conservation laws $D_{TX} = D_X + D_{XX} + D_{XY} + D_{XXY}$ and $D_{TY} = D_Y + D_{YY} + D_{YX} + D_{YYX}$, we arrive at the 2D model:

$$\dot{X} = \frac{\alpha_{xo} + \dfrac{\alpha_{xx}}{K_1 K_X} X^2 + \dfrac{\alpha_{yx}}{K_2 K_\Upsilon} \Upsilon^2 + \dfrac{b_x \cdot R}{K_1 K_2 K_X K_\Upsilon} X^2 \Upsilon^2}{1 + \dfrac{1}{K_1 K_X} X^2 + \dfrac{1}{K_2 K_\Upsilon} \Upsilon^2 + \dfrac{1}{K_1 K_2 K_X K_\Upsilon} X^2 \Upsilon^2} - \gamma_x X + \overline{u}_x,$$

$$\dot{\Upsilon} = \frac{\alpha_{yo} + \dfrac{\alpha_{xy}}{K_X K_3} X^2 + \dfrac{\alpha_{yy}}{K_\Upsilon K_4} \Upsilon^2 + \dfrac{b_y \cdot R}{K_3 K_4 K_X K_\Upsilon} X^2 \Upsilon^2}{1 + \dfrac{1}{K_X K_3} X^2 + \dfrac{1}{K_4 K_\Upsilon} \Upsilon^2 + \dfrac{R}{K_3 K_4 K_X K_\Upsilon} X^2 \Upsilon^2} - \gamma_y \Upsilon + \overline{u}_y,$$

where we have further assumed $K_2/K_5 = K_1/K_6 = K_4/K_7 = K_3/K_8 = R$ and the lumped parameters are equal to:

$\alpha_{xo} = D_{TX}\alpha_x/\eta_x$, $\alpha_{yo} = D_{TY}\alpha_y/\eta_y$, $a_{xx} = D_{TX}\alpha_x^1/\eta_x$, $a_{yx} = D_{TX}\alpha_x^2/\eta_x$, $b_x = D_{TX}\alpha_x^3/\eta_x$, $a_{yy} = D_{TY}\alpha_y^1/\eta_y$, $a_{yx} = D_{TY}\alpha_y^2/\eta_y$, $b_y = D_{TY}\alpha_y^3/\eta_y$, $K_X = d_x/a_x$, $K_Y = d_y/a_y$. Without loss of generality, if we take $R = 1$ and $K_X = K_Y = K_i = 1$ ($i \in \{1, 2, 3, 4\}$), we get the model:

$$\begin{aligned} \dot{X} &= \frac{\alpha_{xo} + a_{xx}X^2 + a_{yx}Y^2 + b_xX^2Y^2}{1 + X^2 + Y^2 + X^2Y^2} - \gamma_x \cdot X + \bar{u}_x \\ &= H_1(X, Y) - \gamma_x X + \bar{u}_x \dot{Y} \\ &= \frac{\alpha_{yo} + a_{xy}X^2 + a_{yy}Y^2 + b_yX^2Y^2}{1 + X^2 + Y^2 + X^2Y^2} - \gamma_y \cdot Y + \bar{u}_y \\ &= H_2(X, Y) - \gamma_y Y + \bar{u}_y \end{aligned} \tag{15}$$

In Eq. 15, the terms $H_1(X, Y)$ and $H_2(X, Y)$ are regulatory functions referred to as Hill functions [39]. In our regulatory functions, the variables X and Y being raised to the power 2 stems from the fact that X and Y were assumed to act as homodimers. In fact, this exponent is more generally referred to as the Hill coefficient n [39], where n is the number of subunits in the multimerized state of the TF (i.e., $n = 1$ for monomers, $n = 2$ for dimers, $n = 3$ for trimers, etc.).

In summary, the assumptions we have made in arriving at the 2D model of our GRN in Eq. 15 are:

Assumption 1: Regulations Take Place at the Transcriptional Level Only: *As shown in Table 1b, the arrows between nodes in the GRN take place through TF protein products binding at the DNA promoters of their targets, where they activate or repress transcription.*

Assumption 2: mRNA Levels Produced via Transcription Reach Steady State Levels Sufficiently Fast: *During reduction of the dynamics from 14D to 2D above, setting* $\dot{m}_X = \dot{m}_Y = 0$ *was predicated on the assumption that mRNA dynamics are sufficiently fast, where sufficiently fast is equivalent to assuming that the decay rates of mRNA are much larger than those of proteins (i.e.,* η_x, $\eta_y \gg \delta_x$, δ_y*). Protein half-lives are typically longer than mRNA half-lives [64], and this is a good assumption when they are longer on the order of one magnitude or more.*

Assumption 3: DNA Binding and Unbinding Rates Are Much Faster than Protein Production and Decay: *Setting the ODEs corresponding to the DNA species to zero during model reduction from 14D to 2D was predicated on the assumption—known as the adiabatic limit [28]—that DNA binding and unbinding rates in Table 1b are much larger than the production and decay rates of the* *protein products, i.e.,* d_i, a_iX, $a_iY \gg \gamma_x, \gamma_y, \kappa_x, \kappa_y$ *for* $i = 1, \ldots, 8$. *In prokaryotic cells, the adiabatic assumption is typically accurate [44, 65], though in eukaryotic cells the time needed to demethylate DNA and unpack chromatin before TFs can bind may mean that the promoter kinetics are slower than assumed [66, 67]. When they become slower than the protein production and decay, the predictions of the reduced model may not be entirely accurate, and it would be preferable to use the full 14D model (i.e., the model without reductions based on QSS assumptions). Nonetheless, the steady states of a reduced model will be in the same location as the steady states of the full model, so we proceed with this assumption to illustrate what a first-pass analysis of the qualitative behavior of a GRN looks like.*

Assumption 4: The Species in Question Exist in Sufficiently High Molecular Counts: *An assumption that has been held throughout this entire ODE modeling scheme is that the species in question exist at sufficiently high molecular counts.*

Now that we have arrived at this reduced ODE model of our GRN, we are ready to represent the Waddington view of cell fate reprogramming from the mathematical perspective.

2.2 A Dynamical Systems View of Cell Fate Reprogramming

By developing a mathematical model for the cell fate reprogramming problem in Subheading 2.1, we can now study its properties using dynamical systems tools. Within a dynamical systems description, a GRN consisting of the n TF species $X_1, X_2, \ldots, X_n$ has a "state" given by the tuple of TF concentrations $\bar{X} = (X_1, X_2, \ldots, X_n)$. The GRN can be described using n ODEs representing the evolution of these concentrations, $\dot{\bar{X}} = \bar{f}(\bar{X})$, in which we use $\dot{\bar{X}}$ for $d\bar{X}/dt$. We refer to the set of points $(S^1, \ldots, S^m)$ in state space such that $\bar{f}(S^i) = 0$ (for $i = 1, \ldots, m$) as the *steady states* of this dynamical system, which can be either stable or unstable [43]. Qualitatively, stable steady states (SSSs) are those that can withstand or damp out small amplitude disturbances and are therefore associated to a phenotype in this context [68–70].

The distinction between a stable and unstable steady state (USS) can be visualized in Fig. 4a, which portrays the states of a GRN as a landscape of wells. In this landscape, SSSs are equivalent to valley-like depressions, while USSs are equivalent to "peaks." On the one hand, under the influence of gravity pointing downward in the figure, any arbitrarily small perturbation applied on a hypothetical ball that is resting on a peak will cause it to permanently lose its position on that peak (and hence that state is unstable). On the other hand, a ball resting in a depression can withstand some level of perturbation and still converge back into that well (and hence that state is stable).

In addition, SSSs have varying degrees of stability, which can in part be captured by the basin of attraction surrounding each SSS, which is the set of states (TF concentrations) starting from which

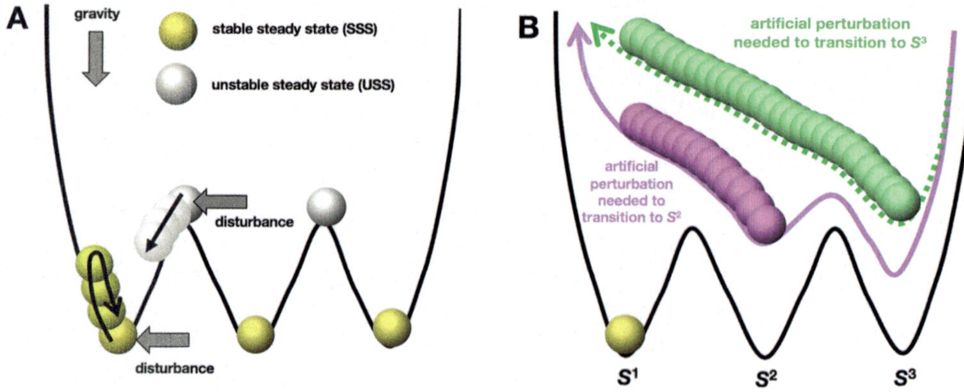

Fig. 4 Visual depiction of stable and unstable steady states. (**a**) Unstable steady states (USSs) are analogous to the "peaks" in a valley landscape, while stable steady states (SSSs) are analogous to well-like depressions in the landscape. If a ball is resting in a SSS, it can withstand some degree of disturbance and still return to the SSS autonomously. However, if it is resting on a peak (USS), then any nonzero disturbance will permanently push it off that peak and hence out of that steady state. (**b**) Cell fate reprogramming is akin to transitioning between wells by altering the landscape through artificial perturbation. If a ball begins at state S^1, it can be forced to transition to S^2 or S^3 by appropriately changing the valley landscape as shown

the system converges autonomously to that SSS. For instance, in Fig. 4a, if the SSSs themselves are the bottommost points in the wells, then the entire well is the basin of attraction of that SSS. Hence in the metaphor of this figure, cell fate reprogramming is akin to enforcing transitions from the basin of attraction of the starting phenotype (one well) to the basin of attraction of the final phenotype (another well). This is accomplished by artificially perturbing (i.e., applying an input to) the endogenous GRN just the right amount such that the landscape pushes the system's state from the basin of attraction of the starting state toward the basin of attraction of the desired final state (Fig. 4b).

In Fig. 5, we depict this reprogramming concept for a prototypical two-dimensional system consisting of two TFs, X and Y. The state space hence consists of concentrations of these TFs: X and Y. In Fig. 5a, when no input is applied, the steady state landscape of the endogenous GRN alone consists of three SSSs: S^1, S^2, and S^3. As introduced above, we view these states as corresponding to distinct phenotypes characterized by the relative concentrations shown and with basins of attraction surrounding each SSS as highlighted in the figure. When an artificial perturbation is introduced for the purpose of reprogramming, the steady state landscape morphs into one that may have a different number of SSSs at different locations (Fig. 5b).

Specifically, the original states S^1, S^2, and S^3 disappear, and two new stable steady states, $S^{2'}$ and $S^{3'}$, arise as depicted. As a result, once the perturbation is applied on a system starting in the location which was originally S^1, the system's state will converge toward $S^{2'}$.

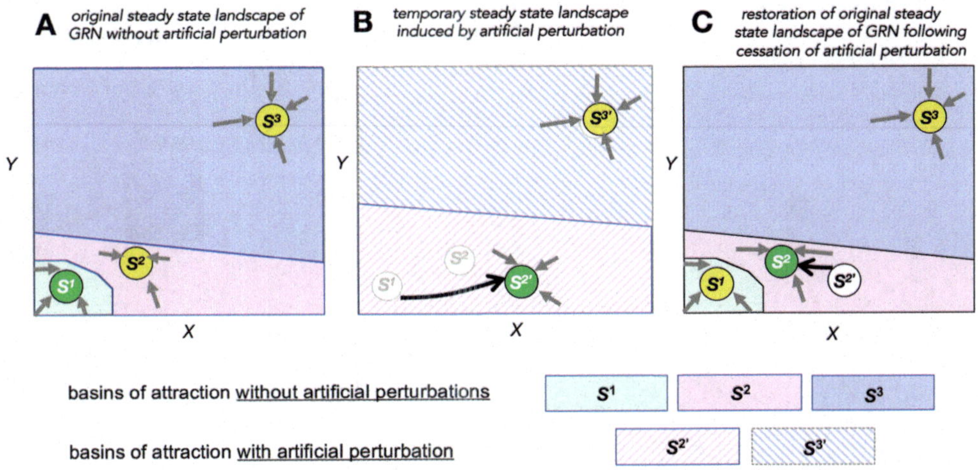

Fig. 5 Cell fate reprogramming as morphing the steady state landscape. (**a**) Without any artificial perturbation, the system possesses three stable steady states (SSSs), S^1, S^2, and S^3, that represent phenotypes characterized by relative concentration levels of the TFs X and Y. The basin of attraction of each SSS (phenotype) is the set of states (TF concentrations) starting from which the system converges autonomously to the corresponding SSS (phenotype), as depicted by the shaded regions. (**b**) When a certain degree of artificial perturbation to the network is applied, the steady state landscape changes shape so as to push the system starting from S^1 to a location $S^{2'}$ in the basin of attraction of the original state S^2 of the system without overexpression. (**c**) Upon removal of the artificial perturbation, the system's state moves from $S^{2'}$ to S^2, and the transition from S^1 to S^2 has been completed

Upon removal of input in Fig. 5c (i.e., completion of the cell fate reprogramming experiment), the system's state now at $S^{2'}$ is found in the basin of attraction of the original system's SSS S^2, and therefore the system's state ultimately converges to S^2. In essence, in a dynamical systems representation, cell fate reprogramming corresponds to morphing the steady state landscape of the GRN through external perturbation such that the current system's state (current phenotype) is pushed into the basin of attraction of the target state (target phenotype). It is in general nontrivial to determine external perturbations that are appropriate to trigger a desired state transition. However, mathematical and computational analysis of the ODE model can reveal useful information in this regard, as we illustrate in the next sections.

To summarize, Fig. 6 juxtaposes the different views of reprogramming at the three levels of abstraction discussed thus far: reprogramming as transitions up and across the Waddington landscape, as transitions in state space, and as an experimental change in the phenotype itself.

2.3 Analysis of Intrinsic Properties of GRN Using Tools from Dynamical Systems

Having introduced the notion of SSS as phenotypes, we provide tools in this section for mathematical characterization of the steady state landscape by continuing with our running example of Fig. 2. In particular, we analyze the intrinsic properties of this GRN by analyzing the steady state landscape of our system in Eq. 15 without

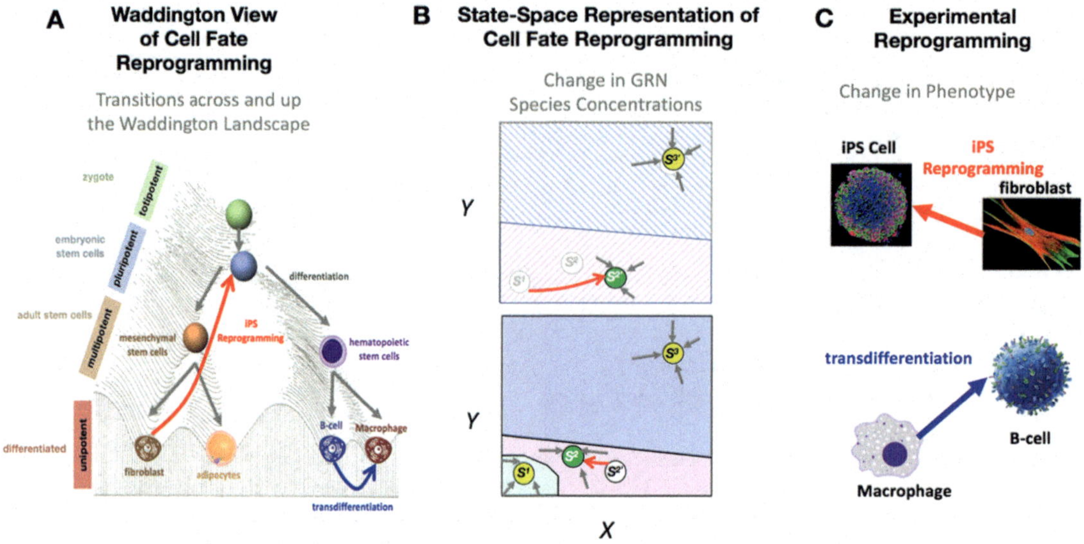

Fig. 6 The reprogramming concept viewed at different levels of abstraction. (**a**) At a high level, cell fate reprogramming is akin to traveling "uphill" in the Waddington landscape. (**b**) Identification of a GRN motif controlling the relevant cell fates allows for the framing of reprogramming as enforcing a transition between stable steady states in the "state space" of an ODE model of that GRN. These stable steady states are characterized by relative concentration levels of TFs in the GRN that in principle map to relative concentration levels seen in the phenotypes being represented. (**c**) Ultimately, the dynamic model of reprogramming is meant to inform improved experimental cell fate reprogramming strategies, which transform one phenotype to another

input (i.e., $\bar{u}_x = \bar{u}_y = 0$). Studying this steady state landscape is of utmost importance, as having a correspondence between the mathematical steady states of an ODE model and the phenotypes being studied in a particular cell fate reprogramming process speaks to the strength of a model and its ability to represent the phenotype switching in question.

In general, the mathematical steady states of an ODE model, regardless of its dimensionality, are the solutions to simultaneously setting the time derivatives of all species in the GRN to zero. In our running example of a GRN in Fig. 2, the mathematical steady states of its 2D model in Eq. 15 are solutions to the equations $dX/dt = dY/dt = 0$. One of the benefits of having a 2D model is that these solutions can be viewed in a 2D-plane, as depicted in Fig. 7a. As shown, setting these time derivatives to zero yields two equations whose solutions are "nullclines" in the X–Y plane. Viewed in this manner, the solution(s) to $dX/dt = dY/dt = 0$, and hence the steady state(s), are given by the intersection(s) of these nullclines.

2.3.1 Stability of Steady States

Once the steady states of a system have been determined, there are different options for determining their stability, depending on the dimensionality of the model. For a system $\dot{\bar{X}} = \bar{f}(\bar{X})$ with

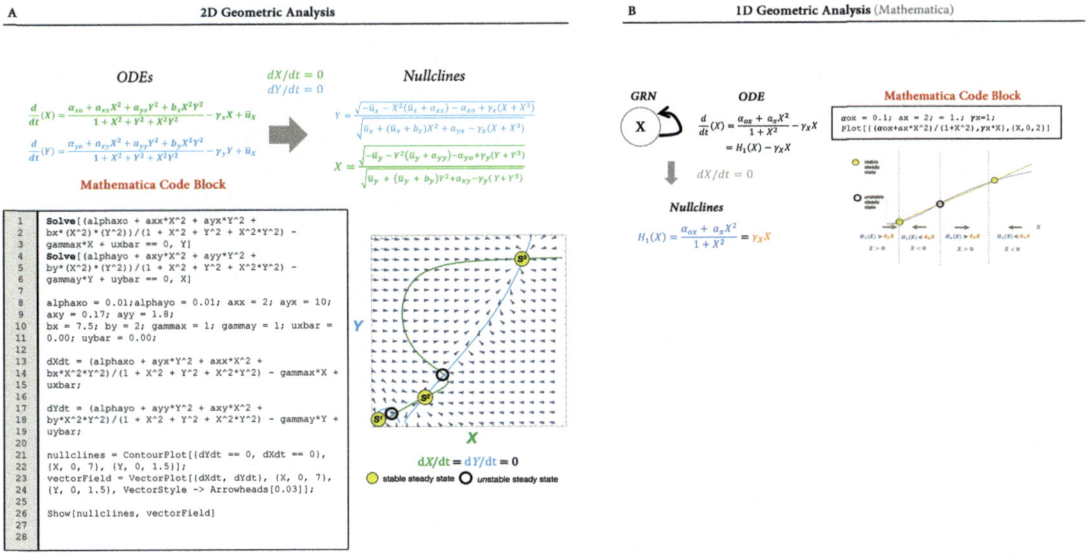

Fig. 7 Steady states of a 2D ODE and their stability (top). One way to find the steady states of our ODE model is to set the time derivatives of each species to zero and solve for the curves in the plane that result. (bottom left) [lines 1–6] The *Solve* function in Mathematica can be used to solve for nullclines if there is an explicit solution. These could then be plotted using the *Plot* function in Mathematica (not shown). [lines 8–11] Parameter value definitions. [lines 13–24] The *ContourPlot* function in Mathematica can also be used to plot nullclines, whether or not there is an explicit solution, as the two ODEs are specified as {*dX*/*dt* == 0, *dY*/*dt* == 0} in the input. The *VectorPlot* function can be used to plot the vector field of the system. The two ODEs are specified as {*dX*/*dt*, *dY*/*dt*} in the input. [line 26] The *Show* function can be used to overlay the nullclines and vector field

$\bar{X} = (X_1, X_2, \ldots, X_n)$ and $\bar{f}(\bar{X}) = (f_1(\bar{X}), \ldots, f_n(\bar{X}))$, the vector $\bar{f}(\bar{X})$ is known as the vector field, which has a different length and direction as a function of $\bar{X}$. Qualitatively, plotting this vector at different points across the state space indicates the local direction and speed at which a point in state space would move under the dynamics of the system. Hence *arrows in a vector field point toward SSSs* while they *point away from USSs* [44]. For systems with dimension higher than 2, plotting the vector field may not be practical, and therefore algebraic tools, such as the Jacobian matrix, are more informative. We illustrate the vector field approach with the 2D model in Eq. 15 and then illustrate the Jacobian matrix approach using a 3D GRN model.

2D Geometric Analysis

With the choice of parameters shown in Fig. 7a, there are five intersections, or steady states.

In the code block of the figure, we use the Mathematica functions `ContourPlot` and `VectorPlot` to simultaneously plot these nullclines and the vector field, respectively. As shown, this allows for a very simple visual inspection of stability at each steady state, and we can see that our system has three SSSs and two USSs. The same

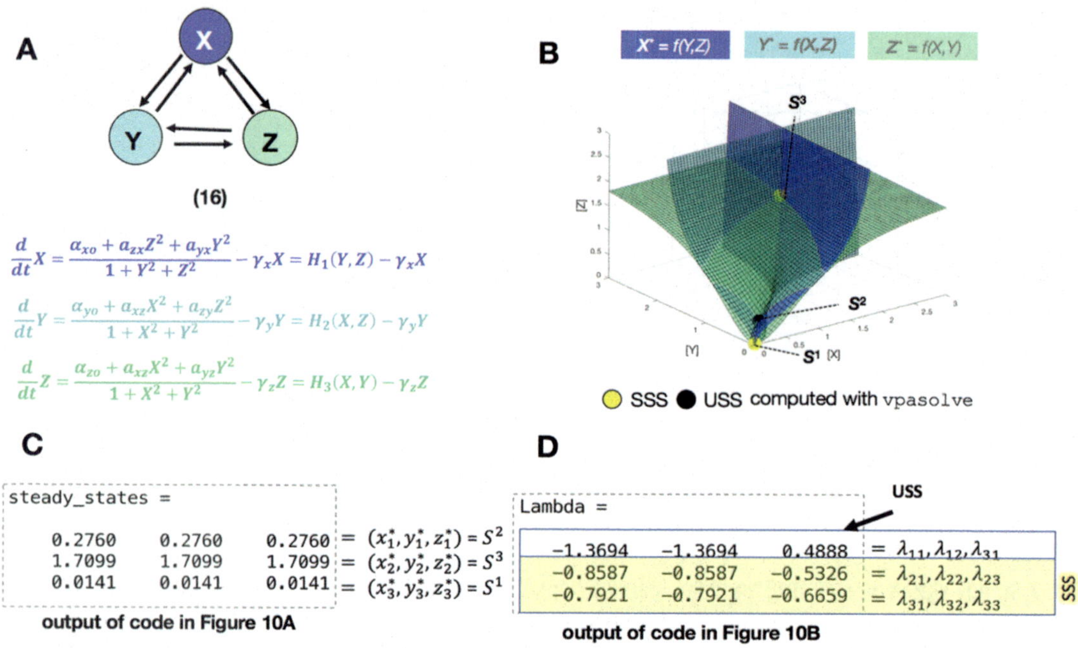

Fig. 8 Steady state landscape analysis of a 3D GRN: (**a**) The ODEs in [18] represent the reduced system that would be obtained by applying the workflow and assumptions of Subheading 2.1 on the GRN shown, with three mutually activating TFs: X, Y, and Z. (**b**) The surfaces X^*, Y^*, Z^* are obtained by setting the time derivatives of X, Y, Z, respectively, equal to zero and solving for each variable. Intersections of all three surfaces represent the steady states of the system, of which there are three as shown. (**c**) These same three steady states can be equivalently computed using vpasolve in MATLAB, as implemented in Fig. 10a. (**d**) Using Jacobian matrices as explained in Fig. 9 and implemented in Fig. 10b, the stability of each steady state can be determined using the eigenvalues of the Jacobian matrix evaluated at that steady state

principles allow for simple visual inspection of stability for 1D systems, an example of which is shown in Fig. 7b.

n-D Computational Analysis

In GRNs with more than two TFs, such as the three-node network shown in Fig. 8a, the reduced ODE system that can be obtained through the workflow from Subheading 2.1 has greater than two dimensions. Equation 16 of Fig. 8a shows the reduced 3D ODE model that can be obtained for the GRN shown. In these cases, the steady state landscape may be inferred using computational techniques outlined in Fig. 9. This workflow is based on standard theory from dynamical systems and linear algebra [43, 71], though in-depth knowledge of these concepts is not required for use of the code we provide.

The workflow is as follows: one can first solve for the steady states of an *n*-D system using an algebraic solver such as `vpasolve` in MATLAB (Fig. 9a). Then, to determine the stability of these steady states, a standard method is to use a Jacobian matrix (that can be found using the function `jacobian` in MATLAB, Fig. 9b),

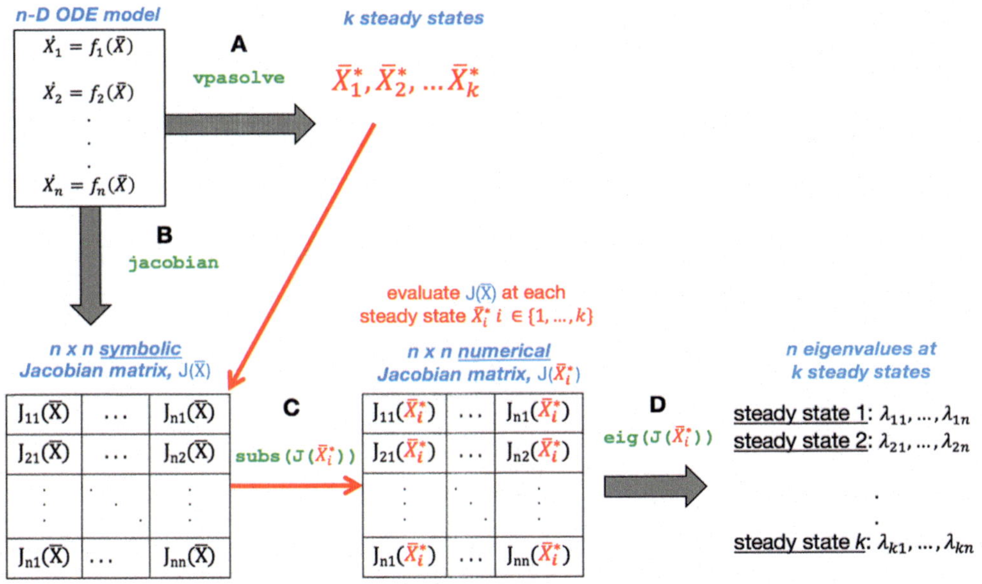

Fig. 9 Using Jacobian matrices to determine the stability of steady states in *n*-D models. (**a**) The steady states of an *n*-D ODE model can be computed using the vpasolve function in MATLAB. (**b**) The Jacobian matrix, *J*, of the *n*-D ODE model can also be computed in symbolic terms using the jacobian function in MATLAB. (**c**, **d**) The outputs of results from (**a**) and (**b**) can be combined to evaluate the symbolic *J* at each of the *k* steady states using the symbolic substitution function subs, and the *n* eigenvalues of each Jacobian matrix can be calculated using the function eig (note: some eigenvalues may be repeated)

which results from linearization of the nonlinear system. The linearization of the system at a steady state of interest provides a good approximation of the system dynamics in the near proximity of the steady state. Since this approximation is linear, we can infer the stability of the linearization by determining the eigenvalues of the Jacobian matrix evaluated at that steady state (Fig. 9c, d). Stability (all eigenvalues with negative real part) or instability (some eigenvalue with positive real part) directly translates into stability or instability of the original system's steady state. Marginal stability of the linearization (when one eigenvalue has zero real part with all others having negative real parts) is inconclusive for the original nonlinear system [44, 72].

In Fig. 8c, d, we show the results of this workflow, the list of steady states and their eigenvalues for the 3D model in Eq. 16. These results indicate that the system possesses two SSSs and one USS. Figure 10a, b implements the MATLAB code used to compute these steady states and eigenvalues, which can be adapted for other systems by simply adding the relevant ODEs and parameters. In Fig. 8b, we also show in 3D how the steady states are distributed in the 3D state space, by plotting the surfaces obtained by setting the time derivatives of X, Y, and Z in Eq. 16 equal to zero (which

A

***n*-D Computational Analysis** - *Find All Steady States* - `surf, vpasolve`

```
1  colormap winter % set coloring for 3D plots
2  a = 1.5; ayx = a; azx = a; azy = a; axy = a; axz = a; ayz = a; % set parameters
3  g = .75; gammaX = g; gammaY = g; gammaZ = g; % decay rates of proteins X,Y,Z
4  alpha0x=0.01;alpha0y=0.01;alpha0z=0.01;%leaky transcription rates of mX,mY,mZ
5  % plot the surface X* = f(Y,Z)
6  dp = 0.05; max = 3;
7  [Y,Z] = meshgrid(0:dp:max,0:dp:max);
8  Xnullcline = (1/gammaZ)*(alpha0x + azx*(Z.^2) + ayx*(Y.^2))./(1+ Y.^2 + Z.^2);
9  C = .25*ones(length(Y),length(Z)); % set color
10 sx = surf(Xnullcline,Y,Z,C); hold on;
11 % plot the surface Y* = f(X,Z)
12 [X,Z] = meshgrid(0:dp:max,0:dp:max);
13 Ynullcline = (1/gammaY)*(alpha0y + azy*(Z.^2) + axy*(X.^2))./(1+ X.^2 + Z.^2);
14 C = 0.55*ones(length(X),length(Z)); % set color
15 sy = surf(X,Ynullcline,Z,C); hold on;
16 % plot the surface Z* = f(X,Y)
17 [X,Y] = meshgrid(0:dp:max,0:dp:max);
18 Znullcline = (1/gammaZ)*(alpha0z + axz*(X.^2) + ayz*(Y.^2))./(1+ X.^2 + Y.^2);
19 C = 0.75*ones(length(X),length(Y));
20 sz = surf(X,Y,Znullcline,C); hold on;
21 % format plots and add axes labels
22 xlabel('[X]','FontSize',15);ylabel('[Y]','FontSize',15);zlabel('[Z]','FontSize',15)
23 alpha(sx,0.4); alpha(sy,0.4); alpha(sz,0.4); % apply transparency on the surfaces
24 % overlay numerically computed solutions on nullclines
25 % (Figure 9A) calculate steady states using vpasolve
26 syms X Y Z % define variables and dynamics
27 dXdt = (alpha0x + azx*(Z.^2) + ayx*(Y.^2))./(1+ Y.^2 + Z.^2) - gammaX*X;
28 dYdt = (alpha0y + azx*(Z.^2) + ayx*(X.^2))./(1+ X.^2 + Z.^2) - gammaY*Y;
29 dZdt = (alpha0z + ayx*(X.^2) + azx*(Y.^2))./(1+ X.^2 + Y.^2) - gammaZ*Z;
30 [sol_X, sol_Y, sol_Z] = vpasolve([dXdt == 0, dYdt == 0, dZdt == 0],...
31     [X, Y, Z],'random',true);
32 % extract indices of real solutions solutions
33 indices1 = find(imag(sol_Y)==0);
34 indices2 = find(imag(sol_Y)==0);
35 indices3 = find(imag(sol_Z)==0);
36 % store real solutions in matrix and save
37 steady_states=double([sol_X(indices1(3:5)),sol_Y(indices2(3:5)), sol_Z(indices3(3:5))])
38 % plot solutions in 3D space (overlay)
39 for i=1:3
40  scatter3(steady_states(i,1),steady_states(i,2),steady_states(i,3),300,'k','filled');
41 hold on;
42 end;
43
```

B

***n*-D: Computational Analysis** - *Determine Stability:* `jacobian, subs, eig`

```
49 % (Figure 9B) find jacobian, in symbolic terms, for 3D system defined above
50 J_sym = jacobian([dXdt,dYdt,dZdt],[X,Y,Z]);
51 Lambda = zeros(size(steady_states)); %create matrix to store eigvals at each SS
52 for i =1:length(steady_states) %loop through steady states
53     x = steady_states(i,1);
54     y = steady_states(i,2);
55     z = steady_states(i,3);
56     %(Figure 9C) evaluate Jacobian matrix at each steady state
57     J_numerical = subs(J_sym,[X,Y,Z],[x,y,z]);
58     %(Figure 9D) compute eigenvalues of Jacobian evaluated at each steady state
59     eigVals = eig(J_numerical);
60     %(store eigenvalues at each SSS)
61     Lambda(i,1) = eigVals(1);
62     Lambda(i,2) = eigVals(2);
63     Lambda(i,3) = eigVals(3);
64 end
65
66 Lambda % display all eigenvalues
```

Fig. 10 Finding steady states. (**a**) Finding steady states using surf and vpasolve in MATLAB: [lines 1–23] Use surf to plot 3D surfaces using the 3D "nullclines" obtained by solving for X, Y, and Z after setting the temporal derivatives of these variables to zero in Eq. 16 of Fig. 8a. [lines 25–38] Use vpasolve to numerically compute steady states of model in Eq. 16. Note that in line 37, we use the solution indices 3–5 since the solutions corresponding to indices 1 and 2 are incorrect (verified manually). Since vpasolve is only a numeric solver, solutions should always be cross-checked by plugging them back into the ODEs. [lines 39–42] Overlay the steady states computed numerically using vpasolve on top of the intersecting surfaces. The steady states returned from this code block, stored in the variable "steady_states," are shown in Fig. 8c. (**b**) Computationally determining the stability of numerically computed steady states. This code block continues from Fig. 10a and implements the Jacobian matrix method outlined in Fig. 9. Output of this code block, a matrix "Lambda" where each row is the set of eigenvalues of a steady state, is shown in Fig. 8d. Documentation for jacobian in MATLAB: https:/www.mathworks.com/help/symbolic/vpasolve.html. Documentation for eig in MATLAB: https:/www.mathworks.com/help/symbolic/eig.html. Documentation for subs in MATLAB: https:/www.mathworks.com/help/symbolic/subs.htm

are the 3D equivalents of nullclines) and seeing that intersections of these surfaces occur at the same points computed by `vpasolve`.

To summarize the Jacobian matrix method of determining stability, for each steady state:

- *Case #1*: if all eigenvalues of the Jacobian matrix evaluated at that steady state have negative real parts, the state is a *stable steady state* for the original nonlinear system.
- *Case #2*: if any eigenvalues of the Jacobian matrix evaluated at that steady state have a positive real part, the state is an *unstable steady state* for the original nonlinear system.
- *Case #3*: if none of the real parts of eigenvalues of the Jacobian matrix evaluated at that steady state are positive, but one or more are zero, the Jacobian linearization method is inconclusive, and other methods must be used.

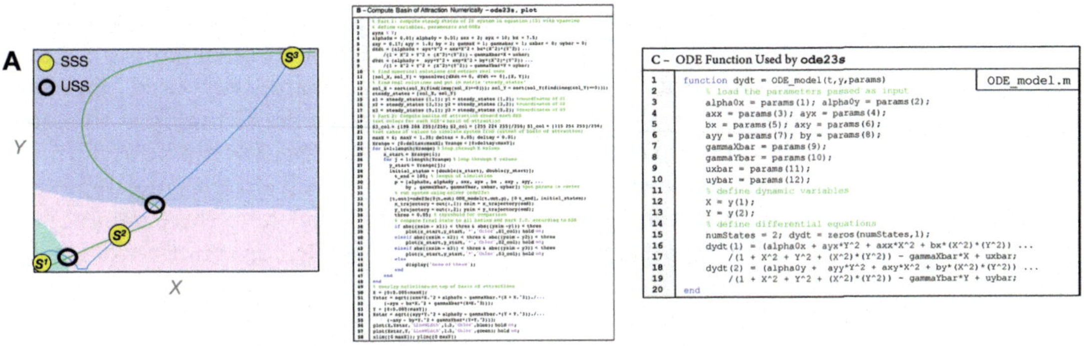

Fig. 11 Basins of attraction for 2D system: (**a**) In MATLAB, the basins of attraction around each SSS can be found numerically by repeatedly simulating the system using one of MATLAB's ODE solvers and starting from a grid of points surrounding the SSSs. (**b**) We implement this numerical calculation of basins of attraction using the ode23s solver to simulate the system starting from each point in the grid. In addition, we overlay nullclines on top of the numerically computed basins using the plot function in MATLAB. (**c**) We define the ODE function "ODE_model" passed to the solver on line (33) of panel **b** in a separate file of the same name. *Documentation for available MATLAB solvers*: mathworks.com/help/matlab/math/choose-an-ode-solver.html

In the rare Case #3, more advanced methods from nonlinear dynamical systems theory, including Lyapunov Theory [73] or the Center Manifold Theorem [74, 75], can be used to investigate stability.

2.3.2 Basins of Attraction

Once the SSSs of a model are identified, determining the basin of attraction around each one can be an informative next step in understanding the landscape. For our running example in Eq. 15, Fig. 11 shows how to compute this basin of attraction numerically in MATLAB. As shown in the figure, the basins of attraction around each of the three SSSs S^1, S^2, and S^3 are the blue-, pink-, and purple-shaded regions, respectively. These represent the sets of concentrations of X and Y starting from which the system converges autonomously to each respective SSS.

2.3.3 Parameter Analysis

Thus far in our study of the steady state landscape, we have taken for granted a certain set of parameters in our model for the 2D GRN, though in reality the values of parameters used can have major consequences on the location, number, and stability of steady states. As discussed in Subheading 2.1, when constructing a biochemical reaction model of a GRN, care should be taken at every step to use parameters from the literature, when available, that most accurately reflect the reactions that take place in reality. However, there will often be large variability on many parameter values, and some reactions may have unknown parameters. It is therefore important to assess how variations in parameters affect the steady state landscape of the GRN.

Here, we describe three different tools that can be used to this end. Specifically, for a single parameter variation at a time, parameter bifurcation analysis is the most commonly used approach [74]. To evaluate the effect of all parameters changing at the same time, local parameter sensitivity analysis [44, 72] can be performed, which provides insight for sufficiently small changes in parameters. This tool can also be used to locally determine the parameters to which the system is more sensitive. Finally, for the effect of changing all parameters at the same time within given ranges, global sensitivity analysis can be performed [76], which is a computationally intensive approach. In the examples that follow, we implement our code in MATLAB for the sake of introducing the concepts. However, several software tools are available, especially for parameter bifurcation analysis, including AUTO (http://indy.cs.concordia.ca/auto/) and Oscill8 (http://oscill8.sourceforge.net/).

Single-Parameter Bifurcations

One way to assess the impact of a given parameter on the steady state landscape is to hold all other parameters constant and plot the location of the steady states as that parameter is changed, with attention given to how the number and stability of steady states change or *bifurcate* [43]. Figure 12 demonstrates this method for the parameter a_{xx} in Eq. 15. The two plots in this figure show the steady states of the system (X^*, Y^*), and their stability, as this *bifurcation parameter* a_{xx} is varied from the values 2 to 2.8, as an example. As shown, the system begins with five steady states at $a_{xx} = 2$ (the nominal value used in Figs. 7 and 11), and it loses two steady states at about $a_{xx} = 2.45$. Stability analysis using the Jacobian matrix method (implemented in the code block of the figure) shows that of the remaining three steady states, two are stable, while one is unstable. In this case, we say that the system goes through a bifurcation at $a_{xx} \approx 2.5$. In particular, for the intermediate stable and unstable steady states, this is a case where the stable and unstable steady states "collide" and annihilate each other, which is known as a saddle-node bifurcation [74]. This is an important insight since, as we have described, SSSs represent phenotypes, and so this model would indicate that at least one phenotype could potentially disappear if this parameter is somehow perturbed experimentally. These notions will become important when we discuss how some external perturbations to ODEs can be modeled by changes in parameters in Subheading 2.4, which will lend insight into possible explanations for how external stimulation during reprogramming changes the steady state landscape.

Local Sensitivity Analysis

A more systematic way of assessing how the steady state landscape changes in response to small parameter perturbations is the use of sensitivity matrices. As depicted, in Fig. 13a, the sensitivity matrix, $S_{X,\Theta}$, of a system with state $X = (X_1, X_2, \ldots, X_n)$ and parameters $\overline{\Theta} = (\theta_1, \theta_2, \ldots, \theta_m)$ at a stable steady state $\bar{X}^*$ and nominal

A

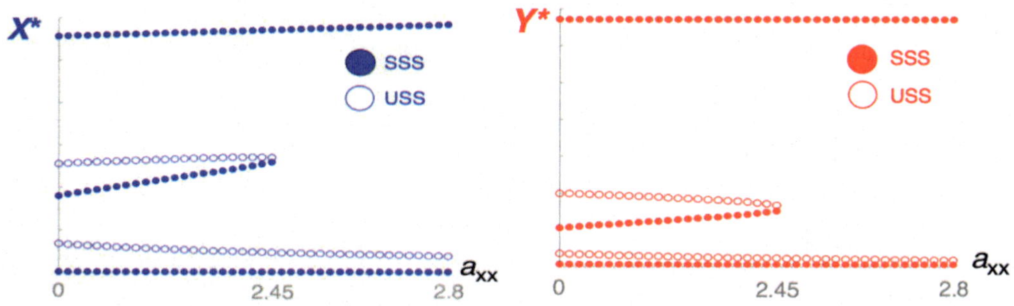

B – Single Parameter Bifurcation

```
1  %% bifurcate along axx
2  axx_list = [2:0.02:2.8]; % parameter range to loop through
3  for i=1:length(axx_list)
4      axx = axx_list(i); % parameters
5      uxbar = 0; uybar = 0; alpha0x = 0.01; alpha0y = 0.01;ayx = 10;
6      bx = 7.5; axy= 0.17; ayy = 1.8; by = 2; gammaX = 1; gammaY = 1;
7      % compute symbolic Jacobian with these parameters
8      syms X Y
9      dXdt = (alpha0x+axx*X^2+ayx*Y^2+bx*(X^2)*(Y^2)) ...
10     /(1+X^2+Y^2+(X^2)*(Y^2))-gammaX*X + uxbar;
11     dYdt = (alpha0y+axy*X^2+ayy*Y^2+by*(X^2)*(Y^2)) ...
12     /(1+X^2+Y^2+(X^2)*(Y^2))-gammaY*Y + uybar;
13     J_sym = jacobian([dXdt,dYdt],[X,Y]);
14     [sol_X, sol_Y] = vpasolve([dXdt==0,dYdt==0],[X, Y]);%numerical solutions
15     % find real solutions
16     sol_X = double(sort(sol_X(find(imag(sol_X)==0))));
17     sol_Y = double(sort(sol_Y(find(imag(sol_Y)==0))));
18     % store solutions in a matrix
19     steady_states = [sol_X, sol_Y];
20     % loop through the steady states just computed, determine stability
21     for j = 1:length(sol_X)
22         x = steady_states(j,1); y = steady_states(j,2);
23         % evaluate jacobian at steady state and calculate eigenvalues
24         J_numerical = double(subs(J_sym,[X,Y],[x,y]));
25         eigVals = eig(J_numerical);
26         anypositives = find(eigVals > 0); % check if any eigenvalues are
27 +
28         if length(anypositives) > 0 % unstable
29             figure(1);  scatter(axx,sol_X(j),80,'b'); hold on;
30             figure(2);  scatter(axx,sol_Y(j),80,'r'); hold on;
31         else % stable
32             figure(1);  scatter(axx,sol_X(j),80,'b','filled'); hold on;
33             figure(2);  scatter(axx,sol_Y(j),80,'r','filled'); hold on;
34         end
35     end
36 end
```

Fig. 12 Bifurcation with parameter a_{xx}: (**a**) Bifurcation plot for the steady state (X^*, Y^*) for the ODE model in Eq. 15. For values of a_{xx} under approximately 2.45, there are a total of five steady states, three of which are stable and two of which are unstable. The system bifurcates above $a_{xx} \approx 2.45$, as a stable and an unstable steady state collide and annihilate each other (also, called a saddle-node bifurcation; Strogatz [43]). (**b**) MATLAB code for looping through a value range for the parameter a_{xx} while holding all other parameters constant (lines 5–6). To bifurcate along any other parameter, define a parameter range in line 1 and adjust lines 4–6 appropriately

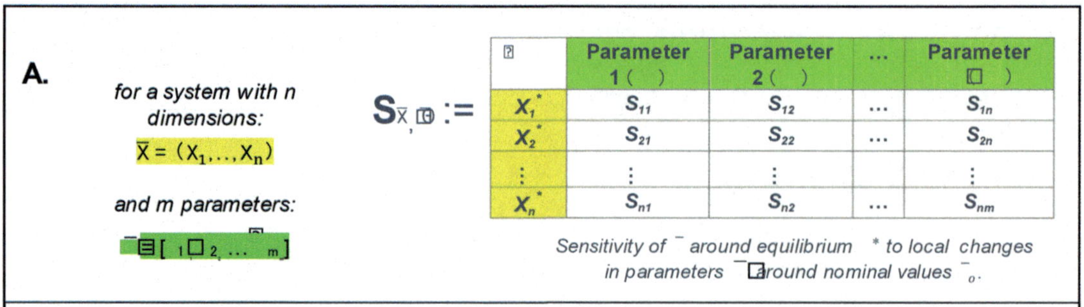

B. Local Sensitivity Analysis – `diff`

```
% Part I: % compute matrix df/dx (f = [f1, f2], x = [X, Y])
% use nominal parameter values
alpha0x = 0.01; alpha0y = 0.01; axx = 2; ayx = 10; bx = 7.5; axy = 0.17; ayy = 1.8;
by = 2; gammaX = 1; gammaY = 1; Dtheta = diag([alpha0x alpha0y axx ayx bx axy ayy
by gammaX gammaY]);
syms X Y % define system
f1=(alpha0x+ayx*Y^2+axx*X^2+bx*(X^2)*(Y^2))/(1+X^2+Y^2+(X^2)*(Y^2))-gammaX*X;
f2=(alpha0y+ayy*Y^2+axy*X^2+by*(X^2)*(Y^2))/(1+X^2+Y^2+(X^2)*(Y^2))-gammaY*Y;
df1dx(X,Y) = diff(f1,X); df1dy(X,Y) = diff(f1,Y);
df2dx(X,Y) = diff(f2,X); df2dy(X,Y) = diff(f2,Y);
% compute all steady states
[sol_X, sol_Y] = vpasolve([f1==0,f2==0],[X, Y]);
sol_X = double(sort(sol_X(find(imag(sol_X)==0))));
sol_Y = double(sort(sol_Y(find(imag(sol_Y)==0))));
% store steady states in matrix, each row is steady state
steady_states = [sol_X, sol_Y]
state_index = 3;%choose steady state by row #. Sensitivity matrix evaluated here.
X_SS = steady_states(state_index,1);
Y_SS = steady_states(state_index,2);
% evaluate df/dx at this steady state
dfdx=[df1dx(X_SS,Y_SS),df1dy(X_SS,Y_SS);df2dx(X_SS,Y_SS),df2dy(X_SS,Y_SS)];
% Part II: compute matrix df/dtheta evaluated at (x,y) = (x_e,y_e)
syms alpha0x alpha0y axx ayx bx axy ayy by gamma1 gamma2
X = X_SS; Y = Y_SS;
f1 = (alpha0x+ayx*Y^2+axx*X^2+bx*(X^2)*(Y^2)) ...
        /(1+X^2+Y^2+(X^2)*(Y^2))-gammaX*X;
f2 = (alpha0y+ayy*Y^2+axy*X^2+by*(X^2)*(Y^2)) ...
        /(1+X^2+Y^2+(X^2)*(Y^2))-gammaY*Y;
% compute derivatives at this steady state
row1= [diff(f1,alpha0x) diff(f1,alpha0y) diff(f1,axx) diff(f1,ayx) diff(f1,bx)
diff(f1,gammaX) diff(f1,gammaY) diff(f1,axy) diff(f1,ayy) diff(f1,by)];
row2 = [diff(f2,alpha0x) diff(f2,alpha0y) diff(f2,axx) diff(f2,ayx) diff(f2,bx)
diff(f2,gammaX) diff(f2,gammaY) diff(f2,axy) diff(f2,ayy) diff(f2,by)];
dfdtheta = [row1; row2];
% compute sensitivity matrix and normalized version
Sxtheta = double(-inv(dfdx)*dfdtheta);
Dx = diag([X, X]);
Sx_theta_normalized = double(inv(Dx)*Sxtheta*Dtheta)
```

Fig. 13 Sensitivity matrices for local sensitivity analysis at steady states and nominal parameters. (**a**) Depiction of the information captured in a sensitivity matrix. (**b**) Code used to implement Eq. 17 for the 2D system in Eq. 15. Documentation for **diff** in MATLAB: https:/www.mathworks.com/help/symbolic/diff.html

parameter values Θ_0 represents the local change in the stable steady state coordinates when small changes are applied to the parameters. The theory behind this method is described in detail in [44], though the general result can be used directly if an ODE model for a GRN is given. For a general n-dimensional system with ODEs described by $\dot{X}_1 = f_1(\bar{X}, \overline{\Theta}), \ldots, \dot{X}_n = f_n(\bar{X}, \overline{\Theta})$, the sensitivity matrix within the vicinity of steady state $\bar{X}^*$ and around a nominal set of parameters Θ_o is:

$$S_{\bar{X},\bar{\Theta}} := \frac{d\bar{X}}{d\bar{\Theta}} = -\left(\frac{\partial \bar{f}}{\partial \bar{X}}\bigg|_{(\bar{X}^*,\bar{\Theta}_o)}\right)^{-1} \frac{\partial \bar{f}}{\partial \bar{\Theta}}\bigg|_{(\bar{X}^*,\bar{\Theta}_o)} \quad (17)$$

where $\bar{f} = [f_1, \ldots, f_n]'$.

In the normalized version of this sensitivity matrix,

$$S^*_{\bar{X},\bar{\Theta}} = \left(D^{\bar{X}^*}\right)^{-1} \cdot S_{\bar{X},\bar{\Theta}} \cdot D^{\bar{\Theta}^*}$$

where $\left(D^{\bar{X}^*}\right)^{-1} = \text{diag}\{\bar{X}^*\}$ and $D^{\bar{\Theta}_o} = \text{diag}\{\bar{\Theta}_o\}$, the entries of this matrix can be compared to each other to assess the parameters to which the system's steady state is most sensitive. In Fig. 13b, a MATLAB code block is provided that carries out this computation for our 2D system in Eq. 15. As seen in the figure, this matrix can be computed at different SSSs (as specified by the parameter "`state_index`").

For the nominal parameters used, the ODE model of Eq. 15 has three SSSs (recall Fig. 7). Using this code, the normalized sensitivity matrix at the SSS S^2 from Fig. 7, for instance, is:

	α_{ox}	α_{ox}	a_{xx}	a_{yx}	b_x	a_{xy}	a_{yy}	b_y	γ_x	γ_y
X^*	0.0033	0.0273	2.1682	0.1437	0.3498	0.0728	0.7706	17.7313	0.1175	0.3814
Y^*	0.0004	0.0070	0.2290	0.0152	0.0370	0.0174	0.1844	4.5339	0.0300	0.0975

A quick glance at this matrix shows that both the X and Y components of S^2 are most sensitive to the parameter b_y compared to the rest of the parameters. Note that the sensitivity matrix can be computed only at steady state/parameter combinations where the Jacobian matrix $d\bar{f}/d\bar{X}$ is non-singular, and therefore cannot be computed at bifurcation points.

Global Sensitivity Analysis

Global sensitivity analysis can offer a wider glimpse into how parameters affect the salient properties of a system. In this type of analysis, Monte Carlo methods [76] can be used to randomly sample several parameters at a time over a sufficiently large number of repetitions. In doing so, any output that one may be interested in as these parameters are varied, such as the number of SSS, their locations, or the dominant eigenvalues associated with them, is observed in the aggregate across a statistically representative set of outputs.

The crucial part of this process is running a sufficient number of repetitions so that the parameter sets used are representative of the entire parameter space such that, in turn, the outputs realized are

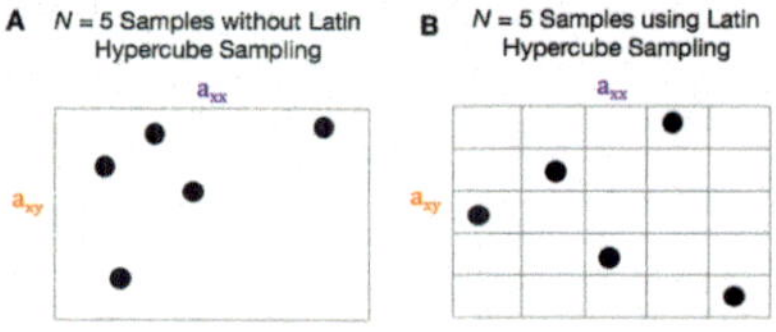

C Latin Hypercube Sampling – lhsnorm

```
N = 1000; % # of samples (partitions in each direction)
% set means and std devs of normal distributionss
mu_axx = 2; mu_ayx = 10; mu_bx = 7.5;
mu_axy = 0.17; mu_ayy = 1.8; mu_by = 2;
mu_list = [mu_axx, mu_ayx, mu_bx, mu_axy, mu_ayy, mu_by];
sigma_axx = 0.2; sigma_ayx = 0.2; sigma_bx = 0.2;
sigma_axy = 0.01; sigma_ayy= 0.2; sigma_by = 0.2;
sigma_list = diag([sigma_axx,sigma_ayx, ...
    sigma_bx,sigma_axy,sigma_ayy,sigma_by])
% generate a latin hypercube sample
hypercube = lhsnorm(mu_list,sigma_list,N)
totals = []; % vector to store # total steady states
for i =1:N % loop through hypercube
    syms X Y; % define variables
    % constant params
    alpha0x = 0.01; alpha0y = 0.01;
    gammaXbar = 1; gammaYbar = 1;
    uxbar = 0; uybar = 0;
    % params sampled from LHS space
    tuple = hypercube(i,:);
    axx = tuple(1); ayx = tuple(2); bx = tuple(3);
    axy = tuple(4); ayy = tuple(5); by = tuple(6);
    % define ODEs
    dXdt =(alpha0x+ayx*Y^2+axx*X^2+bx*(X^2)*(Y^2)) ...
        /(1+X^2+Y^2+(X^2)*(Y^2))-gammaXbar*X + uxbar;
    dYdt =(alpha0y+ayy*Y^2+axy*X^2+ by*(X^2)*(Y^2))...
        /(1+X^2+Y^2+(X^2)*(Y^2))-gammaYbar*Y + uybar;
    % find numerical solutions
    [sol_X,sol_Y]=vpasolve([dXdt == 0, dYdt == 0],[X, Y]);
    % find real solutions & store total # of steady states
    sol_X = sort(sol_X(find(imag(sol_X)==0)));
    sol_Y = sort(sol_Y(find(imag(sol_Y)==0)));
    totals(i) = (length(sol_X)); %add totals to vector
end
histogram(totals) %display totals vector with histogram
```

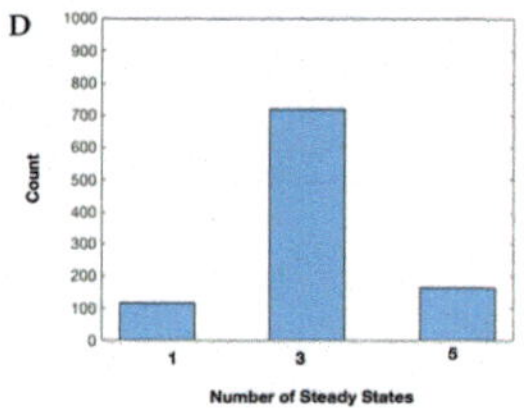

Fig. 14 Global sensitivity analysis using Latin hypercube sampling (LHS). (**a**, **b**) The filling of a 2D parameter space is conceptually shown for five samples. Without LHS, the samples may not efficiently cover the entire parameter space. LHS enforces sampling rules such that each sample must be the only one in its row and column in a grid generated by the partitioning of the parameter space, resulting in a more uniform covering of the sample space while still being random. Partitioning of the parameter space is accomplished by partitioning each individual distribution from which each parameter is sampled. (**c**) Using the lhsnorm function in MATLAB, we implement a Latin hypercube sampling of the parameters a_{xx}, a_{xy}, b_{xy}, a_{yy}, a_{yx}, and b_y of our 2D example in Eq. 15 and calculate the number of total steady states in the model across 1000 repetitions. Each parameter is sampled from a normal distribution centered at its nominal value from Fig. 7, with variances as shown. (**d**) For the distributions used here, the system possesses three steady states most frequently. Note that this could change as the means and variances are changed and that uniform distributions could even be used with the lhsdesign function in MATLAB. Documentation for **lhsnorm** in MATLAB: https:/www.mathworks.com/help/stats/lhsnorm.html. Documentation for **lhsdesign** in MATLAB: https:/www.mathworks.com/help/stats/lhsdesign.html

statistically representative of the output space. As the number of parameters increases, running a sufficient number of repetitions becomes more challenging due to the curse of dimensionality [76].

There are various computational techniques that can be used to control the sample generation process of Monte Carlo methods so that a statistically representative parameter set is formed without an intractable number of samples needed [77]. Latin hypercube sampling (LHS) is one such method that partitions the parameter space into a grid and enforces sampling rules such that every sample resides in a unique row and column [78]. Figure 14 depicts this with a simple example in 2D parameter space. A Latin hypercube is shown with five samples in the parameter spaces a_{xy} and a_{xx}. Without LHS (Fig. 14a), the 2D parameter space might not efficiently cover the space, while with LHS (Fig. 14b), there is some level of spread enforced by the fact that every sample is the only one in the row and column of the grid formed by partitioning the parameter space.

In Fig. 14c, we provide MATLAB code that generates the number of steady states in the model of Eq. 15 as the six parameters

a_{xx}, a_{xy}, b_{xy}, a_{yy}, a_{yx}, and b_y are collectively randomly sampled. The code uses the function `lhsnorm` to generate a 6D normal Latin hypercube, where each parameter is sampled from a normal distribution with the means at nominal values from Fig. 7 and variances as specified in Fig. 14. For $N = 1000$ samples, we see that our model has 3 total steady states around 70% of the time, 5 total steady states about 18% of the time, and 1 total steady state about 12% of the time.

We note that the distributions from which the parameters are being sampled can have a significant difference on the results of a global sensitivity analysis. If sampling from normal distributions, increasing variances of these distributions would include more extreme values away from the nominal means. In addition, the MATLAB function `lhsdesign` offers LHS from uniform distributions, which may be used if those are considered better representative of the parameters in a model. Ultimately, these are decisions that one has to make in accordance with the degree of global sensitivity one is interested in capturing, and the claims made about global sensitivity should be explained in the context of assumptions made on the parameter distributions used.

In summary, the collection of tools we have provided a tutorial for in this section can be used toward understanding the properties of arbitrary GRNs that abide by the assumptions listed at the end of Subheading 2.1. Using the MATLAB and Mathematica implementations of these various tools for our running GRN example, practitioners can adapt the code we provide according to the architecture of their networks.

2.4 Modeling Cell Fate Reprogramming: Response of GRNs to Artificial Perturbations

The dynamical systems tools discussed in Subheading 2.3 were introduced as a means of garnering insight into the intrinsic properties of GRNs. In this section, we use these tools to analyze the response of GRNs to *artificial perturbations* that can capture the manipulations that experimentalists exert during cell fate reprogramming.

In the overexpression reactions of Table 1f for our example GRN in Fig. 2, we modeled these artificial perturbations as constant production terms, u_x and u_y, on the mRNA species m_X and m_Y, respectively. Following reduction of our model to two dimensions in Eq. 15, the protein species X and Y display constant production terms $\bar{u}_x$ and $\bar{u}_y$ (proportional to the original overexpression terms), respectively, in their dynamics. When modeling TF-mediated cell fate reprogramming, these constant overexpression terms are responsible for the change in the steady state landscape in such a way as to induce the desired transition between two SSSs. However, with nonlinear systems such as ours, it is in general nontrivial to determine what types and levels of artificial perturbation can trigger a desired transition.

To begin elucidating the ways in which artificial perturbations might be applied to trigger a certain transition, we can observe the qualitative change in the steady state landscape as we change parameters that correspond to potential artificial perturbations, assuming these stay approximately constant once applied. We will consider ectopic overexpression and enhanced degradation of TFs as the possible artificial perturbations to our model in Eq. 15.

2.4.1 Open-Loop Overexpression

In Fig. 15, we use the nullclines and bifurcation tools discussed in Subheading 2.3 to observe the response of the steady state landscape of the system to increasing values of $\bar{u}_x$ and $\bar{u}_y$. In Fig. 15a, we see that the landscape again begins with three SSSs: S^1, S^2, and S^3. From a control design point of view [79], we can regard prefixed overexpression of TFs as an "open-loop" control strategy, wherein the input of the system (overexpression level) is prefixed at the beginning of the experiment and is not adjusted based on the TF levels throughout the experiment (Fig. 15b). In panels c and d, we note the form in which the nullclines change as a result of increasing $\bar{u}_x$ and $\bar{u}_y$, respectively. And though it is difficult, given the nonlinearity of the system, to pinpoint systematically how the nullclines will change shape or move across the X–Y plane, we can qualitatively see that increasing $\bar{u}_x$ (perturbation #1) causes S^1 to disappear first followed by S^2, while increasing $\bar{u}_y$ (perturbation #2) causes S^2 to disappear first, followed by S^1 (Fig. 15c–f). In either case, sufficiently high overexpression will leave only one SSS, $S^{3'}$, in the proximity of S^3 (i.e., in the basin of attraction of S^3; see bifurcation plots in Fig. 15e, f for $\bar{u}_x$ and $\bar{u}_y$ very high). This is an important insight for the reprogramming strategy. Specifically, if the starting state (phenotype) is S^1 and our objective is to trigger a transition into S^2, overexpression of $\bar{u}_x$ may be a preferable approach to overexpression of $\bar{u}_y$.

In panels g, h of Fig. 15, we simulate a reprogramming experiment using open-loop overexpression with an intermediate level of $\bar{u}_x$ to facilitate a transition from S^1 to S^2. Starting with the system at S^1 (Fig. 15g), we apply $\bar{u}_x = 0.2$ (a level informed by the bifurcation plots of Fig. 15e) and see that the nullclines and vector field change in a fashion that forces the state to transition to $S^{2'}$. Upon removal of this input in Fig. 15h, we see that the system transitions from $S^{2'}$ to S^2, and hence the reprogramming from S^1 to S^2 is complete. We note that even though overexpression was possible with this intermediate level of $\bar{u}_x$, very high overexpression of either TF will not be a successful strategy to triggering transitions from S^1 to S^2 as the system's state will ultimately approach S^3 instead (*see* bifurcation plots in Fig. 15e, f). At best with $\bar{u}_x$, there is a finite window of overexpression that is required for triggering the desired transition, and it may not be experimentally trivial to set the overexpression

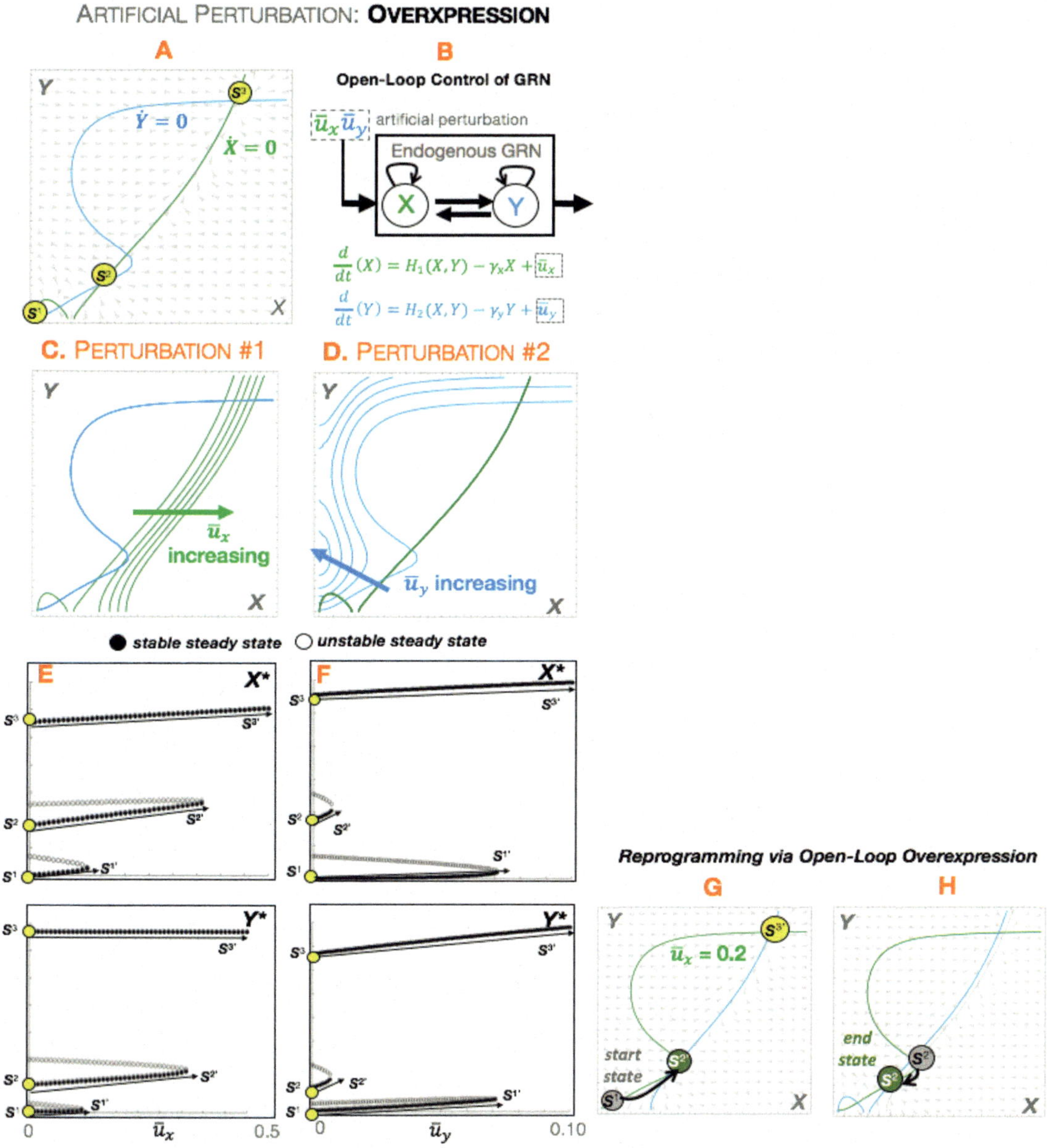

Fig. 15 Educating overexpression strategies in reprogramming. (**a**) In the steady state landscape of our model in [17], there are three SSSs when the model is not artificially perturbed (same parameters used as in Fig. 7). (**b**) From a control systems point of view, we can regard ectopic overexpression of TFs as "open-loop" control in the sense that the perturbation is applied at constant, preset levels ($\bar{u}_X$ and $\bar{u}_Y$) that are not changed in response to how the concentrations of TFs change during the experiment. (**c**) The effect of increasing $\bar{u}_X$ is to cause the nullcline $\dot{X} = 0$ to change in a fashion that causes the steady state S^1 to disappear before S^2 and S^3. (**d**) The effect of increasing $\bar{u}_Y$ is to cause the nullcline $\dot{Y} = 0$ to change in a fashion that causes S_2 to disappear first, followed by the disappearance of S^1 and the persistence of only one SSS in the proximity of S^3. (**e**) Bifurcation plots confirm that S^1, followed by S^2, disappears as $\bar{u}_X$ is increased. (**f**) Bifurcation plots confirm that as $\bar{u}_Y$ is increased, S^2 disappears followed by S^1, with only one state remaining in the proximity of S^3. (**g**) If the system starts at S^1, applying an intermediate level of overexpression, $\bar{u}_X = 0.2$, temporarily changes the landscape such that a transition to $S^{2'}$ is induced. (**h**) Upon removal of this artificial overexpression, the system transitions from $S^{2'}$ to S^2

level precisely in that window. So, overexpression may in practice fail for this example GRN.

Cooperative Monotone Systems: Fundamental Limitations on Reprogrammability

Our model makes some potentially useful predictions about possible reprogramming strategies using open-loop overexpression. In particular, it indicates that there may be a fundamental flaw in these types of perturbations for certain transitions, such as the one from S^1 to S^2. There is a fundamental reason why certain transitions from states characterized by lower concentrations (S^1) to states characterized by higher, but not maximal, concentrations (S^2) cannot be guaranteed with the open-loop overexpression described in Fig. 15. It is because our example GRN of Fig. 2 only has activating arrows between the TFs. At an intuitive level, this means that both genes are always upregulating themselves as well as each other, so stimulating these nodes artificially (via open-loop ectopic overexpression) very easily sets off an upregulation positive feedback cascade that pushes concentrations of both to higher and higher levels so that the system becomes monostable at maximal (i.e., non-intermediate) values. In fact, reprogramming GRNs to intermediate states is a difficult task to achieve in an even more general class of GRNs that can be described as "cooperative monotone" systems [35], of which our example is a type. However, cooperative networks as referred to here should not be confused with cooperative co-binding modes discussed in Subheading 2.1. Hence, it is not possible using these reprogramming paradigms to guarantee that reprogramming from states with lower concentrations to states with intermediate concentrations (such as the transition from S^1 to S^2) will be successful.

Enhanced Degradation

Another way to experimentally perturb our GRN is by adding proteases or microRNA [50] to artificially enhance the degradation rate of TFs X and Y. In our model in Eq. 15, this is akin to increasing the values of the overall decay rates γ_x and γ_y, respectively. In Fig. 16, we repeat the analysis from above, this time analyzing the effect of increasing these decay rates from their nominal values in Fig. 7. As demonstrated by the changing nullclines and bifurcation plots in panels 16C-H, we see that enhancing the degradation rate of γ_x (perturbation #3) causes the SSS S^2 to disappear first, followed by S^3. Likewise, enhancing the degradation rate of γ_y (perturbation #4) causes the SSS S^3 to disappear first and maintains the SSSs in the proximity of S^1 and S^2.

Based on the way increasing degradation rates change the steady state landscape, another perturbation we might consider for use in transitioning from S^1 to S^2 is appropriately combining overexpression with enhanced degradation (perturbations 1 and 4) so that both S^1 and S^3 disappear while keeping a state $S^{2'}$ in the vicinity of S^2. Figure 17 shows the results of simulating this experiment, in which we use the same level of intermediate

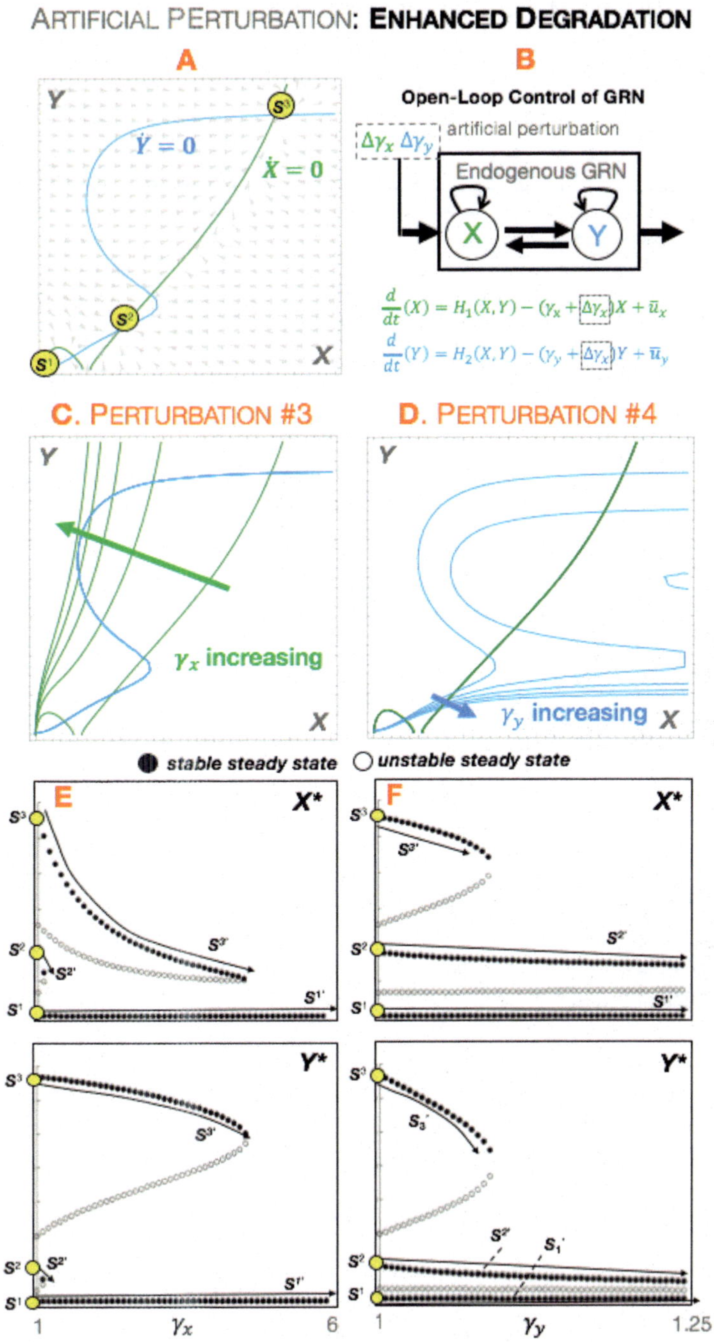

Fig. 16 Educating enhanced degradation strategies in reprogramming. (**a**) In the steady state landscape of our model in [17], there are three SSSs when the model is not artificially perturbed (same parameters used as in Fig. 7). (**b**) We can model enhanced degradation of TFs X and Y via artificial additions of proteases or microRNA as akin to increasing the decay rates $\Delta\gamma_X$ and $\Delta\gamma_Y$, respectively. (**c**) The effect of increasing γ_X is to cause the nullcline $\dot{X} = 0$ to change in a fashion that causes the steady state S^2 to disappear first, followed by S^3. (**d**) The effect of increasing γ_Y is to cause the nullcline $\dot{Y} = 0$ to change in a fashion that causes only S^3 to disappear, with two states in the proximity of S^1 and S^2 remaining. (**e**, **f**) Bifurcation plots confirm these observations

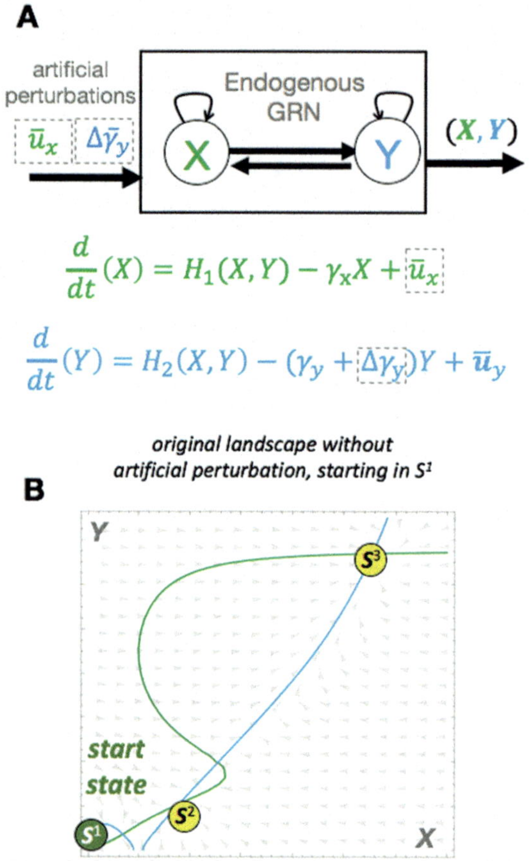

Fig. 17 Reprogramming via combination of overexpression and enhanced degradation: (**a**) Overexpression of TF X and enhanced degradation of TF Y are modeled as the addition of constant term to the dynamics of X ($\bar{u}_x$) and an increased degradation ($\Delta\gamma_y$) in the dynamics of Y, respectively. (**b**) Without artificial perturbations, the landscape possesses the SSSs S^1, S^2, S^3. (**c**) If the system starts at S^1, applying an intermediate level of overexpression, $\bar{u}_x = 0.2$, combined with an increase in degradation of Y, $\Delta\gamma_y = +0.1$, temporarily changes the landscape such that a transition to $S^{2'}$ is induced. (**d**) Upon removal of this artificial overexpression and enhanced degradation, the system transitions from $S^{2'}$ to S^2

overexpression as in the experiment above, in addition to increasing the degradation on TF Y enough to make S^3 disappear (informed by the bifurcation plots of Fig. 16f). This causes the system to transition to a state $S^{2'}$ near S^2 (Fig. 17c), so that upon cessation of both artificial perturbations (Fig. 17d), the system transitions to S^2 and reprogramming is complete.

2.4.2 Closed-Loop Overexpression

The crucial conclusion from these theoretical results is that open-loop overexpression, whether or not it is combined with enhanced degradation, cannot be guaranteed to be a winning strategy for certain reprogramming tasks. To address this, in previous work [35], we have proposed an entirely new paradigm called closed-

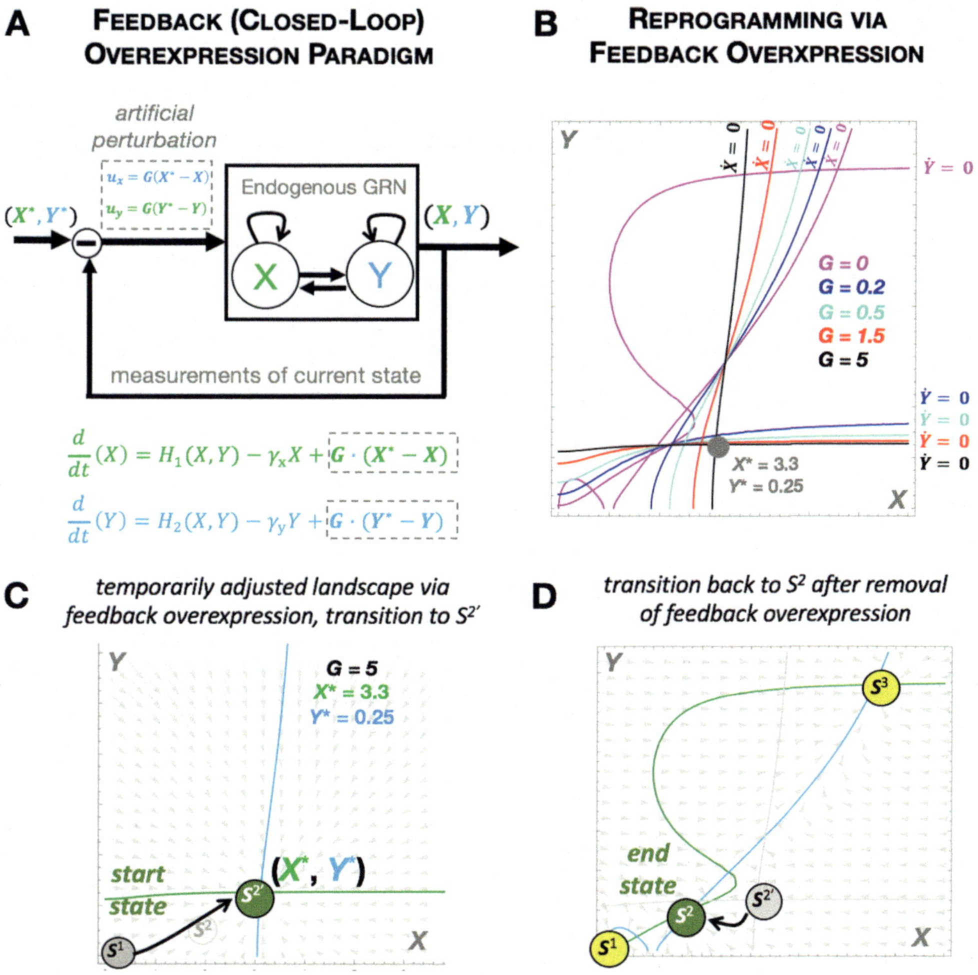

Fig. 18 Reprogramming via closed-loop feedback overexpression. (**a**) In closed-loop overexpression, measurements of the state (X, Y) are taken throughout the experiment, and the overexpression is adjusted in proportion to the distance between the current state (X, Y) and the target state (X^*, Y^*). (**b**) As the values of gain G are increased, the nullclines increasingly morph into straight lines that intersect at the target state. (**c**) If a feedback reprogramming experiment begins in state S^1, and there is high enough gain ($G = 5$), then feedback overexpression can be used to steer the system to any desired point $S^{2'}$ given by the coordinates (X^*, Y^*) in the basin of attraction of S^2. (**d**) Upon removal of feedback overexpression, the system transitions from $S^{2'}$ to S^2

loop or feedback overexpression (Fig. 18a), in which the level of ectopic overexpression is not merely set once at the beginning of an experiment. Rather, if TF X in our system needs to be steered to a target concentration X^* (where X^* may be the concentration seen in a target phenotype in a reprogramming experiment), the actual concentration is measured throughout the experiment, and the level of overexpression or degradation $\bar{u}_x$ is altered in proportion to the distance between the current state and target state, i.e., $\bar{u}_x = G(X^* - X)$. Our theory shows that with high enough values of

G, known as the *gain* of the feedback controller, any starting state can in principle be steered to arbitrary levels X^* of concentrations for each node in the network that is overexpressed via feedback overexpression. This is true of any system regardless of the initial shape of the nullclines, which demonstrates that this strategy is not dependent upon the specific dynamics or parameters of a GRN.

In Fig. 18b, we demonstrate what this means for the nullclines for our model in Eq. 15. As the figure shows, as the values of gain G are increased, the nullclines morph into increasingly straight lines that intersect at the target state (X^*, Y^*). In panels c, d of Fig. 18, we simulate a reprogramming experiment in our running example from S^1 to S^2 via closed-loop overexpression.

2.4.3 Realizing Genetic Feedback Control

In the feedback overexpression paradigm we have presented, which contrasts with the status quo of "open-loop" overexpression paradigms used in most TF-mediated reprogramming experiments, transitions to intermediate states can in fact be theoretically guaranteed (demonstrated in the theory shown in [35]). In addition, we have shown that one way to realize this closed-loop overexpression without requiring continuous measurements of concentration

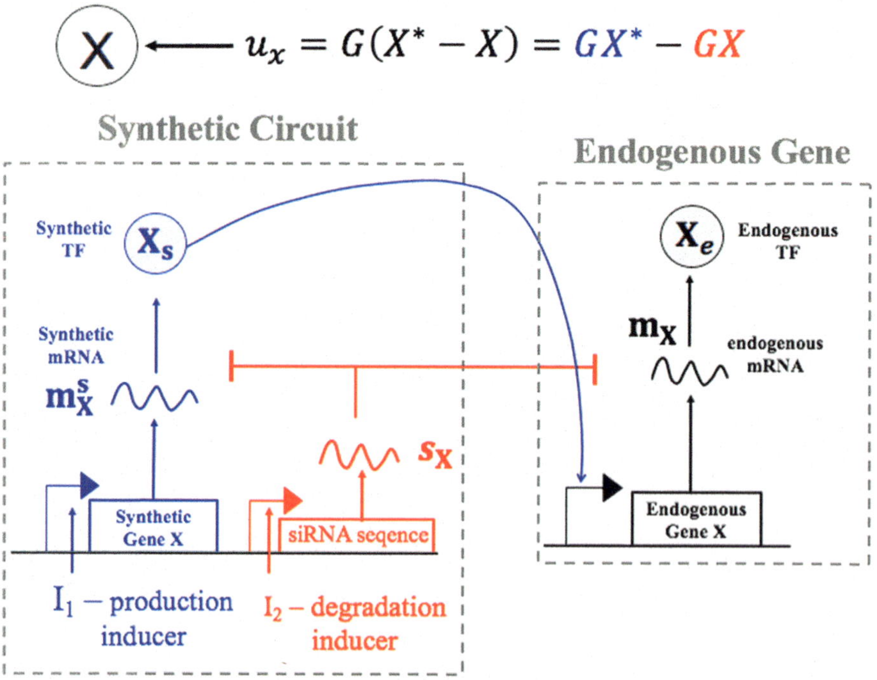

Fig. 19 Realizing feedback overexpression with a synthetic genetic circuit. Feedback overexpression of a TF X can be realized with a synthetic construct consisting of inducer-activated synthetic gene X along with an inducer-activated siRNA sequence that degrades the mRNA of X. With high enough copy number for the synthetic circuit, the right balance of these inducers allows for steering the system's state to arbitrary concentration levels of X, as explained in detail in Del Vecchio et al. [35]

(which may be difficult experimentally) is by the use of a synthetic genetic feedback controller that simultaneously ectopically overexpresses and degrades the mRNA of a species under control using short interfering RNA (siRNA) (Fig. 19). By the appropriate balancing of these two artificial perturbations using inducers that control the expression levels of mRNA using a synthetic gene and the expression levels of the siRNA, the system can be in principle steered to arbitrary concentrations independent of network dynamics and parameters. Moreover, we have shown in detail in [35] that the high gain G needed according to the theory can be achieved by increasing the copy number of the ectopic DNA on which the synthetic circuit is encoded. At an intuitive level, this is tantamount to having a sufficient amount of the synthetic circuit inside the cell being reprogrammed that it dominates the effect of endogenous dynamics and takes over the network to steer it to arbitrary concentration levels.

3 Conclusion

The topics we have discussed in this chapter comprise the basic steps in the computational approach to cell fate reprogramming. At the root of the approach we have described is the existence of a GRN that has been implicated in the cell fate change of interest. Moreover, in order to obtain an ODE model, we have made certain kinetic assumptions about how the species of that GRN interact. By way of a running example of a GRN with two TFs mutually and autoactivating, we have gone through a series of tutorials on methods from dynamical systems theory that can be used to probe the intrinsic properties of GRNs and shown their implementations using Mathematica and MATLAB. These include methods that identify the number and location of stable steady states of a GRN (corresponding to phenotypes) as well as various ways to understand the model's dependence on parameters. The code blocks we have provided can in many cases straightforwardly be adapted to suit practitioners' ODE models corresponding to GRNs with various topologies. We then provided a discussion for how to use these tools to understand GRNs' responses to artificial perturbations, which model the stimulations that experimenters apply to a GRN during cell fate reprogramming. Ultimately, this is the purpose of these models: to understand the response of a GRN in silico to various types of artificial perturbations so that one can inform in vitro and in vivo experiments (*see* Subheadings 2.2–2.4).

Throughout this chapter, the differential equation models we have used were deterministic. As noted in Subheading 2.1, this is because ODE models are an appropriate starting point for a first-pass understanding of the qualitative behavior of GRNs. However, they are only appropriate to describe the mean behavior of species' dynamics when molecular counts are high enough to be stated in terms of concentrations. The next step after understanding GRNs in the deterministic realm is typically to use stochastic differential equation (SDE) models, such as the Chemical Langevin Equation (CLE), which can capture the effects of the prevalent intrinsic and extrinsic noise in biological systems [80, 81]. The Chemical Master Equation (CME) is most appropriate when molecular counts are low, while the Chemical Langevin Equation (CLE) is appropriate when concentration is still a meaningful description of molecular counts. When molecular counts are low, practitioners may find that the GRN behaviors observed in an ODE model are dramatically different from the mean behavior of a stochastic model [48].

Another dimension that we have not considered in this chapter is the epigenetic transformation that might often take place during a cell fate reprogramming process. In the past decade, it has become increasingly clear that the unpacking of genes from their condensed chromatin states and the methylome reprogramming of DNA base pairs on the promoters of GRNs are key events [85–85] that take place during what has typically been assumed to be genetic reprogramming alone. In general, the degree of epigenetic transformation that takes place during a given cell fate reprogramming process should be investigated and potentially included in mathematical models of reprogramming.

References

1. Waddington CH (1957) The strategy of genes. Routledge, New York
2. Mitalipov S, Wolf D (2009) Totipotency, pluripotency and nuclear reprogramming. In: Martin U (ed) Engineering of stem cells. Advances in biochemical engineering/biotechnology, vol vol. 114. Springer, Berlin, Heidelberg
3. Xie H, Ye M, Feng R, Graf T (2004) Stepwise reprogramming of B cells into macrophages. Cell 117(5):663–676
4. Briggs R, King TJ (1952) Transplantation of living nuclei from blastula cells into enucleated frogs' eggs. Proc Natl Acad Sci U S A 38:455–463
5. Gurdon JB, Elsdale TR et al (1958) Sexually mature individuals of *Xenopus laevis* from the transplantation of single somatic nuclei. Nature 182:64–65. https://doi.org/10.1038/182064a0
6. Campbell KH, McWhir J, Ritchie WA, Wilmut I (1996) Sheep cloned by nuclear transfer from a cultured cell line. Nature 380:64–66. https://doi.org/10.1038/380064a0
7. Wakayama T, Perry AC et al (1998) Full-term development of mice from enucleated oocytes injected with cumulus cell nuclei. Nature 394:369–374. https://doi.org/10.1038/28615
8. Tapscott SJ, Davis RL et al (1988) MyoD1: a nuclear phosphoprotein requiring a Myc homology region to convert fibroblasts to myoblasts. Science 21:405–411
9. Davis RL, Weintraub H, Lassar AB (1987) Expression of a single transfected cDNA converts fibroblasts to myoblasts. Cell 51:987–1000
10. Takahashi K, Yamanaka S (2006) Induction of pluripotent stem cells from mouse embryonic and adult fibroblast cultures by defined factors. Cell 126:663–676
11. Radzisheuskaya A, Chia Gle B et al (2013) A defined Oct4 level governs cell state transitions

of pluripotency entry and differentiation into all embryonic lineages. Nat Cell Biol 15 (6):579–590
12. González F, Boué S, Belmonte JC (2011) Methods for making induced pluripotent stem cells: reprogramming àla carte. Nat Rev Genet 12(4):231–242
13. Buganim Y, Faddah DA, Jaenisch R (2013) Mechanisms and models of somatic cell reprogramming. Nat Rev Genet 14(6):427–439. https://doi.org/10.1038/nrg3473
14. David L, Polo JM (2014) Review: phases of reprogramming. Stem Cell Res 12(3):754–761
15. de BW, Zimm R, Brusch L (2013) Transdifferentiation of pancreatic cells by loss of contact-mediated signaling. BMC Syst Biol 7:77. https://doi.org/10.1186/1752-0509-7-77
16. Yao E, Lin C et al (2017) Notch signaling controls transdifferentiation of pulmonary neuroendocrine cells in response to lung injury. Stem Cells 36(3):377–391
17. Malik N, Rao MS (2013) A review of the methods for human ipsc derivation. Methods Mol Biol 997:23–33
18. Schlaeger TM, Daheron L et al (2015) A comparison of non-integrating reprogramming methods. Nat Biotechnol 33:58–63
19. Goh PA, Caxaria S et al (2013) A systematic evaluation of integration free reprogramming methods for deriving clinically relevant patient specific induced pluripotent stem (ips) cells. PLoS One 8(11):e81622. https://doi.org/10.1371/journal.pone.0081622
20. iPS cells 10 years later. Cell 2016;166(6): 1356–1359. https://doi.org/10.1016/j.cell.2016.08.043
21. Zhou Q, Brown J, Kanarek A, Rajagopal J, Melton DA (2008) In vivo reprogramming of adult pancreatic exocrine cells to b-cells. Nature 455:627–632
22. Pang ZP, Yang N et al (2011) Induction of human neuronal cells by defined transcription factors. Nature 476(7359):220–223. https://doi.org/10.1038/nature10202
23. Bussmann LH, Schubert A et al (2009) A robust and highly efficient immune cell reprogramming system. Cell Stem Cell 5:554–566
24. Laiosa CV, Stadtfeld M et al (2006) Reprogramming of committed T cell progenitors to macrophages and dendritic cells by C/EBPα and PU.1 transcription factors. Immunity 25 (5):731–744
25. Vierbuchen T, Ostermeier A et al (2010) Direct conversion of fibroblasts to functional neurons by defined factors. Nature 463:1035–1041
26. Yu U, Lee SH et al (2004) Review: bioinformatics in the post-genome era. J Biochem Mol Biol 37(1):75–82
27. Boyer LA, Lee TI et al (2005) Core transcriptional regulatory circuitry in human embryonic stem cells. Cell 122(6):947–956
28. Zhang B, Wolynes P (2014) Stem cell differentiation as a many body problem. Proc Natl Acad Sci 111:10185–10190
29. Orkin SH, Wang J et al (2008) The transcriptional network controlling pluripotency in ES cells. Cold Spring Harb Symp Quant Biol 73:195–202
30. Chickarmane V, Enver T et al (2009) Computational modeling of the hematopoietic erythroid-myeloid switch reveals insights into cooperativity, priming, and irreversibility. PLoS Comput Biol 5:e1000268
31. Goldfarb AN (2007) Transcriptional control of megakaryocyte development. Oncogene 26 (47):6795–6802
32. Friedman AD (2007) Transcriptional control of granulocyte and monocyte development. Oncogene 26(47):6816–6828
33. Gupta P, Gurudutta GU et al (2009) PU.1 and partners: regulation of haematopoietic stem cell fate in normal and malignant haematopoiesis. J Cell Mol Med 13:4349–4363
34. Chickarmane V, Troein C et al (2006) Transcriptional dynamics of the embryonic stem cell switch. PLoS Comput Biol 2(9):e123. https://doi.org/10.1371/journal.pcbi.0020123
35. Del Vecchio D, Abdallah H et al (2017) A blueprint for a synthetic genetic feedback controller to reprogram cell fate. Cell Syst 4 (1):109–120
36. Abdallah H, Del Vecchio D, Qian Y, Collins JJ (2016) A dynamical model for the low efficiency of induced pluripotent stem cell reprogramming. Paper presented at American Control Conference, Boston, MA, June 2016
37. Olariu V, Lövkvist C et al (2016) Nanog, Oct4 and Tet1 interplay in establishing pluripotency. Sci Rep 6:25438
38. Huang S, Guo YP, May G, Enver T (2007) Bifurcation dynamics in lineage-commitment in bipotent progenitor cells. Dev Biol 305 (2):695–713
39. Santillán M (2008) On the use of the Hill functions in mathematical models of gene regulatory networks. Math Modell Nat Phenom 3

(2):85–97. https://doi.org/10.1051/mmnp:2008056

40. Liew CW, Rand KD et al (2006) Molecular analysis of the interaction between the hematopoietic master transcription factors GATA-1 and PU.1. J Biol Chem 281:28296–28306
41. Tian T, Smith-Miles K (2014) Mathematical modeling of GATA-switching for regulating the differentiation of hematopoietic stem cell. BMC Syst Biol 8(Suppl 1):S8
42. Zhou JX, Brusch L, Huang S (2011) Predicting pancreas cell fate decisions and reprogramming with a hierarchical multi-attractor model. PLoS One 6(3):e14752
43. Strogatz S (2014) Nonlinear dynamics and chaos: with applications to physics, biology, chemistry and engineering (studies in nonlinearity). Westview Press, Boulder, CO
44. Del Vecchio D, Murray RM (2014) Biomolecular feedback systems. Princeton University Press, Boston, MA
45. Gillespie DT (2009) Deterministic limit of stochastic chemical kinetics. J Phys Chem 113 (6):1640–1644. https://doi.org/10.1021/jp806431b
46. Gillespie DT (2000) The chemical langevin equation. J Chem Phys 113(1):297–306
47. Van Kampen NG (1992) Stochastic processes in physics and chemistry, vol 1. North Holland Publishing Co., Amsterdam
48. Al-Radhawi MA, Sontag E, Del Vecchio D (2017) Multi-modality in gene regulatory networks with slow gene binding. arXiv:1705.02330
49. Polynikis A, Hogan SJ, di Bernardo M (2009) Comparing different ODE modelling approaches for gene regulatory networks. J Theor Biol 261(4):511–530
50. Agrawal N, Dasaradhi PV et al (2003) RNA interference: biology, mechanism, and applications. Microbiol Mol Biol Rev 67(4):657–685
51. Berg JM, Tymoczko JL, Stryer L (2002) Biochemistry, 5th edn. W. H. Freeman, New York. Section 10.4, Covalent modification is a means of regulating enzyme activity
52. Gyorgy A, Del Vecchio D (2014) Modular composition of gene transcription networks. PLoS Comput Biol 10(3):e1003486
53. Geertz M, Maerkl SJ (2010) Experimental strategies for studying transcription factor—DNA binding specificities. Brief Funct Genomics 9 (5–6):362–373. https://doi.org/10.1093/bfgp/elq023
54. Horak CE, Snyder M (2002) ChIP-chip: a genomic approach for identifying transcription factor binding sites. Methods Enzymol 350:469–483
55. Park PJ (2009) ChIP-seq: advantages and challenges of a maturing technology. Nat Rev Genet 10:669–680
56. Wang Z, Gerstein M, Snyder M (2009) RNA-Seq: a revolutionary tool for transcriptomics. Nat Rev Genet 10(1):57–63. https://doi.org/10.1038/nrg2484
57. Alberts B, Johnson A, Lewis J et al (2002) Molecular biology of the cell. Garland Science, New York. Available from: https://www.ncbi.nlm.nih.gov/books/NBK26887/
58. Zhou P (2004) Determining protein half-lives. Methods Mol Biol 284:67–77
59. Kuhar MJ (2010) Measuring *levels* of proteins by various technologies: can we learn more by measuring turnover? Biochem Pharmacol 79 (5):665–668. https://doi.org/10.1016/j.bcp.2009.09.029
60. Sezonov G, Joseleau-Petit D et al (2007) *Escherichia coli* physiology in Luria-Bertani broth. J Bacteriol 189(23):8746–8749
61. Cooper GM (2000) The cell: a molecular approach, The eukaryotic cell cycle, 2nd edn. Sinauer Associates, Sunderland, MA. Available from: https://www.ncbi.nlm.nih.gov/books/NBK9876/
62. Takahashi K, Yamanaka S (2016) A decade of transcription factor-mediated reprogramming to pluripotency. Nat Rev Mol Cell Biol 17 (3):183–193. https://doi.org/10.1038/nrm.2016.8
63. Milo R, Jorgensen P et al (2010) BioNumbers—the database of key numbers in molecular and cell biology. Nucleic Acids Res 38: D750–D753
64. Schwanhäusser B, Busse D et al (2011) Global quantification of mammalian gene expression control. Nature 473(7347):337–342. https://doi.org/10.1038/nature10098
65. Alon U (2006) An introduction to systems biology: design principles of biological circuits. CRC Press, Boca Raton, FL
66. Yuan L, Chan GC et al (2016) A role of stochastic phenotype switching in generating mosaic endothelial cell heterogeneity. Nat Commun 7:10160
67. To T-L, Maheshri N (2010) Noise can induce bimodality in positive transcriptional feedback loops without bistability. Science 327 (5969):1142–1145
68. Kauffman S (1973) Control circuits for determination and transdetermination. Science 181:310–318

69. Huang S, Eichler G et al (2005) Cell fates as high-dimensional attractor states of a complex gene regulatory network. Phys Rev Lett 94 (12):128701
70. Huang S (2009) Reprogramming cell fates: reconciling rarity with robustness. BioEssays 31:546–560
71. Strang G (2009) Introduction to linear algebra. Wellesley-Cambridge Press, Wellesley, MA
72. Khalil H (2014) Nonlinear control. Pearson, London
73. Slotine JJ, Li W (1991) Applied nonlinear control. Pearson, London
74. Wiggins S (2003) Introduction to applied nonlinear dynamical systems and chaos. Springer, New York City
75. Carr J (1981) Applications of centre manifold theory. Springer, New York City
76. Saltelli A, Ratto M et al (2008) Global sensitivity analysis: the primer. Wiley-Interscience, Hoboken, NJ
77. McKay M, Beckman RJ et al (1979) A comparison of three methods for selecting values of input variables in the analysis of output from a computer code. Technometrics 21 (2):239–245. https://doi.org/10.2307/1268522
78. Stein M (1987) Large sample properties of simulations using latin hypercube sampling. Technometrics 29(2):143–151. Correction, Vol. 32, p. 367
79. Astrom KJ, Murray RM (2008) Feedback systems: an introduction for scientists and engineers. Princeton University Press, Boston, MA
80. Swain PS, Elowitz MB et al (2002) Intrinsic and extrinsic contributions to stochasticity in gene expression. Proc Natl Acad Sci 99 (20):12795–12800
81. Elowitz MB, Levine AJ et al (2002) Stochastic gene expression in a single cell. Science 297:1183–1186
82. Allis DC, Caparros M-L et al (2015) Epigenetics. Cold Spring Harbor, New York
83. Bagci H, Fisher AG (2013) DNA demethylation in pluripotency and reprogramming: the role of tet proteins and cell division. Cell Stem Cell 13:265–269
84. De Carvalho DD, You JS et al (2010) DNA methylation and cellular reprogramming. Trends Cell Biol 20:609–617
85. Huang K, Fan G (2010) DNA methylation in cell differentiation and reprogramming: an emerging systematic view. Regen Med 5:531–544

Chapter 18

Computational Analysis of Aneuploidy in Pluripotent Stem Cells

Uri Weissbein

Abstract

Due to their unique cellular features, pluripotent stem cells (PSCs) acquire chromosomal aberrations at a rather high frequency during their growth in culture. Analysis of chromosomal integrity should be routinely performed and usually is done at the DNA level of the cells. RNA sequencing (RNA-Seq) has recently become the basic tool for transcriptional studies. Therefore, methods that utilize this already available data to inspect the genomic integrity are very valuable. In this chapter, we provide a practical guide to implement methods of detection of chromosomal aberrations, which are based on RNA-Seq data. The expression-based karyotyping (e-Karyotyping) method is based on global gene expression analysis, while the expressed-SNP-karyotyping (eSNP-Karyotyping) method is based on changes in the ratio between alleles.

Key words Chromosomal aberrations, RNA sequencing, Cytogenetics, Pluripotent stem cells, Allelic bias, Allelic variation, Gene expression

1 Introduction

Stem cells are defined by their self-renewal and differentiation capabilities. PSCs, isolated from the inner cell mass of the blastocyst stage embryos, are the cells that will form the entire organism. When they are grown under defined culture conditions, they can be maintained at a pluripotent state for an indefinite time, which is in contrast to their transient nature during embryonic development.

Cultured pluripotent cells are subjected to genetic changes due to high level of replication stress, weak cell cycle checkpoints, and aberrations in cell division [1, 2]. These genetic changes can be neutral or harmful, but they might also be growth promoting. In these cases, the aberrant cells rapidly take over the culture. Positive selection can be achieved by increasing cell proliferation, reducing cell death, or reducing spontaneous differentiation. Importantly, the recurrent mutations in PSCs are also detected in tumors, such as dominant negative point mutations in P53 [3], duplication of

Patrick Cahan (ed.), *Computational Stem Cell Biology: Methods and Protocols*, Methods in Molecular Biology, vol. 1975, https://doi.org/10.1007/978-1-4939-9224-9_18,

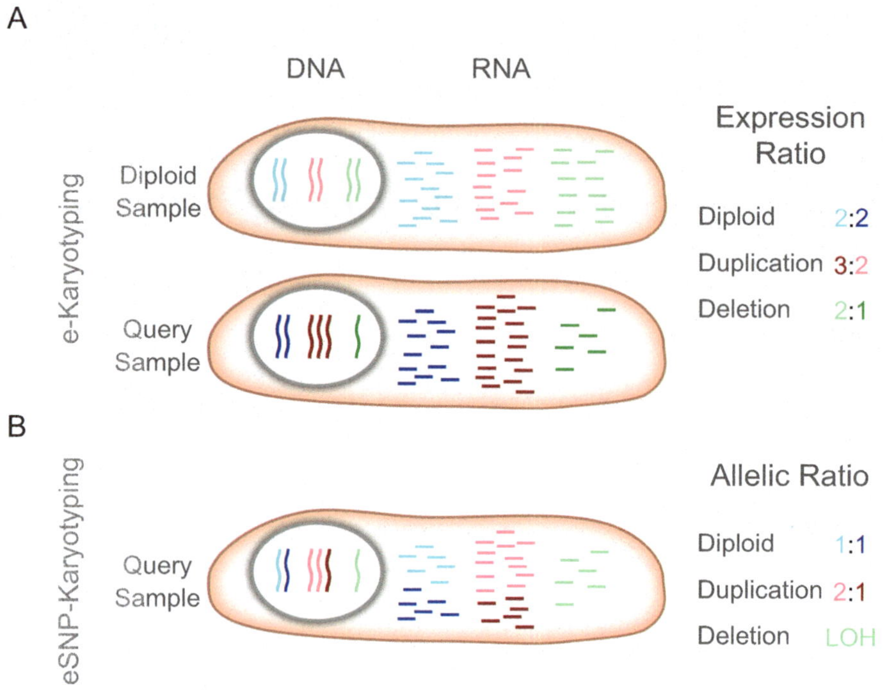

Fig. 1 Schematic representation of the e-Karyotyping (**a**) and eSNP-Karyotyping (**b**) methodologies

chromosome 12 (seen in most germ cell tumors) [4], amplification of the anti-apoptotic gene BCL-XL [5, 6], and more. Thus, these changes are extremely challenging in the context of translational medicine. Moreover, since genetic changes, and especially large chromosomal aberrations, can affect the expression levels of hundreds of genes, they also affect studies related to developmental processes and disease modeling.

There are a variety of tools to detect chromosomal changes. The DNA-based methods rely on analysis of the genetic content and include cytogenetic stainings such as Giemsa or spectral karyotyping (SKY) or molecular methods such as array comparative genomic hybridization (aCGH), single nucleotide polymorphism (SNP) arrays, and whole-genome sequencing (WGS) [7]. However, we have previously introduced a method named e-Karyotyping which deduce genetic changes based on the changes in gene expression levels, analyzed by microarrays [8–10] (Fig. 1a). This method allows identification of chromosomal aberrations by exploiting the correlation between the genetic copy number and gene expression levels. Comparison of gene expression profiles of aberrant and diploid cell lines uncovers large regions with up- or downregulation of many genes correlated with the genetic structural change.

Recently, we adjusted e-Karyotyping to RNA-Seq data, and we have also developed an alternative approach for the detection of chromosomal changes based on RNA-Seq reads [11]. This

Table 1
Comparison between e-Karyotyping and eSNP-Karyotyping

	e-Karyotyping	eSNP-Karyotyping
Small chromosomes	May require parameter adjustments	Works well
Requirement of diploid samples as baseline	Required	Not required
Affected by global gene expression changes	Affected	Not affected
Differentiate between types of LOH	Differentiate	Doesn't differentiate
Complexity of the analysis	Less complex	More complex
Analysis time	About the same time	
Analysis platform	R scripts and CGH-Explorer	R package
Can be applied to microarray data	Yes	No

methodology, named eSNP-Karyotyping, is completely blind to gene expression levels. Instead, it quantifies the ratio between the two chromosomal alleles probing for regions that diverge from the expected 1:1 ratio (Fig. 1b). In case of duplication, the duplicated alleles will be expressed in twice the amount of the other allele. Loss of heterozygosity (LOH, chromosomal deletion, or uniparental disomy) will be identified by large regions of homozygosity (Fig. 1b).

Tools that utilize RNA-Seq information for karyotype inspection are useful since functional studies on gene expression changes are very common. They allow researchers to use the already available sequencing data to examine the cells' genomic integrity. Each one of these methods has its strengths and weaknesses (Table 1). In this chapter, we provide a step-by-step guide to implement both e-Karyotyping and eSNP-Karyotyping from RNA-Seq data.

2 Materials

e-Karyotyping and eSNP-Karyotyping procedures require to acquire a set of tools, files, and R functions as described in the following tables. Most of the analysis (except the CGH-Explorer section) has to be performed in Linux environment. Programs and software are listed in Table 2. Required files and their formats and contents are listed in Table 3. R packages and R commands to install these packages are listed in Table 4.

Table 2
Programs and software

Name	Installation	Comments
TopHat2 Bowtie2 SAMtools	sudo apt install tophat sudo apt install bowtie2 sudo apt install samtools	Linux terminal command
Cufflinks	http://cole-trapnell-lab.github.io/cufflinks/install/	
R	https://www.r-project.org/	RStudio is recommended
GATK	https://software.broadinstitute.org/gatk/download/	Download the file GenomeAnalysisTK.jar
Picard tools	http://broadinstitute.github.io/picard/	Download the file picard.jar
TwoBitToFa	http://hgdownload.soe.ucsc.edu/admin/exe/linux.x86_64/	After downloading twoBitToFa, run the following command: "chmod +x twoBitToFa"
CGH-Explorer	http://heim.ifi.uio.no/bioinf/Projects/CGHExplorer/	Windows software

Table 3
Files

Name	File obtaining	Comments
Whole genome Transcriptome	**Note 1**	FASTA format GTF format
Bowtie2 genome index SAMtools genome index Picard genome dictionary	**Note 2**	Multiple files FAI format DICT format
dbSNP common SNP files	**Note 3**	Separate file for each chromosome, have to be edited using the "Edit_dbSNP_Files" function before use

3 Methods

3.1 Alignment of the Reads to the Reference

The sequencing raw output is a FASTQ file that contains the sequencing reads. After receiving these files, the first step in both the eSNP-Karyotyping and e-Karyotyping analysis is to align the reads to a reference genome, meaning to find the genomic position of each read. In addition to comparing the reads to the genome sequence, an alignment to a transcriptome file is also performed, to allow alignment of reads from spliced exons.

Table 4
R packages

Name	Installation (commands for R)

eSNP-Karyotyping	`install.packages("devtools")`
	`library("devtools")`
	`install_github("BenvenLab/eSNPKaryotyping/eSNPKaryotyping")`
	`library("eSNPKaryotyping")`

gplots	`install.packages("gplots")`
Zoo	`install.packages("zoo")`

The alignment is done on each FASTQ file separately and can be executed with different aligners, such as TopHat2 [12] or STAR [13], which output a file in BAM format. The eSNP-Karyotyping R package contains a function for alignment of FASTQ files with Tophat2. Use the function as described below:

```
Tophat(Directory, Library_Type, Threads, Transcripts_Annotation,
Bowtie_Genome_Index)
```

The function arguments are:

- Directory: The path of the directory where the FASTQ files are located
- Library_Type: Library type: "Paired" or "Single"
- Threads: Number of computer memory threads used for the alignment
- Transcripts_Annotations: Path for the transcript annotation GTF file
- Bowtie_Genome_Index: The base name of the Bowtie genome index files, which is the name of any of the index files up to but not including the first period

To run this function correctly, the FASTQ files of each sample need to be inside a directory called "FASTQ" which resides inside a directory with the name of the sample. The Tophat function creates new directory called BAM in the same path where the FASTQ directory is located. This directory will contain the alignment files. The alignment statistics are available in the "align_summary.txt" file. In case of analysis of publish data that was deposited in the SRA database, *see* **Note 4**.

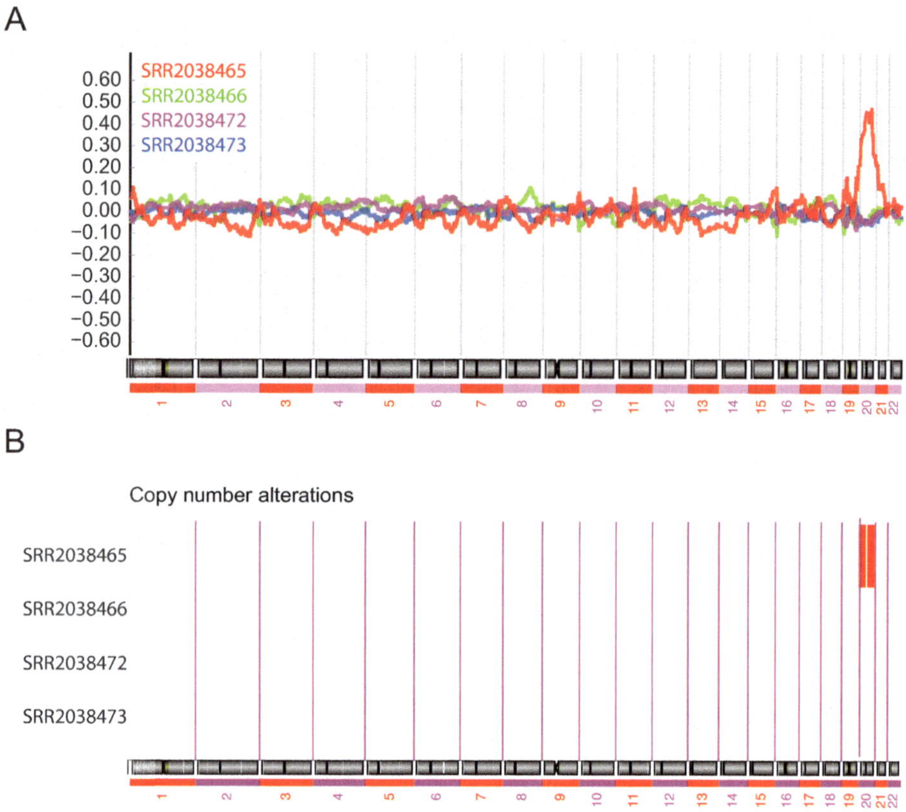

Fig. 2 CGH-Explorer results of three diploid samples and one aberrant sample. (**a**) Moving average plots of all the four samples. (**b**) Results of the PCF algorithm that identifies the trisomy of chromosome 20 in the first sample

3.2 Analysis of Aneuploidy Using e-Karyotyping

3.2.1 Expected Results

The e-Karyotyping methodology for gene expression-based detection of chromosomal aberrations is based on inferring aberrations from alterations in gene expression. In this analysis, for each gene the expression level ratio between each sample and the median of all the samples is calculated. Next the expression ratios are sorted according to their chromosomal location. The graphical output of the e-Karyotyping methodology is a moving average plot of all the expression ratios ordered by the chromosomal location. The ratios are log2 transformed; thus genes from diploid chromosomes are expected to have expression ratio of zero, while trisomies and deletions are expected to have ratios of ~0.58 and −1, respectively (Fig. 2a). The statistical testing used in the analysis is piecewise constant fit (PCF) algorithm (Fig. 2b). Workflow of the analysis is presented in Fig. 3.

3.2.2 Obtaining Gene Expression Levels

Since e-Karyotyping is based on gene expression levels, the first step is to obtain the expression values. This stage can be performed with different bioinformatical tools, such as Cufflinks [14] or HTseq [15], that quantify the number of reads per genes. To obtain

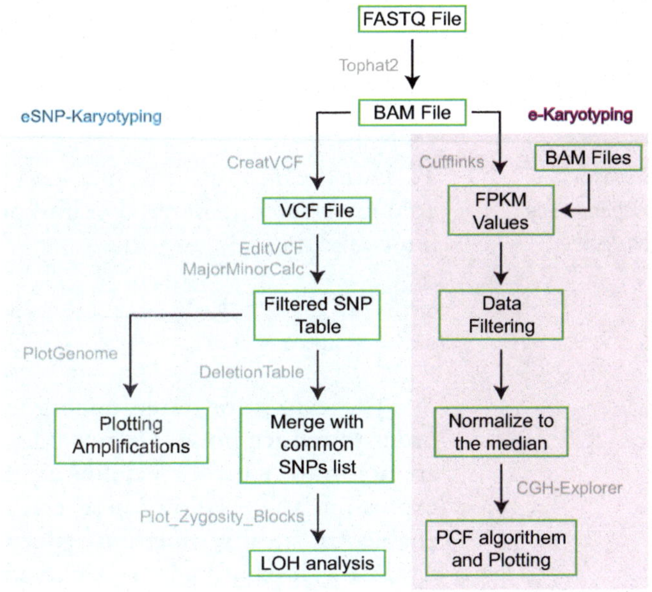

Fig. 3 Workflows of e-Karyotyping and eSNP-Karyotyping methods. In gray are the tools or functions used for each step

fragments per kilobase of transcript per million mapped reads (FPKM) values with Cufflinks, use the following command line in the Linux terminal (this command output files are placed in the current path):

```
cufflinks -p [Number of computer memory threads] -G
[Transcriptome GTF file] [Path to BAM file]
```

e-Karyotyping is based on comparison between multiple samples that most of them have to be diploid. Thus, it is suggested to compare the sample of interest with samples of the same cell lines from different sources or of cell lines that have verified diploid karyotype (*see* **Note 5**). If the alignment output is one file per sample, then the expression values of all the samples should be merged (with R or Excel) into one table that contains the gene ID, gene name, chromosome number, and the gene chromosomal position at the first four columns, followed by columns with the expression values of each sample, as seen in the example below:

```
> head(data)  Gene_ID Gene_Name Chr Start Sample_1 Sample_2 Sample_3 Sample_4
1 NM_000014 A2M 12 9220303 1.373110 1.400730 3.68194 4.82103
2 NM_000015 NAT2 8 18248754 0.130263 0.248107 0.00000 0.00000
3 NM_000016 ACADM 1 76190031 10.740500 13.940700 14.27520 41.02030
4 NM_000017 ACADS 12 121163570 11.824000 16.257300 9.69795 33.58030
5 NM_000018 ACADVL 17 7123149 148.147000 158.987000 124.64800 48.20280
6 NM_000019 ACAT1 11 107992257 10.440800 5.967870 35.40240 12.48530
```

Please change the name of the sex chromosomes to numbers (for example X to 23 and Y to 24 in case of human samples), and remove any genes that are not annotated to chromosomes 1–24 (like mitochondrial genes). Save the table in CSV format.

3.2.3 Processing and Filtering the Gene Expression Table

To start the analysis in R, first load the CSV table, and change the columns that contain number into numeric values using the following commands:

```
data=read.csv("[Path to file]",sep="\t", as.is=T)
data=cbind(data[,1:2],data.matrix(data[-c(1:2)]))
```

The expression table undergoes a set of processing steps in order to reduce noise. First, we discard unexpressed genes as they are non-informative. We define an expression cutoff and we elevate expression value below these threshold to that value. This step enable statistical testing and reduce sequencing noise (usually we use FPKM level of 1). Next, we remove genes that are not expressed (expression value equal to the defined threshold) in high percentage of the samples. Usually, genes that are expressed in less than 80% of the samples are removed. Lastly, the 10% most variable genes are removed from the analysis using calculation of the sum of squares. The following commands perform these three steps:

```
floor=1        # Expression threshold value
expressed=0.8 # Percentage of expressed samples
variable=0.1  # Percentage of most variable genes to remove
data[,-c(1:4)][data[,-c(1:4)]<1]=floor
data=data[rowSums(data[,-c(1:4)]>floor)>=ncol(data[,-
c(1:4)])* expressed,]
row_mean=rowMeans(data.matrix(data[,-c(1:4)]))
distance_mat=sweep(data[,-
c(1:4)],MARGIN=1,STATS=row_mean,"-")
ssq=rowSums(distance_mat^2)
data=data[ssq<quantile(ssq,1-variable),]
```

Then, we compute the median expression value for each gene in all the samples and divide all expression values by this median. For convenience we log2 transform the fold changes and mean-center the values, so that the fold changes of each sample will distribute around the zero. The output is a matrix, called CGH, with the expression fold changes from the median, organized by chromosome number and position.

```
med=apply(data.matrix(data[,-c(1:4)]),1,median)
minus_med=sweep(data[,-c(1:4)],1,med,"/")
minus_med=log(minus_med,2)
minus_med=sweep(minus_med,2,colMeans(minus_med),"-")
CGH=cbind(data[,c(1:4)],minus_med)
CGH=CGH[order(CGH[,3], CGH[,4]),]
```

3.2.4 Visualization and Statistical Testing

For fast visualization in R, run this script which executes moving/rolling average to smooth the data, and plot it:

```
library("zoo")
library("gplots")
plot(1:nrow(CGH),ylim=c(-1,1),col="white")
window=201 # moving average window size
cl=rainbow(ncol(CGH)-4)
i=1
add_line=function(x,window){
 lines(rollmean(x,window),col=cl[i])
 i<<-i+1
}
apply(CGH[,5:ncol(CGH)],2,function(x) add_line(x,window))
abline(v=which(duplicated(round(rollmean(CGH[,3],window)))=
=FALSE),col="grey",lty=2)
```

Better visualization and the statistical testing are performed with CGH-Explorer software [16] (*see* **Note 6**). In order to import the data to CGH-Explorer, the CGH table has to be saved as a tab-delimited TXT file using the following command:

```
write.table(CGH,"[File Path and name].txt",sep="\t",row.names = F)
```

Then, in the CGH-Explorer, go to "File" → "Import Data," and select the file with the karyotyping data. In the "Import of file" window that opens, select the columns with the samples, and press "Add"; choose the columns in the "Clone ID," "Gene names," "Chromosome," and "Position" boxes, and click "OK."

To create a moving average plot, go to "Data" window, select the samples and the chromosomes you wish to draw, and select "Graphs" → "Moving average plot" from the top menu. The window size can be adjusted by clicking "Preferences" → "MA Preferences" and the *y*-axis by choosing "Preferences" → "Vertical plot range." More graphical parameters can be set from the "Graphical parameters" options.

To statistically call aberrations, CGH-Explorer uses the PCF algorithm. Select the samples and chromosomes from the "Data" window, and go to "Aberrations" → "PCF detection" from the top menu. Mark the "Override automatic penalty selection" and

"Detect long CNAs" options, and set the following parameters: "least allowed deviation," 0.15–0.3; "least allowed aberration size," 50–80; "Winsorize at quantile," 0.001; and "Penalty," 8–12. Click "OK," and at the window that appears, select the "Threshold" value of 0.01 (*see* **Note** 7).

At the "piecewise constant fit" window that opens, select "Preferences" → "Plot type," and then select "Aberration plot (single)." This visualization shows the statistically significant regions in each sample. Gains will appear in red and deletions in green.

3.3 Analysis of Genomic Integrity Using eSNP-Karyotyping

3.3.1 Expected Results

eSNP-Karyotyping is a more complex analysis; however it can be easily implemented using the R package that contains the required functions, including the calculations, graphical plotting, and the statistical analysis [11]. Before use, please make sure that all the required files are correctly obtained and all the programs are installed as described in Subheading 2. Here, each sample is analyzed separately without comparison to diploid samples.

The analysis pipeline is separated into duplications and LOH detection. The output of the duplication part of the analysis is a moving median graph of the allelic ratios sorted by the chromosomal position. The statistical test is calculated with a one-tailed *t*-test comparing the SNP major/minor values in each window to the values of the entire genome. The *p*-values are FDR corrected and displayed as color bar above the moving median plot [11] (Fig. 4).

The LOH analysis is performed by merging the detected SNP table with a table of the common SNPs and comparing the distributions of the homozygous and heterozygous positions. Large stretches of homozygosity are considered regions with LOH. The statistical test is performed at the chromosomal arm scale. To obtain *p*-value, the ratio of the number of homozygote to heterozygote SNPs for each chromosome arm is calculated. Then, arms with statistically significant different ratio are recognized by *t*-test. The *p*-values are adjusted for multiple testing using FDR correction. True LOH is considered as an arm with p-value <0.001 and a homozygote to heterozygote SNP ratio five times greater than the ratio of all the autosomal chromosomes. Arm with significant difference in the distribution of the homozygous to heterozygous SNPs will be highlighted in yellow [11] (Fig. 4). Workflow of the analysis is presented in Fig. 3.

3.3.2 Calling SNPs from the Align BAM File

To call genomic positions with sequence variation, we use the GATK haplotype caller tool [17]. This tool output is a variant call format (VCF) file that is saved in the BAM directory and contains all the raw information on the detected variants. The BAM file has to undergo multiple steps of preprocessing before running the haplotype caller tool that are performed automatically by the CreateVCF function:

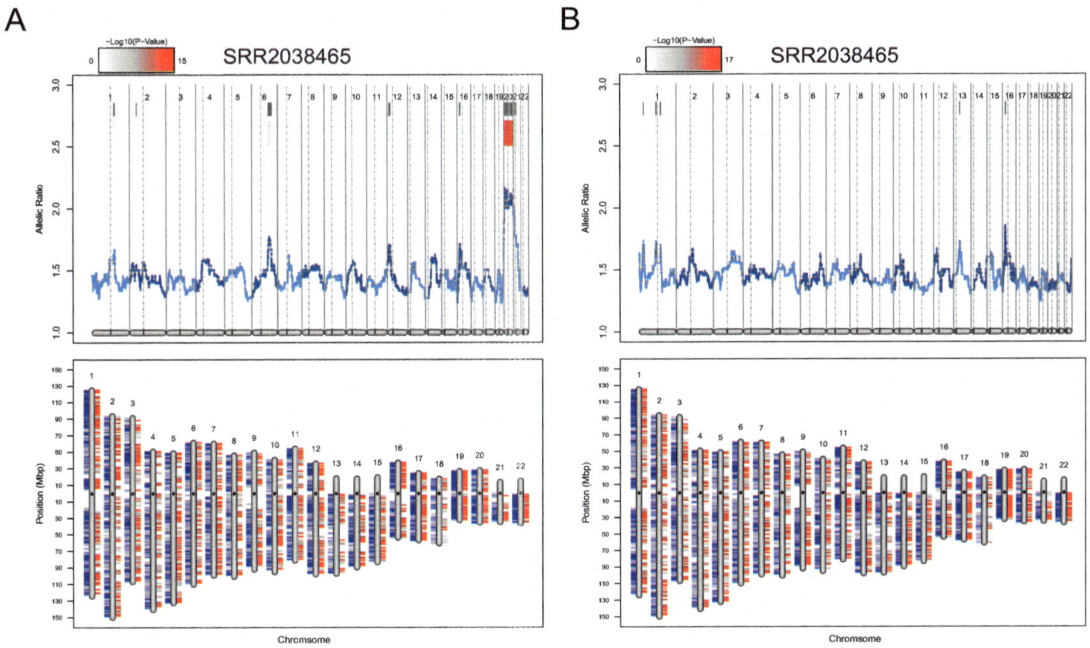

Fig. 4 eSNP-Karyotyping analysis of trisomy 20 (**a**) and diploid (**b**) samples. Upper panels are the output of the duplication analysis, and the bottom panels are the output of the LOH analysis. Homozygous SNPs are shown in blue and heterozygous SNPs in red

```
CreateVCF(Directory, Genome_Fa, Picard_Path, GATK_Path)
```

The function arguments are:

- Directory: The path of the directory where the BAM file is located
- Genome_Fa: Path for the whole-genome FASTQ file
- Picard_Path: Picard directory path, where Picard.jar file is located
- GATK_Path: GATK directory path, where GenomeAnalysisTK.jar file is located

3.3.3 Processing and Filtering the Variant Table

The next stage is to extract the relevant information from the VCF file and to filter the detected variants according to certain criteria. First, run the EditVCF function:

```
table=EditVCF(Directory, Organism)
```

The function arguments are:

- Directory: The path of the BAM directory
- Organism: "Human" or "Mouse"

This function creates file with SNP data at the BAM directory called variantTable.csv (*see* **Note 8**). Next, run the following lines that sort the SNP table according to the chromosomal position:

```
table$chr=as.numeric(table$chr)
table=table[order(table$chr,table$position),]
table=table[table$chr>0,]
```

Lastly, run the MajorMinorCalc function which filters the SNPs according to specified parameters (*see* **Note 9**) and calculates the allelic ratio of each SNP:

```
table2=MajorMinorCalc(Table, minDP, maxDP, minAF)
```

The function arguments are:

- Table: The variable name containing the table of SNPs from previous commands (table)
- minDP: Minimal reading depth required to call SNP, usually set to 10–20
- maxDP: Maximal reading depth required to call SNP, usually set to very high number
- minAF: Minimal minor allele frequency, usually set to 0.2–0.25

The output table contains the frequency of each allele and the ratio between the alleles, as presented in the example below:

```
> head(table2)
chr position AF1 AF2 Major Minor
1 1 14653 0.5333333 0.4666667 1.142857
2 1 16688 0.5454545 0.4545455 1.200000
3 1 135032 0.5098039 0.4901961 1.040000
4 1 137825 0.5357143 0.4642857 1.153846
5 1 138817 0.5625000 0.4375000 1.285714
6 1 494772 0.5505618 0.4494382 1.225000
```

3.3.4 eSNP-Karyotyping Analysis of Chromosomal Duplications

For detection of chromosomal duplications, the package plots the allelic ratio of all the SNPs along the genome. This can be done using the PlotGenome function:

```
PlotGenome(orderedTable, Window, Ylim, PValue, Organism)
```

The function arguments are:

- orderedTable: The variable containing the output of the MajorMinorCalc function
- Window: The window of the moving median plot (odd number), usually 101–201

- Ylim: The plot *y*-axis maximal limit, usually set to 3
- PValue: Determine if *p*-value bar should be calculated and plotted on the graph (slows the analysis). Values of "TRUE" or "FALSE"
- Organism: "Human" or "Mouse"

The *p*-value bar is calculated with a one-tailed *t*-test comparing the SNP major/minor values in each window to the values of the entire genome and correcting for multiple testing using FDR correction (*see* **Note 9**). If necessary, a possible way to refine the analysis is to use only the common SNPs for the analysis (*see* **Note 10**).

3.3.5 eSNP-Karyotyping Analysis of LOH

Before the analysis, make sure that the common SNP files were downloaded and edited as required (*see* **Note 3**), and then run the following function that intersects the observed SNPs with the list of the common SNPs from dbSNP database:

```
Del_tbl=DeletionTable(Directory, Table,
dbSNP_Data_Directory, dbSNP_File_Name, Genome_Fa_dict,
Organism)
```

The function arguments are:

- Directory: The path of the BAM Directory, containing the variantTable.csv file
- Table: The variable containing the output of the MajorMinorCalc function (table2)
- dbSNP_Data_Directory: The path for the directory where the edited dbSNP file is located (created previously by the Edit_dbSNP_Files function)
- dbSNP_File_Name: The edited dbSNP file names, without the chromosome number
- Genome_Fa_dict: Path for the Picard genome dictionary (.dict) file.
- Organism: "Human" or "Mouse"

This function saves the output table in a variable, as well as in a file called Deletions.txt in the BAM directory (*see* **Note 8**).

There are two functions for plotting the distributions of the homozygous and heterozygous SNPs along the chromosome:

```
Plot_Zygosity_Sinle(Table, Organism)
Plot_Zygosity_Blocks(Table, Window, Max, Max2, Organism)
```

The functions arguments are:

- Table: The deletion table containing the output of the DeletionTable function
- Window: The block size in base pairs, usually 1,500,000 (1.5 Mb)
- Max: The number of heterozygous SNP in a block to be colored in full opacity, usually 6
- Max2: The number of homozygous SNP in a block to be colored in full opacity, usually 60
- Organism: "Human" or "Mouse"

These functions plot an ideogram with homozygous SNPs shown in blue to the left of the chromosomes, and heterozygous SNPs are shown in red to the right of the chromosomes (Fig. 4). The Plot_Zygosity_Sinle function plots each variant position on the chromosome ideogram, and it does not perform any statistical testing. Plot_Zygosity_Blocks counts the number of homozygous or heterozygous SNPs in genomic blocks (their size is defined by the window parameter) and plots the data with color gradient that signifies the density of SNPs in each block. Chromosomal arms with significant difference in the distribution of the homozygous to heterozygous SNPs will be highlighted in yellow. To retrieve the number of homozygous and heterozygous SNPs in each chromosome, *see* **Note 11**.

4 Notes

1. *Obtaining the Reference Files*

   For the alignment and analysis, the reference genome, the transcriptome, and the common SNP files have to be in the same format. It is highly recommended to download all the data from the University California Santa Cruz (UCSC) genome browser website [18].

   To download the genome FASTA file, enter the UCSC genome browser website (https://genome.ucsc.edu/index.html), and from the top menu, select "Downloads" → "Genome Data." Select the organism (Human or Mouse), and click "Full data set" under the desired genome assembly (e.g., GRCh38/hg38), and then download the "fa.gz" or "2bit" file. If ".gz" file was downloaded, please extract it by right-clicking → "Extract Here" or by navigating to the directory in the Linux terminal and running "gunzip [filename]" command. If the genome was downloaded in a 2bit format, extract the FASTA file with the TwoBitToFa program using the command "TwoBitToFa [Genome.2bit] [Genome.fa]" in the Linux terminal.

To download the human transcriptome file, select "Tools" → "Table Browser'." Select "Mammal" and "Human" in "Clade" and "Genome" boxes, respectively, and select the genome assembly (should match the genome FASTA file). Next, select the required transcriptome file, for example, select "Genes and Gene Predictions" in "Group," "NCBI RefSeq" in "Track," and "UCSC RefSeq (refGene)" in the table fields. Make sure that "Region" is set to "Genome" and that the "output format" is set to "GTF—gene transfer format (limited)." Then give the file a name in the "Output file" field, and click on "Get output."

To download the mouse transcriptome, select "Mouse" in the "Genome" box, and then the correct genome assembly. Select the desired transcriptome table, for example, select "Genes and Gene Predictions" in "Group," "RefSeq Genes" in "Track" and "refGene" in the table fields. Again, make sure that the output is in GTF file format.

2. *Preparing Genome Indexes*

The eSNP-Karyotyping analysis requires to index the genome by three different programs. To make these indexes, first navigate to the directory where the [genome].fa is stored in the Linux terminal.

To make the bowtie2 genome index, which is required for the genome alignment, navigate to the directory that contains the genome FASTA file, and run the following command:

```
bowtie2-build [Genome filename.fa] [Genome filename without
the extension]
```

To build the SAMtools genome index, run:

```
samtools faidx [Genome filename.fa]
```

To build the Picard genome dictionary file, run:

```
java -jar [path to picard.jar] CreateSequenceDictionary
R=[Genome filename.fa] O=[Genome filename.dict]
```

3. *Obtaining dbSNP Database*

The SNP information can be downloaded from the UCSC table browser tool. In the UCSC homepage, select "Tools" → "Table Browser," and then choose the desired organism and genome assembly. In the "Group" field, select "Variation," in "Track," select "Common SNPs (142)," and in "Table" select "snp142Common." Set the "Output Format" to "GTF—gene transfer format (limited)."

The dbSNP data have to be obtained as separate file for each chromosome. Thus, the SNP information is saved in 24 or 21 files for human or mouse, respectively. To obtain the information on chromosome 1, type "Chr1" in the "Region" field, give a filename that ends with the number 1 in the "Output File" field, and then click "Get Output" to save the file. Repeat this step for all the autosomal chromosomes. For the sex chromosome, type "ChrX" and "ChrY" in the "Region" field, and save them as chromosomes 23 and 24 in human or 20 and 21 in mouse.

Lastly, the downloaded files have to be restructured by the "Edit_dbSNP_Files" function from the eSNP-Karyotyping R package as described below:

```
Edit_dbSNP_Files(Directory, File_Name, Organism)
```

The function arguments are:

(a) Directory: The directory where the files are located, one GTF file per chromosome

(b) File_Name: The file name, without the number of the chromosome

(c) Organism: "Human" or "Mouse"

The function creates a new file for each chromosome, which will be used for the LOH analysis.

4. *Opening SRA File*

In case of analysis of published dataset that was deposited in the SRA database, the sequencing file is compressed to SRA format. Thus, prior to the alignment, it is necessary to extract the sequencing FASTQ file using SRA toolkit [19]. The eSNP-Karyotyping package has a function called "SRAtools" to open the SRA file.

```
OpenSRA(File, Library_Type, SRAPath)
```

The function arguments are:

(a) File: The path to the SRA file

(b) Library_Type: Library type, "Paired" or "Single"

(c) SRAPath: Path for the SRAtools "bin" directory (e.g., SRAPath = "/home/user/SRAtoolkit/sratoolkit.2.5.4/bin/")

Another useful tool in the eSNP-Karyotyping package is the SRAtoSNP function. This function takes a SRA file and performs all the analysis steps automatically. This function runs the MajorMinorCalc function with the following parameters: minDP = 20, maxDP = 10000 and minAF = 0.2.

```
SRAtoSNP(File, Library_Type, TopHat_Threads, Transcript_Anotations, Bowtie_Genome_Index, Genome_FA, SRAPath, Picard_Path, GATK_Path, dbSNP_Data_Directory, dbSNP_File_Name, Genome_Fa_dict, Organism)
```

The function arguments are:

(a) File: The path to the SRA file

(b) Library_Type: Library type, "Paired" or "Single"

(c) TopHat_Threads: Number of computer memory threads used for the alignment

(d) Transcript_Anotations: Path for the transcript annotation GTF file

(e) Bowtie_Genome_Index: The base name of the genome index files, which is the name of any of the index files up to but not including the first period

(f) Genome_FA: Path for the whole-genome FASTA file

(g) SRAPath: Path for the SRAtools "bin" directory

(h) Picard_Path: Picard directory path, where Picard.jar file is located

(i) GATK_Path: GATK directory path, where GenomeAnalysisTK.jar file is located

(j) dbSNP_Data_Directory: The path for the directory where the edited dbSNP file are located

(k) dbSNP_File_Name: The edited dbSNP file names, without the chromosome number

(l) Genome_Fa_dict: Path for the Picard dictionary file created for the genome FASTA file

(m) Organism: "Human" or "Mouse"

5. *Building the e-Karyotyping Database*

The e-Karyotyping analysis is based on comparison of gene expression from multiple samples to infer genetic changes. However, if most/all of the samples contain a certain aberration, then the median baseline will also be aberrant, and the aberration will not be detected. Moreover, in this case, diploid samples will be recognized as having the opposite aberation. Thus, it is essential to obtain many samples from various studies, cell lines, and laboratories together with the samples of interest to minimize this risk. All samples are required to be of the same cell type and to undergo the same data processing. If analyzing differentiated samples, use samples from the same differentiation stage. If some of the samples are very "noisy" and contain multiple aberrations along the entire genome, please remove these outliers.

6. *Notes Regarding the Work with CGH-Explorer Software*

   In order for the CGH-Explorer to run properly, there are few specific points to note:

   (a) Run CGH-Explorer as administrator. To do so right-click on the CGH-Explorer icon → "more" → "Run as administrator" (when using Windows 10).

   (b) Before uploading your CGH file, make sure that the correct genome is selected. In the main menu, click on "Preferences" → "Genome selection," and then select the correct organism.

   (c) If the PCF analysis doesn't work while the moving average plot does, make sure that all chromosomes have multiple genes. In many cases chromosome Y has just a few probes; in these cases delete the chromosome Y probes from the file.

7. *Parameters of CGH-Explorer*

   The PCF parameters need to be adjusted to those that minimize false discovery rates. When applying suitable parameters, diploid samples will be detected as non-aberrant, while the correct chromosomal aberrations will be detected in aberrant samples. Apply different set of parameters until the best parameters are achieved [7]. Detection of aberrations in small chromosomes may require applying different set of parameters (if control samples also show false aberration) [7].

8. *Reloading Previous eSNP-Karyotyping Analysis*

   The output information of the CreateVCF and DeletionTable functions is saved in the BAM directory as variantTable.csv and Deletions.txt, respectively. These are the only files necessary for reanalyzing the data, and they can be loaded using the following R command line:

```
table=read.delim([Path to file],as.is=T)
```

9. *How the Parameters Affect the Detection p-Value*

   The chromosomal duplication analysis plotting is performed by calculating moving median of sliding windows where each window contains a fixed number of allelic ratios, defined by the Window argument in the PlotGenome function. To calculate the p-value of the duplication analysis, the function executes a moving t-test, meaning that each window is compared to the entire genome. Due to the multiple testing, the p-value list is adjusted using FDR correction.

   The parameters defined in the MajorMinorCalc function (minDP, maxDP, minAF) and the size of the moving window will affect the results. The parameters of the MajorMinorCalc define which variants to include in the analysis and which to discard. Thus, they can modulate the amount of noise, in the expense of the number of accepted SNPs. The PlotGenome window size determines the number of variants that are used in the *t*-test; therefore it has a direct impact on the statistical significance. In addition, smaller window size increases the number of statistical testing, thus reducing the statistical power. Since each sequencing experiment is different, we recommend to test different parameters (in our recommended ranges) in order to maximize the detection power.

10. *Refining the eSNP-Karyotyping Duplication Analysis*

    In addition to modifying the variant detection parameters of the different functions, one can improve "noisy" duplication assays by including only the variants that are present in the common SNP list from dbSNP database. This can be done by modifying the output table of the DeletionTable function. The example below takes the output table of the DeletionTable function (Del_tbl) and creates a new table (new) that can be loaded to the PlotGenome function. This new table contains only the common variant.

```
new=Del_tbl[,c(2,3,5,6,7)]
colnames(new)=c("chr","position","AF1","AF2","MajorMinor")
new=new[new$MajorMinor>0,]
PlotGenome(new, 101, 3,"TRUE", "Human")
```

11. *Retrieving the Number of Detected SNPs for Each Chromosome*

    In case of a suspected LOH, one may want to know the number of detected homozygous and heterozygous SNPs in each chromosome. The script below uses the DeletionTable function output (Del_tbl) and builds a table with this information for each chromosome:

```
SNP_Info=cbind(Chr=1:24,
homozygous=table(Del_tbl[Del_tbl$MajorMinor==0,]$chr.x),
heterozygous=table(Del_tbl[Del_tbl$MajorMinor>0,]$chr.x))
SNP_Info=data.frame(SNP_Info)
SNP_Info=cbind(SNP_Info, Ratio= SNP_Info$heterozygous/
SNP_Info$homozygous)
```

References

1. Weissbein U, Benvenisty N, Ben-David U (2014) Genome maintenance in pluripotent stem cells. J Cell Biol 204:153–163
2. Lamm N, Ben-David U, Golan-Lev T et al (2016) Genomic instability in human pluripotent stem cells arises from replicative stress and chromosome condensation defects. Cell Stem Cell 18:253–261
3. Merkle FT, Ghosh S, Kamitaki N et al (2017) Human pluripotent stem cells recurrently acquire and expand dominant negative P53 mutations. Nature 545:229–233

4. Ben-David U, Arad G, Weissbein U et al (2014) Aneuploidy induces profound changes in gene expression, proliferation and tumorigenicity of human pluripotent stem cells. Nat Commun 5:4825
5. Avery S, Hirst AJ, Baker D et al (2013) BCL-XL mediates the strong selective advantage of a 20q11.21 amplification commonly found in human embryonic stem cell cultures. Stem Cell Rep 1:379–386
6. Nguyen HT, Geens M, Mertzanidou A et al (2014) Gain of 20q11.21 in human embryonic stem cells improves cell-survival by increased expression of Bcl-xL. Mol Hum Reprod 20:168–177
7. Ben-David U, Mayshar Y, Benvenisty N (2013) Virtual karyotyping of pluripotent stem cells on the basis of their global gene expression profiles. Nat Protoc 8:989–997
8. Mayshar Y, Ben-David U, Lavon N et al (2010) Identification and classification of chromosomal aberrations in human induced pluripotent stem cells. Cell Stem Cell 7:521–531
9. Ben-David U, Mayshar Y, Benvenisty N (2011) Large-scale analysis reveals acquisition of lineage-specific chromosomal aberrations in human adult stem cells. Cell Stem Cell 9:97–102
10. Weissbein U, Ben-David U, Benvenisty N (2014) Virtual karyotyping reveals greater chromosomal stability in neural cells derived by transdifferentiation than those from stem cells. Cell Stem Cell 15:687–691
11. Weissbein U, Schachter M, Egli D et al (2016) Analysis of chromosomal aberrations and recombination by allelic bias in RNA-Seq. Nat Commun 7:12144
12. Kim D, Pertea G, Trapnell C et al (2013) TopHat2: accurate alignment of transcriptomes in the presence of insertions, deletions and gene fusions. Genome Biol 14:R36
13. Dobin A, Davis CA, Schlesinger F et al (2013) STAR: ultrafast universal RNA-seq aligner. Bioinformatics 29:15–21
14. Trapnell C, Williams BA, Pertea G et al (2010) Transcript assembly and quantification by RNA-Seq reveals unannotated transcripts and isoform switching during cell differentiation. Nat Biotechnol 28:511–515
15. Anders S, Pyl PT, Huber W (2015) HTSeq—a Python framework to work with high-throughput sequencing data. Bioinformatics 31:166–169
16. Lingjaerde OC, Baumbusch LO, Liestol K et al (2005) CGH-Explorer: a program for analysis of array-CGH data. Bioinformatics 21:821–822
17. DePristo MA, Banks E, Poplin R et al (2011) A framework for variation discovery and genotyping using next-generation DNA sequencing data. Nat Genet 43:491–498
18. Kent WJ, Sugnet CW, Furey TS et al (2002) The human genome browser at UCSC. Genome Res 12:996–1006
19. Leinonen R, Sugawara H, Shumway M (2011) The sequence read archive. Nucleic Acids Res 39:2010–2012

Chapter 19

Cell Fate Engineering Tools for iPSC Disease Modeling

Emily K. W. Lo and Patrick Cahan

Abstract

The field of cell fate engineering is contingent on tools that can quantitatively assess the efficacy of cell fate engineering protocols and experiments. CellNet is such a cell fate assessment tool that utilizes network biology to both evaluate and suggest candidate transcriptional regulatory modifications to improve the similarity of an engineered population to its corresponding in vivo target population. CellNet takes in expression profiles in the form of RNA-sequencing data and generates several metrics of cell identity and protocol efficacy. In this chapter, we demonstrate how to (1) preprocess raw RNA-sequencing data to generate an expression matrix, (2) train CellNet using preprocessed expression matrices, and (3) apply CellNet to a query study and interpret its results. We demonstrate the utility of CellNet for analysis of iPSC disease modeling studies, which we evaluate through the lens of cell fate engineering.

Key words Cell fate engineering, Stem cells, Disease modeling, Gene expression profiling, Gene regulatory networks, Computational biology

1 Introduction

1.1 Tools for Cell Fate Engineering

A major subfield of computational stem cell biology addresses the quantitative evaluation of engineered cell identity. Two primary objectives of this subfield are the assessment and improvement of cell fate engineering protocols [1]. The assessment problem refers to a method of discerning whether an engineered cell population recapitulates the properties of an in vivo population. The improvement problem entails determining which (and in what amount) factors, pathways, and transcription factors must be upregulated or downregulated in order to optimize a cell fate engineering protocol. These are nontrivial objectives, as they require, among other things, establishing quantitative metrics of what characteristics constitute a cell or tissue type.

Existing computational tools for evaluating cell identity address both of these problems in the context of different forms of cell fate engineering. These tools conventionally take as input high-throughput expression and/or epigenetic data, apply machine learning methods (e.g., classification, clustering, dimensionality

Patrick Cahan (ed.), *Computational Stem Cell Biology: Methods and Protocols*, Methods in Molecular Biology, vol. 1975, https://doi.org/10.1007/978-1-4939-9224-9_19,

reduction, etc.) to these data, and generate proposed metrics of cell identity.

Most tools specifically address either the assessment problem [2–5] or the improvement problem [6–11] only [12]. The only current tool, to the best of our knowledge, that addresses both the improvement problem and the assessment problem in both stem cells and adult cells is CellNet [13, 14]. CellNet analysis generates three readouts of cell identity: (1) a classification heatmap indicating how indistinguishable a query sample expression profile is from each of the reference cell/tissue types; (2) the gene regulatory network (GRN) status, a metric of the extent to which a cell-/tissue-specific GRN is established in a query sample; and (3) the network influence score (NIS), a list of transcription factors scored by the probability that their expression modulation would improve the desired fate change.

1.2 Cell Fate Engineering Tools in iPSC Disease Modeling

The advent of induced pluripotent stem cell (iPSC) approaches has revolutionized the field of disease modeling. iPSC disease modeling offers several advancements over generic disease modeling approaches. Firstly, it allows us to investigate patient-specific disease mechanisms and assess therapeutic candidates in the C/T type affected by disease, with minimal invasion on the part of the patient. This is particularly necessary when it is infeasible to obtain primary samples of the affected C/T type from the patient, such as in neuronal or cardiac diseases. Secondly, though methods to derive target C/T types from patient-specific adult stem cells exist, iPSC disease modeling has also helped us bypass the issues of limited lineage potential or limited culture expansion potential in adult stem cell approaches [15].

In iPSC disease modeling, the most common approach is to isolate somatic cells (usually dermal fibroblasts, which can be obtained with minimal invasion) from affected patients and healthy controls, reprogram the cells to pluripotency, and then derive the target C/T type through a cell fate engineering protocol. It is therefore essential to be able to verify the fidelity of reprogramming and directed differentiation experiments in iPSC disease modeling studies.

This challenge runs in parallel with the assessment and improvement problems in canonical cell fate engineering. Through the lens of cell fate engineering, the assessment and improvement problems can manifest more specifically as two questions in iPSC disease modeling (Fig. 1a):

1. How does genetic disease affect the efficacy of reprogramming to pluripotency and so-called redifferentiation of iPSCs to target C/T types?
2. Can causal fate-altering mutations in disease be identified using cell fate as a readout?

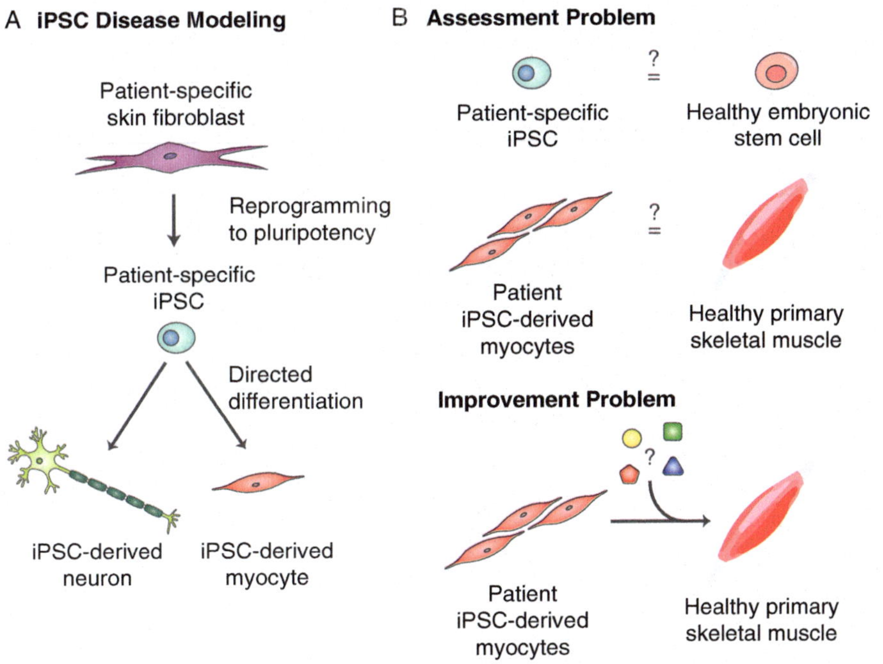

Fig. 1 The assessment and improvement problems in iPSC disease modeling. (**a**) iPSC disease modeling paradigm. (**b**) Assessment and improvement in the context of iPSC disease modeling

This chapter demonstrates three major methods for the application of CellNet to iPSC disease modeling studies:

1. Preprocessing raw RNA-sequencing data to generate an expression matrix.
2. Training CellNet on primary in vivo samples.
3. Applying CellNet to a query iPSC disease modeling study and interpreting results.

2 Materials

2.1 For Preprocessing Raw RNA-Sequencing Data

2.1.1 Hardware

Amazon Web Services (AWS) for cloud computing resources:

- Amazon Machine Image (AMI): CellNet, which includes the following software already installed:
- `screen, R, CellNet, Salmon, cutadapt`
- Instance type: c3.4xlarge (30 GiB RAM, ~$0.796/h) or c3.8xlarge (60 GiB RAM, ~$1.591/h).

Local machine with Linux or UNIX-like (e.g., macOS) operating system or other OS with `ssh` and `scp` clients.

2.1.2 Software and Data

FASTQ files from RNA-sequencing OR SRA IDs for publicly available data

A sample table containing metadata for each FASTQ file (*see* Subheading 3.1 for formatting requirements)

2.2 For CellNet Training

2.2.1 Hardware

Amazon Web Services (AWS) for cloud computing resources:

- Amazon Machine Image (AMI)—CellNet, which includes the following software already installed:
- `screen, R, CellNet, Salmon, cutadapt`
- Instance type: c3.8xlarge (60 GiB RAM, ~$1.591/h) or c5.18xlarge instance (144 GiB RAM, ~$3.06/h), depending on size of training dataset.

Local machine with Linux or UNIX-like (e.g., macOS) operating system or other OS with `ssh` and `scp` clients.

2.2.2 Software and Data

All `sampTab*.R` or `sampTab*.rda` objects generated from preprocessing.

All `expGeneList*.R` or `expGeneList*.rda` objects generated from preprocessing.

2.3 For Application of CellNet to Query Studies

2.3.1 Hardware

Local machine with:

- Linux or UNIX-like (e.g., macOS) operating system or other OS with `ssh` and `scp` clients

  or

- $\geq$8 GB RAM (for analysis on local machine).

2.3.2 Software and Data

R (http://www.r-project.org), Version $\geq$3.2.2.

RStudio (Optional).

CellNet Processor object (`cnProc`) generated from CellNet training.

`sampTab*.R` or `sampTab*.rda` object generated from preprocessing the query study.

`expGeneList*.R` or `expGeneList*.rda` object generated from preprocessing the query study.

Box 1: Glossary of Terms

1. *Amazon Web Services (AWS)**—Amazon Web Services (AWS) is a suite of cloud-based resources that includes data storage (using the Simple Storage Service, or S3) and

(continued)

Box 1 (continued)

computation (using Elastic Compute Cloud, or EC2). AWS provides access to high-performance computing resources without the overhead of setting up and maintaining computer clusters locally. Sign up for AWS at https://aws.amazon.com/console/. Instructions for getting started with AWS can be found at http://docs.aws.amazon.com/AWSEC2/latest/UserGuide/get-set-up-for-amazon-ec2.html. AWS charges users based on increments of services used. For example, the c3.8xlarge instance type currently costs $1.591/h.

2. *Access ID and Secret Access Key**—Security credentials (provided during the AWS sign-up process) for performing account-specific tasks.
3. *Elastic Compute Cloud (EC2)**—AWS's cloud computing service. Further documentation and tutorials available at https://aws.amazon.com/ec2/.
   (a) *Amazon Machine Image (AMI)**—A preserved virtual computer state with specific software and libraries pre-installed, which makes standardized and reproducible analysis feasible.
   (b) *Instance*—A running instantiation of a machine image, for which you can specify operating system and hardware requirements, e.g., RAM, CPU, and storage. Changes made to the instance state while running the instance will not modify the AMI; only saving an AMI or creating a new AMI will preserve the state.
4. *Simple Storage Service (S3)**—AWS's cloud storage service. Further documentation and tutorials available at https://aws.amazon.com/s3/.
   (a) `aws s3 cli`—the S3 command line interface, already installed on all AMIs. Can use commands to transfer files between your instance and S3. More information here https://docs.aws.amazon.com/cli/latest/reference/s3/index.html.
   Useful commands
   - `aws s3 cp`—copy files between EC2 instance and S3 bucket or between S3 buckets.
   - `aws s3 mv`—move files between EC2 instance and S3 bucket or between S3 buckets.
   - `aws s3 rm`—remove files in S3 buckets.

   Useful flags
   - `--include` and `--exclude`—specify pattern matching (differs slightly from UNIX-like pattern matching).

(continued)

Box 1 (continued)

- `--dryrun`—*see* results of command without executing command.
- `--recursive`—include subdirectories in command execution.

5. `ssh`—secure shell.
6. `scp`—secure copy.
7. `screen`—a useful UNIX program that allows you to preserve access to the processes occurring, even if a connection to a remote server (e.g., through `ssh`) is broken
Useful screen commands include:
   (a) `screen --r`—resume screen session.
   (b) Ctrl-a d—detach screen (but leave shell window open).
   (c) `man screen`—open manual page with full usage description.

*AWS-specific term.

3 Methods

3.1 Preprocessing Raw RNA-Sequencing Data

Objective: Convert raw RNA-sequencing data (in the form of FASTQ files) into an expression matrix representing the quantity of expression of every gene or transcript in every sample (sample names as columns, transcript/gene names as rows).

If you have already generated an expression matrix from your RNA-sequencing data, proceed to Subheading 3.2.

This section demonstrates preprocessing for two scenarios: (1) preprocessing FASTQ files already stored on your S3 account and (2) preprocessing publicly available data from SRA.

3.1.1 Preprocessing FASTQ Files Already Stored on S3

This subsection will demonstrate how to preprocess FASTQ files that are already stored on S3. The resulting expression matrix will be downloaded to your local machine.

1. Create a sample metadata table (.csv format) with the following annotations for each sample (column names are case sensitive, but order of columns doesn't matter):
   (a) fname: name of forward read file, e.g., ERR1274906_1.fastq.gz.
   (b) sample_name: desired sample name.
   (c) pe_se: "SINGLE" or "PAIRED".

(d) description1: cell/tissue type.

(e) description2: additional sample annotation (optional).

(f) description3: additional sample annotation (optional).

2. Log in to the AWS console. Search for the CellNet AMI by clicking through "Images" > "AMIs" and then entering "CellNet" into the search field for public images.
3. Launch the CellNet AMI using either a c3.4xlarge (30 GiB RAM, ~$0.796/h) or a c3.8xlarge (60 GiB RAM, ~$1.591/h) instance type, depending on the computing needs of your specific study (e.g., number of samples, the quantity and size of your FASTQ files, etc.).
4. Find the Public DNS (Domain Names System) for your instance by navigating to the EC2 "Instances" page and locating the value in your instance's "Public DNS (IPv4)" column.
5. After the instance has been launched and has completed initialization, launch a terminal (or PuTTY for Windows), and `ssh` into the instance:

```
ssh -i my_aws_private_key.pem ec2-
user@my_instance_public_dns.amazonaws.com
```

6. Launch `screen` (Box 1):

```
screen
```

7. Launch R with superuser privileges and install CellNet:

```
sudo R
library(devtools)
install_github("pcahan1/CellNet", ref="v0.2.3")
q(save = "no")
```

8. Restart R with default privileges and set up disk space for reference transcriptome indices and FASTQ files, define date, and study name:

```
R
library(CellNet) # load installed CellNet package
cn_setup()
mydate <- utils_myDate()
study_name <- "my_study_name"
```

9. Download reference transcriptome indices:

```
# for mouse:
fetch_salmon_indices(species = "mouse")
# for human:
```

```
fetch_salmon_indices(species = "human",
iFile="salmon.index.huan.050316.tgz")
```

10. Read in sample metadata table, define user-specific S3 and study information, and fetch raw FASTQ files from S3:

```
# specify name of S3 bucket in which your FASTQ directory
is:
bucket <- "my_bucket"
# path to directory containing FASTQ files:
# don't include bucket name or leading or trailing slashes
path <- "my_directory/my_FASTQ_directory"
sampTabName <- "my_sample_table.csv" # name of your sample
metadata table
sampTab <- read.csv(sampTabName, as.is = TRUE, strip.white
= TRUE)
# name of column in your sample metadata table with the
FASTQ file names:
fq_column_name <- "fname"
# omit this line if FASTQ files are not gzipped or bzipped:
comp_format <- "gz" # Alternative option is "bz2"
# fetch and decompress FASTQ files:
sampTab <- cn_s3_fetchFastq(bucket, path, sampTab,
fname=fq_column_name, compressed=comp_format)
```

11. Create expression matrix:

```
# for mouse:
expList <- cn_salmon(sampTab)
# for human:
expList <- cn_salmon(sampTab, salmonIndex =
"HS_GRCh38.SalmonIndex.022616", geneTabfname =
"geneToTrans_Homo_sapiens.GRCh38.80.exo_Jul_04_2015.R")
```

12. Save both expression matrix and sample table to home directory:

```
fname <- paste0("~/expList_", study_name, ".rda")
save(expList, file = fname)
fname <-paste0("~/st_", study_name, ".rda")
save(sampTab, file = fname)
```

13. From a different terminal window, `scp` the expression matrix and sample table `.rda` files to your local machine:

```
scp -i my_aws_private_key.pem ec2-
user@my_instance_public_dns.amazonaws.com:~/expList_my_study_name.
```

```
rda my_local_directory
scp -i my_aws_private_key.pem ec2-user@my_instance_public_dns.ama-
zonaws.com:~/st_my_study_name.rda my_local_directory
```

14. ***Important:*** *Terminate your instance when finished.* AWS will continue to charge for compute time until the instance is terminated. This can be done by navigating to the EC2 "Instances" page, selecting your instance, and then clicking through "Actions" > "Instance State" > "Terminate."

3.1.2 Preprocessing Publicly Available Data from SRA

This subsection will demonstrate how to preprocess samples available from SRA. It will automatically download FASTQ files from SRA to a new directory in the S3 path you specify, preprocess the files, and send the resulting expression matrix files to a new directory in the S3 path you specify.

1. For the SRA study of interest, gather the following information:
   (a) SRP # for the study.
   (b) SRR # for each sample.
   (c) GSM # or other names for each sample.
   (d) Whether FASTQ files have single-end reads or paired-end reads.
   (e) Cell/tissue type for each sample.
   (f) Other relevant descriptions for each sample.
   (g) Create a sample metadata table (.csv format) with the following annotations for each sample (column names are case sensitive, but order of columns doesn't matter):
      - sra_id: SRR #.
      - sample_name: usually the GSM #.
      - pe_SE: "SINGLE" or "PAIRED".
      - study_id: SRP #.
      - description1: cell/tissue type.
      - description2: additional sample annotation (optional).
      - description3: additional sample annotation (optional).
2. Log in to the AWS console. Search for the CellNet AMI by clicking through "Images" > "AMIs" and then entering "CellNet" into the search field for public images.
3. Launch the CellNet AMI using either a c3.4xlarge (30 GiB RAM, ~$0.796/h) or a c3.8xlarge (60 GiB RAM, ~$1.591/h) instance type, depending on the computing needs of your specific study (e.g., number of samples, the quantity and size of your FASTQ files, etc.).

4. Find the Public DNS (Domain Names System) for your instance by navigating to the EC2 “Instances” page and locating the value in your instance’s “Public DNS (IPv4)” column.
5. After the instance has been launched and has completed initialization, launch a terminal (or PuTTY for Windows), and `ssh` into the instance:

```
ssh -i my_aws_private_key.pem ec2-
user@my_instance_public_dns.amazonaws.com
```

6. Launch `screen` (Box 1):

```
screen
```

7. Launch R with superuser privileges and install CellNet:

```
sudo R
library(devtools)
install_github("pcahan1/CellNet", ref="v0.2.0")
q(save = "no")
```

8. Restart R with default privileges and set up disk space for reference transcriptome indices and FASTQ files, define date, and study name:

```
R
library(CellNet) # load installed CellNet package
cn_setup()
mydate <- utils_myDate()
study_name <- "my_study_name"
```

9. Download reference transcriptome indices:

```
# for mouse:
fetch_salmon_indices(species = "mouse")
# for human:
fetch_salmon_indices(species = "human",
iFile="salmon.index.human.050316.tgz")
```

10. Read in sample metadata table, define user-specific S3, and study information:

```
# specify name of S3 bucket in which your FASTQ directory
is located:
my_bucket <- "my_bucket"
# path to destination directory where you want your results
stored:
# don't include bucket name or leading or trailing slashes
destPath <- "my_results_directory"
```

```
sampTabName <- "my_sample_table.csv" # name of your sample
metadata table
sampTab <- read.csv(sampTabName, as.is = TRUE, strip.white
= TRUE)
```

11. Create expression matrix:

```
# for mouse:
processSRA(sampTab, bucket = bucket, startDir = destPath,
target = "mouse")
# for human:
processSRA(sampTab, bucket = bucket, startDir = destPath,
target = "human",
gtfFile="/media/ephemeral1/dat/ref/Homo_sapiens.
GRCh38.83.gtf",
geneTabfname="/media/ephemeral1/dat/ref/geneToTrans_Ho-
mo_sapiens.GRCh38.80.exo_Jul_04_2015.R",
salmonIndex="/media/ephemeral1/dat/ref/HS_GRCh38.Salmo-
nIndex.022616")
```

Resulting expression matrix and sample table `.rda` files will be in the specified `destPath` on S3.

12. ***Important:*** *Terminate your instance when finished.* AWS will continue to charge for compute time until the instance is terminated. This can be done by navigating to the EC2 "Instances" page, selecting your instance, and then clicking through "Actions" > "Instance State" > "Terminate."

3.2 CellNet Training

Objective: Generate a new CellNet Processor Object (`cnProc`) from preprocessed training data (Fig. 2a).

If you would like to use an existing trained CellNet Processor Object (`cnProc`), you can download one of several options (https://github.com/pcahan1/CellNet) and proceed to Subheading 3.3.

This subsection assumes that you have already preprocessed and generated expression matrices and sample tables for all of your training datasets.

1. On S3, put all sample tables and expression matrices intended for training into a single directory.
2. Log in to the AWS console. Search for the CellNet AMI by clicking through "Images" > "AMIs" and then entering "CellNet" into the search field for public images.
3. Launch the CellNet AMI using a c3.8xlarge (60 GiB RAM, ~ $1.591/h) instance type. *You may even need an instance type with more RAM for a training dataset with many samples, such as a c5.18xlarge instance (144 GiB RAM, ~$3.06/h).*

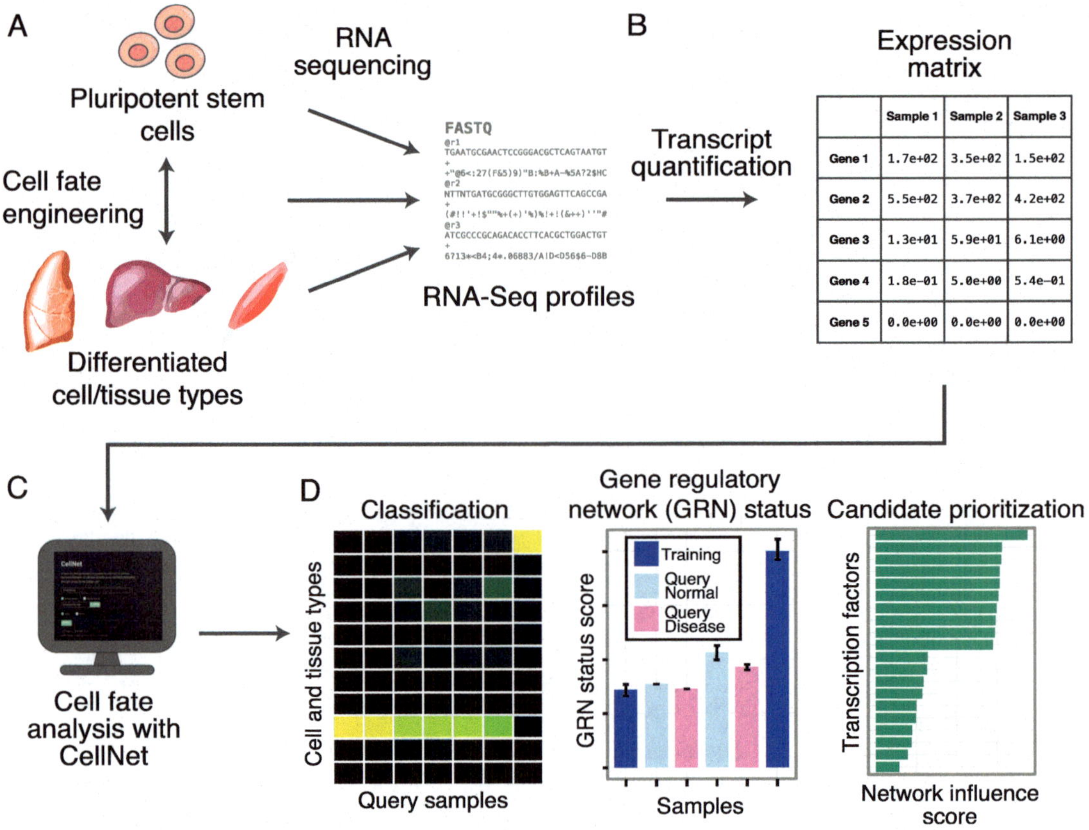

Fig. 2 Cell fate engineering analysis pipeline with RNA-Seq CellNet. (**a**) Cell engineering protocol and expression profiling by RNA-Seq. (**b**) Preprocessing through gene and transcript quantification to generate expression matrix. (**c**) Cell fate analysis with CellNet software. (**d**) Outputs of CellNet: classification heatmap, gene regulatory network status scores, and network influence scores for each sample or treatment group

4. Find the Public DNS (Domain Names System) for your instance by navigating to the EC2 "Instances" page and locating the value in your instance's "Public DNS (IPv4)" column.
5. After the instance has been launched and has completed initialization, launch a terminal (or PuTTY for Windows), and `ssh` into the instance:

```
ssh -i my_aws_private_key.pem ec2-user@my_instance_
public_dns.amazonaws.com
```

6. Launch `screen` (Box 1):

```
screen
```

7. Launch R with superuser privileges and install CellNet:

```
sudo R
library(devtools)
install_github("pcahan1/CellNet", ref="v0.2.0")
q(save = "no")
```

8. Restart R and setup:

```
R
install.packages("dplyr")
library(dplyr)
library(CellNet) # load installed CellNet package
cn_setup()
mydate <- utils_myDate()
# define species:
species <- "mouse"
# or
species <- "human"
```

9. Load sample tables and expression matrices from preprocessing:

```
# specify S3 bucket and path where sample tables and
expression matrices are stored:
bucket <- "my_bucket"
# don't include bucket name, leading or trailing slashes:
path <- "my_path/to/files"
# copy sample tables and expression matrics from S3 to
current directory:
cmd <- paste0("aws s3 cp s3://", bucket, "/", path "/ . --
exclude "*" --include "st*.rda")
system(cmd)
cmd <- paste0("aws s3 cp s3://", bucket, "/", path "/ . --
exclude "*" --include "expList*.rda")
system(cmd)
```

10. Merge sample tables:

```
sampTabNames <- list.files(pattern = "st*.rda")
stAll <- utils_loadObject(sampTabNames[1])
for(i in 2:length(sampTabNames)) {
 add_st <- utils_loadObject(sampTabNames[i])
 stAll <- bind_rows(stAll, add_st)
}
```

`stAll` contains the merged sample table.

11. Merge expression matrices:

```
expListNames <- list.files(pattern = "expList*.rda")
tempExpList <- utils_loadObject(expListNames[1]) # list
expAll <- as.data.frame(tempExpList$normalized) # df
for(i in 2:length(expListNames)) {
  add_expList <- utils_loadObject(expListNames[i]) # list
  add_expListNorm <- as.data.frame(add_expList$normalized) #df
  expAll <- bind_cols(expAll, add_expListNorm) #df
}
expAll <- as.matrix(expAll)
```

`expAll` is the merged expression matrix.

12. Reconstruct C-/T-specific GRNs and train classifier.

```
grnProp <- cn_make_grn(stAll, expAll, species = species)
# assess classifier performance:
classifierPerformance <- cn_splitMakeAssess(stAll, expAll,
grnProp, dLevelStudy = "study_id", dLevelSID= "sra_id")
```

13. Plot classifier performance PR curves and heatmap (Fig. 3a, b):

```
# plot PR curves of classifier performances:
fname <- paste0("classifierPerformance_", species, "_",
mydate, ".pdf")
pdf(file = fname, width = 10, height = 10)
plot_classPRs(classifierPerformance$PRs)
dev.off()
```

14. Generate the CellNet Processor object and save:

```
cnProc <- cn_make_processor(expAll, stAll, grnProp, sidCol
= "sra_id")
fname <- paste0("cnProc_", species, "_", mydate, ".rda")
save(cnProc, file=fname)
```

15. From a different terminal window, `scp` the `.rda` and `.pdf` files to your local machine:

```
scp -i my_aws_private_key.pem ec2-
user@my_instance_public_dns.amazonaws.com:/media/ephemeral0/
analysis/*.rda my_local_directory
scp -i my_aws_private_key.pem ec2-
user@my_instance_public_dns.amazonaws.com:/media/ephemeral0/
analysis/*.pdf my_local_directory
```

16. ***Important:*** *Terminate your instance when finished.* AWS will continue to charge for compute time until the instance is terminated. This can be done by navigating to the EC2

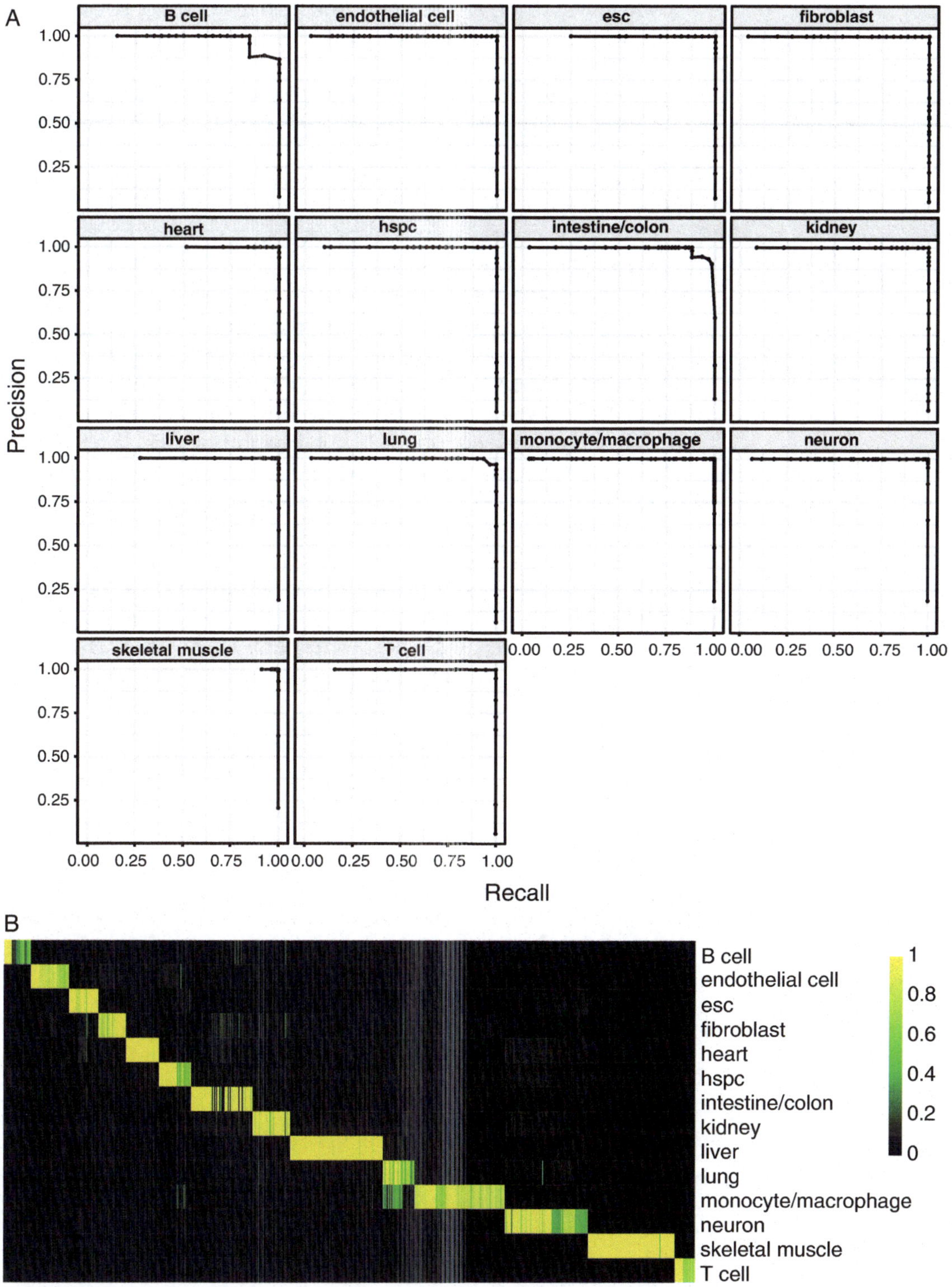

Fig. 3 Assessing CellNet classifier performance. (**a**) Precision-recall curves for each C/T type's binary random forest classifier. (**b**) Classifier performance heatmap for each classifier

"Instances" page, selecting your instance, and then clicking through "Actions" > "Instance State" > "Terminate."

3.3 Application of CellNet to Query Studies

Objective: Apply CellNet to a preprocessed query study, and generate three readouts of cell fate for each sample and/or treatment group in the study.

This procedure can be performed locally or on an EC2 instance.

In this section, we will examine an example iPSC disease modeling study, which explores the contribution of a G296S missense mutation in GATA4 to cardiomyopathy and septal defects. In this study, Ang et al. reprogrammed affected patient and control skin fibroblasts to pluripotency and then directly differentiated the resulting iPSCs to cardiomyocytes. The authors demonstrated that G296S disrupts GATA4 interaction with TBX5, allowing TBX5 to upregulate endothelial genes inappropriately [16].

This subsection assumes that you have an existing CellNet Processor (`cnProc`) object and you have generated your sample metadata table and expression matrix. If you would like to use an existing trained CellNet Processor object (`cnProc`), you can download one of several options here: https://github.com/pcahan1/CellNet.

FASTQ files for this study were obtained from NCBI SRA, Accession ID: SRP082270. The expression matrix was generated in Subheading 3.1.

1. Setup:

```
R
library(devtools)
library(ggplot2)
library(pheatmap)
install_github("pcahan1/CellNet", ref = "v0.2.0") # install
CellNet package
library(CellNet)
```

2. Load CellNet Processor object (`cnProc`), query study expression matrix, and query study sample metadata table:

```
# CellNet processor object
cnProc <- utils_loadObject("cnProc_HS_RS_Jun_20_2017.rda")
# expression matrix:
expGeneList <- utils_loadObject("expGeneList_SRP082270.R")
# sample metadata table
sampTab <- utils_loadObject("sampTab_SRP082270_QC.R")
tissue_types <-
rownames(summary(cnProc$ctGRNs$ctGRNs$geneLists))
```

```
mydate <- utils_myDate()
study_name <- "SRP082270"
```

3. Transform data, apply CellNet, and save Results object (`cnRes`):

```
expQuery<-expGeneList$counts
expQuery<-apply(expQuery, 2, downSampleW, 1e5)
expQuery<-log(1+expQuery)
# apply CellNet and generate CellNet results object
cnResQuery<-cn_apply(expQuery, sampTab, cnProc)
fname<-paste0("cn_", study_name, "_results_nat_prot_",
mydate, ".rda")
save(cnResQuery, file=fname)
```

4. Make plots based on CellNet Results object (Figs. 4 and 5):

```
# plot and save classification heatmap as a PDF
plot_classification_hm(cnResQuery, sampTab, study_name,
mydate)
# define target cell/tissue type for experiment:
target_cell_type <- "heart"
# plot and save GRN status plots as PDFs
grn_status_by_CT(cnResQuery, cnProc, study_name, dlevel=
"description2")
grn_status_by_dlevel(target_cell_type, cnResQuery,
cnProc, study_name, dlevel="description2")
# plot and save Network Influence Score (NIS) plots
# as PDF in current directory:
make_nis_plot(cnResQuery, cnProc, sampTab, target_cell_
type, study_name, mydate)
```

Meta-analysis of Patient-Specific and Control iPSCs and iPSC-Derived Cell Types: CellNet methods can be modified to produce custom heatmaps for a variety of visualization purposes. For example, we have performed a meta-analysis of patient-specific iPSC and iPSC-derived engineered population expression profiles from iPSC disease modeling studies offering publicly available data (Table 1).

5. Load pooled iPSC and iPSC-derived CellNet Results (`cnRes`) objects. Create a data frame representing Column Annotations. Each column in the data frame represents a different attribute of the dataset upon which the samples will be grouped.

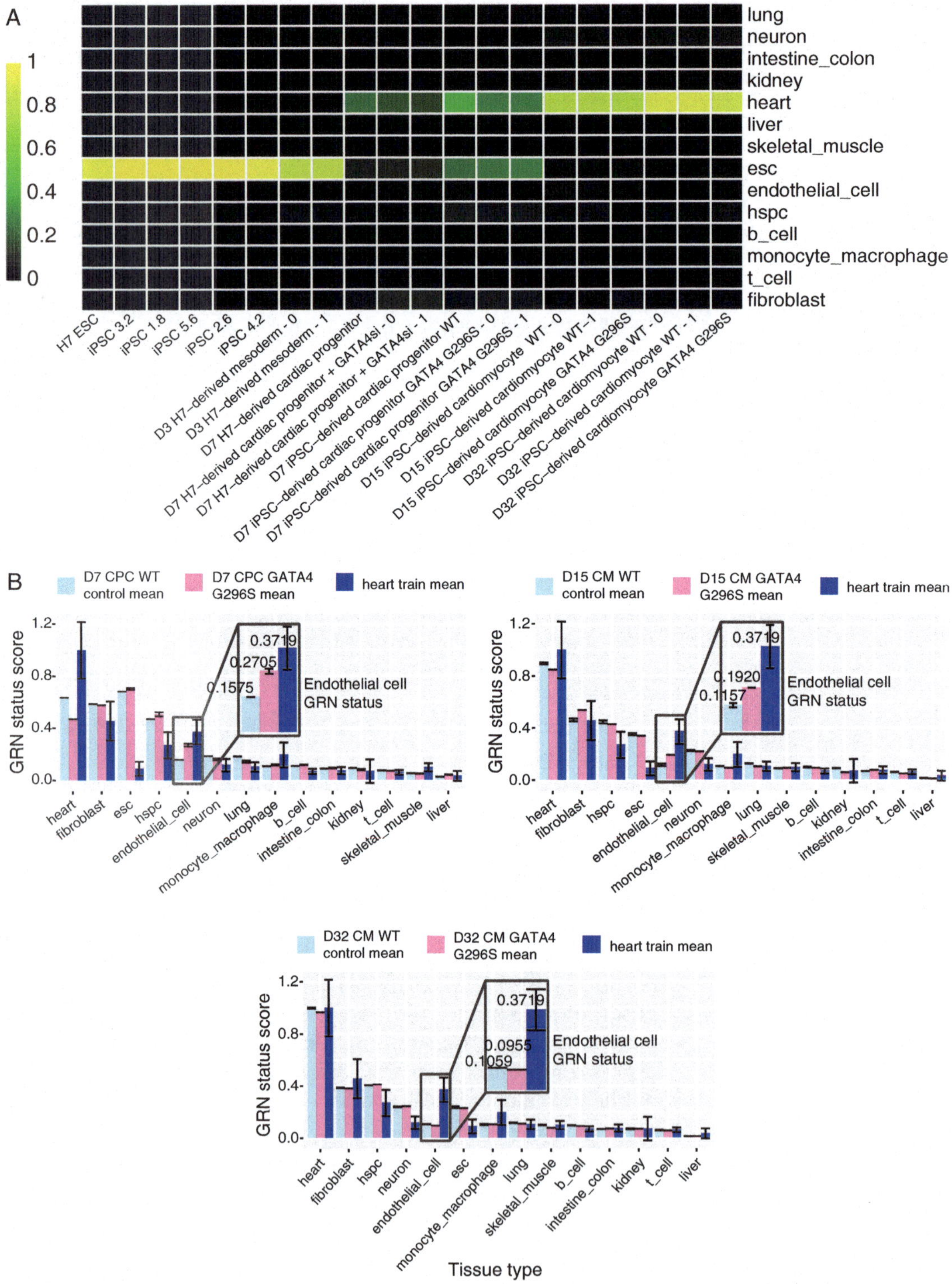

Fig. 4 CellNet classification and GRN status of GATA4 iPSC disease modeling study. (**a**) Classification heatmap of study. (**b**) GRN status for d7, d15, and d32 control and disease treatment groups in study for each C/T type

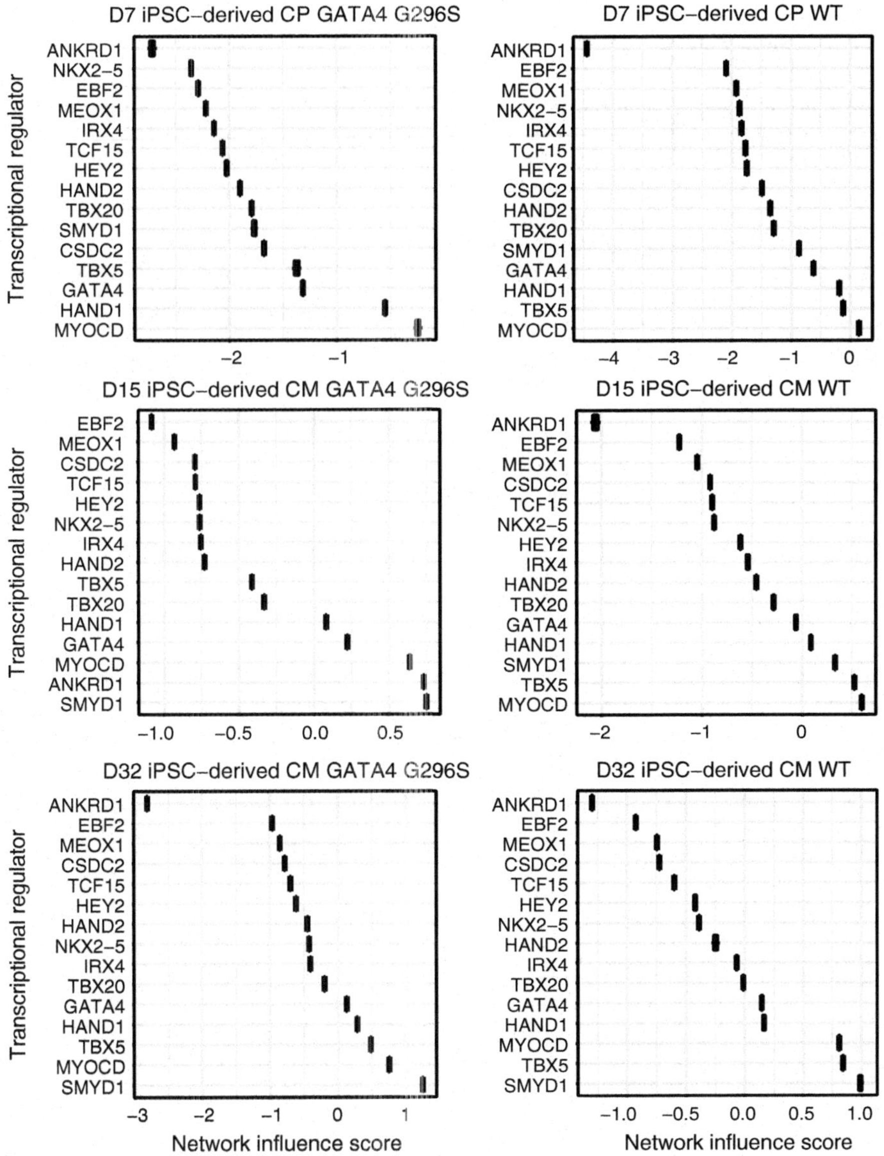

Fig. 5 CellNet network influence score of GATA4 iPSC disease modeling study. Heart network influence score for d7, d15, and d32 control and disease treatment groups

```
cnRes_iPSC <-
utils_loadObject("cn_iPSC_pooled_results_nat_prot_-
Jul_14_2017.R")
sampTab_iPSC <- cnRes_iPSC$stQuery
col_annotations_iPSC <- data.frame(Disease_Status =
factor(sampTab_iPSC$disease_status), Study =
```

Table 1
iPSC disease modeling studies

SRP #	GEO acc. no.	Study name	Disease	Affect. tissue	Ref.
SRP032798	GSE52202	Targeting RNA foci in iPSC-derived motor neurons from ALS patients with C9ORF72 repeat expansion	Amyotrophic lateral sclerosis (ALS)	Neuron	[17]
SRP032928	GSE52249	Modeling and rescuing neurodevelopmental defect of Down syndrome using induced pluripotent stem cells from monozygotic twins discordant for trisomy 21	Trisomy 21 (Down syndrome)	Many	[18]
SRP033078	GSE52431	Human iPSC-based modeling of late-onset disease using progerin-induced aging	Progerin	Neuron	[19]
SRP039348	GSE55504	Domains of genome-wide gene expression dysregulation in Down syndrome	Trisomy 21 (Down syndrome)	Neuron	[20]
SRP042112	GSE57821	Transcriptional dysregulation of synaptic genes in a human iPSC model of major mental disorders	Schizophrenia	Neuron	[21]
SRP042119	GSE57826	Modeling Fanconi anemia pathogenesis and therapeutics using integration-free patient iPSCs	Fanconi anemia	Hematopoietic, endothelial	[22]
SRP043684	GSE58933	Hyperexcitability of neurons generated from patients with bipolar disorder	Bipolar disorder	Neuron	[23]
SRP044749		Gene expression analysis of human-induced pluripotent stem cell-derived neurons carrying copy number variants of chromosome 15q11-q13.1	Autism spectrum disorder (ASD), Angleman syndrome	Neuron	[24]

(continued)

Table 1 (continued)

SRP #	GEO acc. no.	Study name	Disease	Affect. tissue	Ref.
SRP049475	GSE62936	RNA-Seq analysis in hES/iPS cell-derived neuronal samples	Bardet-Biedl syndrome	Neuron	[25]
SRP049593	GSE63055	7q11.23 dosage-dependent dysregulation in the human pluripotent state primes aberrant transcriptional programs in disease-relevant lineages (RNA-Seq)	7q11.23 hemi-deletion (Williams-Beuren syndrome) or microduplication	Neuron	[26]
SRP049676	GSE63161	hiPSCs unravel aberrant TGFβ signaling as an etiology of left ventricular non-compaction	Left ventricular non-compaction	Heart	[27]
SRP050376	GSE63734	Dysregulation of miRNA-9 in a subset of schizophrenia patient-derived neurons	Schizophrenia	Neuron	[28]
SRP050377	GSE63738	Next-generation sequencing facilitates comparisons of control and schizophrenia patient-derived hiPSC-derived NPCs	Schizophrenia	NPCs	[28]
SRP053042	GSE65507	Pluripotent cell models of Fanconi anemia identify the early pathological defect in human hemoangiogenic progenitors	Fanconi anemia	Hematopoietic, endothelial	
SRP053043	GSE65508	Modeling the early phenotype at the neuromuscular junction of spinal muscular atrophy using patient-derived iPSCs	Spinal muscular atrophy (SMA)	Neuron	[29]
SRP056333	GSE67073	RNA sequencing of hiPSC-derived neural crest populations from familial dysautonomia patients	Familial dysautonomia	Neuron	[30]

(continued)

Table 1
(continued)

SRP #	GEO acc. no.	Study name	Disease	Affect. tissue	Ref.
SRP056822	GSE67528	Evidence for proliferation and synaptogenesis impairments in neural cells derived from idiopathic autistic patients	Autism spectrum disorder (ASD)	Neuron	[31]
SRP058640	GSE69175	Genome-wide RNA-Seq of human motor neurons implicates selective ER stress activation in spinal muscular atrophy	Spinal muscular atrophy (SMA)	Neuron	[32]
SRP065036	GSE74201	Genomic analysis reveals disruption of striatal neuronal development and therapeutic targets in a human neural stem cell model of Huntington's disease	Huntington's disease	Neuron	[33]
SRP068017	GSE76490	Transcriptomic analysis of human iPS cells derived from fragile X syndrome patients during neural differentiation	Fragile X	Neuron	[34]
SRP072365	GSE79613	Patient-specific iPSC-derived endothelial cells uncover pathways that protect against pulmonary hypertension in BMPR2 mutation carriers	Familial pulmonary arterial hypertension (FPAH)	Endothelial	[35]
SRP073163	GSE80170	Next-generation sequencing compares effects of microRNA-9 perturbation in control and SZ hiPSC NPCs	Schizophrenia	NPCs	[28]
SRP082270	GSE85623, GSE85631	Disease model of GATA4 mutation reveals transcription factor cooperativity in human cardiogenesis	Cardiac septal defects and cardiomyopathy	Heart	[16]

(continued)

Table 1 (continued)

SRP #	GEO acc. no.	Study name	Disease	Affect. tissue	Ref.
SRP093646	GSE89991	RNA-Seq in PWS iPSC-derived neurons	Prader-Willi syndrome	Neuron	[36]
SRP096788		Gene expression analysis of human spinal motor neurons (MN) differentiated from ALS patient-derived iPSC	ALS	Neuron	[37]
SRP100681	GSE95343	mRNA-Seq of Huntington's disease and control patient iPSC-derived neural cells	Huntington's disease	Neuron	[38]

```
factor(sampTab_iPSC$study_id))
cnRes_derived <- utils_loadObject("cn_iPSC_pooled_
results_nat_prot_Jul_14_2017.R")
sampTab_derived <- cnRes_derived$stQuery
col_annotations_derived <- data.frame(Disease_Status =
factor(sampTab_derived$disease_status), Study =
factor(sampTab_derived$study_id))
```

6. Plot heatmaps (Fig. 6):

```
# reseed random number generator for reproducible colors
set.seed(0)
pdf(file="pooled_iPSC_hm.pdf", width=9, height=5,
onefile=FALSE)
cn_HmClass_disease(cnRes_iPSC, col_annotations_iPSC)
dev.off()
# reseed random number generator
set.seed(0)
pdf(file="pooled_iPSC_derived_hm.pdf", width=9, height=5,
onefile=FALSE)
cn_HmClass_disease(cnRes_derived, col_annotations_derived)
dev.off()
```

4 Notes

1. When trying to login to remove AWS instance, you may receive a message reading "Connection refused." Most likely this indicates that the instance has not finished initializing and that you

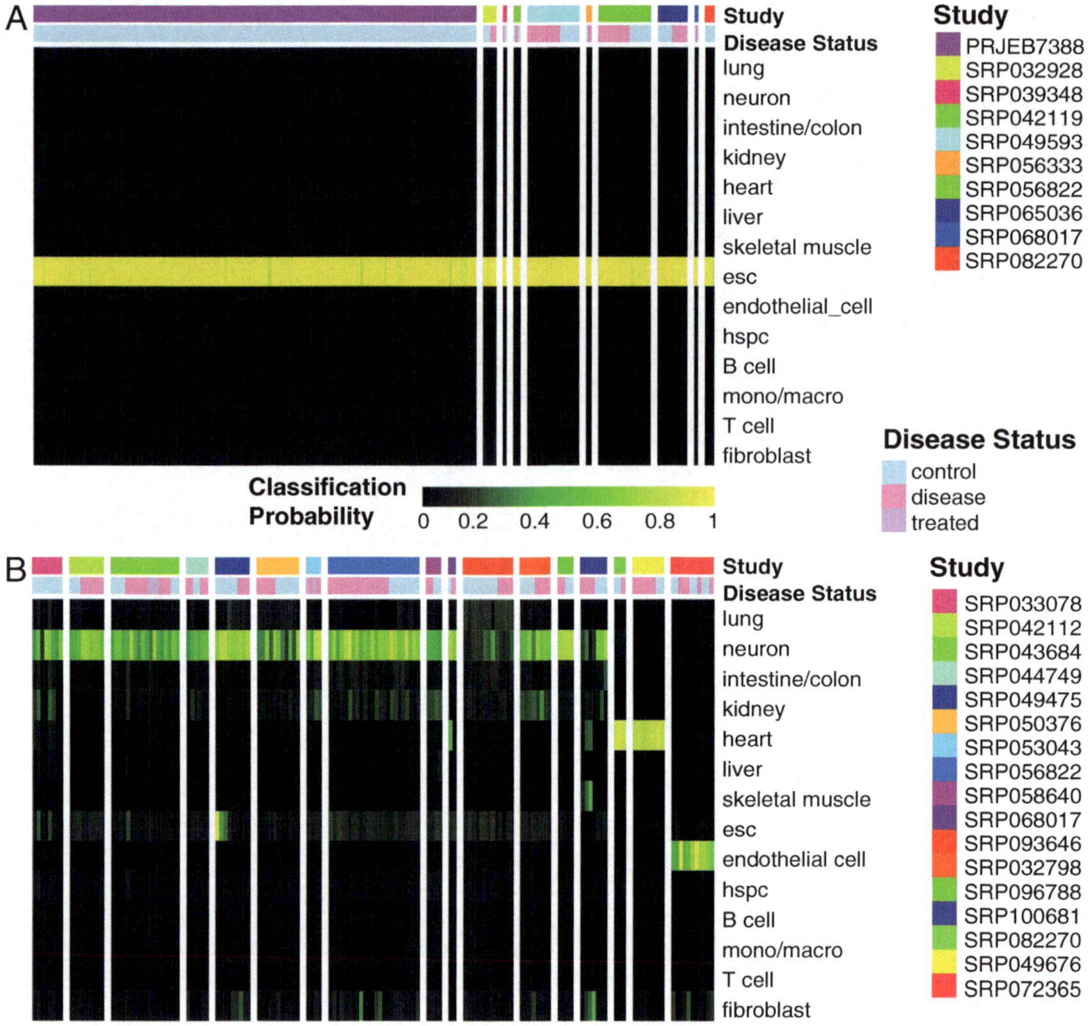

Fig. 6 Classification meta-analysis of iPSC disease modeling studies. (**a**) Classification heatmap of healthy control- and patient-specific iPSCs for several independent iPSC disease modeling studies. (**b**) Classification heatmap of healthy control, patient, and treated iPSC-derived engineered populations for several independent iPSC disease modeling studies. *See* Table 1 for study information

should wait several more minutes and try again. You can check the status of the instance using the AWS Management Console.

2. If you receive a message "Unable to locate credentials," then you are probably still running R as sudo. Quit R and start a new R session as ec2-user. Alternatively, you can run 'aws configure' and input your Access key ID and Secret Access Key when prompted.
3. There are several possible causes of preprocessing failures. First, make sure that you ran `cn_setup()` to set up the required ephemeral drives. Second, it is possible that the image type

launched does not have sufficient RAM or hard disk space to process all of the samples. You can either use an image that has more RAM or add additional storage via the AWS Management Console. Third, the SRA files may be corrupted. You can try to re-fetch the raw data from SRA. If the files are still corrupted, then you will need to contact the study authors to obtain the original raw data.

4. At times, one can encounter difficulties in transferring files to and from S3. This can be caused but incorrect bucket or path specifications, or it may be that you do not have AWS permissions to access the specified S3 resources.
5. GRN reconstruction can fail if there is insufficient RAM. If GRN reconstruction fails, then try launching an instance on an image with more RAM.

References

1. Cahan P (2016) Enabling direct fate conversion with network biology. Nat Genet 48:226–227. https://doi.org/10.1038/ng.3516
2. Tsankov AM, Akopian V, Pop R, Chetty S, Gifford CA, Daheron L, Tsankova NM, Meissner A (2015) A qPCR ScoreCard quantifies the differentiation potential of human pluripotent stem cells. Nat Biotechnol 33:1182–1192. https://doi.org/10.1038/nbt.3387
3. Müller F-J, Schuldt BM, Williams R, Mason D, Altun G, Papapetrou EP, Danner S, Goldmann JE, Herbst A, Schmidt NO, Aldenhoff JB, Laurent LC, Loring JF (2011) A bioinformatic assay for pluripotency in human cells. Nat Methods 8:315–317. https://doi.org/10.1038/nmeth.1580
4. Avior Y, Biancotti JC, Benvenisty N (2015) TeratoScore: assessing the differentiation potential of human pluripotent stem cells by quantitative expression analysis of teratomas. Stem Cell Reports 4:967–974. https://doi.org/10.1016/j.stemcr.2015.05.006
5. Roost MS, van Iperen L, Ariyurek Y, Buermans HP, Arindrarto W, Devalla HD, Passier R, Mummery CL, Carlotti F, de Koning EJP, van Zwet EW, Goeman JJ, Chuva de Sousa Lopes SM (2015) KeyGenes, a tool to probe tissue differentiation using a human fetal transcriptional atlas. Stem Cell Reports 4:1112–1124. https://doi.org/10.1016/j.stemcr.2015.05.002
6. Heinäniemi M, Nykter M, Kramer R, Wienecke-Baldacchino A, Sinkkonen L, Zhou JX, Kreisberg R, Kauffman SA, Huang S, Shmulevich I (2013) Gene-pair expression signatures reveal lineage control. Nat Methods 10:577–583. https://doi.org/10.1038/nmeth.2445
7. D'Alessio AC, Fan ZP, Wert KJ, Baranov P, Cohen MA, Saini JS, Cohick E, Charniga C, Dadon D, Hannett NM, Young MJ, Temple S, Jaenisch R, Lee TI, Young RA (2015) A systematic approach to identify candidate transcription factors that control cell identity. Stem Cell Reports 5:763–775. https://doi.org/10.1016/j.stemcr.2015.09.016
8. Rackham OJL, Firas J, Fang H, Oates ME, Holmes ML, Knaupp AS, Consortium TF, Suzuki H, Nefzger CM, Daub CO, Shin JW, Petretto E, Forrest ARR, Hayashizaki Y, Polo JM, Gough J (2016) A predictive computational framework for direct reprogramming between human cell types. Nat Genet 48:331–335. https://doi.org/10.1038/ng.3487
9. Lang AH, Li H, Collins JJ, Mehta P (2014) Epigenetic landscapes explain partially reprogrammed cells and identify key reprogramming genes. PLOS Comput Biol 10:e1003734. https://doi.org/10.1371/journal.pcbi.1003734
10. Crespo I (2013) A general strategy for cellular reprogramming: the importance of transcription factor cross-repression. Stem Cells—Wiley Online Library. https://stemcellsjournals.onlinelibrary.wiley.com/doi/abs/10.1002/stem.1473. Accessed 30 May 2018
11. Davis FP, Eddy SR (2013) Transcription factors that convert adult cell identity are differentially polycomb repressed. PLOS One 8:

e63407. https://doi.org/10.1371/journal.pone.0063407

12. Bian Q, Cahan P (2016) Computational tools for stem cell biology. Trends Biotechnol 34:993–1009. https://doi.org/10.1016/j.tibtech.2016.05.010
13. Cahan P, Li H, Morris SA, Lummertz da Rocha E, Daley GQ, Collins JJ (2014) CellNet: network biology applied to stem cell engineering. Cell 158:903–915. https://doi.org/10.1016/j.cell.2014.07.020
14. Radley AH, Schwab RM, Tan Y, Kim J, Lo EKW, Cahan P (2017) Assessment of engineered cells using CellNet and RNA-seq. Nat Protoc 12:1089–1102. https://doi.org/10.1038/nprot.2017.022
15. Sterneckert JL, Reinhardt P, Schöler HR (2014) Investigating human disease using stem cell models. Nat Rev Genet 15:625–639. https://doi.org/10.1038/nrg3764
16. Ang Y-S, Rivas RN, Ribeiro AJS, Srivas R, Rivera J, Stone NR, Pratt K, Mohamed TMA, Fu J-D, Spencer CI, Tippens ND, Li M, Narasimha A, Radzinsky E, Moon-Grady AJ, Yu H, Pruitt BL, Snyder MP, Srivastava D (2016) Disease model of GATA4 mutation reveals transcription factor cooperativity in human cardiogenesis. Cell 167:1734–1749.e22. https://doi.org/10.1016/j.cell.2016.11.033
17. Sareen D, O'Rourke JG, Meera P, Muhammad AKMG, Grant S, Simpkinson M, Bell S, Carmona S, Ornelas L, Sahabian A, Gendron T, Petrucelli L, Baughn M, Ravits J, Harms MB, Rigo F, Bennett CF, Otis TS, Svendsen CN, Baloh RH (2013) Targeting RNA foci in iPSC-derived motor neurons from ALS patients with a C9ORF72 repeat expansion. Sci Transl Med 5:208ra149. https://doi.org/10.1126/scitranslmed.3007529
18. Hibaoui Y, Grad I, Letourneau A, Sailani MR, Dahoun S, Santoni FA, Gimelli S, Guipponi M, Pelte MF, Béna F, Antonarakis SE, Feki A (2014) Modelling and rescuing neurodevelopmental defect of Down syndrome using induced pluripotent stem cells from monozygotic twins discordant for trisomy 21. EMBO Mol Med 6:259–277. https://doi.org/10.1002/emmm.201302848
19. Miller JD, Ganat YM, Kishinevsky S, Bowman RL, Liu B, Tu EY, Mandal PK, Vera E, Shim J, Kriks S, Taldone T, Fusaki N, Tomishima MJ, Krainc D, Milner TA, Rossi DJ, Studer L (2013) Human iPSC-based modeling of late-onset disease via progerin-induced aging. Cell Stem Cell 13:691–705. https://doi.org/10.1016/j.stem.2013.11.006
20. Letourneau A, Santoni FA, Bonilla X, Sailani MR, Gonzalez D, Kind J, Chevalier C, Thurman R, Sandstrom RS, Hibaoui Y, Garieri M, Popadin K, Falconnet E, Gagnebin M, Gehrig C, Vannier A, Guipponi M, Farinelli L, Robyr D, Migliavacca E, Borel C, Deutsch S, Feki A, Stamatoyannopoulos JA, Herault Y, van Steensel B, Guigo R, Antonarakis SE (2014) Domains of genome-wide gene expression dysregulation in Down's syndrome. Nature 508:345–350. https://doi.org/10.1038/nature13200
21. Wen Z, Nguyen HN, Guo Z, Lalli MA, Wang X, Su Y, Kim N-S, Yoon K-J, Shin J, Zhang C, Makri G, Nauen D, Yu H, Guzman E, Chiang C-H, Yoritomo N, Kaibuchi K, Zou J, Christian KM, Cheng L, Ross CA, Margolis RL, Chen G, Kosik KS, Song H, Ming G (2014) Synaptic dysregulation in a human iPS cell model of mental disorders. Nature 515:414–418. https://doi.org/10.1038/nature13716
22. Liu G-H, Suzuki K, Li M, Qu J, Montserrat N, Tarantino C, Gu Y, Yi F, Xu X, Zhang W, Ruiz S, Plongthongkum N, Zhang K, Masuda S, Nivet E, Tsunekawa Y, Soligalla RD, Goebl A, Aizawa E, Kim NY, Kim J, Dubova I, Li Y, Ren R, Benner C, del Sol A, Bueren J, Trujillo JP, Surralles J, Cappelli E, Dufour C, Esteban CR, Belmonte JCI (2014) Modelling Fanconi anemia pathogenesis and therapeutics using integration-free patient-derived iPSCs. Nat Commun 5:4330. https://doi.org/10.1038/ncomms5330
23. Mertens J, Wang Q-W, Kim Y, Yu DX, Pham S, Yang B, Zheng Y, Diffenderfer KE, Zhang J, Soltani S, Eames T, Schafer ST, Boyer L, Marchetto MC, Nurnberger JI, Calabrese JR, Ødegaard KJ, McCarthy MJ, Zandi PP, Alba M, Nievergelt CM, Mi S, Brennand KJ, Kelsoe JR, Gage FH, Yao J (2015) Differential responses to lithium in hyperexcitable neurons from patients with bipolar disorder. Nature 527:95–99. https://doi.org/10.1038/nature15526
24. Germain ND, Chen P-F, Plocik AM, Glatt-Deeley H, Brown J, Fink JJ, Bolduc KA, Robinson TM, Levine ES, Reiter LT et al (2014) Gene expression analysis of human induced pluripotent stem cell-derived neurons carrying copy number variants of chromosome 15q11-q13.1. Mol Autism 5:44
25. Wang L, Meece K, Williams DJ, Lo KA, Zimmer M, Heinrich G, Martin Carli J, Leduc CA, Sun L, Zeltser LM, Freeby M,

Goland R, Tsang SH, Wardlaw SL, Egli D, Leibel RL (2015) Differentiation of hypothalamic-like neurons from human pluripotent stem cells. J Clin Invest 125:796–808. https://doi.org/10.1172/JCI79220

26. Adamo A, Atashpaz S, Germain P-L, Zanella M, D'Agostino G, Albertin V, Chenoweth J, Micale L, Fusco C, Unger C, Augello B, Palumbo O, Hamilton B, Carella M, Donti E, Pruneri G, Selicorni A, Biamino E, Prontera P, McKay R, Merla G, Testa G (2014) 7q11.23 dosage-dependent dysregulation in human pluripotent stem cells affects transcriptional programs in disease-relevant lineages. Nat Genet 47:132–141. https://doi.org/10.1038/ng.3169
27. Kodo K, Ong S-G, Jahanbani F, Termglinchan V, Hirono K, InanlooRahatloo K, Ebert AD, Shukla P, Abilez OJ, Churko JM, Karakikes I, Jung G, Ichida F, Wu SM, Snyder MP, Bernstein D, Wu JC (2016) iPSC-derived cardiomyocytes reveal abnormal TGF-β signalling in left ventricular non-compaction cardiomyopathy. Nat Cell Biol 18:1031–1042. https://doi.org/10.1038/ncb3411
28. Topol A, Zhu S, Hartley B, English J, Hauberg M, Tran N, Rittenhouse C, Simone A, Ruderfer D, Johnson J, Readhead B, Hadas Y, Gochman P, Wang Y-C, Shah H, Cagney G, Rapoport J, Gage F, Dudley J, Sklar P, Mattheisen M, Cotter D, Fang G, Brennand K (2016) Dysregulation of miRNA-9 in a subset of schizophrenia patient-derived neural progenitor cells. Cell Rep 15:1024–1036. https://doi.org/10.1016/j.celrep.2016.03.090
29. Yoshida M, Kitaoka S, Egawa N, Yamane M, Ikeda R, Tsukita K, Amano N, Watanabe A, Morimoto M, Takahashi J, Hosoi H, Nakahata T, Inoue H, Saito M (2015) Modeling the early phenotype at the neuromuscular junction of spinal muscular atrophy using patient-derived iPSCs. Stem Cell Reports 4:561–568. https://doi.org/10.1016/j.stemcr.2015.02.010
30. Zeltner N, Fattahi F, Dubois NC, Saurat N, Lafaille F, Shang L, Zimmer B, Tchieu J, Soliman MA, Lee G, Casanova J-L, Studer L (2016) Capturing the biology of disease severity in a PSC-based model of familial dysautonomia. Nat Med 22:1421–1427. https://doi.org/10.1038/nm.4220
31. Marchetto MC, Belinson H, Tian Y, Freitas BC, Fu C, Vadodaria KC, Beltrao-Braga PC, Trujillo CA, Mendes APD, Padmanabhan K, Nunez Y, Ou J, Ghosh H, Wright R, Brennand KJ, Pierce K, Eichenfield L, Pramparo T, Eyler LT, Barnes CC, Courchesne E, Geschwind DH, Gage FH, Wynshaw-Boris A, Muotri AR (2017) Altered proliferation and networks in neural cells derived from idiopathic autistic individuals. Mol Psychiatry 22:820–835. https://doi.org/10.1038/mp.2016.95
32. Ng S-Y, Soh B, Rodriguez-Muela N, Hendrickson D, Price F, Rinn J, Rubin L (2015) Genome-wide RNA-Seq of human motor neurons implicates selective ER stress activation in spinal muscular atrophy. Cell Stem Cell 17:569–584. https://doi.org/10.1016/j.stem.2015.08.003
33. Ring K, An M, Zhang N, O'Brien R, Ramos E, Gao F, Atwood R, Bailus B, Melov S, Mooney S, Coppola G, Ellerby L (2015) Genomic analysis reveals disruption of striatal neuronal development and therapeutic targets in human Huntington's disease neural stem cells. Stem Cell Reports 5:1023–1038. https://doi.org/10.1016/j.stemcr.2015.11.005
34. Lu P, Chen X, Feng Y, Zeng Q, Jiang C, Zhu X, Fan G, Xue Z (2016) Integrated transcriptome analysis of human iPS cells derived from a fragile X syndrome patient during neuronal differentiation. Sci China Life Sci 59:1093–1105. https://doi.org/10.1007/s11427-016-0194-6
35. Gu M, Shao N-Y, Sa S, Li D, Termglinchan V, Ameen M, Karakikes I, Sosa G, Grubert F, Lee J, Cao A, Taylor S, Ma Y, Zhao Z, Chappell J, Hamid R, Austin ED, Gold JD, Wu JC, Snyder MP, Rabinovitch M (2017) Patient-specific iPSC-derived endothelial cells uncover pathways that protect against pulmonary hypertension in BMPR2 mutation carriers. Cell Stem Cell 20:490–504.e5. https://doi.org/10.1016/j.stem.2016.08.019
36. Burnett LC, LeDuc CA, Sulsona CR, Paull D, Eddiry S, Levy B, Salles JP, Tauber M, Driscoll DJ, Egli D, Leibel RL (2016) Induced pluripotent stem cells (iPSC) created from skin fibroblasts of patients with Prader-Willi syndrome (PWS) retain the molecular signature of PWS. Stem Cell Res 17:526–530. https://doi.org/10.1016/j.scr.2016.08.008
37. Bhinge A, Namboori SC, Zhang X, VanDongen AMJ, Stanton LW (2017) Genetic correction of SOD1 mutant iPSCs reveals ERK and JNK activated AP1 as a driver of neurodegeneration in amyotrophic lateral sclerosis. Stem Cell Reports 8:856–869. https://doi.org/10.1016/j.stemcr.2017.02.019
38. Lim RG, Salazar LL, Wilton DK, King AR, Stocksdale JT, Sharifabad D, Lau AL, Stevens B, Reidling JC, Winokur ST, Casale MS, Thompson LM, Pardo M, Díaz-Barriga

AGG, Straccia M, Sanders P, Alberch J, Canals JM, Kaye JA, Dunlap M, Jo L, May H, Mount E, Anderson-Bergman C, Haston K, Finkbeiner S, Kedaigle AJ, Gipson TA, Yildirim F, Ng CW, Milani P, Housman DE, Fraenkel E, Allen ND, Kemp PJ, Atwal RS, Biagioli M, Gusella JF, MacDonald ME, Akimov SS, Arbez N, Stewart J, Ross CA, Mattis VB, Tom CM, Ornelas L, Sahabian A, Lenaeus L, Mandefro B, Sareen D, Svendsen CN (2017) Developmental alterations in Huntington's disease neural cells and pharmacological rescue in cells and mice. Nat Neurosci 20:648–660. https://doi.org/10.1038/nn.4532

INDEX

Patrick Cahan (ed.), *Computational Stem Cell Biology: Methods and Protocols*, Methods in Molecular Biology, vol. 1975, https://doi.org/10.1007/978-1-4939-9224-9, © Springer Science+Business Media, LLC, part of Springer Nature 2019

H

I

L

M

N

P

Q

R

S

T

U

W

Printed by Printforce, the Netherlands